ANALYZING MULTIVARIATE DATA

ANALYZING MULTIVARIATE DATA

Norman Cliff

University of Southern California

HBJ Harcourt Brace Jovanovich, Publishers

and its subsidiary, Academic Press

San Diego New York Chicago Austin Washington, D.C.

London Sydney Tokyo Toronto

ISBN: 0-15-502704-2

Library of Congress Catalog Card Number: 86-80760

Printed in the United States of America

To Rosemary

PREFACE

This book attempts to fill the need for a less technical yet sophisticated text, one that considers the computer's role in multivariate analysis. My main motive for writing this book was the realization that, while many behavioral scientists were performing multivariate analyses, few textbooks provided a conceptual understanding of what goes on in multivariate analysis, particularly in a way that relates to the questions the investgator might have in mind. There also seemed to be a gap between the methods described in the textbooks and the statistical computer packages that are so widely used.

Although the level of mathematics has been kept basic, this text contains many formulas and an occasional derivation. The purpose of the mathematics is either to show the connections among methods or to reveal what is going to affect some statistical method. Although some matrix algebra is used, considerable space is taken in the introductory chapters to familiarize readers with matrix notation and to connect it with common statistical formulas.

Parts of this book have been used by a number of my graduate students, most of whom are studying psychology but a few of whom are studying other behavioral sciences, education, or management. Most students who use this text should have completed one or two previous statistics courses, although the text has been used successfully by some students with no prior background in statistics. It assumes no special mathematics preparation, although it does assume a willingness to accept mathematical notation and some manipulation of symbols.

The book describes the most common statistical approaches that can be taken when more than two variables are considered simultaneously. It covers multiple regression and its generalization as the general linear model, including the analysis of variance and covariance, discriminant function analysis, multivariate analysis of variance, canonical correlation analysis, and the various forms of factor analysis. These are the most common topics for the majority of textbooks having the word "multivariate" in their titles.

In multiple regression, there are a number of predictors (independent variables) and a criterion (dependent variable). The goal of the investigator includes (1) finding a set of weights for combining the independent variable into a composite in order to predict or explain the dependent variables, (2) assessing the accuracy of the prediction, and (3) making statistical inferences about the weights and the accuracy.

In the precomputer era, analysis of variance and covariance had an enormous advantage over multiple regression. Analyses could be carried out quite readily using mechanical hand calculators, except in the most complex cases. It was well known in the theoretical statistical literature that there was a direct mathematical connection between analysis of variance and covariance and multiple regression. However, learning to apply one tended to preclude learning to apply the other. In recent years, there has been more widespread appreciation of the relationship between these two areas. Analyses of variance and covariance are special types of multiple regression in which the predictors and the methods of analysis take on special forms.

In regression and analysis of variance, a single dependent variable is the focus of analysis. When considering multiple dependent variables simultaneously, one uses canonical correlation analysis, discriminant function analysis, and multivariate analysis of variance and covariance. Canonical correlation analysis is a generalized form of multiple regression, and discriminant analysis and multivariate analysis of variance are generalized forms of analysis of variance.

The various forms of factor analysis come from a somewhat separate tradition. In factor analysis, the variables are not separated into two camps, dependent and independent. Instead, one attempts to understand the whole collection of variables, largely through the supposition and interpretation of underlying variates that explain the observed variables.

All of these methods are related. They all, in one way or another, focus on forming *linear,* that is, weighted, combinations of variables to form new variables. These weighted combinations are formed so as to maximize or minimize a variance or correlation. Thus, properties of variances and covariances of linear combinations are central to this book.

This central theme largely explains why areas such as multidimensional scaling, cluster analysis, and log-linear analysis of frequency data have been omitted from this text. None of these specialized areas fits readily into the weighted-combination-of-variables framework.

A somewhat related area of study is the analysis of covariance structures, which is perhaps the most widely expanding form of multivariate analysis. This text touches on only one aspect of that—confirmatory factor analysis. Additional background would be needed in order to treat covariance structures in a more complete way.

Beginning in Chapter 9, the book relates statistical methods to three common statistical packages: BMDP, SAS, and SPSS^x. At the end of each chapter, a section on computer applications describes how the methods of each chapter can be implemented using these systems. These general descriptions are by no means a substitute for a careful reading of the manuals for the various systems, however.

The text also contains many detailed tables, or exhibits, that trace the calculations for analysis. The calculations are simplified as much as possible for these rather complex procedures. To understand the procedures better, it is helpful to go through the exhibits step by step. Each chapter also contains problem sets and their corresponding answers. Again, calculations are kept as simple as possible for the particular method being illustrated. Most can be carried out on a pocket calculator that includes a square-root calculation and a memory.

It will probably take more than a single semester to cover all the material included in this book. Chapters 1 through 12 could be the basis for a semester course in regression and the general linear model. The remaining chapters could be the basis for a more strictly multivariate course. If necessary, the easiest chapters to omit are 10, 11, 12, and 15.

Many readers come to a statistical book with the attitude that statistical analysis is a dreary business. On the contrary, one of the most interesting parts of any study is figuring out the appropriate statistical analyses and then sitting down with the printouts to see what in fact did happen. I hope this book encourages readers to enjoy this process, as well as guiding them in correct and valid analysis and inference.

A book like this depends on the efforts of many people. It benefited from the suggestions of several reviewers, including William I. Sauser, Jr., of Auburn University, Douglas K. Spiegel of the University of South Dakota, Lester C. Shine II of Texas A & M University, Douglas Herrmann of Hamilton College, and John C. Loehlin of the University of Texas, Austin. A number of typists have been patient and accurate with my manuscript, most recently Ms. Nana Sadamura. The students who attended my classes have also been instrumental, working with earlier typed versions and offering suggestions and corrections. To all these people I express my gratitude and appreciation.

Norman Cliff

CONTENTS

7 Partial Correlation and Stepwise Regression 108

8 Statistical Inference in Multivariate Statistics 134

ANALYZING MULTIVARIATE DATA

1

Elements of Matrix Algebra for Statistical Applications

DEFINITION OF DATA MATRICES

The use of multivariate statistics implies the existence of a matrix of data. From the original data matrix, one or more other matrices may need to be constructed. Although the statistical formulas of multivariate statistics can be stated using the more-or-less familiar expressions involving summation signs and subscripted variables, these expressions become very cumbersome. So, the shorthand of matrix algebra is needed. It is therefore important to become familiar with the manipulation of matrices and with the translation of formulas into and out of matrix notation. Both this chapter and the next will attempt to provide background in matrix development and description of multivariate statistical methods. A few other aspects of matrix algebra are introduced in later chapters as they become necessary.

A data matrix may be defined simply as a *rectangular table of numbers* on which it is legitimate to perform matrix algebra. Some sets of numbers that are not in a rectangular table can be arranged so that they do form such a table, or, more often, they can be treated symbolically as if they do. Alternatively, there are some tables of numbers on which it makes no sense to perform matrix algebra. Whether or not a set of numbers can be treated as a matrix ultimately depends on whether it can be legitimately considered to conform to the algebraic axioms of matrix theory, and whether considering it as a matrix serves any useful purpose.

If a table of numbers is to be considered a matrix, it must be arranged in an orderly fashion. A necessary characteristic is that any number (or value of a variable) that is part of a matrix has a tag that specifies which row and column of the matrix it belongs to. For example, x_{ij} is the value (or number) that belongs in row i and column j of matrix X. Conversely, there is a value of x that goes with every row and column combination of the matrix. Moreover, the indices i and j

must have a meaning, and the meaning must be consistent. If X is a matrix of data, the index j often refers to which *variable* is involved, and the index i tells which person or other experimental unit is referred to. In a matrix, all rows have the same number of entries and all columns have the same number of entries. The entries may be zero, but the row and column designations cannot be empty. So, to be a matrix, each entry in the table must occupy a definite row and column, and all the entries must be filled.

The requirement that a matrix be complete is the reason why missing data is a major concern in multivariate statistics. If a matrix has missing entries, then it is not really a matrix, and matrix mathematics applies only approximately at best.

Thus, the following sets of numbers could be arranged as matrices: the set of scores of each of 271 people on 43 tests; the number of responses of a subject observed under all combinations of four stimulus intensities and three durations of food deprivation; the per capita income, percentage of owner-occupied homes, ratio of persons to dwelling units, and average number of years of education of people living in each of the cities having populations of more than 10,000; and the number of messages sent from individual i to individual j. In each of these examples, the data numbers belong in a particular row and column designation (cell) and, given a row and column designation, we know the information (the value of the variable) that belongs in it. In the terminology of analysis of variance, the row variable and the column variable must be *crossed* for the table to be a matrix.

There are a few instances in which rectangular tables should not be treated as matrices. For example, the list of data from k groups of subjects, each group having n subjects, cannot ordinarily be treated as a matrix with k columns *if* there are different subjects in each group. This is because the row-and-column defining sets are not crossed; whatever determines which entry goes in which row is not necessarily the same from column to column. Within any column, the data (subjects) can be arbitrarily reordered. If, on the other hand, each row of the data came from the same subject, we could not arbitrarily rearrange the order within a single column. Such a table is a matrix; if we rearrange the order, we must make the same rearrangement in all the columns. This illustrates the requirement that there be a single property that determines which row a matrix element belongs in (for example, the subject who made the response) and which column the entry belongs in (for example, the condition under which the response was observed). Sometimes the same property may define both rows and columns.

We see that a matrix is a table, provided that certain logical conditions for its entries are met. This is only the most minimal restriction, however. In addition, certain kinds of arithmetic may be performed on matrices. The numbers or abstract variables representing numbers in a matrix may be added and subtracted, multiplied by pure numbers, and multiplied together in a particular way. These arithmetic operations are discussed later in this chapter.

In order to be able to refer to single numbers as well as matrices, appropriate terms are needed. A single number, or variable whose value is a single number, is called a *scalar*. The number 2 is a scalar, as is the number pi or the gross national product of the United States. A matrix having a single row or column, or any other one-dimensional list of numbers, is called a *vector*. The numbers (3, 5, 7) are a vector—so is the list of scores on a test, or all the scores of a person on several tests. We can think of a matrix as a two-dimensional array, a vector as a one-dimensional array, and a scalar as a zero-dimensional array. Indeed, some computer languages use these concepts. In this text, we will not consider arrays of more than two-dimensions. Two is hard enough!

MATRIX OPERATIONS

Notational Conventions

Matrices come in all sizes and degrees of rectangularity. They may have one to an infinite number of rows and columns. They may have equal numbers of rows and columns, in which case they are referred to as *square matrices*.

The size of a matrix is referred to as its *order*, and is given as a pair of numbers, the first being the number of rows: two by three, m by n, 1×2, $r \times 4$, 47×243, and so on. A matrix presented as a table is usually enclosed in large brackets, as if there were a danger of the entries escaping (see Exhibit 1-1).

It is customary to use a single capital letter to stand for a matrix: these letters often appear in boldface, although boldface notation will not be used here. The letters G, X, O, P, and Z all stand for matrices. A few *italic* uppercase letters are used for other purposes. For example, R is the multiple correlation. For a given matrix, A or R, the entries are referred to by using the same letter in lowercase. This symbol is subscripted to designate the row and column to which it belongs, with the row index again coming first. Thus, a_{23} refers to the entry in the second row and third column of matrix A. A comma may be used between the subscripts to avoid ambiguity: $a_{4,17}$. Most frequently, however, one refers to an unspecified or generalized element such as a_{ij}, x_{pq}, z_{km}, and so on.

Another method of designating a matrix is the use of a generalized element, either instead of or in addition to the capital letter. Here, the symbol is conventionally enclosed in double vertical lines: $\|a_{ij}\|$, $\|r_{pq}\|$, $\|z_{km}\|$. A rather complete definition of a particular matrix may be given in the following manner:

$$X = \|x_{ij}\| \qquad (i = 1, 2, \ldots, n;\ j = 1, 2, \ldots, m)$$

$$X = \begin{bmatrix} x_{11} & x_{12} & \cdots & x_{1n} \\ x_{21} & x_{22} & \cdots & x_{2n} \\ \vdots & \vdots & \vdots\vdots & \vdots \\ x_{m1} & x_{m2} & \cdots & x_{mn} \end{bmatrix} \quad P = \begin{bmatrix} p_{11} & p_{12} & \cdots & p_{1r} \\ p_{21} & p_{22} & \cdots & p_{2r} \\ \vdots & \vdots & \vdots\vdots & \vdots \\ p_{r1} & p_{r2} & \cdots & p_{rr} \end{bmatrix} \quad Z = \begin{bmatrix} z_{11} & z_{12} & \cdots & z_{1,243} \\ z_{21} & z_{22} & \cdots & z_{2,243} \\ \vdots & \vdots & \vdots\vdots & \vdots \\ z_{47,1} & z_{47,2} & \cdots & z_{47,243} \end{bmatrix}$$

$$m \times n \qquad\qquad r \times r \qquad\qquad 47 \times 243$$

$$A = \begin{bmatrix} a_{11} & a_{12} & a_{13} \\ a_{21} & a_{22} & a_{23} \end{bmatrix} \qquad \Theta = [\theta_{11}\theta_{12}]$$

$$2 \times 3 \qquad\qquad 1 \times 2$$

Exhibit 1-1 Examples of matrices.

This means that the matrix X has typical element x_{ij} where the index i runs from one to n (n rows) and j runs from one to m (m columns).

Lowercase boldface letters refer to vectors, with **a**, **q**, **x**, and so on, referring to vectors arranged as a column matrix, and $\mathbf{a}'$, $\mathbf{q}'$, $\mathbf{x}'$ referring to a row matrix. Occasionally, uppercase letters are used to refer to a matrix that happens to have only one row or column.

There are a number of different kinds of matrices, and some have special names that describe their appearance or notational characteristics. Already noted was the square matrix, which has the same number of rows as columns. A *rectangular* matrix is one that does not necessarily have the same number of rows as columns. An important part of a square matrix is its *main diagonal.* This refers to the elements running from the upper left to lower right corner whose row subscript and column subscript are equal. For example, a_{11}, a_{22}, or, more abstractly, a_{jj}. A special kind of square matrix is a *symmetric* matrix, in which the symmetrically placed elements are equal: $a_{12} = a_{21}$; $a_{35} = a_{53}$. Thus, if $a_{jk} = a_{kj}$, for all j and k, the matrix A is symmetric. Symmetric matrices are common in statistical applications; a matrix of correlation coefficients, for example, is symmetric.

A special kind of symmetric matrix is one in which all the elements *except* those on the main diagonal are equal to zero. Such a matrix is called a *diagonal* matrix. Diagonal matrices have special properties and serve some special functions in matrix algebra. A matrix that is all zeros on just one side of the diagonal is called *triangular.* If the zeros are above the diagonal, it is called *lower* triangular; that is, if $t_{pq} = 0$ for all $q > p$, the matrix T is lower triangular. If the zeros are below the diagonal, the matrix is *upper* triangular. In either case, the diagonal elements are not assumed to be zero, although conceivably some or all the diagonal entries might happen to be zero.

In addition to their differences in appearance, these various kinds of matrices also have different mathematical properties, which will be considered later. The mathematical properties of other special kinds of matrices also lead to different terms; a few of these terms will also be addressed later on.

Exhibit 1-2 illustrates some of the different types of matrices discussed so far.

Addition and Subtraction

Addition and subtraction of matrices takes place in the usual manner, but must be done element by element. If $\mathrm{C} = \mathrm{A} + \mathrm{B}$, $c_{ij} = a_{ij} + b_{ij}$ for all i and j. If $\mathrm{D} = \mathrm{A} - \mathrm{B}$, $d_{ij} = a_{ij} - b_{ij}$ for all i and j. Here are some examples:

$$\begin{bmatrix} 2 & 1 \\ 4 & 7 \end{bmatrix} + \begin{bmatrix} 1 & 0 \\ 2 & 3 \end{bmatrix} = \begin{bmatrix} 2+1 & 1+0 \\ 4+2 & 7+3 \end{bmatrix} = \begin{bmatrix} 3 & 1 \\ 6 & 10 \end{bmatrix}$$

$$\begin{bmatrix} -4 & 2 \\ 1 & 0 \\ 1 & -1 \end{bmatrix} + \begin{bmatrix} 5 & 1 \\ -2 & 1 \\ 2 & -2 \end{bmatrix} = \begin{bmatrix} -4+5 & 2+1 \\ 1-2 & 0+1 \\ 1+2 & -1-2 \end{bmatrix} = \begin{bmatrix} 1 & 3 \\ -1 & 1 \\ 3 & -3 \end{bmatrix}$$

$$\begin{bmatrix} 3 & -1 \\ -1 & 2 \end{bmatrix} - \begin{bmatrix} -2 & -9 \\ 1 & 4 \end{bmatrix} = \begin{bmatrix} 3-(-2) & -1-(-9) \\ -1-1 & 2-4 \end{bmatrix} = \begin{bmatrix} 5 & 8 \\ -2 & -2 \end{bmatrix}$$

Symmetric Matrices

$$\begin{bmatrix} 1 & 2 & 3 \\ 2 & 4 & 5 \\ 3 & 5 & 6 \end{bmatrix} \quad \begin{bmatrix} 1.00 & .45 & .26 & .11 \\ .45 & 1.00 & .84 & .23 \\ .26 & .84 & 1.00 & .52 \\ .11 & .23 & .52 & 1.00 \end{bmatrix} \quad \begin{bmatrix} 2.10 & .98 & -.77 \\ .98 & 4.13 & 1.25 \\ -.77 & 1.25 & 5.00 \end{bmatrix}$$

Diagonal Matrices

$$\begin{bmatrix} 1 & 0 & 0 \\ 0 & 2 & 0 \\ 0 & 0 & 3 \end{bmatrix} \quad \begin{bmatrix} .64 & 0 & 0 \\ 0 & 1.15 & 0 \\ 0 & 0 & 2.11 \end{bmatrix} \quad \begin{bmatrix} a_{11} & 0 & 0 \\ 0 & a_{22} & 0 \\ 0 & 0 & a_{33} \end{bmatrix}$$

Triangular Matrices

$$\underset{\text{(lower)}}{\begin{bmatrix} 1 & 0 & 0 \\ 2 & 3 & 0 \\ 4 & 5 & 6 \end{bmatrix}} \quad \underset{\text{(upper)}}{\begin{bmatrix} 2 & 4 & 6 \\ 0 & 8 & 10 \\ 0 & 0 & 0 \end{bmatrix}} \quad \underset{\text{(lower)}}{\begin{bmatrix} t_{11} & 0 & 0 \\ t_{21} & t_{22} & 0 \\ t_{31} & t_{32} & t_{33} \end{bmatrix}}$$

Exhibit 1-2 Examples of different types of matrices.

Note that *only matrices of the same size may be added or subtracted.* Such matrices are of the *same order.*

As in ordinary addition, the addition of matrices is commutative and associative. This means that

$$\text{(1)} \quad A + B = B + A$$

$$\text{(2)} \quad A + B + C = (A + B) + C = A + (B + C)$$

for any matrices A, B, and C, that can legitimately be added together. The first axiom is called the *commutative law,* which states that the sequence in which things (in this case, matrices) are added does not affect the sum; this is analogous to $3 + 7 = 7 + 3$ or $x + y = y + x$. The second axiom is the *associative law,* which states that if three or more things are added, "subtotals" between adjacent terms may be formed in any way and added together without changing the sum. In combination, the two laws may be interpreted as saying that matrices may be added together in any order or combination without affecting the sum.

Like ordinary subtraction, matrix subtraction is *distributive:*

$$A - (B + C) = A - B - C$$

Also, for any matrix A there exists its negative, $-A$, whose elements are the negatives of the elements of A. So, as in ordinary algebra, subtraction may be translated into the addition of the negative of a matrix. The following rule parallels the ordinary subtraction of scalars:

$$\mathrm{A} - (\mathrm{B} - \mathrm{C}) = \mathrm{A} - \mathrm{B} - (-\mathrm{C}) = \mathrm{A} - \mathrm{B} + \mathrm{C}$$

This rule is a combination of the distributive rule for subtraction and the existence of the negative of a matrix. When working with ordinary numbers, we make frequent use of zero. Its prime utility is that zero added to any quantity leaves the quantity unchanged:

$$2 + 0 = 2 \qquad x + 0 = x$$

Corresponding to the ordinary zero is the zero or *null* matrix, a matrix consisting entirely of zeros. Because the addition of matrices is the addition of the corresponding elements, adding the zero matrix to any matrix leaves it unchanged:

$$\begin{bmatrix} 2 & 1 \\ 4 & -3 \end{bmatrix} + \begin{bmatrix} 0 & 0 \\ 0 & 0 \end{bmatrix} = \begin{bmatrix} 2+0 & 1+0 \\ 4+0 & -3+0 \end{bmatrix} = \begin{bmatrix} 2 & 1 \\ 4 & -3 \end{bmatrix}$$

This may be stated more generally as

$$\mathrm{A} + 0 = 0 + \mathrm{A} = \mathrm{A} - 0 = \mathrm{A}$$

for any matrix A. Here 0 refers to the null matrix, a matrix of zeros.

In algebra, zero is used to solve equations. For example, if $2 + x = 3$, $x = 1$. Solving equations of this type may be so routine that one forgets that it involves the following series of intermediate steps:

$$\begin{aligned} 2 + x &= 3 \\ (-2) + 2 + x &= (-2) + 3 \\ 0 + x &= 1 \\ x &= 1 \end{aligned}$$

The zero matrix is similarly used in solving matrix equations. Suppose we have the equation $\mathrm{A} + \mathrm{X} = \mathrm{B}$, in which the matrices are

$$\underset{\mathrm{A}}{\begin{bmatrix} 2 & 4 \\ 1 & -3 \end{bmatrix}} + \underset{\mathrm{X}}{\begin{bmatrix} x_{11} & x_{12} \\ x_{21} & x_{22} \end{bmatrix}} = \underset{\mathrm{B}}{\begin{bmatrix} 5 & 2 \\ 1 & -2 \end{bmatrix}}$$

Note the way the subscripts are arranged on entries of matrix X. This equation may be solved by adding $-A$ to both sides:

$$\begin{bmatrix} -2 & -4 \\ -1 & 3 \end{bmatrix} + \begin{bmatrix} 2 & 4 \\ 1 & -3 \end{bmatrix} + \begin{bmatrix} x_{11} & x_{12} \\ x_{21} & x_{22} \end{bmatrix} = \begin{bmatrix} -2 & -4 \\ -1 & 3 \end{bmatrix} + \begin{bmatrix} 5 & 2 \\ 1 & -2 \end{bmatrix}$$

$$-A \quad + \quad A \quad + \quad X \quad = \quad -A \quad + \quad B$$

$$\begin{bmatrix} 0 & 0 \\ 0 & 0 \end{bmatrix} + \begin{bmatrix} x_{11} & x_{12} \\ x_{21} & x_{22} \end{bmatrix} = \begin{bmatrix} 5-2 & 2-4 \\ 1-1 & -2+3 \end{bmatrix}$$

$$0 \quad + \quad X \quad = \quad (-A + B)$$

$$\begin{bmatrix} x_{11} & x_{12} \\ x_{21} & x_{22} \end{bmatrix} = \begin{bmatrix} 3 & -2 \\ 0 & 1 \end{bmatrix}$$

$$X \quad = (-A + B)$$

Thus, matrix equations of this simple additive type are readily solved.

This equation implies that $x_{11} = 3$; $x_{12} = 2$; $x_{21} = 0$; and $x_{22} = 1$. Note that it is as if we had an equation for each entry in X ($2 + x_{11} = 5$, and so on) and solved each of these equations for x_{ij}. This illustrates an additional rule: *Two matrices are equal if and only if each of their elements is equal.*

We see that the addition and subtraction of matrices parallels ordinary addition and subtraction and obeys the following rules:

1. Matrices are added and subtracted in the same way as scalars.
2. The negative of any matrix exists.
3. There is a matrix consisting entirely of zeros called the null matrix.
4. Equations of the type $A + X = B$ can be solved.

The fact that matrices consist of many elements and may be of a variety of different sizes makes the following two qualifications necessary:

5. Matrices can be added only if they have the same dimensions (are of the same order).
6. Matrices are equal if and only if each of their elements is equal.

Multiplication

There are two types of multiplication involving matrices. The simpler kind is used when we wish to multiply all the entries in a matrix by the same number, such as 2, .10, 1.65, 3, or some unspecified constant k or n. Multiplying by a constant includes multiplying by a reciprocal, 1/2, $1/n$, that is, dividing the entries in a matrix by a constant. Multiplying or dividing all the entries in a matrix by a constant is called *scalar multiplication.* Scalar multiplication is symbolized by letting a lowercase letter stand for the constant: kA means "multiply each entry in A by the constant k." The k can come first or second: kA and Ak mean the same thing. Dividing by a constant is shown as multiplication by a reciprocal, for example $(1/n)$X or n^{-1}X both mean "divide every element of X by n." Thus, *for every scalar k and matrix* X, the scalar product $k\mathrm{X} = \mathrm{V} = \mathrm{X}k$ *exists, where* $v_{ij} = kx_{ij}$ *for every i and j.*

Matrix multiplication involves a different and more complex operation. For example, when we say C = AB, every element of C is represented as follows:

$$c_{jk} = \sum_i a_{ji} b_{ik} \tag{1.1}$$

where $i = 1$ to n; $j = 1$ to m; $k = 1$ to p. Equation 1.1 is fundamental to the material presented in later chapters, so it needs to become familiar. Here, A is an m by n matrix and B is n by p, according to the limits given on the right of Equation 1.1. The element in row j and column k of the product matrix C is found by taking *row* j of A (the left member of the product) and *column* k of B (the right member) and computing the sum of products of corresponding entries.

For example, the product of A and B is computed as follows:

$$\underset{\text{A}}{\begin{bmatrix} 2 & 3 \\ 1 & -1 \end{bmatrix}} \underset{\text{B}}{\begin{bmatrix} 1 & 3 & -3 \\ 2 & 0 & 2 \end{bmatrix}} \underset{=}{=} \underset{\text{C}}{\begin{bmatrix} 8 & 6 & 0 \\ -1 & 3 & -5 \end{bmatrix}}$$

$$\begin{bmatrix} ② & ③ \\ 1 & -1 \end{bmatrix} \begin{bmatrix} ① & 3 & -3 \\ ② & 0 & 2 \end{bmatrix} = \begin{bmatrix} 2 \times 1 + 3 \times 2 & 2 \times 3 + 3 \times 0 & 2 \times -3 + 3 \times 2 \\ 1 \times 1 + -1 \times 2 & 1 \times 3 + -1 \times 0 & 1 \times -3 + -1 \times 2 \end{bmatrix}$$

It may help to visualize the matrix on the left as being turned on its side:

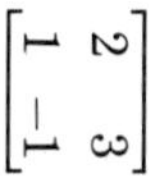

$$\begin{bmatrix} 1 & 3 & -3 \\ 2 & 0 & 2 \end{bmatrix}$$

Here are some more examples; follow through each step.

(a) $$\begin{bmatrix} 3 & 0 \\ 0 & 2 \end{bmatrix} \begin{bmatrix} 1 \\ 2 \end{bmatrix} = \begin{bmatrix} 3 \times 1 + 0 \times 2 \\ 0 \times 1 + 2 \times 2 \end{bmatrix} = \begin{bmatrix} 3 \\ 4 \end{bmatrix}$$

(b) $$\begin{bmatrix} 4 & 1 \\ -1 & -1 \\ 0 & 1 \end{bmatrix} \begin{bmatrix} 3 & 0 & 4 \\ 1 & 2 & -1 \end{bmatrix} = \begin{bmatrix} 13 & 2 & 15 \\ -4 & -2 & -3 \\ 1 & 2 & -1 \end{bmatrix}$$

(c) $\begin{bmatrix} 0 & 2 \\ -1 & 4 \end{bmatrix} \begin{bmatrix} 1 & 0 \\ 0 & 1 \end{bmatrix} = \begin{bmatrix} 0 & 2 \\ -1 & 4 \end{bmatrix}$

(d) $\begin{bmatrix} a & b \\ c & d \end{bmatrix} \begin{bmatrix} e & c \\ f & g \end{bmatrix} = \begin{bmatrix} ae + bf & ac + bg \\ ce + df & c^2 + dg \end{bmatrix}$

Now solve the following examples:*

(e) $\begin{bmatrix} 2 & 1 \\ 1 & 3 \end{bmatrix} \begin{bmatrix} 4 & 1 \\ 0 & 2 \end{bmatrix} =$

(f) $\begin{bmatrix} 2 & 3 & 1 \\ 1 & 1 & 0 \end{bmatrix} \begin{bmatrix} 3 & 0 & 0 \\ 0 & 1 & 0 \\ 0 & 0 & 2 \end{bmatrix} =$

Remember that the *order* of a matrix is the number of its rows and columns. To be multiplied, the matrix on the *left* must have *as many columns* as the matrix on the right has *rows*. Also, when multiplying an m by n matrix by an n by p matrix, the product matrix must be m by p because m is the number of *rows* of the matrix on the left and p is the number of *columns* of the matrix on the right. This can be verified with all the examples presented so far. Now find the order of the product matrix in the following example:

(g) $\begin{bmatrix} 2 & 1 & 7 \\ 4 & -1 & 0 \end{bmatrix} \begin{bmatrix} 1 & 0 & 4 \\ -1 & -1 & 1 \\ 2 & 3 & -2 \end{bmatrix}$

Which of the following are not proper matrix products?†

(h) $\begin{bmatrix} 2 & 1 & 0 \end{bmatrix} \begin{bmatrix} 3 & 2 \\ 1 & -1 \end{bmatrix}$

(i) $\begin{bmatrix} 4 & 4 \\ 2 & 1 \end{bmatrix} \begin{bmatrix} 3 & 1 & 1 \\ -1 & 0 & -1 \end{bmatrix}$

---

*Answers: (e) $\begin{bmatrix} 8 & 4 \\ 4 & 7 \end{bmatrix}$; (f) $\begin{bmatrix} 6 & 3 & 2 \\ 3 & 1 & 0 \end{bmatrix}$; (g) 2 rows by 3 columns.

†Answers: (h), (j), and (k) are improper.

(j) $\begin{bmatrix} 2 & 2 \end{bmatrix} \begin{bmatrix} 1 \\ 4 \\ 7 \end{bmatrix}$

(k) $\begin{bmatrix} x_{11} & x_{12} & x_{13} \\ x_{21} & x_{22} & x_{23} \end{bmatrix} \begin{bmatrix} a_{11} & a_{12} \\ a_{21} & a_{22} \end{bmatrix}$

Let us return to the general example to review the correspondence between the scalar and matrix formulas and the limitations that are implied for legitimate matrix multiplication:

$$c_{jk} = \sum_i a_{ji} b_{ik}$$

What is most important is keeping track of the subscripts. The first example shows the dimensions of the matrices labeled with the appropriate subscripts:

$$\overset{\quad i}{j\begin{bmatrix} 2 & 3 \\ 1 & -1 \end{bmatrix}} \; \overset{\quad k}{i\begin{bmatrix} 1 & 3 & -3 \\ 2 & 0 & 2 \end{bmatrix}} = \overset{\quad k}{j\begin{bmatrix} 8 & 6 & 0 \\ -1 & 3 & -5 \end{bmatrix}}$$

In the scalar formula, the elements of the product were written in the same order they appeared in the matrix multiplication, a_{ji} followed by b_{ik}. Thus, the *second* subscript of the *first* member of the product is i, the same as the *first* subscript of the *second* member of the product. Furthermore, this is the subscript that is summed over. These are universal characteristics of concrete matrix multiplication, in which the matrices stand for actual tables of numbers. The two elements must share a subscript in common, and this must be the subscript that is summed over. When the elements of the scalar formula are in the same order as the matrix multiplication, the first subscript of the left member defines the row of the product element and the second subscript of the right member defines the column of the product element.* That is,

$$\sum a_{3i} b_{i2} = c_{32}$$

This is very important because it is the key to translating statistical formulas into matrix notation, and vice versa. Note that the column index of the left matrix (called the *pre*multiplier) must *refer to the same set* as the row index of the right matrix (called the *post*multiplier). That is, they must both refer to the same set of variables, persons, dimensions, and so on, if the multiplication is to make sense. It is not enough that the *number* of *columns* of the premultiplier be the same as the *number* of *rows* of the postmultiplier; the left column index must refer to the same set as the right row index.

*Avoid trying to memorize the principles in this section in written form; it is much better to remember these things visually. Not only is the visual memory more efficient than verbal memory, it is also much less susceptible to interference.

If one instructs a computer to multiply matrices of the wrong size, it will balk, perhaps even print a supercilious comment. If one tells it to multiply two matrices that are the right size but are defined by the wrong indices, it will readily oblige, even though the answer will be nonsense.

Consider the following matrix multiplication:

$$\begin{bmatrix} 4 & 1 & 1 \end{bmatrix} \begin{bmatrix} 2 \\ 3 \\ 1 \end{bmatrix} = [12]$$

Note what happens when we multiply these matrices in the opposite order:

$$\begin{bmatrix} 2 \\ 3 \\ 1 \end{bmatrix} \begin{bmatrix} 4 & 1 & 1 \end{bmatrix} = \begin{bmatrix} 8 & 2 & 2 \\ 12 & 3 & 3 \\ 4 & 1 & 1 \end{bmatrix}$$

The matrix on the left has three rows, and the one on the right has three columns, making the product 3 by 3 rows. Compare the product below to (e) on page 9 in which the same matrices are multiplied in the opposite order.

$$\begin{bmatrix} 4 & 1 \\ 0 & 2 \end{bmatrix} \begin{bmatrix} 2 & 1 \\ 1 & 3 \end{bmatrix} = \begin{bmatrix} 9 & 7 \\ 2 & 6 \end{bmatrix}$$

This example shows that matrix multiplication is not *in general* commutative, AB ≠ BA. Some matrices have this property. We take note of these exceptions as the occasion arises.

Matrix multiplication is *associative,* however, just like ordinary multiplication:

$$ABC = A(BC) = (AB)C$$

That is, when multiplying three or more matrices together, it does not matter which subproducts we multiply first, as long as they are between adjacent matrices. For example, A, B, and C could be multiplied as follows:

$$\underset{A}{\begin{bmatrix} 2 & 4 \\ 1 & -1 \end{bmatrix}} \underset{B}{\begin{bmatrix} 3 & 4 \\ 0 & 2 \end{bmatrix}} \underset{C}{\begin{bmatrix} 1 & 2 \\ 1 & 0 \end{bmatrix}} = \underset{(AB)}{\begin{bmatrix} 6 & 16 \\ 3 & 2 \end{bmatrix}} \underset{C}{\begin{bmatrix} 1 & 2 \\ 1 & 0 \end{bmatrix}} = \underset{(ABC)}{\begin{bmatrix} 22 & 12 \\ 5 & 6 \end{bmatrix}}$$

Alternatively, B and C could be multiplied first:

$$\underset{A}{\begin{bmatrix} 2 & 4 \\ 1 & -1 \end{bmatrix}} \underset{B}{\begin{bmatrix} 3 & 4 \\ 0 & 2 \end{bmatrix}} \underset{C}{\begin{bmatrix} 1 & 2 \\ 1 & 0 \end{bmatrix}} = \underset{A}{\begin{bmatrix} 2 & 4 \\ 1 & -1 \end{bmatrix}} \underset{(BC)}{\begin{bmatrix} 7 & 6 \\ 2 & 0 \end{bmatrix}} = \underset{(ABC)}{\begin{bmatrix} 22 & 12 \\ 5 & 6 \end{bmatrix}}$$

This property is useful for simplifying both algebra and arithmetic in many applications.

In ordinary algebra or arithmetic, one frequently makes use of the *distributive* law as follows:

$$2(4 + 7) = 2(4) + 2(7); \quad 8(6) + 8(4) = 8(6 + 4); \quad (b + c)d = bd + cd$$

Like the associative law, the distributive law applies to matrices as well as other algebraic systems. *The product of a matrix with the sum of two matrices is the sum of its products with each:*

$$\mathrm{Q(M + Z) = QM + QZ}$$

$$\mathrm{(R + S)V = RV + SV}$$

The converse of this relationship is also useful; one has frequent occasion to substitute a sum of products like those on the right side of the above equations with the equivalent form on the left. This is called "factoring out." Note, however, that for QM + ZQ the failure of the commutative law *prevents* factoring out Q, even though it appears twice, except in special cases in which the multiplication is commutative.

For our purposes, the basic properties of matrix multiplication can be summarized as follows:

1. If AB = C, where $\|a_{ji}\|$ is m by n, and $\|b_{ik}\|$ is n by p, then $\|c_{jk}\|$ is m by p,

$$c_{jk} = \sum_i a_{ji} b_{ik}$$

(Refer to Equation 1.1.) This describes the operation of matrix multiplication, which has the following properties:

2. ABC = A(BC) = (AB)C (associative law)

3. S(T + U) = ST + SU (distributive law)

4. AB ≠ BA (Except in special cases, multiplication is noncommutative.)

Transposing

A matrix operation that is important for representing many calculations but which has no direct counterpart in ordinary arithmetic is called *transposing*. Every matrix A has its transpose, denoted A′, or in some cases, A^t. The transpose A′ is the same as A, except that each row of A is a column of A′ as follows:

$$\mathrm{A} = \begin{bmatrix} 3 & 1 & 2 \\ -1 & 2 & 4 \end{bmatrix} \qquad \begin{bmatrix} 3 & -1 \\ 1 & 2 \\ 2 & 4 \end{bmatrix} = \mathrm{A}'$$

The matrix on the right is the transpose of the one on the left.

The transpose of a transpose is the original matrix. Thus the transpose of A′ is A, which is true for all matrices:

$$(A')' = A$$

The purpose of the transpose operation is to arrange a matrix so that a particular multiplication can be carried out. A common operation in statistical applications is to compute the sums of squares and products. Suppose $X = \|x_{ij}\|$ represents the scores of n persons on p variables, and that $n = 4$ and $p = 3$.

$$X = \begin{bmatrix} 3 & 1 & 1 \\ 1 & 1 & 0 \\ 1 & 2 & 2 \\ 3 & 2 & 1 \end{bmatrix}$$

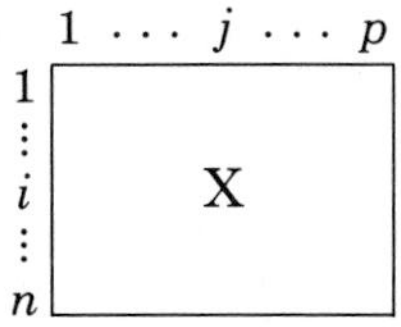

Now suppose we transpose X and multiply X′X (note the order):

$$X'X = \begin{bmatrix} 3 & 1 & 1 & 3 \\ 1 & 1 & 2 & 2 \\ 1 & 0 & 2 & 1 \end{bmatrix} \begin{bmatrix} 3 & 1 & 1 \\ 1 & 1 & 0 \\ 1 & 2 & 2 \\ 3 & 2 & 1 \end{bmatrix}$$

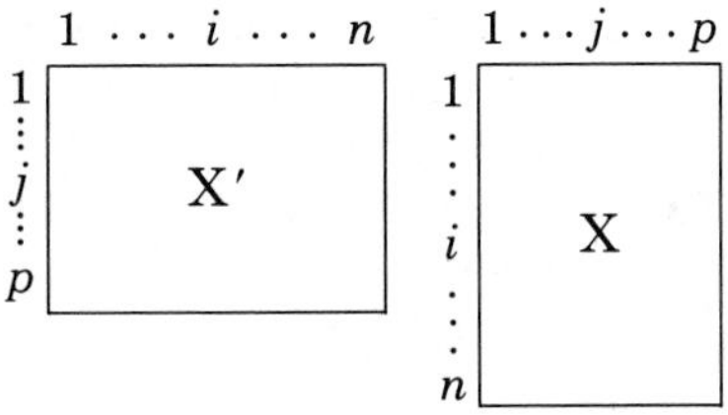

According to Equation 1.1,

$$X'X = \begin{bmatrix} 20 & 12 & 8 \\ 12 & 10 & 7 \\ 8 & 7 & 6 \end{bmatrix} \begin{bmatrix} \Sigma_i x_{i1}^2 & \Sigma_i x_{i1}x_{i2} & \Sigma_i x_{i1}x_{i3} \\ \Sigma_i x_{i2}x_{i1} & \Sigma_i x_{i2}^2 & \Sigma_i x_{i2}x_{i3} \\ \Sigma_i x_{i3}x_{i1} & \Sigma_i x_{i3}x_{i2} & \Sigma_i x_{i3}^2 \end{bmatrix}$$

Next, let $T = X'X$ and use k as an alternate subscript for variables. The general expression is $t_{jk} = \Sigma_i x_{ij}x_{ik}$ in which t_{jk} is a sum of products if j and k are different and a sum of squares if they are the same.

In the example, it is clear that $t_{12} = t_{21}$; $t_{13} = t_{31}$; and $t_{23} = t_{32}$, and in general, $t_{jk} = t_{kj}$ when $T = X'X$. Thus, T is an example of a symmetric matrix. We will deal with matrix products involving transposes in more detail in Chapter 2.

Multiplying Special Kinds of Matrices

So far we have considered matrix multiplication as a general procedure. When this multiplication is applied to special kinds of matrices, it may turn out to have special properties. Naturally enough, these types of matrices with special properties may be given their own names.

Consider the following example in which the matrix on the left side of the equation is the same as the one on the far right, even though the latter has been multiplied by a matrix:

$$\begin{bmatrix} 0 & 2 \\ -1 & 4 \end{bmatrix} = \begin{bmatrix} 1 & 0 \\ 0 & 1 \end{bmatrix} \begin{bmatrix} 0 & 2 \\ -1 & 4 \end{bmatrix}$$
$$\mathrm{A} \quad = \quad \mathrm{I} \quad \mathrm{A}$$

The same result would occur if we postmultiplied instead of premultiplied by matrix I. This matrix, which has no effect on A, is called the *identity* matrix. In general, the identity matrix has a row of ones along the main diagonal (that is, the elements where the row index and the column index are equal) and zeros everywhere else. *The letter I is reserved for this matrix.* Its property may be stated as follows:

There is a matrix I such that for any matrix A, AI = IA = A.

Thus, I has the same property that the number 1 has in arithmetic. It is *multiplicative identity,* whereas zero and the null matrix are additive identities. The identity matrix need not be thought of as having a particular size—that is why we speak of *the* identity rather than *an* identity matrix. It is as big as it needs to be in any given context. The identity matrix is extremely important to the solution of matrix equations. In statistical applications, it is perhaps more important than the null matrix.

Suppose we multiply the following matrices:

$$\begin{bmatrix} 2 & 0 \\ 0 & 2 \end{bmatrix} \begin{bmatrix} 4 & 1 \\ 3 & -2 \end{bmatrix} = \begin{bmatrix} 8 & 2 \\ 6 & -4 \end{bmatrix}$$
$$\mathrm{C} \quad \mathrm{W} \quad = \quad \mathrm{V}$$

Remember scalar multiplication, multiplying every entry by the same constant. Because each entry of V is just twice that of W, we see that scalar multiplication is like multiplying by a matrix that was similar to C in that it was all zero except it had a constant along the main diagonal.

This special kind of matrix multiplication *is* commutative, because $c\mathrm{W} = \mathrm{W}c$ for all scalars c and matrices W. Multiplication by the identity is multiplication by a scalar matrix where $c = 1$.

In example (a) (p. 8), the left matrix, the premultiplier, has nonzero numbers in the main diagonal only, like the postmultiplier of (f) (p. 9). Thus they resemble scalar matrices. However, because the main diagonal entries are not the same number, they are referred to as *diagonal* matrices:

If D (n by n) has entry $d_{ij} = 0$ when $i \neq j$, then D is a diagonal matrix.

Multiplication by a diagonal matrix has a very specific effect; compare the product from (f) (p. 9) with the premultiplier. Now, compare the product in (a) with the postmultiplier. The principles of this operation may be stated as follows: When a matrix A is *postmultiplied* by a diagonal matrix D, the entries in each *column* of A are multiplied by the corresponding diagonal entry of D. If B = AD, $b_{ij} = a_{ij}d_{jj}$. When A is *premultiplied* by a diagonal matrix, the entries in each *row* of A are multiplied by the corresponding diagonal entry of D. So, if C = DA, $c_{ij} = d_{ii}\, a_{ij}$. The reason is that in applying the general multiplication formula, $c_{ij} = \Sigma_k\, d_{ik} a_{kj}$, all the entries in column j are multiplied by zero *except* the ith. When we want to multiply each column by a different constant, we express it in matrices as postmultiplication by a diagonal matrix.

Sometimes the elements of a diagonal matrix are referred to with only a single subscript, since the second one is redundant. Therefore, it is important to know whether a scalar expression involving a variable with a single subscript refers to a diagonal matrix or to a vector. This can be determined by examining any summations that are involved. If there is a sum over the singly subscripted variable, it implies multiplication by a vector. If there is no sum over the singly subscripted variable, it implies multiplication by a diagonal matrix.

SUMMARY

A matrix is a rectangular table of numbers, or symbols of numbers, with certain specialized properties. The indices (subscripts) of an element x_{ij} of matrix X refer to particular sets, and the meaning of one index must stay the same, whatever the value of the other. A certain kind of arithmetic may be performed on matrices, including addition, subtraction, and a special kind of multiplication. A variety of descriptive terms specify matrices with particular forms.

Matrix operations are similar in many respects to their ordinary scalar counterparts. Addition and subtraction of matrices take place by adding (subtracting) the corresponding elements of the matrices. Matrix multiplication is more complicated, taking the special form shown in Equation 1.1. An operation that is unique to matrices is transposing; the transpose of a matrix has the same elements, but the rows of the matrix become the columns of the transpose.

Addition and subtraction of matrices are associative and commutative; multiplication of matrices is associative but not commutative, except in special cases. Multiplication is distributive over addition and subtraction. The null matrix, consisting of all zeros, serves as an additive identity (the same as a scalar zero). The identity matrix—a square matrix that is all zeros except for the main diagonal, which is all unities—serves as a multiplicative identity.

COMPUTER APPLICATIONS

Many computer installations have packages that carry out matrix algebra in a very literal fashion. The user inputs certain matrices, and then writes equations using them in a very close counterpart to matrix equations. The computer does the rest. Two of the most useful examples are Speakeasy, which has the advantage of being interactive, and the PROC MATRIX aspect of SAS. Using such a system is quite beneficial in familiarizing oneself with matrix operations.

PROBLEMS

Several matrices are defined below. Perform the following operations, indicating any that are impossible to perform.

$$\underset{A}{\begin{bmatrix} 2 & 1 \\ 2 & 4 \end{bmatrix}} \quad \underset{B}{\begin{bmatrix} 1 & -1 \\ 2 & 4 \end{bmatrix}} \quad \underset{C}{\begin{bmatrix} 4 & 0 \\ 0 & 9 \end{bmatrix}} \quad \underset{D}{\begin{bmatrix} 2 & 6 & 10 \\ 4 & 8 & 12 \end{bmatrix}} \quad \underset{E}{\begin{bmatrix} .8 & -.2 \\ -.6 & .4 \end{bmatrix}} \quad \underset{F}{[1 \ \ 1 \ \ 1]} \quad \underset{G}{[3 \ \ 2]}$$

1. A + B 2. B + A. What is its relation to Problem 1? 3. B + D
4. A + B + C. Perform the operations in two ways.
5. A′ 6. C′. What can you say about it? 7. (B′)′ 8. AB
9. AC. Describe this product. 10. BA 11. CA. Describe this product.
12. AD 13. DD′. What is in the main diagonal of DD′?
14. D′D. What is in the main diagonal of D′D?
15. DF′. Describe this product.
16. GD 17. AE 18. GAC. Perform the operations in two ways.
19. EA 20. (AB)′ 21. B′A′. What is its relation to Problem 20?

Use the following matrices to perform the operations described in problems 22–26.

$$\underset{H}{\begin{bmatrix} h_{11} & h_{12} \\ h_{21} & h_{22} \\ h_{31} & h_{32} \end{bmatrix}} \quad \underset{J}{\begin{bmatrix} j_{11} & j_{12} \\ j_{21} & j_{22} \end{bmatrix}} \quad \underset{K}{\begin{bmatrix} k_{11} & 0 \\ 0 & k_{22} \end{bmatrix}} \quad \underset{M}{\begin{bmatrix} m_{11} & 0 & 0 \\ 0 & m_{22} & 0 \\ 0 & 0 & m_{33} \end{bmatrix}}$$

22. H′H; HH′. Express the elements of the products both using Σ signs and as complete sums: $h_{11}^2 + h_{12}^2$, and so on.
23. HJ 24. HK, JK, KJ, and K′J. K′J = ? 25. MH and H′M 26. J + K; J − K

ANSWERS

1. $\begin{bmatrix} 3 & 0 \\ 4 & 8 \end{bmatrix}$ 2. Same as Problem 1. 3. Impossible.

4. $\begin{bmatrix} 3 & 0 \\ 4 & 8 \end{bmatrix} + \begin{bmatrix} 4 & 0 \\ 0 & 9 \end{bmatrix} = \begin{bmatrix} 2 & 1 \\ 2 & 4 \end{bmatrix} + \begin{bmatrix} 5 & -1 \\ 2 & 13 \end{bmatrix} = \begin{bmatrix} 7 & 0 \\ 4 & 17 \end{bmatrix}$

5. $\begin{bmatrix} 2 & 2 \\ 1 & 4 \end{bmatrix}$ 6. $C' = C$ 7. $(B')' = B$ 8. $\begin{bmatrix} 4 & 2 \\ 10 & 14 \end{bmatrix}$

9. $\begin{bmatrix} 8 & 9 \\ 8 & 36 \end{bmatrix}$ Each column of A has been multiplied by the corresponding entry of C. 10. $\begin{bmatrix} 0 & -3 \\ 12 & 18 \end{bmatrix}$

11. $\begin{bmatrix} 8 & 4 \\ 18 & 36 \end{bmatrix}$ Each row of A is multiplied by the corresponding entry of C.

12. $\begin{bmatrix} 8 & 20 & 32 \\ 20 & 44 & 68 \end{bmatrix}$ 13. $\begin{bmatrix} 140 & 176 \\ 176 & 224 \end{bmatrix}$ The sum of squares of the row.

14. $\begin{bmatrix} 20 & 44 & 68 \\ 44 & 100 & 156 \\ 68 & 156 & 244 \end{bmatrix}$ The sum of squares of the column.

15. $\begin{bmatrix} 18 \\ 24 \end{bmatrix}$ The sum of each row. 16. $[14 \quad 34 \quad 54]$ 17. $\begin{bmatrix} 1.0 & 0 \\ -0.8 & 1.2 \end{bmatrix}$

18. $[3 \quad 2] \begin{bmatrix} 8 & 9 \\ 8 & 36 \end{bmatrix} = [10 \quad 11] \begin{bmatrix} 4 & 0 \\ 0 & 9 \end{bmatrix} = [40 \quad 99]$

19. $\begin{bmatrix} 1.2 & 0 \\ 0.4 & 1.0 \end{bmatrix}$ 20. $\begin{bmatrix} 4 & 10 \\ 2 & 14 \end{bmatrix}$ 21. Same as Problem 20.

22. $$H'H = \begin{bmatrix} h_{11}^2 + h_{21}^2 + h_{31}^2 & h_{11}h_{12} + h_{21}h_{22} + h_{31}h_{32} \\ h_{12}h_{11} + h_{22}h_{21} + h_{32}h_{31} & h_{12}^2 + h_{22}^2 + h_{23}^2 \end{bmatrix} = \begin{bmatrix} \Sigma_i h_{i1}^2 & \Sigma_i h_{i1}h_{i2} \\ \Sigma_i h_{i2}h_{i1} & \Sigma_i h_{i2}^2 \end{bmatrix}$$

$$HH' = \begin{bmatrix} h_{11}^2 + h_{12}^2 & h_{11}h_{21} + h_{12}h_{22} & h_{11}h_{31} + h_{12}h_{32} \\ h_{21}h_{11} + h_{22}h_{12} & h_{21}^2 + h_{22}^2 & h_{21}h_{31} + h_{22}h_{32} \\ h_{31}h_{11} + h_{32}h_{12} & h_{31}h_{21} + h_{32}h_{22} & h_{31}^2 + h_{32}^2 \end{bmatrix}$$

$$= \begin{bmatrix} \Sigma_j h_{1j}^2 & \Sigma_j h_{1j}h_{2j} & \Sigma_j h_{1j}h_{3j} \\ \Sigma_j h_{2j}h_{1j} & \Sigma_j h_{2j}^2 & \Sigma_j h_{2j}h_{3j} \\ \Sigma_j h_{3j}h_{1j} & \Sigma_j h_{3j}h_{2j} & \Sigma_j h_{3j}^2 \end{bmatrix}$$

23. $$HJ = \begin{bmatrix} h_{11}j_{11} + h_{12}j_{21} & h_{11}j_{12} + h_{12}h_{22} \\ h_{21}j_{11} + h_{22}j_{21} & h_{21}j_{12} + h_{22}j_{22} \\ h_{31}j_{11} + h_{32}j_{21} & h_{31}j_{12} + h_{32}j_{22} \end{bmatrix} = \begin{bmatrix} \Sigma_v h_{1v}j_{v1} & \Sigma_v h_{1v}j_{v2} \\ \Sigma_v h_{2v}j_{v1} & \Sigma_v h_{2v}j_{v2} \\ \Sigma_v h_{3v}j_{v1} & \Sigma_v h_{3v}j_{v2} \end{bmatrix}$$

v is being used as a subscript here instead of j to avoid duplication.

24. $\mathrm{HK} = \begin{bmatrix} h_{11}k_{11} & h_{12}k_{22} \\ h_{21}k_{11} & h_{22}k_{22} \\ h_{31}k_{11} & h_{32}k_{22} \end{bmatrix}$ $\mathrm{JK} = \begin{bmatrix} j_{11}k_{11} & j_{12}k_{22} \\ j_{21}k_{11} & j_{22}k_{22} \end{bmatrix}$

$\mathrm{KJ} = \begin{bmatrix} j_{11}k_{11} & j_{12}k_{11} \\ j_{21}k_{22} & j_{22}k_{22} \end{bmatrix}$ K′J = KJ because K is symmetric.

25. $\mathrm{MH} = \begin{bmatrix} m_{11}h_{11} & m_{11}h_{12} \\ m_{22}h_{21} & m_{22}h_{22} \\ m_{33}h_{31} & m_{33}h_{32} \end{bmatrix}$ $\mathrm{H'M} = \begin{bmatrix} h_{11}m_{11} & h_{21}m_{22} & h_{31}m_{33} \\ h_{12}m_{11} & h_{22}m_{22} & h_{32}m_{33} \end{bmatrix}$

H′M = (MH)′ because of the rule about the transpose of a product and because M is symmetric.

26. $\mathrm{J} + \mathrm{K} = \begin{bmatrix} j_{11} + k_{11} & j_{12} \\ j_{21} & j_{22} + k_{22} \end{bmatrix}$ $\mathrm{J} - \mathrm{K} = \begin{bmatrix} j_{11} - k_{11} & j_{12} \\ j_{21} & j_{22} - k_{22} \end{bmatrix}$

READINGS

Ayres, F. (1962). *Theory and problems of matrices.* New York: Schaum, pp. 1–12.

Green, P. E., and J. D. Carroll (1976). *Mathematical tools for applied multivariate analysis.* New York: Academic Press, pp. 26–50.

Horst, P. (1963). *Matrix algebra for social scientists.* New York: Holt, Rinehart, and Winston, pp. 1–61; 99–215. Most texts in multivariate statistics contain a brief overview of matrix operations.

2

Statistical Formulas in Matrix Form

TRANSLATING FORMULAS INTO MATRIX NOTATION

General Multiplication

Even when one has a clear understanding of matrix operations, it is not necessarily obvious how one goes from a scalar formula to a matrix formula, or vice versa. The keys to the process of disentangling a scalar formula usually center on paying close attention to the subscripts and looking for certain patterns in the summation signs.

For matrix procedures to apply, the variables must be subscripted, and the meaning and limits of the subscripts must be specified. For example, the variable x_{ij} has two indices, i and j, and these indices have limits. For x_{ij} there is, implicitly or explicitly, a specification such as $i = 1, 2, \ldots, n$; $j = 1, 2, \ldots, p$. Matrix X is n by p. Sometimes two different subscripts refer to the same index. Suppose we refer to r_{jk}; $j, k = 1, 2, \ldots, p$. The r_{jk} might refer to the correlation between variables j and k. If so, we would have a square matrix of order p by p, consisting of the correlations of all possible pairs of variables, and j and k would refer to the same set of variables. Sometimes, to save letters, j' is used as the alternate subscript. If the same subscript is used twice, for example, r_{jj}, it refers specifically to the elements of the matrix that are in the main diagonal. (In a correlation matrix, these entries are the correlations of a variable with itself, and they are all 1.00.) In any event, it is important to examine the subscripts of variables and the limits specified for the subscripts which define the matrices involved.

Most instances of translating formulas involve recognizing special cases of multiplication. Several of these were encountered in Chapter 1, but we will see them again here and note instances in which they apply in statistical equations.

The basic multiplication equation is

$$c_{jk} = \sum_i a_{ji} b_{ik} \qquad (2.1)$$

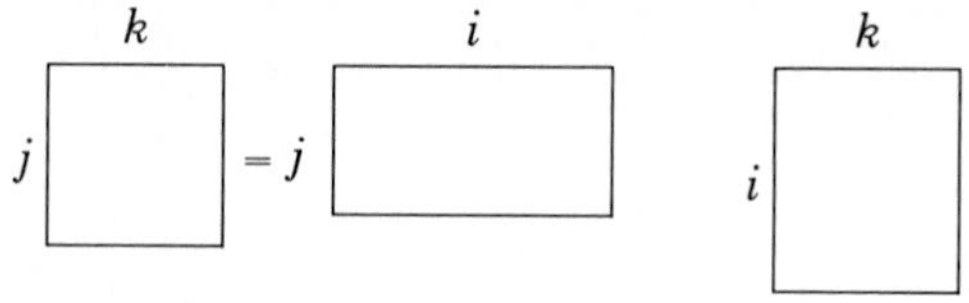

(see Equation 1.1). This is expressed in matrices as C = AB. Note the way the subscripts are used. The element in row j (first subscript) and column k (second subscript) of the product matrix is equal to the sum over i (second subscript of the *pre*multiplier) and the first subscript of the *post*multiplier) of the products of a_{ji} (row j) and b_{ik} (column k). Do not be confused if the scalar expression is stated with the terms in the reverse order: $c_{jk} = \Sigma_i b_{ik} a_{ji}$, because in multiplying *numbers* it does not matter which letter comes first. Both $\Sigma_i a_{ji} b_{ik}$ and $\Sigma_i b_{ik} a_{ji}$ are expressed as AB, although it is less confusing to use the former.

Special Cases

Equation 2.1 applies to all matrix multiplication, although special types of matrices yield different looking scalar formulas. Some of the more common examples are presented below.

Matrix Times Vector Sometimes one member of a product has only one subscript. This often means multiplication of a matrix times a vector:

$$y_i = \sum_j x_{ij} b_j \qquad (2.2)$$

In matrix notation this is

$$\mathbf{y} = \mathrm{X}\mathbf{b} \qquad (2.3)$$

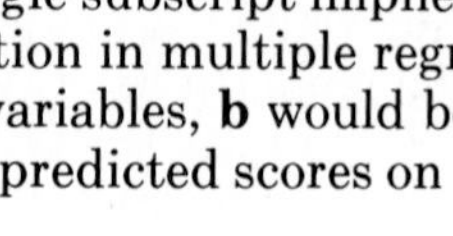

which means matrix $\|x_{ij}\|$ is being *post*multiplied by vector **b**. Because the subscript summed over is the one that refers to columns of X, the vector comes second. The fact that $\mathbf{b}_j$ has only a single subscript implies that it is a vector.

We could perform the above operation in multiple regression, where X would be the matrix of scores on predictor variables, **b** would be a vector of regression weights, and $\hat{\mathbf{y}}$ would be the vector of predicted scores on the dependent variable y. For example:

$$\begin{bmatrix} x_{11} & x_{12} \\ x_{21} & x_{22} \\ x_{31} & x_{32} \end{bmatrix} \begin{bmatrix} b_1 \\ b_2 \end{bmatrix} = \begin{bmatrix} x_{11}b_1 + x_{12}b_2 \\ x_{21}b_1 + x_{22}b_2 \\ x_{31}b_1 + x_{32}b_2 \end{bmatrix} = \begin{bmatrix} \Sigma_j x_{1j} b_j \\ \Sigma_j x_{2j} b_j \\ \Sigma_j \;\; {}_{3j} \;\; {}_j \end{bmatrix}$$

Note the correspondence between subscripts of the product terms.

If the singly subscripted variable was the row index of the matrix, and the subscript was summed over, then the vector comes first:

$$u_i = \sum_i x_{ij} g_i \tag{2.4}$$

$$\mathbf{u} = \mathbf{g}'\mathrm{X} \tag{2.5}$$

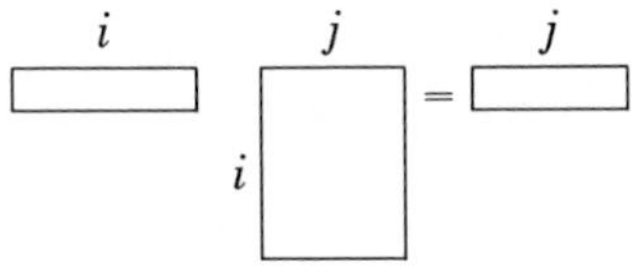

Here the vector is written as a transpose because it has to be a row instead of the conventional column. Below is an example:

$$[g_1 \quad g_2 \quad g_3]\begin{bmatrix} x_{11} & x_{12} \\ x_{21} & x_{22} \\ x_{31} & x_{32} \end{bmatrix} = [g_1x_{11} + g_2x_{21} + g_3x_{31} \quad g_1x_{12} + g_2x_{22} + g_3x_{32}]$$

$$= [\Sigma_i \, g_i x_{i1} \quad \Sigma_i \, g_i x_{i2}]$$

Matrix Summation The previous forms are employed when we want to find weighted sums. What about simple sums such as $\Sigma_j x_{ij}$? To represent this, we need the vector that consists of all ones, **1**. Suppose we multiply a matrix by this vector. Multiplying numbers by 1.0 and then adding them is the same as just adding them, so $x_{i.}$, which is $\Sigma_j x_{ij}$, (when a variable is summed over, its index is replaced by a dot), can be represented as

$$\mathbf{x}_{i.} = \mathrm{X}\mathbf{1} \tag{2.6}$$

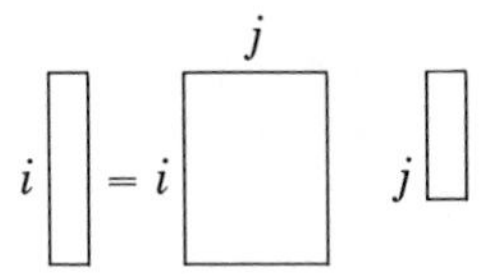

This operation could be used to find each person's total score over several parts of a test or over several trials.

If we wish to sum over rows, the **1** must be transposed and come first. When $x_{.j} = \Sigma_i \, x_{ij}$,

$$\mathbf{1}'\mathrm{X} = \boldsymbol{x}_j \tag{2.7}$$

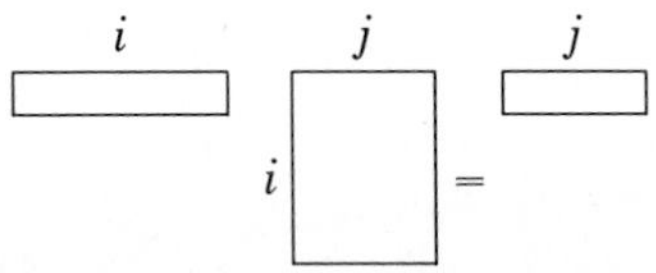

To sum over the whole matrix, we would use both **1** and **1**′. This might be used as a step toward computing the grand mean over both subjects (rows) and trials (columns) where $x_{..} = \Sigma_i \Sigma_j x_{ij}$:

$$x_{..} = \mathbf{1}'\mathbf{X}\mathbf{1} \qquad (2.8)$$

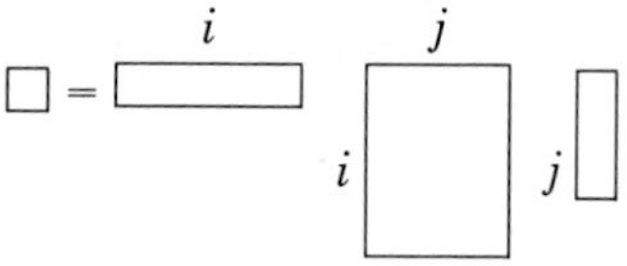

Because $x_{..}$ is a single number, it is not shown in boldface in Equation 2.8.

Multiplying or Dividing by Constants As we saw in Chapter 1, dividing or multiplying every element of a matrix by a constant is symbolized $k\mathrm{X}$ or $\mathrm{X}k$. If we wish to make it clear that we are dividing by a particular constant, we may make use of the exponent -1, as in $n^{-1}\mathrm{X}$ or $\mathrm{X}n^{-1}$.

Multiplying different columns of a matrix by different constants is accomplished by postmultiplying by a diagonal matrix. For example, $x_{ij}d_j$ is shown in matrix notation as

$$\mathrm{Z} = \mathrm{XD} \qquad (2.9)$$

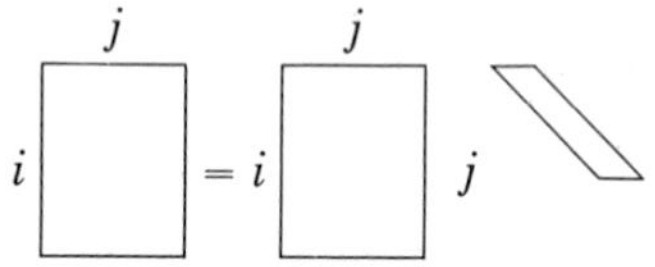

$$\begin{bmatrix} x_{11} & x_{12} \\ x_{21} & x_{22} \\ x_{31} & x_{32} \end{bmatrix} \begin{bmatrix} d_1 & 0 \\ 0 & d_2 \end{bmatrix} = \begin{bmatrix} x_{11}d_1 & x_{12}d_2 \\ x_{21}d_1 & x_{22}d_2 \\ x_{31}d_1 & x_{32}d_2 \end{bmatrix}$$

The scalar expression could be either $x_{ij}d_j$ or d_jx_{ij}, but the matrix expression in both cases is XD because the column subscript of x_{ij} is matched by the subscript of d_j.

Operations like that shown in Equation 2.9 are carried out whenever we wish to rescale all the variables for some reason. The most common case occurs in converting to standard scores. The d_j then become the reciprocals of standard deviations of the respective variables.

To multiply *rows* by different constants, the diagonal matrix is a premultiplier. The scalar expression $q_{ij} = u_ix_{ij}$ translates as

$$\mathrm{Q} = \mathrm{UX} \qquad (2.10)$$

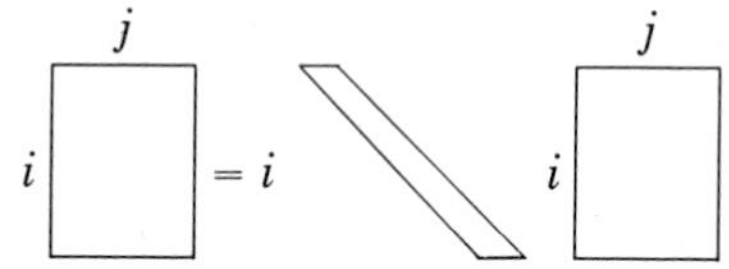

This expression may be used if we wish to normalize the rows of a matrix to have a constant sum of squares.

Multiplying a matrix by a diagonal matrix can be discriminated from the similar-looking expression for multiplying by a vector. The key is the absence of a summation sign. When a summation is used, it means matrix-times-vector; when it is not used, it means multiplication by a diagonal matrix. Below are examples of multiplication by diagonal matrices:

$$\begin{bmatrix} x_{11} & x_{12} \\ x_{21} & x_{22} \\ x_{31} & x_{32} \end{bmatrix} \begin{bmatrix} d_1 & 0 \\ 0 & d_2 \end{bmatrix} = \begin{bmatrix} x_{11}d_1 & x_{12}d_2 \\ x_{21}d_1 & x_{22}d_2 \\ x_{31}d_1 & x_{32}d_2 \end{bmatrix}$$

$$\begin{bmatrix} u_1 & 0 & 0 \\ 0 & u_2 & 0 \\ 0 & 0 & u_3 \end{bmatrix} \begin{bmatrix} x_{11} & x_{12} \\ x_{21} & x_{22} \\ x_{31} & x_{32} \end{bmatrix} = \begin{bmatrix} u_1x_{11} & u_1x_{12} \\ u_2x_{21} & u_2x_{22} \\ u_3x_{31} & u_3x_{32} \end{bmatrix}$$

Products Involving Transposes In general, scalar expressions corresponding to matrix multiplication have a subscript that is summed over. It comes first in one term and second in the other. When the summed-over subscript appears in the same position in both terms, this means that one of the matrices must be transposed. For example, in $t_{jj'} = \Sigma_i x_{ij}x_{ij'}$, where j and j' are alternate subscripts for the same set, X is being premultiplied by its own transpose:

$$\mathrm{T} = \mathrm{X'X}$$

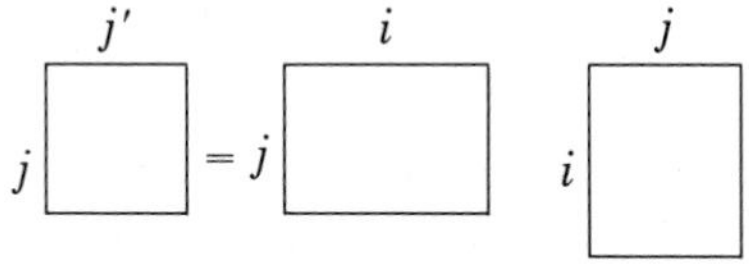

(2.11)

Transposing X and making it a premultiplier has the effect of putting the i subscript in the right place so we can get a sum of products over it. Such an example was given in Chapter 1.

We may also wish to sum the products over the column subscript. That is, when $r_{jj'} = \Sigma_m f_{jm}f_{j'm}$, the transpose is the postmultiplier:

$$\mathrm{R} = \mathrm{FF'}$$

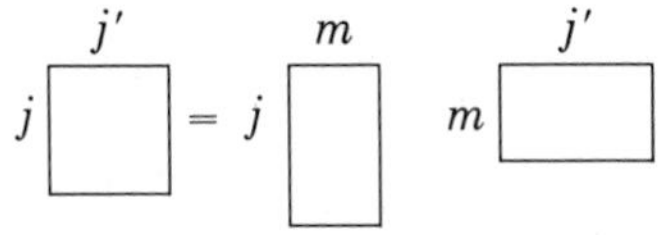

(2.12)

A matrix multiplication involving a transpose may also involve two different matrices. For example, $c_{jk} = \Sigma_i a_{ij}b_{ik}$ would translate as C = A′B. The A is transposed and comes first because its subscript, j, is first, specifying the row of C.

The expressions used in products involving transposes seem to violate the general principle of matrix multiplication that says that the subscript that is summed over must come first in one term and second in the other (see Equation 2.1). Actually, the necessity of being consistent with the equation makes it necessary to transpose one matrix. For example, transposing X and using it as a premultiplier of X itself makes i the column label, matching the row label of X. Now corresponding elements of X are matched together to calculate the sum of products.

Perhaps the easiest approach is just to remember that when the common subscript comes first, the transpose comes first, and when the common subscript comes second, the transpose comes second.

More than Two Terms

The last section illustrates the main cases involving the multiplication of two matrices. However, we must also be prepared to disentangle expressions having three or four terms. Here it is necessary to break down the expressions into parts, identify each part as one of the cases already discussed, and then put the parts back together. Often there is more than one way to break down an expression. Seeing alternatives may be useful in demonstrating the equivalence of two or more different expressions.

Perhaps the simplest expression involving more than two terms is

$$p_{rs} = q_{rs} a_r b_s \tag{2.13}$$

This involves premultiplication and postmultiplication by diagonal matrices:

$$\mathrm{P} = \mathrm{AQB} \tag{2.14}$$

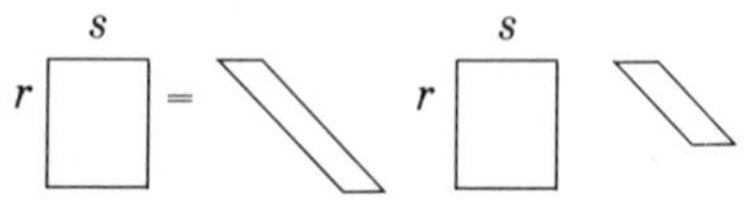

The positions of the matrices are determined by which subscripts match. Because the symbol a is used twice, the expression $q_{rs} a_r a_s$ would only make sense if Q were square and if it was pre- *and* postmultiplied by the diagonal matrix A. The same symbol is never used to represent different elements in the same expression.

Another common type of expression is

$$v = \sum_j \sum_k w_j w_k m_{jk} \tag{2.15}$$

The matrix notation for this expression would be

$$v = \mathbf{w}'\mathrm{M}\mathbf{w} \tag{2.16}$$

where M is square and $\mathbf{w}$ is a vector. Remember that a term whose subscript is *not involved* in a summation may be taken outside it, because such terms are constant with respect to that summation. Thus Equation 2.15 could be written as

$$v = \sum_j w_j \sum_k w_k m_{jk} \tag{2.17}$$

The part of the equation to the right of the second summation can be recognized as $\mathrm{M}\mathbf{w}$, which may suggest the final product more readily:

$$v = \mathbf{w}'(\mathrm{M}\mathbf{w}) \tag{2.18}$$

Note that the product is a scalar, not a matrix.

An elaboration on the multiplication by two diagonal matrices is a matrix multiplied by its transpose, each multiplied by diagonal matrices. The following is an example:

$$g_{pq} = d_p d_q \sum_m a_{mp} a_{mq} \tag{2.19}$$

The part of the equation to the right of the summation sign can be recognized as A′A, which must be pre- and postmultiplied by a diagonal matrix D:

$$\mathrm{G} = \mathrm{DA'AD} \tag{2.20}$$

If the d's were on the right of the summation sign in the scalar expression, their subscripts would not be involved in the summation and could be taken outside it.

One final type of expression should be noted. This expression has the most complex appearance of all, because it includes three summation signs and four terms. In translating, we make use of the following summation principle: *Where two or more summations exist, they may be carried out in any order.*

For example, suppose we have

$$t = \sum_i \sum_j \sum_k x_{ij} x_{ik} w_j w_k \tag{2.21}$$

The order in which we write the summation signs is arbitrary. The equation could just as well be written as

$$t = \sum_j \sum_k \sum_i x_{ij} x_{ik} w_j w_k$$

Because w_j and w_k do not have i subscripts, they may be brought *outside* their summation sign, and w_j may be brought outside the summation sign involving k, giving us

$$t = \sum_j w_j \sum_k w_k \sum_i x_{ij} x_{ik} \tag{2.22}$$

The expression on the far right should be recognizable as the matrix-by-transpose product X′X. If we give this part of the product the nickname M, it will resemble Equation 2.18;

$$t = \mathbf{w}'(\mathrm{X'X})\mathbf{w} \tag{2.23}$$

It is possible to go about this in a different way. Leaving the summations in their original order, we could have

$$t = \sum_i \sum_j x_{ij} w_j \sum_k x_{ik} w_k \tag{2.24}$$

The summed product $\Sigma_j x_{ij} w_j$ is matrix-by-vector, and the similar term must be its transpose. For the time being we will give them the nicknames $\mathbf{y}'$ and $\mathbf{y}$, where $\mathbf{y}$ is a column vector, $\mathrm{X}\mathbf{w}$. The summation over i is telling us to take $\mathbf{y}'\mathbf{y}$, the sum of products between these. Therefore,

$$t = (\mathrm{X}\mathbf{w})'\mathrm{X}\mathbf{w} \tag{2.25}$$

Because of the way matrix multiplication is carried out, there is a general principle that *the transpose of a product is the product of the transposes in the reverse order.* Applying this principle to $(\mathrm{X}\mathbf{w})'$, we find that

$$t = \mathbf{w}'\mathrm{X}'\mathrm{X}\mathbf{w}$$

The parentheses in Equation 2.23 serve no function except to indicate which multiplication we intend to carry out first. When they are removed, we can see that the expression is equivalent to the equation above.

The preceding examples of matrix multiplication represent the majority of cases encountered in statistical applications. Two or three additional matrix operations will be introduced in later chapters as necessary. In the next section, we will learn how scalar statistical expressions are formulated using matrices.

MATRIX MULTIPLICATION EXAMPLES

Learning to use matrices is like learning any second language. We must first learn to translate between the language we know and the new one. This is a two-sided proposition: we need to know how to translate scalar formulas into matrix formulas *and* vice versa. The serious student aspires to be able to think in the new language, but even then, translation is necessary for communication purposes.

To translate a matrix product into a scalar formula, we must know the scalar notation of the elements of the factors, including the subscripts and their limits. From this we can deduce the formula that is represented and the subscripts and limits of the product. For example, in

$$\mathrm{PQ} = \mathrm{M}$$

suppose the elements of P and Q are p_{gh}, $g = 1, r$; $h = 1, s$; and g_{hi}, $i = 1, t$. (Note the appearance and position of h in each.) After making the following diagram

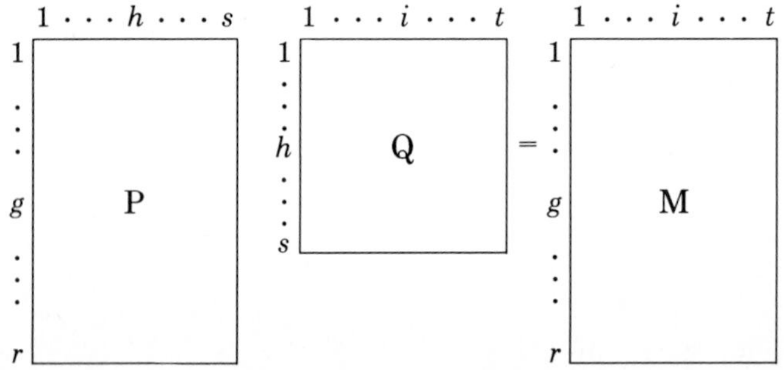

we deduce that the scalar formula for the elements of M must be

$$m_{gi} = \sum_h p_{gh} q_{hi}$$

This is just like Equation 2.1, except that different letters are used.

The order of multiplication is always important, but for equations with diagonal matrices, it is essential in translating. For example, the expression

$$\text{A} = \text{DP}$$

where P is as before and D is a diagonal matrix, must mean

$$a_{gh} = d_g p_{gh}$$

(We could use d_{gg} instead of d_g.) The diagonal term d_g shares the g (left) index of f_{gh} because the diagonal term is on the left of P. Therefore, each of *its* rows is modifying the corresponding row of P. If the diagonal were on the right, it would share the column index for a similar reason.

Triple and quadruple multiplications look scary, but they can be handled by taking two adjacent matrices, finding out what product they imply, and then combining this product with the one that is next to the first pair, before continuing. For example, in the expression

$$\text{DPP}'\text{D} = \text{T}$$

where D and P are defined as above, one can start with DP. We just saw that an element of DP is $d_g p_{gh}$. Now we need to think about P′D and conclude that it is the same thing, realizing that it is arranged differently, as a transpose:

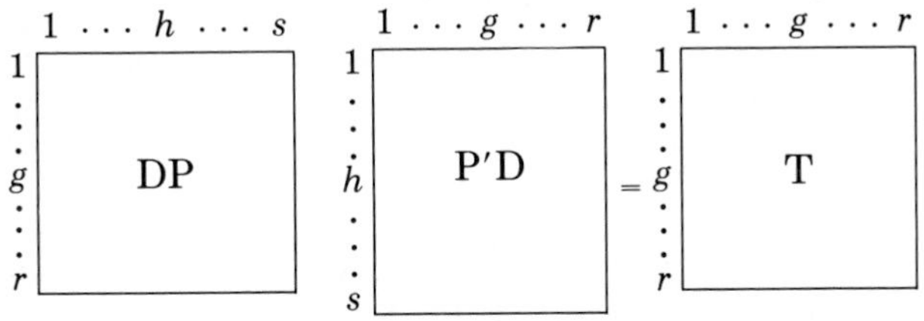

Thus, the subscript that will be summed over is h. In general,

$$t_{gg'} = \sum_h p_{gh} p_{g'h} d_g d_{g'}$$

but on the diagonal, where $g = g'$,

$$t_{gg} = \sum_h d_g^2 p_{gh}^2$$

We could have started in the middle, focusing on P′P, where the product is $\Sigma_h\, p_{gh} p_{g'h}$ as before. Let us call these terms $f_{gg'}$ for the moment. Then the effect of the D's is to multiply each element of the initial product by the d for its column *and* the d for its row: $d_g d_{g'} f_{gg'}$. Substituting back the summation gives $d_g d_{g'} = \Sigma_h p_{gh} p_{g'h}$. This differs from Equation 2.25 in that the d's are to the left of the summation, but because the subscripts of the d's are not involved in the summation, they may be taken outside the summation, giving $d_g d_{g'} \Sigma_h p_{gh} p_{g'h}$ just as before.

STATISTICAL FORMULAS IN MATRIX FORM

Means, Deviation Scores, and Standard Scores

We should now be able to translate some common statistical computations into matrix form. Let $X = \|x_{ij}\|$, $i = 1, 2, \ldots, n$; $j = 1, 2, \ldots, p$; X may consist, for example, of the scores of n persons on p measures. How do we compute the mean of each of the measures? The mean of measure j is $\bar{x}_{\cdot j} = \Sigma_i x_{ij}/n$. The summation sign is over rows, $\mathbf{1}'X$, and involves multiplication by the constant $1/n$. Therefore, if $\bar{\mathbf{x}}'_{\cdot j}$ stands for the row vector of means, we can use the following formula:

$$\bar{\mathbf{x}}'_{\cdot j} = \mathbf{1}'Xn^{-1} \tag{2.26}$$

A drawing may be helpful:

$$
\begin{array}{cccc}
j & i & j & j \\
\boxed{\bar{x}_{\cdot 1}\ \bar{x}_{\cdot 2}\ \ldots\ \bar{x}_p} = & \boxed{1\ \ 1\ \ \ldots\ \ 1} & \boxed{\begin{matrix} x_{11} & x_{12} & \cdots & x_{1p} \\ x_{21} & x_{22} & \cdots & x_{2p} \\ \cdot & \cdot & & \cdot \\ \cdot & \cdot & & \cdot \\ x_{n1} & x_{n2} & & x_{np} \end{matrix}} & \boxed{\begin{matrix} 1/n & 0 & \cdots & 0 \\ 0 & 1/n & & \cdot \\ \cdot & & \ddots & \cdot \\ 0 & \cdots & & 1/n \end{matrix}} \\
\bar{\mathbf{x}}'_{\cdot j} & \mathbf{1}' & X & n^{-1}
\end{array}
$$

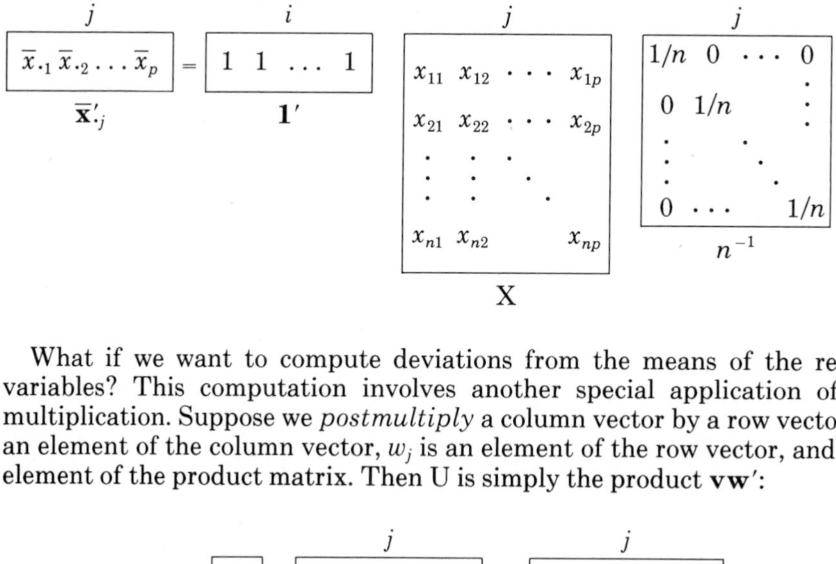

What if we want to compute deviations from the means of the respective variables? This computation involves another special application of matrix multiplication. Suppose we *postmultiply* a column vector by a row vector. If v_i is an element of the column vector, w_j is an element of the row vector, and u_{ij} is an element of the product matrix. Then U is simply the product $\mathbf{vw}'$:

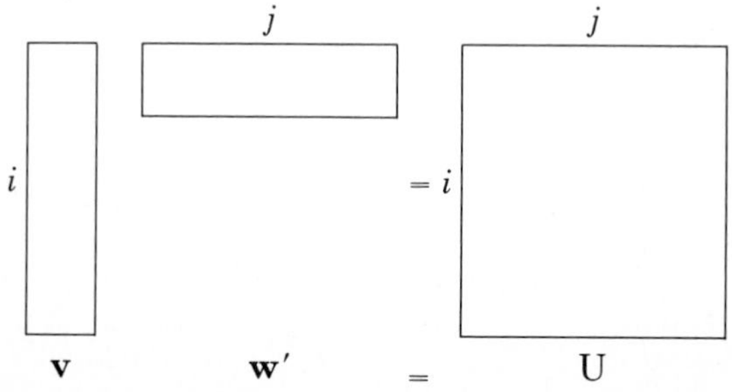

The product of a column vector postmultiplied by a row vector is sometimes referred to as the *outer product* of the two vectors. This general principle is used to construct a matrix $\overline{X}_{\cdot j}$ in which each column j is a constant $\overline{x}_{\cdot j}$

$$\overline{X}_{\cdot j} = \mathbf{1}\overline{\mathbf{x}}'_{\cdot j} \qquad (2.27)$$

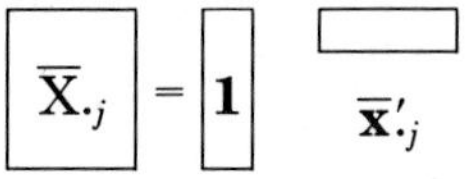

This column of n ones *post*multiplied by a row consisting of the p means gives an $n \times p$ matrix, each column having the corresponding mean $\overline{x}_{\cdot j}$.

Now we can use this to transform X to deviation score form. Defining Y as the matrix of deviation scores, we have the formula

$$y_{ij} = x_{ij} - \overline{x}_{\cdot j} \qquad (2.28)$$

expressed in matrices as

$$\mathrm{Y} = \mathrm{X} - \mathbf{1}\overline{\mathbf{x}}_{\cdot j} \qquad (2.29)$$

or, by further substitution, $\mathbf{1}'\mathrm{X}\,n^{-1}$ for $\overline{\mathbf{x}}_{\cdot j}$

$$\mathrm{Y} = \mathrm{X} - \mathbf{1}\mathbf{1}'\mathrm{X}n^{-1} \qquad (2.30)$$

Now let us see what we can do with Y. Suppose D_s^{-1} is a diagonal matrix whose entries are the reciprocals of the standard deviations s_j of the corresponding variables, that is, $d_{jj} = 1/s_j$. Now postmultiply Y by D_s^{-1} to give Z:

$$\mathrm{Z} = \mathrm{Y}\mathrm{D}_s^{-1} \qquad (2.31)$$

What effect does this have? Each deviation score will be divided by the appropriate standard deviation, making Z a matrix of standard scores. They relate directly back to the original X matrix by substituting Equation 2.30 for Y:

$$\mathrm{Z} = (\mathrm{X} - \mathbf{1}\mathbf{1}'\mathrm{X}n^{-1})\mathrm{D}_s^{-1} \qquad (2.32)$$

Variance-Covariance Matrices

The matrix of deviation scores Y can be used to compute the variances and covariances of the variables. Recall that s_j^2, the variance of variable j, is defined as

$$s_j^2 = \frac{\sum_i y_{ij}^2}{n - 1} \qquad (2.33)$$

Similarly, the covariance s_{jk} between two variables j and k is defined as

$$s_{jk} = \frac{\sum_i y_{ij}y_{ik}}{n - 1} \qquad (2.34)$$

Recall what happens when we premultiply Y by its transpose Y′, yielding the matrix T:

$$T = Y'Y$$

For a 3 × 2 Y matrix:

$$\begin{bmatrix} y_{11} & y_{21} & y_{31} \\ y_{12} & y_{22} & y_{32} \end{bmatrix} \begin{bmatrix} y_{11} & y_{12} \\ y_{21} & y_{22} \\ y_{31} & y_{32} \end{bmatrix} = \begin{bmatrix} \Sigma\, y_{i1}^2 & \Sigma\, y_{i1}y_{i2} \\ \Sigma\, y_{i2}y_{i1} & \Sigma\, y_{i2}^2 \end{bmatrix}$$

An element of T will be defined as

$$t_{jk} = \sum_i y_{ji}' y_{ik}$$

but $y_{ji}' = y_{ij}$, so t_{jk} is exactly the numerator of s_{jk}. If we define S as a matrix of variances and covariances between variables (referred to as the *variance-covariance matrix,* or simply the *covariance matrix,* which has the variances down the main diagonal and covariances everywhere else, the covariance matrix S is expressed as

$$S = Y'Y\,(n - 1)^{-1} \tag{2.35}$$

In the future we will be using x_{ij} to stand for x in *either* raw or deviation score form, making it clear from context which is meant. The letter y will be used for some other variable or set of variables. Right here, we must use Y this way.

Correlation Matrices

The standard score matrix Z can be used to calculate the matrix of correlations between the variables. The standard score formula for the correlation matrix is

$$r_{jk} = \frac{\sum_i z_{ij} z_{ik}}{n - 1} \tag{2.36}$$

Using the same device that was used for the covariance matrix, we can convert to the matrix formula

$$R = Z'Z(n - 1)^{-1} \tag{2.37}$$

This formula can be related back to the deviation scores as shown below. Recall that the transpose of a product is the product of the transposes in the reverse order. Because $Z = YD_s^{-1}$, $Z' = (D_s^{-1})'Y'$. But because D_s^{-1} is a diagonal matrix, it is symmetric, and the same as its transpose: $D_s^{-1} = (D_s^{-1})'$. If we substitute for Z′ and Z, the correlation matrix becomes

$$R = (D_s^{-1}Y')\,(YD_s^{-1})\,(n - 1)^{-1}$$

The associative law allows us to transfer parentheses in a matrix product:

$$R = D_s^{-1}(Y'Y)D_s^{-1}(n - 1)^{-1}$$

The fact that the scalar $(n - 1)^{-1}$ can be moved gives us

$$R = D_s^{-1}(Y'Y)\,(n - 1)^{-1}D_s^{-1}$$

The middle part of this expression is the covariance matrix S, making an alternate matrix expression for R

$$R = D_s^{-1}SD_s^{-1} \tag{2.38}$$

The scalar notation for this formula is

$$r_{jk} = \frac{s_{jk}}{s_j s_k} \tag{2.39}$$

or, alternatively,

$$r_{jk} = \frac{\Sigma\,(x_{ij} - \overline{x}_{\cdot j})\,(x_{ik} - \overline{x}_{\cdot k})}{(n - 1)\,s_j s_k} \tag{2.40}$$

A more efficient way to compute R, rather than converting each deviation score to a standard score, would be to use Equation 2.40, as illustrated in Exhibit 2-1. Although we could further substitute for Y its definition in terms of X, this would complicate rather than simplify the processes.

The series of steps discussed above will be used in subsequent chapters to illustrate the relationships among statistical formulas, or to deduce the consequences of changing a matrix from one form to another. Most of these manipulations will be variations on what has been done here. For example, we may substitute one form of a matrix for another, equivalent form, as we substituted YD_s^{-1} for Z. We will also make frequent use of the associative rule, as in $(D_s^{-1}Y')(YD_s^{-1}) = D_s^{-1}(Y'Y)D_s^{-1}$. We use the associative rule when the new form of the product has some status as a particular matrix that we want to recognize. We will also make use of the principle that $(AB)' = B'A'$ and that certain matrices are symmetric. Additional principles like these will be introduced in later chapters as needed.

$$\mathbf{G} = \mathbf{Y'Y} = \begin{bmatrix} y_{11} & y_{21} & y_{i1} & \cdots & y_{1n} \\ y_{12} & y_{22} & y_{i2} & \cdots & y_{2n} \\ y_{13} & y_{23} & y_{i3} & \cdots & y_{3n} \end{bmatrix} \begin{bmatrix} y_{11} & y_{12} & y_{13} \\ y_{21} & y_{22} & y_{23} \\ \vdots & \vdots & \vdots \\ y_{i1} & y_{i2} & y_{i3} \\ y_{n1} & y_{n2} & y_{n3} \end{bmatrix}$$

$$\mathbf{G} = \qquad \mathbf{Y'} \qquad \mathbf{Y}$$

$$\mathbf{G} = \begin{bmatrix} \Sigma y_{i1}^2 & \Sigma y_{i1}y_{i2} & \Sigma y_{i1}y_{i3} \\ \Sigma y_{i2}y_{i1} & \Sigma y_{i2}^2 & \Sigma y_{i2}y_{i3} \\ \Sigma y_{i3}y_{i1} & \Sigma y_{i3}y_{i2} & \Sigma y_{i3}^2 \end{bmatrix}$$

$$\mathbf{S} = \mathbf{G}(n-1)^{-1} = \begin{bmatrix} \Sigma y_{i1}^2 & \Sigma y_{i1}y_{i2} & \Sigma y_{i1}y_{i3} \\ \Sigma y_{i2}y_{i1} & \Sigma y_{i2}^2 & \Sigma y_{i2}y_{i3} \\ \Sigma y_{i3}y_{i1} & \Sigma y_{i3}y_{i2} & \Sigma y_{i3}^2 \end{bmatrix} \begin{bmatrix} \frac{1}{n-1} & 0 & 0 \\ 0 & \frac{1}{n-1} & 0 \\ 0 & 0 & \frac{1}{n-1} \end{bmatrix}$$

$$\mathbf{S} = \qquad \mathbf{G} \qquad (n-1)^{-1}$$

$$\mathbf{S} = \begin{bmatrix} \frac{\Sigma y_{i1}^2}{n-1} & \frac{\Sigma y_{i1}y_{i2}}{n-1} & \frac{\Sigma y_{i1}y_{i3}}{n-1} \\ \frac{\Sigma y_{i2}y_{i1}}{n-1} & \frac{\Sigma y_{i2}^2}{n-1} & \frac{\Sigma y_{i2}y_{i3}}{n-1} \\ \frac{\Sigma y_{i3}y_{i1}}{n-1} & \frac{\Sigma y_{i3}y_{i2}}{n-1} & \frac{\Sigma y_{i3}^2}{n-1} \end{bmatrix}$$

$$\mathbf{R} = \mathbf{D}_s^{-1}\mathbf{S}\mathbf{D}_s^{-1} = \begin{bmatrix} \frac{1}{s_1} & 0 & 0 \\ 0 & \frac{1}{s_2} & 0 \\ 0 & 0 & \frac{1}{s_3} \end{bmatrix} \begin{bmatrix} \frac{\Sigma y_{i1}^2}{n-1} & \frac{\Sigma y_{i1}y_{i2}}{n-1} & \frac{\Sigma y_{i1}y_{i3}}{n-1} \\ \frac{\Sigma y_{i2}y_{i1}}{n-1} & \frac{\Sigma y_{i2}^2}{n-1} & \frac{\Sigma y_{i2}y_{i3}}{n-1} \\ \frac{\Sigma y_{i3}y_{i1}}{n-1} & \frac{\Sigma y_{i3}y_{i2}}{n-1} & \frac{\Sigma y_{i3}^2}{n-1} \end{bmatrix} \begin{bmatrix} \frac{1}{s_1} & 0 & 0 \\ 0 & \frac{1}{s_2} & 0 \\ 0 & 0 & \frac{1}{s_3} \end{bmatrix}$$

$$\mathbf{R} = \qquad \mathbf{D}_s^{-1} \qquad \mathbf{S} \qquad \mathbf{D}_s^{-1}$$

Exhibit 2-1 Computation of the correlation matrix for $m = 3$ variables starting from the deviation score matrix. Note the subscripts in Y′. The matrix G is the sum of squares and cross-products matrix which is divided by $n - 1$ to give the covariance matrix, S. This in turn is pre- and then postmultiplied by the diagonal matrix of reciprocals of standard deviations to give R.

$$R = DSD = \begin{bmatrix} \frac{\Sigma y_{i1}^2}{(n-1)s_1} & \frac{\Sigma y_{i1}y_{i2}}{(n-1)s_1} & \frac{\Sigma y_{i1}y_{i3}}{(n-1)s_1} \\ \frac{\Sigma y_{i2}y_{i1}}{(n-1)s_2} & \frac{\Sigma y_{i2}^2}{(n-1)s_2} & \frac{\Sigma y_{i2}y_{i3}}{(n-1)s_2} \\ \frac{\Sigma y_{i3}y_{i1}}{(n-1)s_3} & \frac{\Sigma y_{i3}y_{i2}}{(n-1)s_3} & \frac{\Sigma y_{i3}^2}{(n-1)s_3} \end{bmatrix} \begin{bmatrix} \frac{1}{s_1} & 0 & 0 \\ 0 & \frac{1}{s_2} & 0 \\ 0 & 0 & \frac{1}{s_3} \end{bmatrix}$$

$$R = \qquad (D_s^{-1}S) \qquad D_s^{-1}$$

$$R = D_s^{-1}SD_s^{-1} = \begin{bmatrix} \frac{\Sigma y_{i1}^2}{(n-1)s_1^2} & \frac{\Sigma y_{i1}y_{i2}}{(n-1)s_1s_2} & \frac{\Sigma y_{i1}y_{i3}}{(n-1)s_1s_2} \\ \frac{\Sigma y_{i2}y_{i1}}{(n-1)s_2s_1} & \frac{\Sigma y_{i2}^2}{(n-1)s_2^2} & \frac{\Sigma y_{i2}y_{i3}}{(n-1)s_2s_3} \\ \frac{\Sigma y_{i3}y_{i1}}{(n-1)s_3s_1} & \frac{\Sigma y_{i3}y_{i2}}{(n-1)s_3s_2} & \frac{\Sigma y_{i3}^2}{(n-1)s_3^2} \end{bmatrix} = \begin{bmatrix} 1 & r_{12} & r_{13} \\ r_{21} & 1 & r_{23} \\ r_{31} & r_{32} & 1 \end{bmatrix}$$

$$R = \qquad D_s^{-1}SD_s^{-1} \qquad = \qquad R$$

Exhibit 2-1 (continued)

SUMMARY

In this chapter we have accomplished two things. First, we amplified the previous chapter's discussion of matrix multiplication, calling attention to the way in which subscripts in a scalar formula define how a particular matrix multiplication is carried out. Second, we connected this to the ways in which some special kinds of matrices operate, for example, matrix transposes, diagonal matrices, vectors, and so on. Some of the common statistical formulas were presented in matrix form. In matrix notation, the vector of means of a set of variables x_{ij} is expressed in Equation 2.26 as

$$\overline{\mathbf{x}}_{\cdot j} = \mathbf{1}'Xn^{-1}$$

The deviation score matrix **Y** is shown in Equation 2.30:

$$Y = X - \mathbf{11}'Xn^{-1}$$

The variance-covariance matrix is shown in Equation 2.35:

$$S = Y'Y(n-1)^{-1}$$

The matrix of standard scores Z is shown in Equation 2.31:

$$Z = YD_s^{-1}$$

where D_s^{-1} is a diagonal matrix of reciprocals of standard deviations. The correlation matrix is shown in Equation 2.37:

$$R = Z'Z(n-1)^{-1}$$

and also in Equation 2.38:

$$R = D_s^{-1} S D_s^{-1}$$

PROBLEMS

Express the following as matrix operations. Include a diagram of the matrices.

1. $\Sigma_i x_{ij} x_{ik}$ $\quad i = 1, n;\ j, k = 1, p$
2. $\Sigma_i x_{ij} y_{ki}$ $\quad i = 1, n;\ j = 1, p;\ k = 1, q$
3. $\Sigma_i x_{ij} y_{ik}$ $\quad i = 1, n;\ j = 1, p;\ k = 1, q$
4. $\Sigma_j x_{ij} y_j$ $\quad i = 1, n;\ j = 1, p$
5. $\Sigma_q a_{pq} b_{qr}$ $\quad q = 1, m;\ p = 1, s;\ r = 1, t$
6. $\Sigma_i x_{ij}$ $\quad i = 1, n;\ j = 1, p$
7. $x_{ij} s_j$ $\quad i = 1, n;\ j = 1, p$
8. $x_{ij} s_i$ same limits as Problem 7
9. $\Sigma_j x_{ij} w_j$ same limits as Problem 7
10. $a_{pq} v_q$ $\quad p = 1, s;\ q = 1, t$
11. $a_{pq} v_p$ same limits as Problem 10
12. $\Sigma_q a_{pq} v_q$ same limits as Problem 10
13. $\Sigma_q a_{pq}$ same limits as Problem 10
14. $\Sigma_p \Sigma_q a_{pq}$ same limits as Problem 10
15. $x_{ij} + c_j$ $\quad i = 1, n;\ j = 1, p$
16. $x_{ij} - c$ same limits as Problem 15
17. $\Sigma_i x_{ij} x_{ik}/n$ $\quad i = 1, n;\ j, k = 1, p$
18. $\Sigma_i x_{ij} x_{ik}/n s_j s_k$ same limits as Problem 17
19. $\Sigma_p a_{pq} a_{pr}/s v_q v_r$ $\quad p = 1, s;\ q, r = 1, t$
20. $\Sigma_i \Sigma_j x_{ij} b_j y_i$ $\quad i = 1, n;\ j = 1, p$
21. $\Sigma_i \Sigma_j \Sigma_k x_{ij} b_j x_{ik} b_k$ same limits as Problem 20
22. Let $z_{ij} = x_{ij}/s_j$. Using matrices, show that $r_{jk} = \Sigma\, z_{ij} z_{ik}/(n-1)$ is equivalent to $r_{jk} = \Sigma x_{ij} x_{ik}/s_j s_k (n-1)$ if $s_j^2 = \Sigma\,(x_{ij} - \bar{x}_{\cdot j})^2/(n-1)$

Translate the following matrix products and specifications into the corresponding scalar expression. For example, the answer to Problem 23 is $g_{ik} = \Sigma_j a_{ij} b_{jk}$. It is helpful to diagram the matrices involved, labeling the dimensions of each with the relevant subscript and limits.

Product	*Elements*	*Limits*
23. AB = G	a_{ij}, b_{jk}	$i = 1, n; j = 1, p; k = 1, m$
24. C′D = X	c_{pq}, d_{pr}	$p = 1, s; q = 1, t; r = 1, u$
25. C′D′ = Y	c_{pq}, d_{rp}	same limits as Problem 24
26. CD′ = Z	c_{pq}, d_{rq}	same limits as Problem 24
27. R**s** = M	r_{jk}, s_k, **s** vector	$j = 1, p' k = 1, r$
28. **s**′R′ = N	same	Problems 28 through 34 same limits as Problem 27
29. **s**′R = Q	r_{jk}, s_j, **s** vector	
30. RS = P	r_{jk}, s_k, S diagonal	
31. SR′ = V	same	
32. SR = W	r_{jk}, s_j, S diagonal	
33. R**1** = T	r_{jk}, **1** vector	
34. 1′R = M	same	
35. **vw**′ = S	v_i, w_j, **v**, **w** vector	$i = 1, n; j = 1, p$
36. **w**′**v** = z	v_i, w_i, **v**, **w** vector	$i = 1, n$
37. DGD = H	g_{pq}, d_p, D diagonal	$p, q = 1, m$
38. **v**G**v** = s	g_{pq}, v_p, **v** vector	same limits as Problem 37
39. **1**′G**1** = x	g_{pq}, **1** vector	same limits as Problem 37
40. A′A**w** = **p**	a_{ij}, w_j, **w** vector	$i = 1, n; j = 1, p$
41. **w**′A′A**w** = u	same	same limits as Problem 40
42. **v**′B′A**w** = q	a_{ij}, b_{ik}, v_k, w_j, **v**, **w** vectors	$i = 1, n; j = 1, p; k = 1, r$
43. DA′AD = K	a_{ij}, d_j, D diagonal	$i = 1, n; j = 1, p$
44. EAA′E = C	a_{ij}, e_i, E diagonal	same limits as Problem 43
45. EB′AD = M	a_{ij}, b_{ik}, E, D diagonal	$i = 1, n; j = 1, p; k = 1, r$

ANSWERS

1. X′X 2. YX 3. Y′X or X′Y

4. X**y** 5. AB 6. **1**′X

7. XS, S diagonal 8. SX, S diagonal 9. X**w**

10. AV, V diagonal 11. VA 12. A**v**

13. A**1** 14. **1**′A**1** 15. X + **1c**′

16. X − c**11**′, or X − **c1**′, or X − **1c**′, where **c** = $(c, c, \ldots, c)$

17. n^{-1}X′X 18. n^{-1}SX′XS, S diagonal 19. s^{-1}VA′AV, V diagonal

20. **y**′X**b** 21. **b**′X′X**b**

22. $\Sigma\, x_{ij}x_{ik}/n = n^{-1}$X′X; then R = n^{-1}SX′XS, S diagonal; $R = n^{-1}$(SX′) (XS). Z = XS; R = n^{-1}Z′Z = n^{-1}(XS)′ (XS) = n^{-1}(SX′) (XS), so the identity is there

23. $g_{ik} = \Sigma_j a_{ij} b_{jk}$ 24. $x_{qr} = \Sigma_p c_{pq} d_{pr}$ 25. $y_{qr} = \Sigma_p c_{pq} d_{rp}$

26. $z_{pr} = \Sigma_q c_{pq} d_{rq}$ 27. $m_j = \Sigma_k r_{jk} s_k$

28. $n_j = \Sigma_k r_{jk} s_k$

29. $q_k = \Sigma_j r_{jk} s_j$ 30. $p_{jk} = r_{jk} s_k$ 31. $v_{kj} = r_{jk} s_k$

32. $w_{jk} = r_{jk} s_j$ 33. $t_j = \Sigma_k r_{jk}$ 34. $m_k = \Sigma_j r_{jk}$ 35. $s_{ij} = v_i w_j$

36. $z = \Sigma_i v_i w_i$ (z is a scalar) 37. $h_{pq} = d_p g_{pq} d_q$

38. $s = \Sigma_p \Sigma_q v_p v_q g_{pq}$ 39. $x = \Sigma_p \Sigma_q g_{pq}$ 40. $p_k = \Sigma_i \Sigma_j a_{ij} a_{ik} w_j$

41. $u = \Sigma_i \Sigma_j \Sigma_k a_{ij} a_{ik} w_j w_k$ 42. $\Sigma_i \Sigma_k b_{ik} v_k \Sigma_j a_{ij} w_j$

43. $k_{jj'} = \Sigma_i a_{ij} a_{ij} d_j d_{j'}$ (j' is the same index as j; some other letter also could be used.)

44. $c_{ii'} = \Sigma_j a_{ij} a_{i'j} e_i e_{i'}$ 45. $m_{kj} = \Sigma_i a_{ij} b_{ik} d_j e_k$

READINGS

Green, P. E., and J. D. Carroll (1976). *Mathematical tools for applied multivariate analysis.* New York: Academic Press, pp. 69–74.

Horst, P. (1963). *Matrix algebra for social scientists.* New York: Holt, Rinehart, and Winston, pp. 268–305.

3

Vectors

NATURE OF VECTORS

Vectors in a Vector Space

Earlier, we defined a vector as a one-dimensional array of numbers, conventionally displayed as a one-column matrix. A matrix can be thought of as representing a collection of vectors with each column of the matrix representing one vector. Alternatively, each row can be considered as a transposed vector, and in fact, there is something to be gained by thinking of it both ways. This duality provides an extra avenue for understanding some aspects of multivariate statistics.

It is best to begin by considering vectors entirely geometrically. We can depict them as arrows starting from a common origin but going off in different directions for different distances (see Figure 3-1). This chapter will deal with collections of vectors—one, two, five, or a thousand of them.

Figure 3-1 shows four sets of vectors, with each arrow representing a vector. First, we will consider how many dimensions are required to contain the sets of vectors. Figure 3-1 (a) apparently requires two dimensions for the three vectors, whereas (b) needs only one because all three vectors lie within the same line. Figures 3-1 (c) and (d) also require two dimensions. Now imagine that the page is a transparent screen and that the arrows in (a) are coming out of the page or extending back from the page to varying degrees in three-dimensional space. Imagine the same thing for Figure 3-1 (c) and (d). What if the page is still a transparent screen but the arrows in (a) are arranged in a fan, coming out from the bottom of the page at a certain angle? Then the vectors would constitute only a two-dimensional space in the sense that we could pluck the fan from the page and put it in an envelope. So, a number of vectors can exist in a three-dimensional space, although they may require only two dimensions. The three superimposed lines in (b) require only one dimension, although they exist in the two-dimensional space on the page. While we might imagine *folding* the fan to make it one-dimensional and putting it in a tube, this is inadmissible because it violates the angular separations between the vectors.

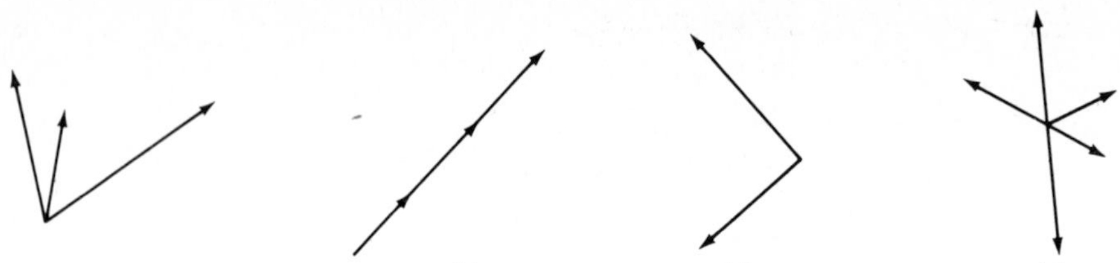

(a) (b) (c) (d)

Figure 3-1 Configurations of vectors.

The principle here is that, although we might imagine a set of vectors as existing in a space of a certain dimensionality, they may require only a smaller number of dimensions. A single vector is one-dimensional; whereas two vectors are two-dimensional (at most), and three vectors are three-dimensional. These are upper limits; fewer dimensions may suffice for a given set. These principles may be considered abstractly, extending to spaces of more than three dimensions, although admittedly the visualization becomes difficult.

Now imagine a vector space as having *any* dimensionality—two, three, ten, or simply p-dimensions. In any dimensionality, there is an infinite number of possible vectors radiating out from an origin, and the vectors are of all possible lengths. In a p-dimensional space, p of the vectors may be selected as a *basis* for the space. These can be any p, provided they span the whole space and do not fall into a subspace of lower dimensionality; for example, if $p = 3$, we need three basis vectors which cannot lie in a plane. All the possible vectors in the space can be expressed as weighted combinations of these basis vectors. The numbers that specify the weights are definitions of their respective vectors.

Because there is an infinite number of possible vectors, these weights can be any number—positive, negative, large, small, or even zero. The numbers in a given matrix can be thought of as expressing the coefficients that define some of these vectors. If a matrix is n by p, each column can be thought of as corresponding to one basis vector, and each row can be thought of as one of the vectors to be defined. The numbers in the matrix will then represent the coefficients for defining the row vectors as a particular combination of the basis vectors. For example, in

$$\begin{array}{c} \\ \mathbf{a}_1 \\ \mathbf{a}_2 \\ \mathbf{a}_3 \end{array} \begin{array}{c} \begin{array}{cc} \mathbf{b}_1 & \mathbf{b}_2 \end{array} \\ \begin{bmatrix} 2 & 1 \\ -1 & 4 \\ 0.2 & 0 \end{bmatrix} \end{array}$$

there are three vectors (rows), $\mathbf{a}_1$, $\mathbf{a}_2$, and $\mathbf{a}_3$, expressed with respect to two basis vectors, $\mathbf{b}_1$ and $\mathbf{b}_2$. In this matrix, $\mathbf{a}_1 = 2\mathbf{b}_1 + \mathbf{b}_2$; $\mathbf{a}_2 = -\mathbf{b}_1 + 4\mathbf{b}_2$; $\mathbf{a}_3 = .2\mathbf{b}_1$. Whenever we conceive of a matrix, it may be thought of in this fashion, expressing how to define a set of n vectors with respect to a p-dimensional space.

We also need to be able to think of a matrix *conversely,* with the *rows* representing the n basis vectors and the numbers describing how to make the p-column vectors. Occasionally we will need to alternate between the two ways of considering the matrix.

A geometrical application to the basis vectors and the coefficients in matrices is offered in textbooks on matrix algebra. Although we do not consider geometric applications here, one of these geometric ideas deserves mention. We already said that any set of p vectors in a p-dimensional space can act as the basis, as long as it does not fall in a subspace. However, a number of concepts and operations in applied work later are simplified if the p vectors are chosen so that they are arranged at right angles to each other and are of equal length. Such a basis is called *orthonormal.*

An important concept in considering vectors is that of a *transformation.* It is often desirable, particularly in statistical applications, to change from one set of basis vectors to another. This means that if the coefficients of a set of vectors are expressed numerically in a matrix, these numbers have to be changed so that they refer to the new basis. Fortunately, there is a systematic relation between the coefficients that express a set of vectors in terms of one basis and the coefficients that express these same vectors with respect to the new one. If the rows of a matrix A are the coefficients for the old basis, then the new coefficients would be $B = AT$, where T is a square matrix that expresses how to construct the new basis out of the old one.

A special kind of transformation is an orthonormal transformation. It represents a special situation in which the old basis vectors are obtained from the new ones by a rigid rotation. When the transformation matrix has this special property, then $TT' = T'T = I$, the identity matrix.

Rank and Dimensionality

After receiving a multiple regression problem for analysis, a computer occasionally declines to perform the analysis, offering the excuse, "Matrix not of full rank." On reading a journal report on a factor analysis, one may encounter the assertion, "The rank of the correlation matrix was five." The term "rank" refers to the numerical counterpart of the dimensionality of the space occupied by the vectors represented in a given matrix. Rank is an exact mathematical concept, however, and the two quotes about rank, particularly the latter, may refer to an "approximate" rank or dimensionality.

If a particular matrix has, for example, three columns, and we consider each column as referring to a basis vector, the basis is three-dimensional. Suppose we have three or more vectors given as rows of a matrix and defined by coefficients on the basis. It could be that the three vectors lie in a fan, as suggested earlier, or even in a single line. Then they do not constitute a space of "full dimensionality." This can be determined by the nature of the numbers representing these vectors, that is, from the matrix of coefficients. *The dimensionality of the given vectors can be determined by finding the rank of the matrix.*

Rank and Linear Dependence

The rank of a matrix is related to the concept of using a basis to construct other vectors in the space. If some of the vectors in a matrix can be used to construct all the others, then this subset is in fact an alternative basis for the whole set. The number of vectors (rows) of the matrix required to construct the complete set is the rank of the matrix. Consider the following matrix:

$$\begin{bmatrix} 1 & 2 & 3 \\ 3 & 6 & 9 \end{bmatrix}$$

Let the first row define the vector $\mathbf{x}_1 = (1, 2, 3)$ and the second row the vector $\mathbf{x}_2 = (3, 6, 9)$. Then, we can see that $\mathbf{x}_2 = 3\mathbf{x}_1$, or alternatively, $\mathbf{x}_1 = (1/3)\mathbf{x}_2$. Consider the following matrix in which the rows define vectors $\mathbf{x}_1$, $\mathbf{x}_2$, and $\mathbf{x}_3$.

$$\begin{bmatrix} 1 & 2 & 3 \\ 1 & 0 & 1 \\ 3 & 4 & 7 \end{bmatrix}$$

Although it may not be obvious, $2\mathbf{x}_1 + \mathbf{x}_2 = \mathbf{x}_3$; $-2\mathbf{x}_1 + \mathbf{x}_3 = \mathbf{x}_2$; and $-\frac{1}{2}\mathbf{x}_2 + \frac{1}{2}\mathbf{x}_3 = \mathbf{x}_1$. That is, given any two of the vectors, we can construct the third. Because coefficients applied to any two vectors make the third, two vectors are a sufficient basis for constructing the whole set of three.

In general, if a set of k vectors are such that any one may be constructed by weighting the $k - 1$ others, the vectors are said to be *linearly dependent,* and the corresponding matrix is not of *full rank.* If this is not possible, then the vectors are *linearly independent,* and the matrix is of full rank, Note that "independence" has a different meaning in this context than "statistical independence."

The formal criterion of linear dependence can be stated somewhat differently. For the two vectors in the first matrix, we can restate it as $3\mathbf{x}_1 + (-1)\mathbf{x}_2 = 0$. For the three vectors in the second matrix, we can restate it as $2\mathbf{x}_1 + 1\mathbf{x}_2 + (-1)\mathbf{x}_3 = 0$.

> Given a set of vectors $\mathbf{x}_1, \mathbf{x}_2, \ldots, \mathbf{x}_k$, if there is a set of coefficients $c_1, c_2, \ldots, c_k$, which are not all zero, such that $c_1\mathbf{x}_1 + c_2\mathbf{x}_2, \ldots + c_k\mathbf{x}_k = 0$, then the vectors are *linearly dependent.*

If the rows of a square matrix are linearly dependent, so are the columns, and vice versa. If the number of rows of a matrix *exceeds* the number of columns, the rows are linearly dependent. So, if the matrix is n by p, we can take the first p rows as a basis, and construct the remainder from them. Alternatively, if $p > n$, we could take the first n columns and make the remaining $p - n$ out of them. *This proves that the rank of an* n *by* p *matrix is, at most, the lesser of* n *and* p.

Rank and the Number of Linearly Dependent Rows or Columns

So far, all we have is a method for finding out if the rank of a matrix is less than its order. But how does one find out what the rank actually is? *The rank of a matrix is the largest number of rows (columns) that are linearly independent.*

The numerical process of finding the rank of a matrix can be tedious, although computer programs for carrying out the process are quite simple and widely available. We will not be concerned with how to do this here.

Remember that linear dependence was stated in terms of the sum of weighted vectors equal to zero. The rank of a matrix can be stated in a related way as follows:

> The rank of a matrix is r if and only if r rows or columns $\mathbf{x}_j$ can be found such that $\Sigma_j c_j \mathbf{x}_j \neq 0$ for *any* set of coefficients c_j (except all zeros), but for all sets of $r + 1$ or more rows or columns, there is a set of coefficients such that $\Sigma_j c_j \mathbf{x}_j = 0$.

Therefore, at least one set of r rows or columns is linearly independent, but all larger sets are linearly dependent.

As noted earlier, the order of a matrix limits its rank, the maximum rank being equal to the number of rows or columns, whichever is fewer. Rank may be reduced further if there are linear dependencies in the data. An example of linear dependence is the inclusion of a pretest, x_1, a posttest, x_2, and the gain score, $d = x_2 - x_1$. Then $(1)(x_2) + (-1)(x_1) + (-1)d = 0$, which is clearly an example of linear dependence. Linear dependence can also occur with "ipsatized" scores—sets of scores in which every person has the same total score. This happens with some personality inventories, experimental tasks, and artificial matrices. If any of these sets of variables occur in a data matrix, the rank of the matrix will be less than its order.

Another source of rank reduction deserves note: The rank of a matrix is at most the lesser of n and p. Ordinarily, data analyses are carried out on matrices in which the number of observations (n) is greater than the number of variables (p), so the rank is governed by the latter. If this is not the case, the rank is at most n, not p. When deviation scores are calculated as a step toward computing the covariances among the variables, the rank is reduced to at most $n - 1$. Because the deviation scores on each variable sum to zero, this is an example of $\Sigma_i c_i x_i = 0$, with each $c_i = 1.0$. If $n > p$, this effect does not occur because p is limiting the rank.

Another very important principle that should be noted is that *the rank of a matrix product is at most the lesser of the ranks of the matrices multiplied together*. That is, matrix multiplication never increases rank; multiplying a rank two matrix by a rank ten matrix gives at most a rank two matrix. This principle may occur when $n \leqq p$ and a covariance matrix is computed. As we have seen, the rank of the deviation score matrix and the covariance matrix is then $n - 1$. That is, if a score matrix is 10 persons by 50 variables, the rank of the covariance and correlation matrices is at most nine, even though the matrix is 50 by 50.

Approximate Rank

So far in this discussion the rank of a matrix has been an exact integer quantity, which is what the usage is in mathematics. Many references to rank in the context of data matrices refer to an "approximate" rank, however. We might speak of a matrix as being "barely" rank four, or "almost" linearly dependent. In the strict sense, linear dependence is a qualitative condition; it is either present or it is not. We might as well speak of being "almost pregnant" as of "almost linearly dependent." In dealing with data, however, it may well be a matter of interest if we can "almost" construct one vector out of the rest. After all, a Japanese fan, no matter how exquisite, is three-dimensional, not two. So are the pages in this book, even though two of the dimensions are much broader than the other. In the same sense, a collection of vectors and their numerical representation in a matrix may be very thin in one or more dimensions, and barely require that dimension to contain it. It is in this sense that "approximate" rank, or dimensionality, is used.

Some readers may have encountered the term *eigenvalue,* or equivalent terms such as *characteristic root, latent root,* or *singular value.* These terms will be discussed in later chapters. For now, we will define eigenvalues as particular numbers associated with a given matrix. The exact rank of a square matrix is the *number of its eigenvalues which are nonzero.* Having eigenvalues close to zero indicates that there are dimensions in which the vectors represented in the matrix are "thin." Thus eigenvalues are used to determine both the exact and approximate rank of a matrix. A 5×5 matrix with one zero eigenvalue, one very small one, and three large ones would have an exact rank of four and an approximate rank of three.

QUANTITATIVE GEOMETRIC CONCEPTS

Angles Between Vectors

Some of the quantitative aspects of geometry have application in multivariate statistics, particularly those having to do with angles. For any two vectors, $\mathbf{x}$ and $\mathbf{y}$, there is an angle between them, θ_{xy}. This angle is zero if the vectors are colinear, 180° or π radians if the vectors are directed oppositely, and 90° or $\pi/2$ radians if one vector is orthogonal (at right angles) to the other.

The angles among several vectors are obviously not independent of each other. Given θ_{xy} and θ_{xz}, there are limitations on θ_{yz}. In two-space, in fact, either $\theta_{yz} = \theta_{xy} + \theta_{xz}$ or $\theta_{yz} = |\theta_{xy} - \theta_{xz}|$. In three-space, the limitations are not as stringent, although θ_{yz} must be *between* these two limits, $(\theta_{xy} + \theta_{xz}) \geqq \theta_{yz} \geqq |\theta_{xy} - \theta_{xz}|$. We can visualize this restriction as follows. Draw three lines that intersect at the same point on a transparent or thin sheet of paper. Label the middle line x and the other two y and z, and fold the paper along the x line as shown in Figure 3-2. When the paper is flat, we have $\theta_{yz} = \theta_{zy} + \theta_{xz}$. When the paper is folded over and laid flat, we have $\theta_{yz} = |\theta_{xy} - \theta_{xz}|$. In both of these cases, the three vectors fall in

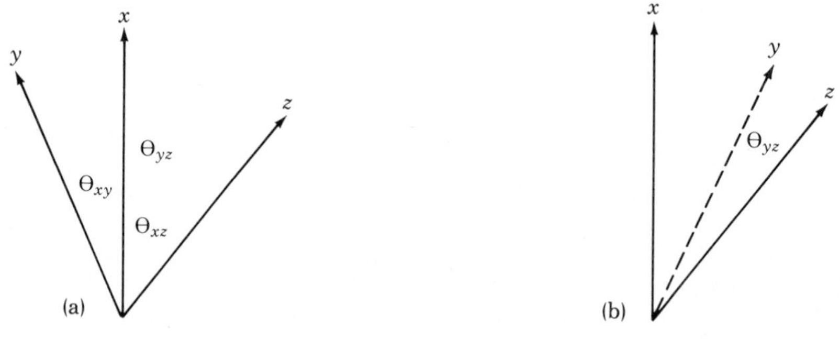

Figure 3-2 If figure (a) is cut out and folded along x, the angle θ_{yz} can take any value from that in figure (a) to that in (b).

two dimensions. When the paper is partially folded, the angle θ_{yz} is between these limits.

If we know all three angles, then this fact determines the positions of the three vectors relative to each other. When we have more vectors, and thus the possibility of more than three dimensions, there are parallel restrictions. These are more complicated to state and virtually impossible to visualize.

Cosines of Angles

Except for casual purposes, one rarely has cause to deal with the angles themselves. For our purposes, virtually all angular relations are expressed in terms of the *cosines* of the angles. The cosine of an angle may be familiar from trigonometry or calculus.

In this book, we only need to recall a few of the numerical and algebraic properties of cosines. The cosine of an angle is a number between -1.0 and 1.0. The function $\cos \theta_{xy}$ expresses the angle θ_{xy} between two vectors, **x** and **y**. Cos $\theta_{xy} = 1.0$ when the angle is zero, that is, when the vectors are colinear. Cos θ_{xy} equals zero when the angle is 90°, that is, when the vectors are orthogonal, and -1.0 when the angle is 180°, that is, when their directions are opposite. The relationship between cosines and angles is curvilinear; cos 60 degrees = .5, cos 45 degrees = .707. The pleasant shape in Figure 3-3 represents the complete relationship.

Another important property of cosines is that the squared cosine of an angle plus the squared cosine of its complementary angle is equal to 1.0. This is expressed as

$$\cos^2 \theta + \cos^2 (90° - \theta) = 1.0$$

For our purposes, this means that if **x** and **y** are orthogonal and **z** is some third vector in their plane, $\cos^2 \theta_{xz} + \cos^2 \theta_{yz} = 1.0$.

We need to be able to consider a large number of vectors simultaneously. If the cosines among a number of vectors are all near 1.0, it means that they are tightly bunched into a bundle, like the arrows grasped by the eagle in the Great Seal of the United States. If they are all close to zero, the vectors are nearly at right angles to each other. For the vectors to be mutually nearly orthogonal, there must be about as many dimensions as vectors.

The most useful type of basis for a collection of vectors is an orthonormal one in which any given vector may be described almost completely by the cosines of its angles with the basis vectors. We would easily be able to express the direction of any vector with respect to the basis and know its direction. If one of the cosines of a vector with the basis is 1.0 and the rest are zero, it means that the vector is colinear with one of the basis vectors. If all the vectors have the same length, then knowing their cosines with the basis vectors specifies them completely. If we have k vectors, and a $k + 1$st vector has cosines of zero with all, we know the vector is orthogonal to all other vectors and that its inclusion with them will require one more dimension than the other vectors have.

The sum of squares of the cosines of the angles that a vector makes with an orthogonal basis is equal to 1.0. This is the generalization of the relationship discussed above concerning the squared cosine of an angle plus the squared

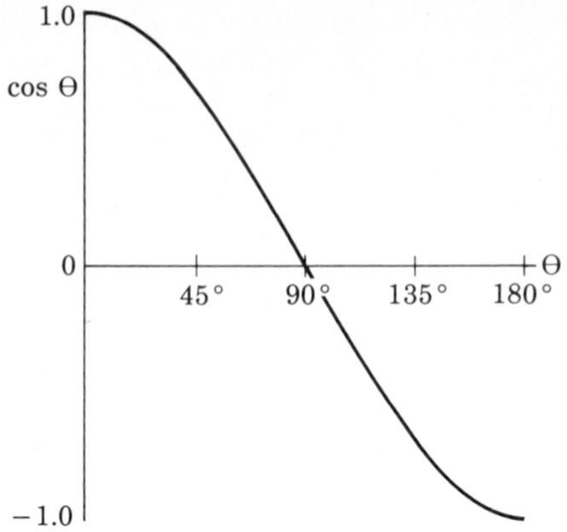

Figure 3-3 Cosine θ for $\theta = 0$ to 180°.

cosine of its complement. This relationship is summarized as follows:

$$\sum_m \cos^2 \theta_{mj} = 1.0 \tag{3.1}$$

where the $\mathbf{x}_j$, $j = 1, p$ is an orthogonal basis set of vectors and $\mathbf{x}_j$ is any vector in that space.

Scalar Products, Length, and Distance

If $\mathbf{x}$ is a vector expressed in terms of coefficients on an orthonormal basis, then the *length* of the vector is the square root of the sum of squared coefficients. In two-space, this is a straightforward application of the Pythagorean theorem. A two-stage application of the theorem shows that it is true in three-space as well, as illustrated in Figure 3-4. Repeated applications of the theorem as many times as necessary would show that it is true for any dimensionality. In general,

> If $\mathbf{x}$ is a vector with coefficients $(x_1, x_2, \ldots, x_j, \ldots, x_p)$ on an orthonormal basis, then the *length*, ℓ_x, of the vector $\mathbf{x}$ is

$$\ell_x = \sqrt{\sum_j x_j^2} \tag{3.2}$$

If a vector is defined with respect to an orthonormal basis, knowing the cosines of its angles with the basis vectors and its length defines it completely.

A *scalar product* of two vectors that are expressed in terms of their coefficients is the sum of products of coefficients:

$$t_{xy} = \sum_j x_j y_j \tag{3.3}$$

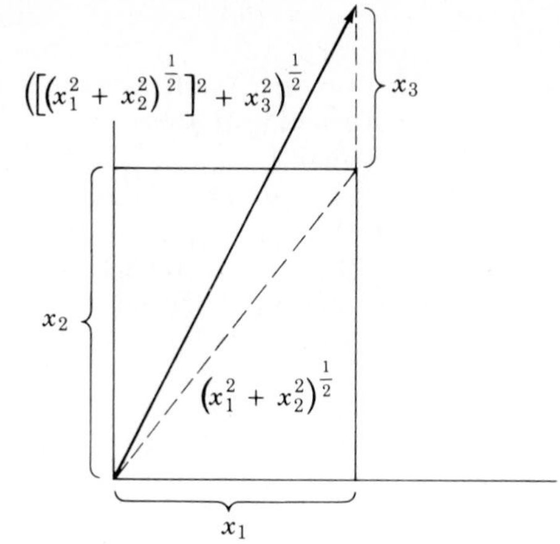

Figure 3-4 Length of a vector in three dimensions as an extension of the Pythagorean theorem. If (x_1, x_2, x_3) are the coordinates of the vector, x is the hypotenuse of a right triangle whose legs are x_3 and $(x_1^2 + x_2^2)^{1/2}$.

Note that this is the same as the basic operation in matrix multiplication. When we multiply a matrix A by a matrix B, we find the scalar products between each of the vectors that are the *rows* of A with each of the vectors that are the *columns* of B.

The term "scalar product" comes from the fact that the product of the two multicomponent vectors is a single number, a scalar. In some areas of study, the equivalent terms *inner* product and *dot* product are used. From the definition of length given earlier, we can see that the length of a vector is the square root of its scalar product with itself.

The scalar product of two vectors is related to the cosine of the angle between them and their respective lengths:

$$t_{xy} = \ell_x \ell_y \cos \theta_{xy} \tag{3.4}$$

Conversely, the cosine may be expressed in terms of the scalar product and the lengths of the vectors:

$$\cos \theta_{xy} = \frac{t_{xy}}{\ell_x \ell_y} \tag{3.5}$$

Replacing t_{xy} and the two lengths by their definitional formulas gives the cosine

$$\cos \theta_{xy} = \frac{\sum_j x_j y_j}{\sqrt{\sum_j x_j^2 \sum_j y_j^2}} \tag{3.6}$$

The right-hand side of Equation 3.6 may look familiar. It is one of the formulas for the correlation between x and y, the one that applies if x and y are in deviation score form. This interpretation is quite direct and general. *The Pearson-product moment correlation is the cosine of the angle between the vectors representing the two variables* in deviation score form.

The distance between two points representing the termini of two vectors is also related to these quantities. Consider two, $\mathbf{a}_1$ and $\mathbf{a}_2$, in a space with orthonormal basis vectors $\mathbf{x}_1$ and $\mathbf{x}_2$, so $\mathbf{a}_i$ has coefficients a_{11} and a_{12}, and $\mathbf{a}_2$ has coefficients a_{21} and a_{22}. Again, by application of the Pythagorean theorem, the distance between them is

$$\sqrt{(a_{11} - a_{21})^2 + (a_{12} - a_{22})^2}$$

as illustrated in Figure 3-5. The same principle applies to higher dimensionalities.

The following general definition applies to the distance between two points or vectors $\mathbf{a}_i$ and $\mathbf{a}_j$:

$$d_{ik} = \sqrt{\sum_j (a_{ij} - a_{kj})^2} \tag{3.7}$$

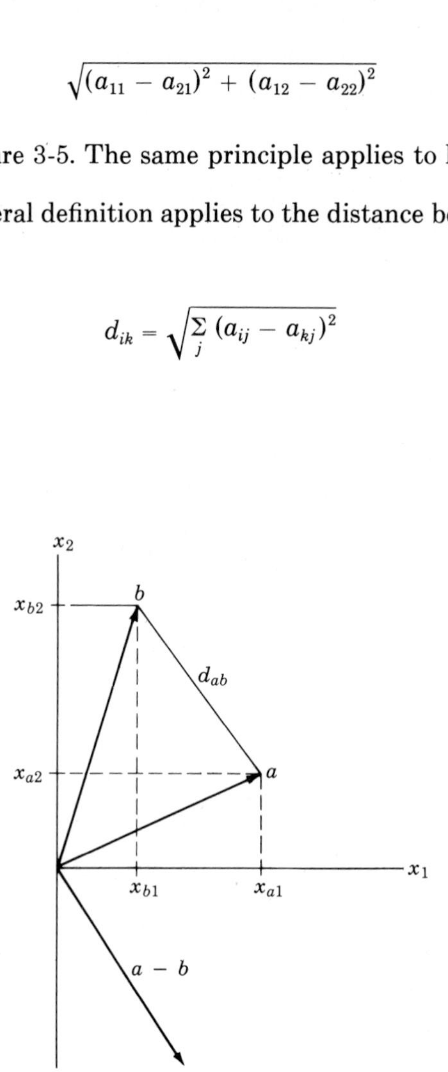

Figure 3-5 Distance and the Pythagorean theorem. The distance between vectors **a** and **b** is the hypotenuse of a right triangle with legs $(x_{a1} - x_{b1})$ and $(x_{a2} - x_{b2})$. This is also the length of the vector **a** − **b**.

This equation can be rearranged and expressed as two other useful equations:

$$d_{ik} = \sqrt{\sum_j a_{ij}^2 + \sum_j a_{kj}^2 - 2 \sum_j a_{ij}a_{kj}} \tag{3.8}$$

$$d_{ik} = \sqrt{\ell_i^2 + \ell_k^2 - 2\ell_i\ell_k \cos \theta_{ik}} \tag{3.9}$$

Note the resemblance of Equation 3.9 to the standard deviation of a difference score. The concept of distance, then, applies not only to one dimension (along a line), two dimensions, (a plane) and three dimensions (ordinary space), but to higher dimensionalities as well. If you have trouble imagining distances in more than three dimensions, don't try. Just work with the equations.

Therefore, a number of properties of vectors or pairs of vectors can be expressed in terms of the coefficients, including the length of a vector (Equation 3.2), the scalar product between two vectors (Equation 3.3), the cosine of the angle between them (Equation 3.5), and the distance between them (Equation 3.7). Typically in our applications, the only definitions for vectors are numerical, given by or derived from a data matrix. Thus we make extensive use of these numerically defined properties, and frequently carry out the matrix operations that correspond to them.

SUMMARY

Most of our work with multivariate data is numerical, but it is useful to have the geometrical concepts of vectors as an alternate way of thinking about multivariate data. Vectors can be considered as directed line segments emanating from a particular origin. Important vector concepts include the dimensionality of the space containing them and the idea of selecting a subset of the vectors to form a *basis* for the space. There is one basis vector for each dimension.

The numerical entries in a matrix can be thought of as the coefficients that tell how to construct a set of vectors using a particular basis. A set of vectors is *linearly dependent* if there are more vectors than dimensions to their configuration, and the *rank* of a matrix is the dimensionality of the set of vectors represented in it.

Several other geometrical concepts were introduced in this chapter: the cosine of the angle between two vectors, vector length, and the distance between vectors.

PROBLEMS

1. What is the rank of each of the following matrices? (If there are linear dependencies, they are simple ones.)

$$\underset{A}{\begin{bmatrix} 2 & 1 \\ -4 & -2 \end{bmatrix}} \quad \underset{B}{\begin{bmatrix} 3 & 2 \\ 4 & 6 \end{bmatrix}} \quad \underset{C}{\begin{bmatrix} 1 & 1 & 0 \\ 2 & 3 & 4 \\ 1 & 2 & 4 \end{bmatrix}} \quad \underset{D}{\begin{bmatrix} 1 & 0 & 1 \\ 0 & 1 & 1 \\ 1 & 1 & 0 \end{bmatrix}}$$

2. What is the maximum possible rank of the following matrices A, B, C, D, and E, as judged by their order and the presence of simple linear dependencies?

$$\underset{\text{A}}{\begin{bmatrix} 1 & 0 & 1 \\ 1 & 1 & 0 \\ 0 & 1 & 1 \\ 0 & 1 & 0 \\ 0 & 0 & 1 \end{bmatrix}} \quad \underset{\text{B}}{\begin{bmatrix} 1 & 3 & -4 \\ 2 & -1 & -1 \\ 1 & 0 & -1 \\ 4 & -2 & -2 \\ -1 & 2 & -1 \end{bmatrix}} \quad \underset{\text{C}}{\begin{bmatrix} 1 & 2 \\ 2 & 4 \\ 3 & 6 \\ 4 & 8 \end{bmatrix}} \quad \underset{\text{D}}{\begin{bmatrix} 3 & 1 & 2 \\ 2 & 4 & 1 \end{bmatrix}} \quad \underset{\text{E}}{\begin{bmatrix} 4 & 0 & 3 & 0 & 1 \\ 2 & -1 & 0 & 3 & 0 \end{bmatrix}}$$

3. Given your answers to Problem 2, what is the maximum rank of the product A′B? CD? EA? E′D? AB′?

4. If $\theta_{12} = 30°$ and $\theta_{23} = 110°$, what are the limits for θ_{13}? What does it mean if θ_{13} is at one of these limits?

5. The following three vectors are expressed with respect to an orthonormal basis:
   $\mathbf{a} = (2, 1, -1)'$
   $\mathbf{b} = (1, -3, 2)'$
   $\mathbf{c} = (0, 7, -5)'$
   a. What is the *scalar product* $\mathbf{a}'\,\mathbf{b}$?
   b. What is the *length* of $\mathbf{a}$? Of $\mathbf{b}$?
   c. What is $\cos \theta_{ab}$?
   d. What is the dimensionality of the space defined by $\mathbf{a}$, $\mathbf{b}$, and $\mathbf{c}$?

6. Plot the preceding three vectors in all three pairs of coordinate axes.

ANSWERS

1. A: 1; B: 2; C: 2 (row 1 − row 2 + row 3 = 0, or col. 1 − col. 2 + ¼ col. 3 = 0); D: 3.

2. A: 3; B: 2; C: 1; D: 2; E: 2.

3. A′B: 2; CD: 1; EA: 2; E′D: 2; AB′: 2.

4. $80° \le \theta_{13} \le 140°$. At the limits, the configuration is two-dimensional.

5. $a \circ b = -3$; $|a| = \sqrt{6}$; $|b| = \sqrt{14}$; $\cos \theta_{ab} = -.327$. Dimensionality is 2 ($a - 2b + c = 0$).

READINGS

Ayres, F. (1962). *Theory and problems of matrices.* New York: Schaum, pp. 39; 85–93; 100–109.

Green, P. E. and J. D. Carroll (1976). *Mathematical tools for applied multivariate analysis.* New York: Academic Press, pp. 77–124, 167–75.

Mills, G. (1969). *Introduction to linear algebra: A primer for social scientists.* Chicago: Aldine, pp. 25–41, 80–101.

4

Variances and Covariances of Linear Combinations

WEIGHTS AND LINEAR COMBINATIONS

Types of Weights

Multivariate statistics is based on forming linear combinations of variables and then examining their variances and covariances. In multiple regression, a set of weights is found to form a linear combination of the p predictors so as to come as close as possible to the values of the criterion variable y. The various forms of factor analysis are used to define factors as linear combinations of the variables and then reverse the process and describe the variables as linear combinations of the factors. When we perform an analysis of variance on a given set of data we are simply using different methods to take linear combinations of the different means—methods that correspond to finding the variances and covariances of different linear combinations of the group-membership "dummy variables."

Discriminant function analysis and multivariate analysis of variance seek linear combinations of variables that maximize the ratio of between-group variance to within-group variance. Canonical correlation analysis is a system for finding sets of weights for two sets of variables so that the covariance between the two combinations is maximal while their variances are held at 1.0. As a general principle, the various formal forms of multivariate statistics seek linear combinations of variables that have maximal or optimal variances and covariances in one sense or another.

Linear combinations of variables are also formed on an a priori basis. We might wonder after doing a multiple regression what happens to the error sum of squares when we use some simple, integral weights instead of the regression weights—for example, using all unit weights and just adding up the predictor scores. We might want to find out the effect of choosing a different form of factor analysis, or the effect of including gain scores in a battery of measures, and so on. Both these and many other kinds of statistical activities are simply ways of finding out about the effects of taking certain linear combinations of the variables. Thus there is considerable benefit in understanding the general aspects of this process. It is also useful to gain familiarity with the general principles of linear combinations before moving on to the specifics. A "linear combination" of variables means that a new variable is formed as a weighted sum of other variables. For example,

$$y_i = w_1x_{i1} + w_2w_{i2} + \cdot\cdot\cdot + w_px_{ip} \tag{4.1}$$

Such a weighted combination is called "linear" because the variables are involved in their simple, linear form; there are no powers, products, logs, sines, and so on, of the x's in the combination. The main concern is the general case in which there are any number of variables and the weights can be anything, but situations involving special kinds of weights are very common. Their places in the general scheme may be obscure unless they are pointed out.

The simplest weighted combination is an "unweighted" combination. If y is the simple sum of several x's, $y_i = x_{i1} + x_{i2} + \cdot\cdot\cdot + x_{ip}$, this simply means that all the weights are one, $w_1 = w_2 = \cdot\cdot\cdot = w_p = 1$. If all the x's are not in the sum, this means their weights are zero. Another important special case is the difference score or gain score, $d_i = x_{i2} - x_{i1}$, such as post-test minus pretest. In this case, there are only two variables to be weighted, and their weights are -1.0 and 1.0, respectively. An instructor may decide that the two midterms and final in a course should be weighted 30–30–40, and assign final grades on the basis of a weighted total, $.30x_{i1} + .30x_{i2} + .40x_{i3}$, for the two midterms and final, respectively. So, variables may receive a variety of a priori weights, often simple ones.

Weight Vectors and Score Matrices

By now the translation of the formula for a linear combination into matrix notation should be relatively easy. If the x variables are arranged in an n by p matrix, then

$$\mathbf{y} = \mathbf{X}\mathbf{w}_1 \tag{4.2}$$

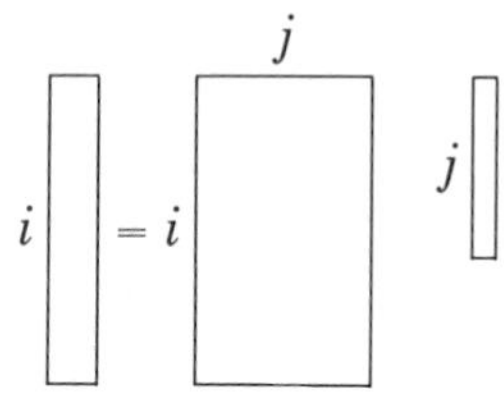

where $\mathbf{w}_1$ is a column vector of weights. If $\mathbf{y}$ is a simple sum, then $\mathbf{w}_1$ contains all unities. Common examples of unit weights include ordinary tests in which the columns of X refer to different items or parts and y is a total score on the test. In the case of a difference score, the weight vector would be -1 and 1, for the two variables involved, the rest would be zeros. In general, if only some variables play a part in a particular combination, the remainder have weights of zero. Whatever the weights are, they go in $\mathbf{w}_1$; whatever variables are or *might be* weighted, go in X. $\mathrm{X}\mathbf{w}_1$ will consistently be used here instead of $\mathbf{w}_1'\mathrm{X}'$, in accord with our general bias toward working with matrices in which variables define the columns.

The score matrix X can be anything that is worth recording. Most often the variables will be raw or deviation scores on continuous variables, but they may also be dichotomous variables, such as test items scored 1 or 0, 1 or -1, or even 1 or $-1/4$, as in formula scores. Also dichotomous, or frequently so, are the values used to code categorical variables, as in analysis of variance.

Most often the variables weighted are in raw or deviation score form, although they may sometimes be standard scores (z scores). It is important to be clear about which form the weights are intended to apply to—that is, whether raw or standard scores are to be summed, or difference scores are to be used. If weights are shown, it must be made clear which type they are.

It is quite straightforward to convert raw to standard score weights and vice versa. All that is needed are the standard deviations. For example, suppose we wish to weight standard scores equally but do not wish to go to the trouble of computing the standard scores. What is desired is that $\mathbf{y} = \mathbf{Z1}$, where $\mathbf{1}$ is a column vector of ones. Because $\mathrm{Z} = \mathrm{XD}_s^{-1}$, where D_s^{-1} is a diagonal matrix with the reciprocals of the standard deviations, $\mathrm{Y} = \mathrm{XD}_s^{-1}\mathbf{1}$. Inserting parentheses points out that $\mathrm{Y} = \mathrm{X}\ (\mathrm{D}_s^{-1}\mathbf{1})$. $\mathrm{D}_s^{-1}\mathbf{1}$ will be a vector of reciprocals of standard deviations. This shows that if we wish to weight standard scores equally, we should weight raw scores according to the reciprocals of the standard deviations; that is, $\mathbf{w}_1 = \mathrm{D}_s^{-1}\mathbf{1}$.

This principle has practical applications. If we want to go from raw to standard score weights, for example, we divide by the standard deviations of the variable; to go in reverse, we multiply.

STATISTICS OF LINEAR COMBINATIONS

Means

If $\mathbf{y} = \mathrm{X}\mathbf{w}$, the mean of y, $\bar{y}$, is computed in the usual way, $\mathbf{1}'\mathbf{y}n^{-1}$. Simple substitution yields $\mathbf{1}'\mathrm{X}\mathbf{w}(n)^{-1}$. Because n^{-1} is a scalar, it may be moved and represented by a diagonal matrix with entries $1/n$:

$$\bar{y} = (\mathbf{1}'\mathrm{X}n^{-1})\,\mathbf{w} \qquad (4.3)$$

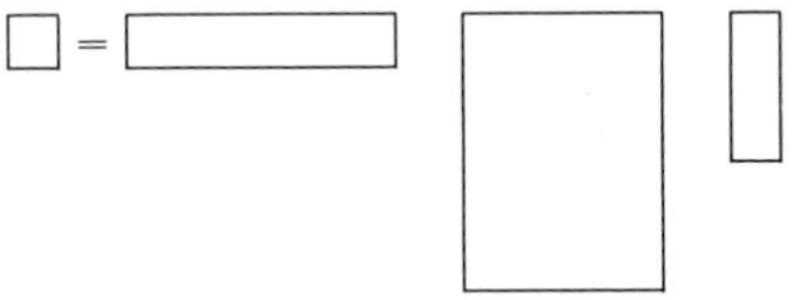

The part of the equation in parentheses is the row vector of means of the x's, which is denoted $\bar{\mathbf{x}}$. So,

$$\bar{y} = \bar{\mathbf{x}}'\mathbf{w} \qquad (4.4)$$

This shows that the mean of a weighted combination is the weighted combination of the means. Note that if X is in deviation form, or if the matrix is a standard score matrix, the mean of each variable is zero, and the mean of the weighted combination is also zero.

This also allows us to assume that we are dealing with matrices of deviation rather than raw scores when later variances, covariances, and correlations are considered. This is because these statistics are unaffected by means. We can make the assumption of deviation scores because the notations and derivations are simplified. One of the attractions of statistics is that we can sometimes assume what is most convenient and be sure there will be no unpleasant and unexpected consequences. Statistics is simpler than life in this respect.

Variances of Linear Combinations

If $\mathbf{y} = \mathrm{X}\mathbf{w}$, what is the variance of $\mathbf{y}$? By definition,

$$s_y^2 = \frac{1}{n-1} \Sigma y_i^2 \tag{4.5}$$

which can be expressed in matrices as

$$s_y^2 = (n-1)^{-1}\, \mathbf{y}'\mathbf{y} \tag{4.6}$$

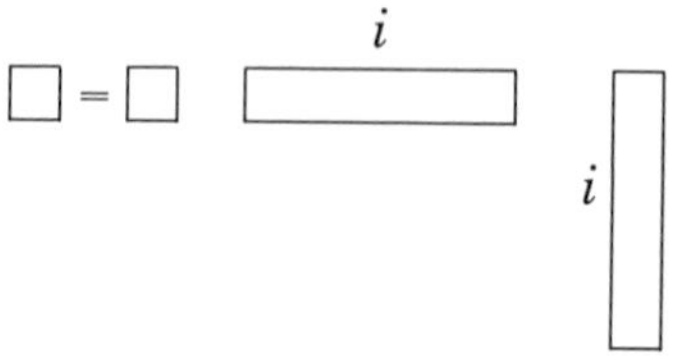

Direct substitution shows that s_y^2 can be found from the variance-covariance matrix of the x's as follows:

$$s_y^2 = (n-1)^{-1}\, (\mathrm{X}\mathbf{w})'\mathrm{X}\mathbf{w}$$

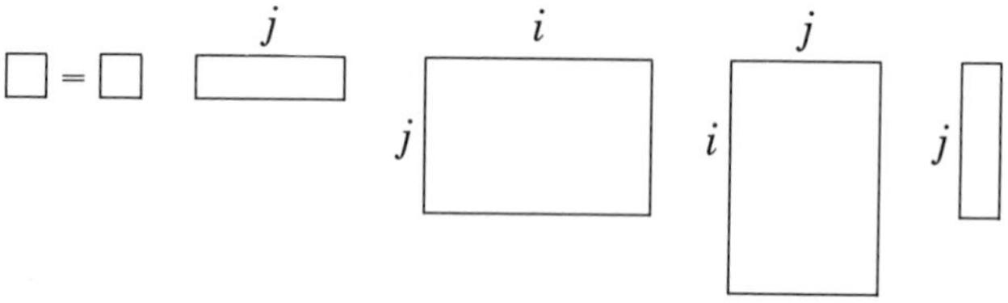

(4.7)

because relocating the parentheses and moving the scalar $(n-1)^{-1}$ leads to

$$s_y^2 = \mathbf{w}'\,[(n-1)^{-1}\,\mathrm{X}'\mathrm{X}]\mathbf{w} \tag{4.8}$$

and the middle portion is recognizable as the covariance matrix

$$s_y^2 = \mathbf{w}'\mathrm{S}\mathbf{w} \tag{4.9}$$

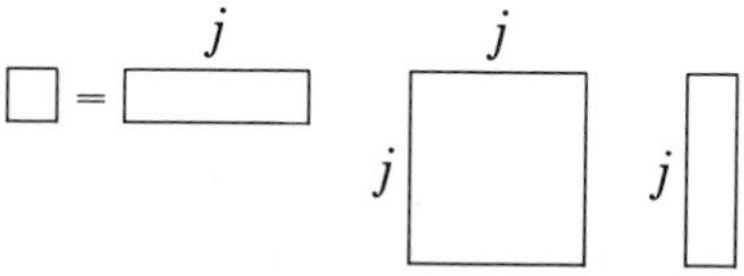

This means that the variance of the linear combination can be computed by pre- and postmultiplying the variance-covariance matrix of the constituent variables by the weight vector. The linear combination scores need not be computed at all to get the variance of the new variable. We will use this method many times over.

In scalar form, the above formula takes on the more forbidding form

$$s_y^2 = \sum_j \sum_k w_j w_k s_{jk} \tag{4.10}$$

Remember that s_{jj} is the variance of j, and that j and k are alternate subscripts that refer to variables.

If the weights are to be applied to the standard scores, the only difference is that now it is the correlation matrix that is pre- and postmultiplied by the weights. The linear combination of standard scores is not itself a standard score unless it happens that the weights are chosen in such a way as to give the combination unit variance.

An important special case of weighting is the one in which unit weights are used so the linear combination is the simple sum. Here, the variance of the combination is simply the sum of all the elements of the variance-covariance or correlation matrix, whichever is appropriate. We will refer to this from time to time in future chapters.

Exhibit 4-1 illustrates the principles introduced in this section, computing the variance of the linear combination from the scores themselves and also from the variance-covariance matrix.

Covariances and Correlations of Linear Combinations

The process used in deducing the variance of a linear combination from the variance-covariance matrix can also be used in the case of covariances. First, the covariance of the linear combination with some other single variable will be considered, and then, the covariance between two linear combinations.

In the first case, we will define y as the linear combination variable, and describe its covariance with some variable x_j:

$$s_{yj} = \sum_i y_i x_{ij}/(n-1)$$

According to the equations used here, we have

$$s_{yj} = (n-1)^{-1}\mathbf{y}'\mathbf{x}_j$$

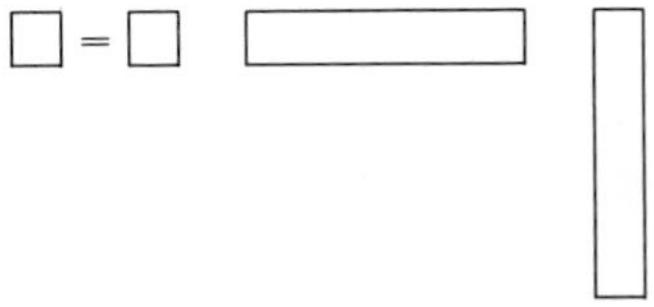

where $\mathbf{x}_j$ is subscripted to show that it is a column vector consisting of the single

(a)

$$\underset{\mathbf{X}}{\begin{bmatrix} -3 & 1 & 0 & 1 \\ -1 & 2 & -3 & -4 \\ 2 & 0 & 1 & 4 \\ 3 & 3 & 2 & 0 \\ 0 & 2 & -2 & 2 \\ 2 & -1 & 2 & -2 \\ -1 & 3 & -1 & -2 \\ -1 & 0 & 2 & 1 \\ -2 & -6 & 2 & -3 \\ 1 & -4 & -3 & 3 \end{bmatrix}} \underset{\mathbf{w}}{\begin{bmatrix} 2 \\ 1 \\ -1 \\ 3 \end{bmatrix}} \underset{=}{=} \underset{\mathbf{y}}{\begin{bmatrix} -2 \\ -9 \\ 15 \\ 7 \\ 10 \\ -5 \\ -4 \\ -1 \\ -21 \\ 10 \end{bmatrix}}$$

(b)

$$\underset{\frac{1}{n-1}}{\tfrac{1}{9}} \underset{\mathbf{y}'}{\begin{bmatrix} -2 & -9 & 15 & 7 & 10 & -5 & -4 & -1 & -21 & 10 \end{bmatrix}} \underset{\mathbf{y}}{\begin{bmatrix} -2 \\ -9 \\ 15 \\ 7 \\ 10 \\ -5 \\ -4 \\ -1 \\ -21 \\ 10 \end{bmatrix}} \underset{= s_y^2}{\begin{matrix} = \tfrac{1}{9}(1042) \\ = 115.78 \end{matrix}}$$

(c)

$$\underset{(n-1)^{-1}}{\begin{bmatrix} 1/9 & 0 & 0 & 0 \\ 0 & 1/9 & 0 & 0 \\ 0 & 0 & 1/9 & 0 \\ 0 & 0 & 0 & 1/9 \end{bmatrix}} \underset{\mathbf{X}'}{\begin{bmatrix} -3 & -1 & 2 & 3 & 0 & 2 & -1 & -1 & -2 & 1 \\ 1 & 2 & 0 & 3 & 2 & -1 & 3 & 0 & -6 & -4 \\ 0 & -3 & 1 & 2 & -2 & 2 & -1 & 2 & 2 & -3 \\ 1 & -4 & 4 & 0 & 2 & -2 & -2 & 1 & -3 & 3 \end{bmatrix}} \underset{\mathbf{X}'}{\begin{bmatrix} -3 & 1 & 0 & 1 \\ -1 & 2 & -3 & -4 \\ 2 & 0 & 1 & 4 \\ 3 & 3 & 2 & 0 \\ 0 & 2 & -2 & 2 \\ 2 & -1 & 2 & -2 \\ -1 & 3 & -1 & -2 \\ -1 & 0 & 2 & 1 \\ -2 & -6 & 2 & -3 \\ 1 & -4 & -3 & 3 \end{bmatrix}} \underset{=}{=}$$

Exhibit 4-1 (continued)

$$\begin{bmatrix} 3.78 & 0.78 & 0.78 & 1.67 \\ 0.78 & 8.89 & -1.00 & -0.11 \\ 0.78 & -1.00 & 4.44 & -0.33 \\ 1.67 & -0.11 & -0.33 & 7.11 \end{bmatrix}$$
$$\mathrm{S}$$

(d)
$$\underset{\mathbf{w}'}{[2 \quad 1 \quad -1 \quad 3]} \underset{\mathrm{S}}{\begin{bmatrix} 3.78 & 0.78 & 0.78 & 1.67 \\ 0.78 & 8.89 & -1.00 & -0.11 \\ 0.78 & -1.00 & 4.44 & -0.33 \\ 1.67 & -0.11 & -0.33 & 7.11 \end{bmatrix}} \underset{\mathbf{w}}{\begin{bmatrix} 2 \\ 1 \\ -1 \\ 3 \end{bmatrix}}$$

$$= [12.57 \quad 11.12 \quad -4.87 \quad 24.89] \begin{bmatrix} 2 \\ 1 \\ -1 \\ 3 \end{bmatrix} = 115.80$$

$$= \qquad (\mathbf{w}'\mathrm{S})\mathbf{w} \qquad = s_y^2$$

Exhibit 4-1 Two equivalent ways of computing the variance of a linear combination. (a) Illustration of computation of weighted combination; (b) Matrix computation of variance of combination; (c) Computation of variance-covariance matrix of X, NX′X = S; (d) Two-step illustration of computing s_y^2 from variance-covariance matrix, $\mathbf{w}'\mathrm{S}\mathbf{w} = (\mathbf{w}'\mathrm{S})\mathbf{w} = s_y^2$. Note that in (a) the mean of each column of X is zero, and so is the mean of y.

variable $\mathbf{x}_j$. Substitution for $\mathbf{y}$ in the matrix equation gives

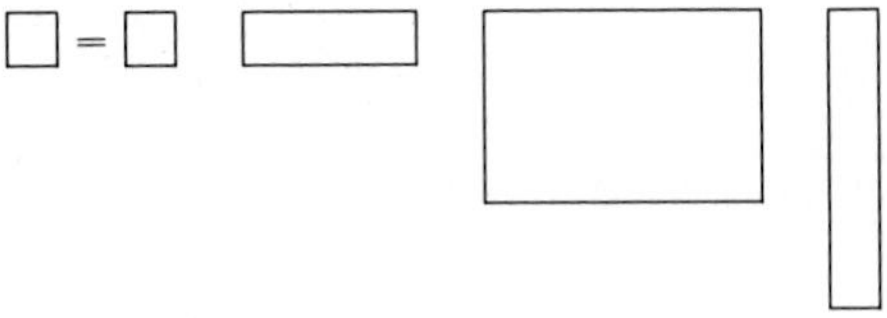

$$s_{yj} = (n - 1)^{-1}\mathbf{w}'\mathrm{X}'\mathbf{x}_j$$

Because the scalar may be moved, this in turn gives

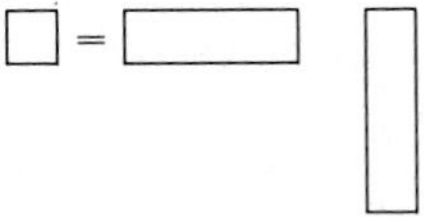

$$s_{yj} = \mathbf{w}'\mathbf{s}_j \tag{4.11}$$

where $\mathbf{s}_j$ is the vector of covariances of all the variables with x_j. That is, the covariance of the linear combination with variable j is the same linear combination of the individual covariances of the constituent variables with variable x_j. This calculation is illustrated in Exhibit 4-2. As a scalar formula, Equation 4.11 is

$$s_{yj} = \sum_j w_j s_{jk}/(n - 1)$$

$$\underset{(n-1)^{-1}}{\tfrac{1}{9}}\ \underset{\mathbf{w}}{[1\ \ -1\ \ 2\ \ 0]}\ \underset{\mathrm{X}'}{\begin{bmatrix} -3 & -1 & 2 & 3 & 0 & 2 & -1 & -1 & -2 & 1 \\ 1 & 2 & 0 & 3 & 2 & -1 & 3 & 0 & -6 & -4 \\ 0 & -3 & 1 & 2 & -2 & 2 & -1 & 2 & 2 & -3 \\ 1 & -4 & 4 & 0 & 2 & -2 & -2 & 1 & -3 & 3 \end{bmatrix}}\ \underset{\mathrm{X}_4}{\begin{bmatrix} 1 \\ -4 \\ 4 \\ 0 \\ 2 \\ -2 \\ -2 \\ 1 \\ -3 \\ 3 \end{bmatrix}}$$

$$= \underset{\mathbf{w}}{[1\ \ -1\ \ 2\ \ 0]}\ \underset{\mathbf{s}_4}{\begin{bmatrix} 1.67 \\ -0.11 \\ -0.33 \\ 7.11 \end{bmatrix}} = 1.11$$

Exhibit 4-2 Covariance of a linear combination with a variable. Variables are the same as in Exhibit 4-1, so $\mathbf{s}_j$ is the fourth column of the S matrix in part (c) of that exhibit.

The corresponding correlation between variable x_j and the weighted combination can be found by using the expression for the variance of y, in Equation 4.9:

$$r_{yj} = \frac{\mathbf{w}'\mathbf{s}_j}{s_j\,(\mathbf{w}'\mathrm{S}\mathbf{w})^{1/2}} \tag{4.12}$$

Here, the s_j in the denominator is simply the standard deviation of x_j.

Covariances Between Linear Combinations

Determining the covariance between two linear combinations, or seeing its dependence on the component variances and covariances, can be done in much the same way as in the preceding section. For example, let $\mathbf{w}$ represent one set of weights and $\mathbf{v}$ the other, and y_w and y_v represent the respective linear combinations. The covariance between these two is

$$s_{wv} = (n-1)^{-1}\mathbf{y}_w{}'\mathbf{y}_v{}' \tag{4.13}$$

which, by substitution, and by moving the scalar $(n-1)^{-1}$, gives

$$s_{wv} = \mathbf{w}'\,[(n-1)^{-1}\mathrm{X}'\mathrm{X}]\,\mathbf{v}$$

and

$$s_{wv} = \mathbf{w}'\mathrm{S}_{xx}\mathbf{v} \tag{4.14}$$

Once again, the part in parentheses is the covariance matrix of the x's.

The scalar formula is

$$s_{wv} = \sum_j \sum_k w_j v_k s_{jk} \tag{4.15}$$

Therefore, the covariance between two linear combinations of the variables can be found by premultiplying the covariance matrix by one set of weights and postmultiplying the result by the other weights. Which set is which is arbitrary because s_{wv} is a single number. Taking the transpose of both sides has no effect, and the positions of **w** and **v** could be interchanged.

This expression applies even if one is getting a covariance of a weighted combination of p variables in one set (X_w) and a weighted combination of q variables in another (X_v). The two sets of variables are simply arranged side-by-side as a larger matrix which is called X. Then **w** has weights for the variables in X_w followed by q zeros for the variables in X_v. Similarly, **v** has p weights of zero followed by the weights for the variables in X_v. If there are p variables in X_w, and q in X_v, $w_j = 0$ for $j > p$, and $v_j = 0$ for $j < p + 1$. The formula will then still hold. These general procedures are illustrated in Exhibit 4-3.

Equation 4.14 is readily converted into the correlation between two weighted combinations by dividing by the two standard deviations obtained from Equation 4.9:

$$r_{wv} = \frac{\mathbf{w}'\mathrm{S}_{xx}\mathbf{v}}{(\mathbf{w}'\mathrm{S}_{xx}\mathbf{w})^{1/2}(\mathbf{v}'\mathrm{S}_{xx}\mathbf{v})^{1/2}} \tag{4.16}$$

The scalar form of Equations 4.17 is rather forbidding:

$$r_{wv} = \frac{\sum_j\sum_k w_j v_k s_{jk}}{\left(\sum_j \sum_k w_j w_k s_{jk}\right)^{1/2} \left(\sum_j \sum_k v_j v_k s_j k\right)^{1/2}} \tag{4.17}$$

These expressions for the covariances or correlations are critical in making connections among various multivariate statistical methods. In several later chapters, it will be pointed out that one or another method revolves around the variance, covariance, or correlation of some weighted sum or sums. It is merely a matter of realizing what the weights are.

The material here has been presented in terms of raw score weights, but is equally relevant if the weights are to be applied to standard scores. If w and/or v are standard score weights, then we would need to substitute correlations or correlation matrices in the equations wherever there are covariances or covariance matrices.

(a)

$$\underset{\mathbf{w}'}{\begin{bmatrix}1 & -1 & 2 & 0\end{bmatrix}}\underset{\mathrm{S}}{\begin{bmatrix}3.78 & 0.78 & 0.78 & 1.67\\0.78 & 8.89 & -1.00 & -0.11\\0.78 & -1.00 & 4.44 & -0.33\\1.67 & -0.11 & -0.33 & 7.11\end{bmatrix}}\underset{\mathbf{v}}{\begin{bmatrix}1\\1\\1\\-1\end{bmatrix}}$$

$$= \begin{bmatrix}1 & -1 & 2 & 0\end{bmatrix}\begin{bmatrix}3.67\\8.78\\4.55\\-5.88\end{bmatrix} = 3.99$$

$$= \mathbf{w}'(\mathrm{S}\mathbf{v}) = s_{vw}$$

(b)

$$\underset{\mathbf{w}'}{\begin{bmatrix}1 & -1 & 2 & 0\end{bmatrix}}\underset{\mathrm{S}}{\begin{bmatrix}3.78 & 0.78 & 0.78 & 1.67\\0.78 & 8.89 & -1.00 & -0.11\\0.78 & -1.00 & 4.44 & -0.33\\1.67 & -0.11 & -0.33 & 7.11\end{bmatrix}}\underset{\mathbf{w}}{\begin{bmatrix}1\\-1\\2\\0\end{bmatrix}}$$

$$= \begin{bmatrix}4.56 & 10.11 & 10.66 & 1.12\end{bmatrix}\begin{bmatrix}1\\-1\\2\\0\end{bmatrix} = 35.99$$

$$= (\mathbf{w}'\mathrm{S})\mathbf{w} = s_w^2$$

(c)

$$\underset{\mathbf{v}'}{\begin{bmatrix}1 & 1 & 1 & -1\end{bmatrix}}\underset{\mathrm{S}}{\begin{bmatrix}3.78 & 0.78 & 0.78 & 1.67\\0.78 & 8.89 & -1.00 & -0.11\\0.78 & -1.00 & 4.44 & -0.33\\1.67 & -0.11 & -0.33 & 7.11\end{bmatrix}}\underset{\mathbf{v}}{\begin{bmatrix}1\\1\\1\\-1\end{bmatrix}}$$

$$= \begin{bmatrix}3.67 & 8.78 & 4.55 & -5.88\end{bmatrix}\begin{bmatrix}1\\1\\1\\-1\end{bmatrix} = 22.88$$

$$= (\mathbf{v}'\mathrm{S})\mathbf{v} = s_v^2$$

(d)

$$\underset{\mathrm{W}'}{\begin{bmatrix}1 & -1 & 2 & 0\\1 & 1 & 1 & -1\end{bmatrix}}\underset{\mathrm{S}}{\begin{bmatrix}3.78 & 0.78 & 0.78 & 1.67\\0.78 & 8.89 & -1.00 & -0.11\\0.78 & -1.00 & 4.44 & -0.33\\1.67 & -0.11 & -0.33 & 7.11\end{bmatrix}}\underset{\mathrm{W}}{\begin{bmatrix}1 & 1\\-1 & 1\\2 & 1\\0 & -1\end{bmatrix}} = \underset{\mathrm{S}_w}{\begin{bmatrix}35.99 & 3.99\\3.99 & 22.88\end{bmatrix}}$$

Exhibit 4-3 (continued)

$$\text{(e)}\quad \begin{bmatrix} .1667 & 0 \\ 0 & .2091 \end{bmatrix} \begin{bmatrix} 35.99 & 3.99 \\ 3.99 & 22.88 \end{bmatrix} \begin{bmatrix} .1667 & 0 \\ 0 & .2091 \end{bmatrix} = \begin{bmatrix} 1.000 & .139 \\ .139 & 1.000 \end{bmatrix}$$

$$\mathrm{D}_s^{-1} \qquad \mathrm{S}_w \qquad \mathrm{D}_s^{-1} \quad = \quad \mathrm{R}_w$$

Exhibit 4-3 Computation of covariances, variances, and correlations of linear combinations. (a) Covariance of X_w and X_v; (b) and (c) Variances of X_w and X_v; (d) $\mathbf{w}$ and $\mathbf{v}$ are arranged as the first and second rows, respectively, of a matrix W, which is used to calculate the variance-covariance matrix of the linear combination. In (a), (b), and (c) a two-step calculation is illustrated, employing the associative rule. The calculations also could have been $(\mathbf{v}'\mathrm{S})\mathbf{v}$, $\mathbf{w}'(\mathrm{S}\mathbf{w})$, and $\mathbf{v}'(\mathrm{S}\mathbf{v})$; (e) The matrix of correlations of the weighted combinations is calculated.

Variances and Covariances among Several Linear Combinations

In a variety of multivariate statistical contexts, one needs to be able to express the variances and covariances among a number of linear combinations, not just one or two. This can be done by a relatively straightforward extension of what has been shown so far. Let W now stand for a whole matrix of weights for making r different weighted combinations. Each column of W then has as its elements the numbers that define a particular linear combination, and W is $p \times r$ for r different linear combinations. Therefore, $\mathrm{Y} = \mathrm{XW}$ will correspondingly be an $n \times r$ matrix of the scores of n observations on the r linear combinations. The variance-covariance matrix for the linear combinations will be

$$s_{yy} = (n-1)^{-1}\mathrm{Y}'\mathrm{Y} \qquad (4.18)$$

which, by substitution and moving the scalar $(n-1)^{-1}$, gives

$$s_{yy} = \mathrm{W}'(n-1)^{-1}\mathrm{X}'\mathrm{XW}$$

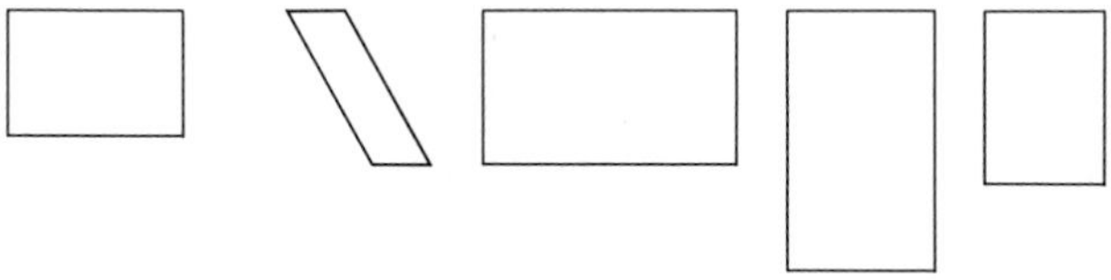

The part of the equation between W′ and W is the matrix of covariances among the X's, so

$$s_{yy} = \mathrm{W}'\mathrm{S}_{xx}\mathrm{W} \qquad (4.19)$$

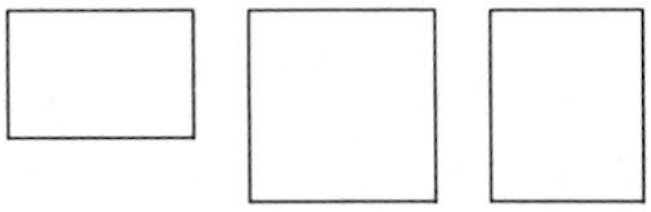

Thus we deduce the complete matrix of covariances among the combinations by pre- and postmultiplying the covariance matrix of the components by the weight matrix.

In S_{yy}, the diagonal elements are the variances of the linear combinations and the off-diagonal elements are the covariances between one linear combination and another. This, and the corresponding expressions involving correlations, is widely used in multivariate statistics. Exhibit 4-3(d) illustrates these calculations. When there are several weighted combinations, we can convert Equation 4.19 to a correlation matrix. This is done by pre- and postmultiplying the equation by the diagonal matrix containing the reciprocals of the standard deviations of the y's:

$$R_{yy} = D_y^{-1} S_{yy} D_y^{-1} = D_y^{-1} W' S_{xx} W D_y^{-1} \tag{4.20}$$

where D_y^{-1} contains the reciprocals of the standard deviations of the combinations s_y. Remember that the s_y^2 is in the diagonal of S_{yy}. The correlation matrix may include some instances of correlations of single variables with more complicated sums. Just remember that a single variable is a "weighted sum" when it has a weight of 1.0 and that the remaining variables are all zero.

The Difference Score and Its Correlations

The equations for covariances and correlations of weighted sums have a variety of applications throughout multivariate analysis. We will make use of them shortly in our discussion of multiple regression. These equations are also useful for testing the effects of various weighting systems that may suggest themselves on intuitive grounds.

A common example is the simple difference score. It can be used to represent changes of various kinds, reflecting learning or the effects of various treatments. But the difference score is one in which one variable, x_1, is weighted -1.0 and the other, x_2, is weighed $+1.0$. According to Equation 4.11, the covariance of the difference score with some other variable will be the difference between the covariances of x_1 and x_2 with that variable.

This situation is interesting when the "other" variable is actually x_1 or x_2, for now one of the "covariances" is actually a variance. We would label the difference score as d, $s_{d2} = s_2^2 - s_{12}$. This is almost bound to be positive unless the correlation r_{12} is high and s_1^2 is greater than s_2^2. On the other hand, s_{d1} is $s_{12} - s_1^2$, which is almost bound to be negative. When these covariances are converted to correlations, their signs remain the same, and their magnitudes can be quite high. They have, however, no valid substantive interpretation because they just depend on these basic statistics of the variables used in the difference score. So, one has to avoid making what seems superficially to be the obvious interpretations. For example, if x_1 is a pretest and x_2 is a posttest, one cannot conclude that "those who knew the most gained the least," based on the fact that r_{d1} is negative.

The covariance of d with some third variable, x_3, also depends on its covariances with x_1 and x_2 in a simple way: $s_{d3} = s_{23} - s_{13}$. Thus, in all these cases, the covariances—and therefore the correlations of a difference score—depend in simple ways on the variances and covariances of its two components. The signs and sizes of these covariances and correlations contain no information beyond the relationships between the sizes of the covariances and variances of the original variables. Substantive interpretations of correlations of difference scores are correspondingly weak.

These equations for the covariances of difference scores can readily be converted into equations for the corresponding correlations. Dividing these equations by the standard deviations and rearranging some terms gives

$$r_{d2} = \frac{1 - r_{12}\left(\dfrac{s_1}{s_2}\right)}{\left[1 + \dfrac{s_1^2}{s_2^2} - 2r_{12}\left(\dfrac{s_1}{s_2}\right)\right]^{1/2}} \tag{4.21}$$

$$r_{d1} = \frac{r_{12}\left(\dfrac{s_2}{s_1}\right) - 1}{\left[1 + \dfrac{s_2^2}{s_1^2} - 2r_{12}\left(\dfrac{s_2}{s_1}\right)\right]^{1/2}} \tag{4.22}$$

$$r_{d3} = \frac{r_{23}\left(\dfrac{s_2}{s_1}\right) - r_{13}}{\left(1 + \dfrac{s_2^2}{s_1^2} - 2r_{12}\dfrac{s_2}{s_1}\right)^{1/2}} \tag{4.23}$$

The point here is that nothing new is learned by correlating difference scores with observed variables because such correlations depend in rather simple ways on the correlations and standard deviations of the observed variables. Attempts to interpret them substantively can be misleading.

Covariances Between Simple Sums

Suppose we define two composites by forming the simple sums of two sets of variables, that is, two total scores. As we have seen, the variances of such sums are the sums of the variances of the components in the totals and of all the covariances among them. Now think what the covariance between the two totals represents. Suppose all the variables in one set come first, and all those in the other set come second. We would like to know about the covariance between the two totals.

This can be found by introducing the two weight vectors. The first vector, **w**, has weights of 1.0 for the first p entries and zero for the rest. The second, **v**, is the reverse. Now we premultiply the supermatrix by one vector (**w**′, for example), and postmultiply by **v**. Note that the pre- and postmultiplication is simply summing the covariances in the section of S that is S_{wv}, the section that contains covariances between members of different sets. The variances of the totals are just the sums of all the terms in S_{ww} and S_{vv}, respectively, that is, all the variances and covariances in the individual sets. Therefore, the correlation between the two totals depends predominately on how closely the average covariance between sets resembles the average covariance within sets, and partly on the sizes of the covariances relative to the variances and on the number of components. For example, suppose there are two variables in each set:

$$\mathbf{v}' = [1.0 \quad 1.0 \quad 0 \quad 0\,]$$
$$\mathbf{w}' = [\,0 \quad 0 \quad 1.0 \quad 1.0]$$

Applying these vectors to the variance-covariance matrix will show that only the four boxed "between set" covariance terms receive nonzero weights in the covariance $s_{y_v y_w}$:

$$\begin{bmatrix} s_1^2 & s_{12} & \boxed{s_{13} \quad s_{14}} \\ s_{12} & s_2^2 & \boxed{s_{23} \quad s_{24}} \\ s_{13} & s_{23} & s_3^2 \quad s_{34} \\ s_{14}^2 & s_{24}^2 & s_{34}^2 \quad s_4^2 \end{bmatrix}$$

These ideas have applications in several areas of psychometric theory. For example, they form the basis for the reliability coefficient called Kuder-Richardson 20, and its generalization, Cronbach's Alpha. They are also used to derive the general form of the Spearman Brown Prophecy Formula, which estimates the reliability of a test if it is lengthened or shortened.

When all the covariances of a linear combination are zero, the variance of the weighted combination (Equation 4.10) reduces to $\Sigma_j w_j^2 s_j^2$. That is, the variance of the combination is the sum of products of the components' variances with their own *squared* weights. It is worth noting further that when all the weights are 1.0 or -1.0, or a mixture of the two, as in simple sum *or* difference scores, the equation reduces to the simple *sum* of the variances.

SUMMARY

This brief chapter introduces some operations and relationships that will be used many times over in subsequent chapters. These operations include the deduction of the variances and covariances of weighted combinations from the weights and the covariance matrix of the components.

In particular, it was shown that the variance of a composite can be computed by pre- and postmultiplying S_{xx} by the weight vector (Equation 4.9). The covariance between two weighted combinations is the covariance matrix premultiplied by one weight vector and postmultiplied by the other (Equation 4.14). The completely general case in which one obtains the variances and covariances among several sets of weights, is handled by putting all the sets of weights into a weight matrix W and using Equation 4.22.

The concept of weights is completely general and various devices can be used to fit special cases into the concept, such as single variables, simple sums, simple sums of subsets, differences, and so on. The specification of the weights that define a special kind of score can be used, in conjunction with the covariance matrix for the components, to assess the relationships among such composites without actually computing the scores. A useful device here is the supermatrix, which is a matrix divided into submatrices.

PROBLEMS

For this and subsequent exercises, extensive calculations may be carried out on a terminal using Speakeasy, Minitab, APL, and so on.

1. Given the following X matrix and the four weight vectors $\mathbf{w}_c$ ($c = 1, 4$), compute the weighted combinations $Y = XW$, where W is the matrix of the $\mathbf{w}_c$. Verify that $\bar{\mathbf{y}} = \bar{\mathbf{x}}W$.

$$\underset{X}{\begin{bmatrix} 3 & 2 & 3 & -1 \\ 2 & -1 & 1 & -1 \\ -1 & -1 & -1 & 1 \\ -3 & 1 & -3 & 3 \\ -1 & -1 & 0 & -2 \end{bmatrix}} \quad \underset{\mathbf{w}_1}{\begin{bmatrix} 1 \\ 1 \\ 1 \\ 1 \end{bmatrix}} \quad \underset{\mathbf{w}_2}{\begin{bmatrix} 1 \\ 1 \\ 0 \\ 0 \end{bmatrix}} \quad \underset{\mathbf{w}_3}{\begin{bmatrix} 0 \\ 0 \\ 2 \\ 1 \end{bmatrix}} \quad \underset{\mathbf{w}_4}{\begin{bmatrix} 1 \\ -1 \\ 0 \\ 0 \end{bmatrix}}$$

2. Compute the variances of the y_c by the usual formula for the variance.

3. Compute the covariance matrix $S_{xx} = (n - 1)^{-1}X'X$.

4. Compute the variances of the four y_c by $\mathbf{w}_c' S_x \mathbf{w}_c$. Verify that they agree with Problem 2.

5. Compute the covariances of the four y_c with x_1 by the usual formula and verify that they are the weighted sums of the corresponding covariances.

6. Compute the covariance of y_2 with y_3, using the scores on the y's themselves and the covariances among the x's.

7. Compute the rest of S_{yy}, the y covariance matrix, by means of $W'S_{xx}W$.

8. What common types of composite scores correspond to y_1, y_2, and y_4?

9. What is special about the covariance between y_2 and y_3?

10. What raw score weights are required here to make equal *standard* score weights?

11. Given the following correlation matrix, what is the correlation between the sum of z_1 and z_2 and the sum of z_3 and z_4?

$$\begin{bmatrix} 1.00 & .62 & .20 & .42 \\ & 1.00 & .01 & .63 \\ & & 1.00 & -.02 \\ & & & 1.00 \end{bmatrix}$$

ANSWERS

1. $$XW = \begin{bmatrix} 7 & 5 & 5 & 1 \\ 1 & 1 & 1 & 3 \\ -2 & -2 & -1 & 0 \\ -2 & -2 & -3 & -4 \\ -4 & -2 & -2 & 0 \end{bmatrix} = Y$$

$$\bar{\mathbf{x}} = [\,0 \quad 0 \quad 0 \quad 0\,]$$
$$\bar{\mathbf{y}} = [\,0 \quad 0 \quad 0 \quad 0\,]$$
$$\bar{\mathbf{x}}W = [\,0 \quad 0 \quad 0 \quad 0\,]$$

2. $s^2_{y1} = 18.5$; $s^2_{y2} = 9.5$; $s^2_{y3} = 10.0$; $s^2_{y4} = 6.5$

3. $$S_{xx} = \begin{bmatrix} 6.0 & .75 & 5.25 & -3.25 \\ .75 & 2.00 & .75 & .75 \\ 5.25 & .75 & 5.00 & -3.50 \\ -3.25 & .75 & -3.50 & 4.00 \end{bmatrix}$$

4. $\mathbf{w}_1{}'\mathrm{S}_{xx} = [8.75 \quad 4.25 \quad 7.50 \quad -2.00]$; $s^2_{y1} = 18.5$
   $\mathbf{w}_2{}'\mathrm{S}_{xx} = [6.75 \quad 2.75 \quad 6.00 \quad -2.50]$; $s^2_{y2} = 9.5$
   $\mathbf{w}_3{}'\mathrm{S}_{xx} = [7.25 \quad 2.25 \quad 6.50 \quad -3.00]$; $s^2_{y3} = 10.0$
   $\mathbf{w}_4{}'\mathrm{S}_{xx} = [5.25 \quad -1.25 \quad 4.50 \quad -4.00]$; $s^2_{y4} = 6.5$

5. $s_{11'} = 8.75$; $s_{12'} = 6.75$; $s_{13'} = 7.25$; $s_{14'} = 5.25$

6. $s_{2'3'} = 9.50$. From Problem 4, $\mathbf{w}_2{}'\mathrm{S}_{xx} = [6.75 \quad 2.75 \quad 6.00 \quad -2.50]$; $\mathbf{w}_2{}'\mathrm{S}_{xx}\mathbf{w}_3 = 0 \times 6.75 + 0 \times 2.75 + 2 \times 6.0 + 1 \times (-2.50) = 9.50$

7. Using again the results of Problem 4 as $W'S_x$:

$$W'S_{xx}W = \begin{bmatrix} 18.5 & 13.00 & 13.00 & 4.50 \\ 13.00 & 9.50 & 9.50 & 4.00 \\ 13.00 & 9.50 & 10.00 & 5.00 \\ 4.50 & 4.00 & 5.00 & 6.50 \end{bmatrix}$$

8. The simple sum of all variables is y_1. The sum of the first two only is y_2. The difference or gain (loss) score is y_4.

9. The covariance is between a weighted combination of variables from disjoint sets, that is, x_1 and x_2 versus x_3 and x_4.

10. $w_1^* = [.408, .707, .447, .500]$

11. $$r = \frac{1.26}{\sqrt{3.24}\,\sqrt{1.96}} = .50.$$

READINGS

Mulaik, S. A. (1972). *The foundations of factor analysis.* New York: McGraw-Hill, pp. 61–76.

5

The Inverse

NATURE OF INVERSES

Definition and Use of the Inverse

In Chapter 1 it was noted that the identity matrix is a generalization of the number one in that AI = A just as $a \times 1 = a$. It is also true that any number times its reciprocal equals one—for example, $2 \times 1/2 = 1$. Note that in the following matrix multiplication,

$$\begin{bmatrix} .4 & .2 \\ -.1 & .2 \end{bmatrix} \begin{bmatrix} 2 & -2 \\ 1 & 4 \end{bmatrix} = \begin{bmatrix} 1.0 & 0 \\ 0 & 1.0 \end{bmatrix}$$
$$\text{A} \qquad\qquad \text{B} \quad = \quad \text{C}$$

a matrix A multiplied by another matrix B gives the identity matrix: B is analogous to the reciprocal or inverse of A, and vice versa. Such matrix relationships are referred to by either term. More commonly, however, B is called the *inverse* of A, denoted as A^{-1}. Specifically, *if* $AB = I$, *then* B *is the inverse of* A, A^{-1}, *and* A *is the inverse of* B, B^{-1}.

Inverses are just as necessary for solving matrix equations as reciprocals are for solving simple ones. Suppose we have the simple equation

$$bx = a$$

This is solved by the following series of steps, even though they would not usually be detailed:

$$\left(\frac{1}{b}\right)(b)x = \frac{1}{b}a$$

$$1x = \frac{1}{b}a$$

$$x = \frac{1}{b}a$$

Sometimes, $1/b$ may be designated as b^{-1}.

The analogous matrix equation has a similar solution:

$$\mathrm{BX} = \mathrm{A}$$

$$\mathrm{B}^{-1}\mathrm{BX} = \mathrm{B}^{-1}\mathrm{A}$$

$$\mathrm{IX} = \mathrm{B}^{-1}\mathrm{A}$$

$$\mathrm{X} = \mathrm{B}^{-1}\mathrm{A}$$

In this case, X could be a whole matrix of coefficients. But more commonly, it is a vector. For example, consider the following equations:

$$2x_1 - 2x_2 = 3$$

$$x_1 + 4x_2 = -1$$

In matrices, this is

$$\begin{bmatrix} 2 & -2 \\ 1 & 4 \end{bmatrix} \begin{bmatrix} x_1 \\ x_2 \end{bmatrix} = \begin{bmatrix} 3 \\ -1 \end{bmatrix}$$

Knowing the inverse of this matrix from the formula of the preceding example, the following steps can be carried out:

$$\begin{bmatrix} .4 & .2 \\ -.1 & .2 \end{bmatrix} \begin{bmatrix} 2 & -2 \\ 1 & 4 \end{bmatrix} \begin{bmatrix} x_1 \\ x_2 \end{bmatrix} = \begin{bmatrix} .4 & .2 \\ -.1 & .2 \end{bmatrix} \begin{bmatrix} 3 \\ -1 \end{bmatrix}$$

$$\begin{bmatrix} 1.0 & 0 \\ 0 & 1.0 \end{bmatrix} \begin{bmatrix} x_1 \\ x_2 \end{bmatrix} = \begin{bmatrix} 1.0 \\ -.5 \end{bmatrix}$$

$$\begin{bmatrix} x_1 \\ x_2 \end{bmatrix} = \begin{bmatrix} 1.0 \\ -.5 \end{bmatrix}$$

If a simple equation like this is expressed in scalar form, it might be $a_i = \Sigma_j c_{ij} x_j$, where we wish to solve for the x_j''s.

Conditions for Inverses

Not all matrices have inverses. The truth of this statement is the source of a good bit of aggravation to people who work with matrices. You may recall an injunction from algebra never to divide by zero, and that nonsense results when someone inadvertently does divide by zero in an equation. Attempting to use the inverse of a matrix when none exists is the generalized form of dividing by zero, and it too provides nonsense—only now the nonsense may be multidimensional.

A matrix that has an inverse is called *nonsingular,* and one that does not is called *singular.* Only square matrices of full rank are *nonsingular.* So, no rectangular matrix has a true inverse.* In addition, any square matrix that is not of full rank is also singular.

Properties of Inverses

If A^{-1} is the inverse of A, $A^{-1}A = AA^{-1} = I$, that is, multiplication of a matrix by its own inverse is commutative—which is an exception to the general rule about noncommutativity of matrix multiplication. However, in solving equations such as $CB = A$, and in other such circumstances, one must remember to multiply on the appropriate side of the expression, $C^{-1}CB = C^{-1}A$, not AC^{-1}. It may even be impossible to carry out the multiplication AC^{-1}.

If two matrices both have inverses, then the inverse of their product is the product of their inverses in the reverse order:

$$(PQ)^{-1} = Q^{-1}P^{-1} \tag{5.1}$$

The necessity of this principle can be seen as follows. Suppose a matrix $A = PQ$. Then if $A^{-1} = Q^{-1}P^{-1}$, we have

$$Q^{-1}P^{-1}PQ = Q^{-1}IQ = I$$

Both P and Q are eliminated, whereas $P^{-1}Q^{-1}PQ$ cannot be simplified because $Q^{-1}P \neq I$.

Our earlier statement regarding the rank of a product being at most the lesser of the ranks of the factors can be made somewhat more specific: If A is n by p ($p \leq n$), rank r, and B is q by n ($q \leq p \leq n$), rank s, with $s \leq r$, *the rank of the product BA is at most s and at least $s - (p - r)$.* This may not seem like much help at first, but in the special case where $p = q = r = s$—that is, where both matrices are of the same full rank—it is useful. In this case the (square) product will be of full rank also and will have an inverse. This applies in a number of statistical situations in which we want the inverse of the product of two rectangular matrices that are of full rank.

Changes of Basis

The first application of inverses, then, allows for solving equations like $A\mathbf{x} = \mathbf{c}$ for $\mathbf{x}$. We will have occasion to do this in the context of regression, solving $S_{xx}\mathbf{b} = \mathbf{s}_{xy}$ for the regression weights $\mathbf{b}$. There are, of course, more complex equations—

*Rectangular matrices can have an inverse of a limited kind.

indeed, a whole body of mathematics—where inverses are needed. Most of this is irrelevant for our purposes, although we do need inverses when we want to shift from one basis for a set of vectors to another. In the multivariate statistical context, we can think of this as a *transformation* from one score matrix into another. Suppose the linear composite $\mathbf{Xw} = \mathbf{y}$ has been made, where X is an n by p score matrix, $\mathbf{w}$ is a p-vector, and $\mathbf{y}$ is an n-vector. Now suppose that X is a derived score matrix that was made up from some original score matrix Q by means of the transformation T: $X = QT$. The reason for the transformation might be that X is somehow more convenient or interpretable than Q. (This principle is used in analysis of variance.)

Now, if we decided instead to express $\mathbf{y}$ in terms of Q rather than X—go back to basics, as it were—what weights $\mathbf{v}$ are needed instead of $\mathbf{w}$ to give us $\mathbf{y}$ from Q? By simple substitution, $(QT)\,\mathbf{w} = \mathbf{y} = Q(T\mathbf{w})$. If we set $T\mathbf{w} = \mathbf{v}$ and applied $\mathbf{v}$ to Q, it would be the same as using $\mathbf{w}$ on X. That is, if we obtain X from Q by means of the transformation T, we would apply the same transformation to $\mathbf{w}$ to find out the weights to apply to the original matrix Q.

In solving the above problem, no inverse is needed because it is a simple identity. This system works for any T whatsoever, even one that is singular. The inverse is needed to go the other way, however. Suppose we had already found that $Q\mathbf{v} = \mathbf{y}$, but now wished to change Q into X because X is the more convenient or prettier form. What weights are needed to apply to X if we are still to get $\mathbf{y}$ from it? If $QT = X$, then $XT^{-1} = Q$, provided T is nonsingular. Now substituting for Q we have $XT^{-1}\mathbf{v} = X(T^{-1}\mathbf{v}) = \mathbf{y}$. If $T^{-1}\mathbf{v} = \mathbf{w}$, obviously this is the same $\mathbf{w}$ we had before since $T\mathbf{w} = \mathbf{v}$ and $\mathbf{w} = T^{-1}\mathbf{v}$. Note, however, that this works *only* if T is nonsingular.

Several aspects of the preceding problem are worth noting. One is the straightforward fact that if we have a weight vector to be applied to a matrix of scores and wish to transform the scores, we can readily find out what weights to apply to the transformed scores by using the inverse of the transformation on the original set of weights. The second aspect worth noting is that we obtain the same result, $\mathbf{y}$, whether we use X or Q, *provided* they are linked by a nonsingular transformation T. This is an important and far-reaching principle, as we will frequently be switching from one form of a score matrix to another in order to have X in a form that leads to a certain kind of simplicity of interpretation or that has a simpler computational form. These ideas come up repeatedly in the general linear model interpretation of analysis of variance and in factor analysis. Thus we have another reason for becoming familiar with inverses.

FINDING INVERSES AND RANKS

Special Cases

Computing inverses is a chore, one best left to the computer or some other uncomplaining and meticulously accurate assistant. However, it is important to know the inverses of some special matrices because we will refer to them here.

One of the simplest inverses is that of a diagonal matrix. If D is a diagonal matrix with diagonal entries $d_{jj} \neq 0$, then D^{-1} is also a diagonal matrix with entries $1/d_{jj}$. Note that if any diagonal terms are zero, the matrix would not be of full rank, which means the matrix is singular.

Another matrix that has a simple inverse is the *orthogonal transformation,* which is particularly important in the area of changes of basis. An orthogonal (or more strictly, an orthonormal) transformation is a matrix whose transpose is its inverse: $T^{-1} = T'$. A simple example is

$$\underset{T}{\begin{bmatrix} .8 & -.6 \\ .6 & .8 \end{bmatrix}} \underset{T'}{\begin{bmatrix} .8 & .6 \\ -.6 & .8 \end{bmatrix}} = \underset{I}{\begin{bmatrix} 1.0 & 0 \\ 0 & 1.0 \end{bmatrix}}$$

A more complicated one is

$$\begin{bmatrix} .985 & .101 & -.134 \\ .119 & .148 & .981 \\ .119 & -.983 & .133 \end{bmatrix} \begin{bmatrix} .985 & .119 & .119 \\ .101 & .148 & -.983 \\ -.134 & .981 & .133 \end{bmatrix} = \begin{bmatrix} 1.0 & 0 & 0 \\ 0 & 1.0 & 0 \\ 0 & 0 & 1.0 \end{bmatrix}$$

The process works both ways: $TT' = I$ and $T'T = I$. Another way of stating this is that in an orthogonal transformation, the sum of squared entries in every row and column is 1.0 and the sum of products between rows and between columns is zero. An orthogonal transformation may be thought of as representing a change of basis when the basis vectors are orthonormal and the transformation is rigidly rotating the basis while the other vectors in the space remain stationary. Alternatively, we can think of the vectors as rotating rigidly in a body while the basis stays in place. Both are equivalent, but the former is more typical, and usually corresponds more closely to the use of transformations.

Frequently, we may refer to an explicit inverse in examples, but usually the size can be kept to 2×2. So that we have it to refer to, the form of the inverse of the general 2×2 matrix is

$$A^{-1} = \begin{bmatrix} a_{11} & a_{12} \\ a_{21} & a_{22} \end{bmatrix}^{-1} = \begin{bmatrix} \dfrac{a_{22}}{g} & \dfrac{-a_{12}}{g} \\ \dfrac{-a_{21}}{g} & \dfrac{a_{11}}{g} \end{bmatrix} \tag{5.2}$$

where $g = a_{11}a_{22} - a_{12}a_{21}$. The denominator, g, is called the *determinant* of the matrix. If the denominator is zero, the matrix is singular. The fact that the two matrices in Equation 5.2 are inverses of each other can be verified by multiplying them together.

Symmetric matrices have simpler inverses, and the inverse of a correlation matrix is simpler still. By substitution, Equation 5.2 shows that

$$\begin{bmatrix} 1.0 & r_{12} \\ r_{12} & 1.0 \end{bmatrix}^{-1} = \begin{bmatrix} \dfrac{1}{1 - r_{12}^2} & \dfrac{-r_{12}}{1 - r_{12}^2} \\ \dfrac{-r_{12}}{1 - r_{12}^2} & \dfrac{1}{1 - r_{12}^2} \end{bmatrix} \tag{5.3}$$

From time to time the inverses of other simple matrices will be used.

Where there is no fairly simple algebraic form for an inverse, someone must compute it. This is a formidable task and one to which considerable energy has been devoted by both applied mathematicians and computer specialists. There are several possible approaches, particularly if the matrix is symmetric, but we will not go into them here. They can be found in texts on matrix algebra.

PROBLEMS

1. Verify that $A = B^{-1}$ by computing AB and BA

$$A = \begin{bmatrix} 3 & -1 \\ 1 & -2 \end{bmatrix} \qquad B = \begin{bmatrix} .4 & -.2 \\ .2 & -.6 \end{bmatrix}$$

2. Verify the inverse relationship between the following matrices:

$$\begin{bmatrix} 2 & 0 & 0 \\ 3 & 3 & 3 \\ 0 & -2 & -1 \end{bmatrix} \begin{bmatrix} \frac{1}{2} & 0 & 0 \\ \frac{1}{2} & -\frac{1}{3} & -1 \\ -1 & \frac{2}{3} & 1 \end{bmatrix} =$$

3. For the following matrix C, find C^{-1} and show that $(AC)^{-1} = C^{-1}B$, using A and B from Problem 1. (Multiply AC by $C^{-1}B$.)

$$C = \begin{bmatrix} 1.0 & .4 \\ .5 & 1.0 \end{bmatrix}$$

4. Solve the following equations.
   (a) $3x_1 - x_2 = 1 \qquad x_1 - 2x_2 = 3$
   (b) $\mathbf{x}'C = [-1 \;\; 1]$, where C is from Problem 3.

5. Solve the following equations for X, assuming that all the relevant matrices are nonsingular.
   (a) $GX = M$ (b) $XG = P$ (c) $RSX = T$
   (d) $RXS^{-1} = V$ (e) $MX = P + Q$ (f) $MXN = V + W$

6. Find the inverses of the following matrices without using Equation 5.2.

$$T = \begin{bmatrix} .60 & -.80 \\ .80 & .60 \end{bmatrix} \qquad V = \begin{bmatrix} .96 & .28 \\ .28 & -.96 \end{bmatrix}$$

7. Find the inverses of the following correlation matrices:

$$R_1 = \begin{bmatrix} 1.0 & .5 \\ .5 & 1.0 \end{bmatrix} \qquad R_2 = \begin{bmatrix} 1.0 & -.8 \\ -.8 & 1.0 \end{bmatrix}$$

ANSWERS

Problems 1 and 2 are self-verifying.

3. $C^{-1} = \begin{bmatrix} 1.25 & -.5 \\ -.625 & 1.25 \end{bmatrix}$ $AC = \begin{bmatrix} 2.50 & .2 \\ 0 & -1.6 \end{bmatrix}$ $(AC)^{-1} = \begin{bmatrix} .400 & .050 \\ 0 & -.625 \end{bmatrix}$

   $C^{-1}B = \begin{bmatrix} .400 & .050 \\ 0 & -.625 \end{bmatrix}$

4. (a) $x_1 = 0; x_2 = -1.0$ (b) $x' = -1.875 \quad 1.750.$

5. (a) $X = G^{-1}M$ (b) $X = PG^{-1}$ (c) $X = S^{-1}R^{-1}T$
   (d) $X = R^{-1}VS$ (e) $X = M^{-1}(P + Q)$ (f) $X = M^{-1}(U + W)N^{-1}$

6. These are orthonormal matrices. $T^{-1} = T'$; $V^{-1} = V'$, and $V' = V$ because V is symmetric.

7. $\begin{bmatrix} 1.333 & -.667 \\ -.667 & 1.333 \end{bmatrix}$ $\begin{bmatrix} 2.778 & 2.222 \\ 2.222 & 2.778 \end{bmatrix}$

READINGS

Green, P. E., and J. D. Carroll (1976). *Mathematical tools for applied multivariate analysis.* New York: Academic Press, pp. 127–90.

Horst, P. (1963). *Matrix algebra for social scientists.* New York: Holt, Rinehart, and Winston, pp. 406–19, 434–65.

6

Regression

LINEAR PREDICTION

Height and Shoe Size

In a typical mystery story, the famous detective, Sherlock Chan, examines a footprint outside a library window and remarks, "The intruder was a man, 5′7″ tall, married to a pediatrician." When the intruder is apprehended, these "deductions" prove to be true. The detective turns away the admiring, incredulous comments of his assistant with the classic, "I have my methods, Watson." One of Chan's methods might be linear regression. It is possible to make a pretty good prediction of a person's height from his or her shoe size. It is also possible to guess the person's sex from the shape of the print—that problem has been saved for a later chapter. The inference about the pediatrician is the result of a flash of intuition possible only for a genius—a genius with an author on his side, that is.

If we take a miscellaneous group of men, record each person's shoe size and height, and then make a scatter plot with shoe size as the horizontal axis and height as the vertical, the result might look like that shown in Figure 6-1. It is apparent that there is a pretty strong tendency for larger feet to go with greater height. (Next time you see a basketball player up close, look at his feet.) Furthermore, the rate of increase is pretty consistent throughout the range of the variables. On the average, an increase of one shoe size corresponds to an increase of about two inches in height. Thus, according to the fictitious data shown in Figure 6-1, a shoe size of nine might lead to a height of 69 inches. Note, however, that there is likely to be an error of a couple of inches in this estimate, an aspect traditionally ignored by fictitious detectives. (In fiction everything is neat; this may lead us to suspect fiction whenever the statistics are too neat.)

In Figure 6-1, a straight line has been drawn through the scattered points in order to represent the relationship between the two variables as closely as possible. The line serves two functions. First, it simplifies the prediction process. Sherlock Chan needs only to remember $h = 2s + 62$ instead of the height that corresponds to each shoe size. Second, if the data represent a situation with

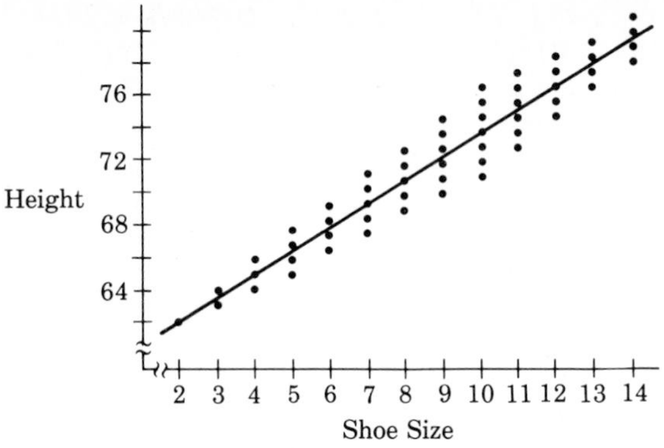

Figure 6-1 Height plotted against shoe size.

well-understood variables, as is often the case with scientific data, the line may express the functional dependence of one variable on the other.

Linear relationships between variables are used in a variety of areas, including economic forecasting, predicting school grades from test scores, the dependence of discrimination reaction time on the amount of information in a stimulus display, and the estimation of family size from education level. Sometimes a relationship may not be linear in the observed variables but becomes so for a suitable transformation of them. The judged loudness of a sound, for example, is linear with respect to the physical intensity of the sound if the sound is measured in log units; the world-record times for all the running races are a linear function of their distances *if* both variables are transformed to log units. Moreover, when relationships are not exactly linear, the linear relationship often serves as an adequate first approximation. Therefore, it is worthwhile to find out about straight lines and how to fit them to data.

Equation of a Straight Line

Any straight line in a two-dimensional vector space, that is, a plane, needs only two parameters to characterize it. Such data is traditionally displayed in Cartesian coordinates like the data for shoes and heights shown in Figure 6-1. Typically, an x-axis is defined horizontally and a y-axis vertically, and appropriate units are marked off on each. In terms of the x, y coordinate system, any straight line can be defined by an equation of the form $y = b_1x + b_0$, where b_1 and b_0 are parameters of the line in question.

The b_1 parameter represents the slope of the line. A positive slope indicates that y increases as x increases, whereas a negative slope indicates that y decreases as x increases. Large slope values indicate that the slope is steep, and small slope values indicate that it is gentle. The b_0 parameter is called the intercept, indicating the vertical position of a line of a given slope. It is called the intercept because it is the value of y when $x = 0$, that is, the point on the y axis where the line "intercepts" it.

The figures below show positive, zero, and negative slopes in the first three examples, while the fourth example shows lines corresponding to the same slope but different intercepts.

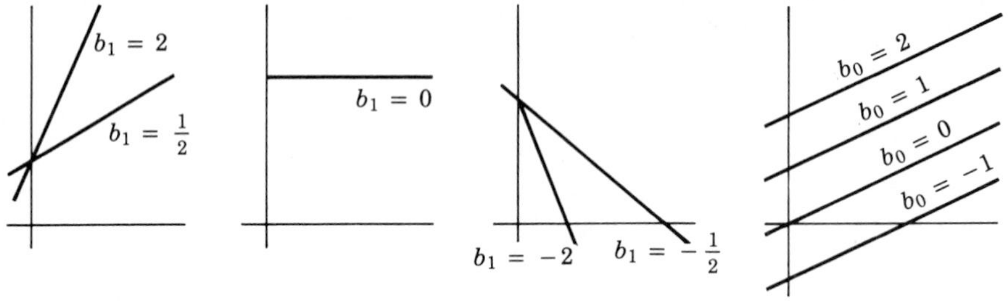

A few other things about b_1 and b_0 are worth noting. First, for any two pairs of values of x and y that satisfy the equation of the line, for example, x_1, y_1 and x_2, y_2, $b_1 = (y_1 - y_2)/(x_1 - x_2)$. For example, in the equation $y = 0.5x + 3$, the two pairs of values 2, 4 and 10, 8 show that $(4 - 8)/(2 - 10) = -4/-8 = 0.5$. This is illustrated as follows:

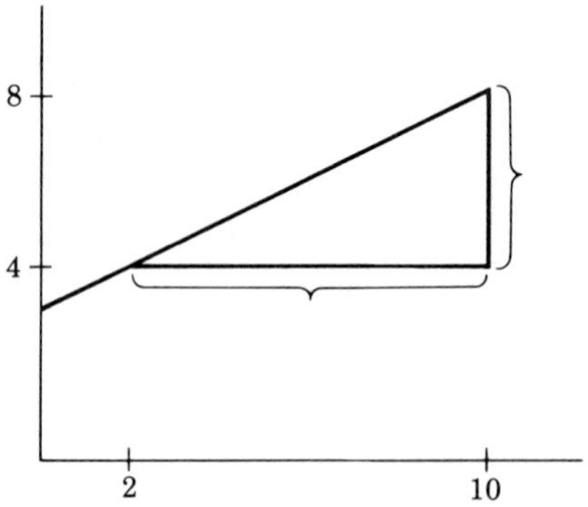

That is, b_1 is the change in y for a unit change in x. In fact, in elementary calculus we learn that the derivative of y with respect to x is b_1, and to say that a relation is linear is to say that a given amount of change in one variable always leads to the proportional amount of change in the other, no matter where the change occurs on the scale.

If y is expressed as a linear function of x, then x can be expressed as a linear function of y. All we have to do is solve the former equation for x. If $y = b_1x + b_0$, then $x = b_1'y + b_0$, where $b_1' = 1/b_1$ and $b_0' = -b_0/b_1$.

When real data is used, the relation is never exact, so we must make allowance for error in the relation of y and x. So, an empirical straight-line relation between a single dependent variable y and a single independent variable x is expressed as follows:

$$y_i = b_1 x_i + b_0 + e_i \tag{6.1}$$

This says that any value of y, for example, the ith, is assumed to be a linear function of the corresponding value of x, with slope coefficient b_1 and intercept b_0, plus an error e_i. A predicted value for y, $\hat{y}$, is correspondingly

$$\hat{y}_i = b_1 x_i + b_0 \tag{6.2}$$

Later we will use more than one independent variable, necessitating a subscript for variables on b and a second subscript on x. There will also be more than one dependent variable. This will necessitate b having two subscripts, one designating predictor and one the predictee, and b_0 will have a second subscript to tell which dependent variable it belongs to. In a very general case we have

$$y_{ik} = b_{jk} x_{ij} + b_{0k} + l_{ik} \tag{6.3}$$

This lets us know that we are using the coefficient, b_{jk}, for predicting y_k from x_i and adding the constant appropriate for y_k. Such doubly subscripted coefficients will be discussed in Chapter 13.

Fitting a Straight Line

Because this book is concerned with data analysis rather than mathematics, the b coefficients are unknown, and have to be estimated from the data. There are a variety of ways of doing this, but only two will be examined here. The first is the graphical method. To do this, we begin by plotting the pairs of values in x, y coordinates, as was done with heights and shoe sizes in Figure 6-1. *This is the single most important step in any data analysis no matter how complex, and should always be carried out;* more is usually learned from this step than from any other single thing. Not only does it reveal the nature and closeness of the relationship between the variables, but also the existence of outlying, wild, or possibly erroneous values can easily be noted, as can the existence of clumps of points, and anomalies in the distributions of the variables. This and other interesting information will not be revealed by a statistical analysis of a numerical type.

Such a diagram is called a "scatterplot." Perhaps the main danger in its use is the tendency to see more than is there—to focus on configurations of points which seem to suggest more than one type of relation, special subgroups of subjects, mystic signs, and so on. Therefore, a little restraint may be in order. If necessary, we can randomly divide the data in half and plot the two sets separately. Anything that does not show up in both graphs can be regarded as an accident.

We may find that plotting points is a tedious pastime. However, once the data are in a form suitable for computer analysis, statistical analysis packages can be used to do the plotting in any of a variety of different forms.

Once the data have been plotted and we have decided to try to fit the data with a straight line, a best-fitting line can be estimated visually. In regression analysis, the term "best-fitting" means that we want to minimize the sum of squared deviations in predicting one variable from the other, traditionally y from x. That is, we define an error, e_i as $y_i - \hat{y}_i$ such that

$$\Sigma e_i^2 = \Sigma(y_i - b_1 x_i - b_0)^2 \tag{6.4}$$

We want to choose b and b_0 in such a way as to make Σe_i^2 as small as possible. This is the *least squares* criterion, which minimizes the sum of the squared deviations of the points from the line *measured in the vertical direction*. In most cases, we can fit the point by eye and do quite well. In the long run, however, an analytic method is preferred, and will be discussed next.

Imagine a club of statisticians that meets weekly in 1790, before the technique was discovered. At each meeting, one of the main entertainments is to take a dozen or so points scattered on an x, y plot and have each member try to fit the points visually with a straight line. The club secretary measures the vertical deviations with a ruler and calculates the error function above (Equation 6.4) for each of the members. The one with the lowest score gets a free dinner at the next week's meeting. Suppose one particular member wins three weeks in a row. This calls into play a timeworn statistical maxim, "One is an accident; two is a coincidence; three is a trend." The other members press the winner for an explanation. Has he bribed the secretary? Consulted an astrologer? Threatened with expulsion, he reluctantly reveals his secret. He has found a mathematical solution that enables him to find the best possible coefficients!

This solution is the same one we will use here. The slope and intercept that give the minimum sum of squared deviations are

$$b_1 = \frac{s_{xy}}{s_x^2} \tag{6.5}$$

and

$$b_0 = \overline{y} - b_1\overline{x} \tag{6.6}$$

So, the best way to minimize Σe^2 is to compute the means and variances of y and x and the covariance between them and use the indicated combinations of statistics to compute b_1 and b_0.

The equation for b_1 may be recognized as closely related to the correlation between x and y:

$$b_1 = r_{xy}\,(s_y/s_x) \tag{6.7}$$

These values (see Equations 6.5, 6.6, and 6.7) will always give a smaller sum of squared errors than any other possible values. Two kinds of arguments can be used to demonstrate this. One is partly algebraic and partly heuristic and the other involves elementary calculus. We will consider the algebraic-heuristic argument first, and then treat the calculus proof in a separate section. The line of reasoning begins as follows. Whatever the value of b_1, the value of b_0 must be chosen so as to make the mean error zero: $\overline{e}_. = 0$. That is, the average difference between $\hat{y}_i$ and y_i must be zero. This may *seem* like a good idea in the sense that we do not want to make systematic over- or under-predictions, but it can also be proved algebraically that the sum of squared errors (Equation 6.4) is smallest if $\bar{y}. = \bar{\hat{y}}$. To say that the mean error is zero is the same as saying the sum is zero:

$$\Sigma e_i = 0 \tag{6.8}$$

The second part of the argument goes as follows: If *errors* e_i are correlated with the x_i, then to some extent they should be predictable. If they are predictable, then we could adjust b_1 so as to predict them. That is, improvement in b_1, leading to a reduction in the sum of squared errors could take place unless the *errors are uncorrelated with the predictors.*

This means that if the correlation *is* zero, no further improvement can take place, and the best value has been found. This provides our second requirement. Because the correlation r_{xe} is the covariance s_{xe} divided by the two standard deviations, saying that the correlation must be zero means that the covariance must be zero. We know that

$$s_{xe} = \frac{\Sigma(x_i - \bar{x})(e_i - \bar{e})}{n - 1} \tag{6.9}$$

But we just finished requiring that $\bar{e} = 0$. In that special case, Equation 6.9 is the same as

$$s_{xe} = \frac{\Sigma x_i e_i}{n - 1} \tag{6.10}$$

because $\Sigma(x_i - x)e_i = \Sigma x_i e_i - \Sigma \bar{x} e_i = \Sigma x_i e_i - \bar{x}\Sigma e_i$, and $\Sigma e_i = 0$. We can multiply both sides of Equation 6.10 by $n - 1$, and use it to express our second requirement on the errors:

$$\Sigma x_i e_i = 0 \tag{6.11}$$

Equations 6.8 and 6.11 specify the circumstances under which b_1 and b_0 minimize the sum of squared deviations from the regression line. The reason for expressing them in these particular ways is that they very readily generalize to the case in which we have more than one predictor, and lend themselves to a matrix expression.

Based on the principles of translating equations into matrix expressions, we can see that Equation 6.8 is the same as $\mathbf{1}'\mathbf{e} = 0$, and that Equation 6.11 is the same as $\mathbf{x}'\mathbf{e} = 0$, where $\mathbf{x}$ and $\mathbf{e}$ are the vectors consisting of the x_i and the e_i, respectively. Now we put $\mathbf{x}$ and 1 together to form an $n \times 2$ matrix that we will call X_+:

$$X_+ = \begin{bmatrix} x_1 & 1 \\ x_2 & 1 \\ \vdots & \vdots \\ x_n & 1 \end{bmatrix} \tag{6.12}$$

(Hereafter, we will use X_+ to represent an $n \times p + 1$ matrix where the last column is a constant of ones.) Both Equations 6.8 and 6.11 may be expressed as the following single matrix equation, one in which we will be able to solve for the b's:

$$X_+'\mathbf{e} = 0 \tag{6.13}$$

If Equation 6.13 is true, we have found the best values for b_1 and b_0. To see how to find the values, we can express the errors as a vector equation

$$\mathbf{e} = \mathbf{y} - \hat{\mathbf{y}} \tag{6.14}$$

and make a matrix translation of $\hat{y}_i = b_1 x_i + b_0$ into

$$\hat{\mathbf{y}} = X_+\mathbf{b}_+ \tag{6.15}$$

where $\mathbf{b}_+$ is the vector $(b_1, b_0)'$. We substitute this into Equation 6.14 and get

$$\mathbf{e} = \mathbf{y} - X_+\mathbf{b}_+ \tag{6.16}$$

and substitute Equation 6.16 into Equation 6.13, leaving

$$X_+'(\mathbf{y} - X_+\mathbf{b}_+) = 0 \tag{6.17}$$

When this is multiplied out and rearranged it becomes

$$X_+'\mathbf{y} = (X_+'X_+)\mathbf{b}_+ \tag{6.18}$$

Because X_+ must be of full column rank, unless all the x's are 0, $X_+'X_+$ will have an inverse, and we can solve for b_+ by multiplying both sides by $(X_+'X_+)^{-1}$:

$$\mathbf{b}_+ = (X_+'X_+)^{-1}X_+'\mathbf{y} \tag{6.19}$$

Equation 6.19 is one of the basic solution equations in multiple regression; it extends to multiple predictors, as we will see, and it is the simplest, most convenient way of expressing the solution for the regression coefficients in many contexts.

We will use our knowledge of inverses to determine the expression for the solution with single predictor:

$$X_+'\mathbf{y} = \begin{bmatrix} \Sigma x_i y_i \\ \Sigma y_i \end{bmatrix}$$

and

$$X_+'X_+ = \begin{bmatrix} \Sigma x_i^2 & \Sigma x_i \\ \Sigma x_i & n \end{bmatrix} \tag{6.20}$$

so Equation 6.19 is

$$\begin{bmatrix} b_1 \\ b_0 \end{bmatrix} = \begin{bmatrix} \Sigma x_i^2 & \Sigma x_i \\ \Sigma x_i & n \end{bmatrix}^{-1} \begin{bmatrix} \Sigma x_i y_i \\ \Sigma y_i \end{bmatrix}$$

According to Equation 5.2,

$$(\mathrm{X}_+'\mathrm{X}_+)^{-1} = \begin{bmatrix} \dfrac{n}{n\Sigma x_i^2 - (\Sigma x_i)^2} & \dfrac{-\Sigma x_i}{n\Sigma x_i^2 - (\Sigma x_i)^2} \\ \dfrac{-\Sigma x_i}{n\Sigma x^2 - (\Sigma x_i)^2} & \dfrac{\Sigma x_i^2}{n\Sigma x_i^2 - (\Sigma x_i)^2} \end{bmatrix} \tag{6.21}$$

Multiplying together $(\mathrm{X}_+'\mathrm{X}_+)^{-1}$ (Equation 6.21), and $\mathrm{X}_+'\mathbf{y}$ (Equation 6.20) gives

$$\mathbf{b}_+ = \begin{bmatrix} \dfrac{n\Sigma x_i y_i - \Sigma x_i \Sigma y_i}{n\Sigma x_i^2 - (\Sigma x_i)^2} \\ \dfrac{-\Sigma x_i \Sigma x_i y_i + \Sigma x_i^2 \Sigma y_i}{n\Sigma x_i^2 - (\Sigma x_i)^2} \end{bmatrix} = \begin{bmatrix} b_1 \\ b_0 \end{bmatrix} \tag{6.22}$$

If the numerator and denominator of the expression for b_1 are each divided by $n(n - 1)$, Equation 6.22 may be recognized as Equation 6.5. It may be hard to believe, but the b_0 term here is the same as that in Equation 6.6.

A numerical example may be useful at this point. Below are values of y and x for $n = 3$. For later use, the column of ones is included in the X_+ matrix, but only the first column of X represents the values of the predictor x.

$$\underset{\mathbf{y}}{\begin{bmatrix} 6 \\ 9 \\ 18 \end{bmatrix}} \quad \underset{\mathrm{X}_+}{\begin{bmatrix} 2 & 1 \\ 3 & 1 \\ 7 & 1 \end{bmatrix}}$$

A scatterplot of y as a function of x is shown in Figure 6-2. Note that the points lie very close to a straight line. The standard statistics relevant to Equations 6.5 and 6.6 are given below.

	x	y
mean	4.00	11.00
s^2	7.00	39.00
s	2.646	6.245
r_{yx}	.9986	
s_{yx}	16.500	

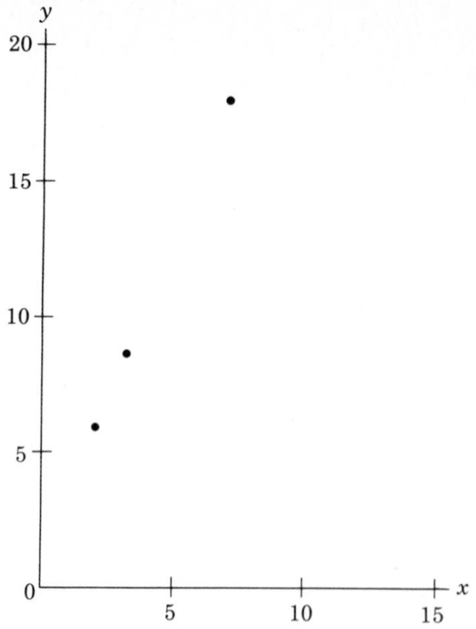

Figure 6-2 Illustrative points plotted, y vertically and x horizontally. These are used to show how a line is fitted by least squares. The fitted line is shown in Figure 6-3.

According to Equation 6.5,

$$b_1 = 16.5/7.00 = 2.357$$

and according to Equation 6.6,

$$b_0 = 11.00 - (2.357)\,(4.00) = 11.00 - 9.428 = 1.572$$

Below we have calculated $\mathbf{X}_+'\mathbf{X}_+$ and $\mathbf{X}_+'\mathbf{y}$ and displayed the components of Equation 6.19. Note, however, that the inverse of $\mathbf{X}_+'\mathbf{X}_+$ has not yet been found.

$$\begin{bmatrix} b_1 \\ b_0 \end{bmatrix} = \begin{bmatrix} 62 & 12 \\ 12 & 3 \end{bmatrix}^{-1} \begin{bmatrix} 165 \\ 33 \end{bmatrix}$$

$$\mathbf{b}_+ = (\mathbf{X}_+'\mathbf{X}_+)^{-1}\,\mathbf{X}_+'\mathbf{y}$$

Now we have found the inverse. Compare the identity of the various terms in the inverse below with Equation 6.21.

$$\begin{bmatrix} b_1 \\ b_0 \end{bmatrix} = \begin{bmatrix} 3/42 & -12/42 \\ -12/42 & 62/42 \end{bmatrix} \begin{bmatrix} 165 \\ 33 \end{bmatrix}$$

$$\mathbf{b}_+ = (\mathbf{X}_+'\mathbf{X}_+)^{-1}\mathbf{X}_+'\mathbf{y}$$

Carrying out the final multiplication gives

$$\begin{bmatrix} b_1 \\ b_0 \end{bmatrix} = \begin{bmatrix} 2.357 \\ 1.571 \end{bmatrix}$$

These values, obtained by means of the matrix Equation 6.9, are the same as those from Equations 6.4 and 6.5.*

Deviations from Regression

If we apply the least-squares parameters to the x values, we obtain the estimated or predicted values of y, $\hat{y}$. These can be compared to the actual values in order to get the deviations from regression. When squared and summed, this will give the value of the error function. If these operations are carried out for the above example, we obtain the following results:

x_i	y_i	$\hat{y}_i$	$y_i - \hat{y}_i$	$(y_i - \hat{y}_i)^2$
2	6	6.286	−.286	.081
3	9	8.643	.357	.128
7	18	18.071	−.071	.005
			Σ .001	.214

The errors are fairly small compared to the variation in y. They may seem large, given the size of the correlation (.9986), but this is actually the magnitude to be expected. This will be discussed further in later chapters, but for the moment we will concentrate on the methods used to arrive at these figures and their meaning. In Figure 6-3 the regression line has been drawn to the points of Figure 6-2.

It should be noted that the errors have the properties they are specified to have: Their sum is zero, as are their covariance and sum of products with x.

It is not necessary to compute the $\hat{y}_i$ in order to get their sum of squares. Because

$$\mathbf{e} = \mathbf{y} - \mathrm{X}_+\mathbf{b}_+$$

(see Equation 6.16), the Σe_i^2 is

$$\mathbf{e}'\mathbf{e} = (\mathbf{y}' - \mathbf{b}_+'\mathrm{X}_+')\,(\mathbf{y} - \mathrm{X}_+\mathbf{b}_+) \tag{6.23}$$

Multiplying this out part way gives

$$\mathbf{e}'\mathbf{e} = \mathbf{y}'(\mathbf{y} - \mathrm{X}_+\mathbf{b}_+) - \mathbf{b}_+'\mathrm{X}_+'\,(\mathbf{y} - \mathrm{X}_+\mathbf{b}_+) \tag{6.24}$$

---

*At various points in the examples in this book, there will be small discrepancies when two numbers are said to be equal. This is due to the effects of rounding.

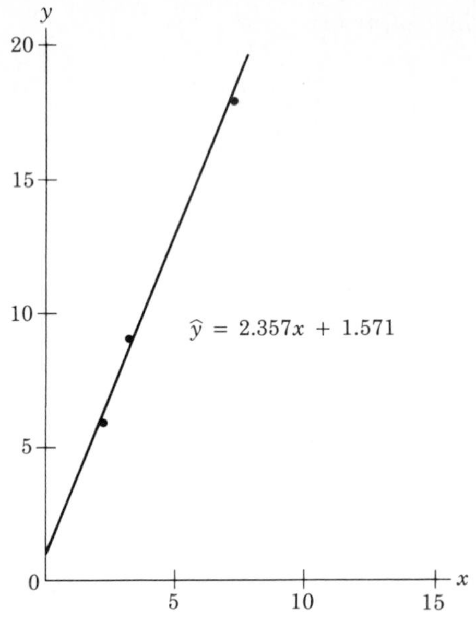

Figure 6-3 Points of Figure 6-2 with best-fitting straight line. Note that y_1, the upper right point, is slightly below the line, whereas y_2 is above it and y_3 is virtually on it, consistent with their calculated error deviations.

But the right most term is **e**, and we know from earlier discussion that the product of X_+ and **e** is zero, so

$$\mathbf{e}'\mathbf{e} = \mathbf{y}'\mathbf{y} - \mathbf{y}'X_+\mathbf{b}_+ \tag{6.25}$$

Thus, we can get the sum of squared errors by first getting the sum of squared y's and subtracting $(X_+'\mathbf{y})'b_+$. The latter in this case is

$$[165 \quad 33] \begin{bmatrix} 2.357 \\ 1.571 \end{bmatrix} = 440.786$$

Subtracting it from Σy_i^2, which was 441, gives .214, as it should.

One further aspect of $\mathbf{y}'X\mathbf{b}$ should be noted. It represents the sum of squares of the *predicted* $\hat{y}$'s, $\Sigma\hat{y}^2$. These properties of $\mathbf{y}'X\mathbf{b}$ are significant because they generalize to the multiple regression case.

Standard Score Regression

Suppose that both variables x and y were in standard score form, that is, with a mean of zero and standard deviation of one. What would the slope and intercept be then? We could probably find the answer using standard formulas. A more interesting approach, however, is to find out what would happen if we made the appropriate redefinitions of X_+ and **y**. Suppose we insert z_x in X_+ in place of x

and insert z_y in **y** in place of y, where $z_x = (x - \bar{x})/s_x$ and $z_y = (y - \bar{y})/s_y$. Then

$$\mathbf{X}_+'\mathbf{X}_+ = \begin{bmatrix} \Sigma z_x^2 & 0 \\ 0 & n \end{bmatrix}$$

and

$$\mathbf{X}_+'\mathbf{y} = \begin{bmatrix} \Sigma z_x z_y \\ 0 \end{bmatrix}$$

The zeros appear because Σz_x and Σz_y equal zero, that is, standard scores have zero means. Thus, the formula for the inverse of a diagonal matrix gives

$$(\mathbf{X}_+'\mathbf{X}_+)^{-1} = \begin{bmatrix} \dfrac{1}{\Sigma z_x^2} & 0 \\ 0 & \dfrac{1}{n} \end{bmatrix}$$

and multiplying out Equation 6.19 gives

$$\mathbf{b}_+ = \begin{bmatrix} \dfrac{\Sigma z_x z_y}{\Sigma z_x^2} \\ 0 \end{bmatrix}$$

If we divide the numerator and denominator of the first entry by $n - 1$, we get $r_{xy}/1$, or r_{xy}, because of the nature of z-scores. This shows that if x and y are in standard score form, the slope is r_{xy} and the intercept is zero. These conclusions are consistent with Equations 6.5 and 6.6 because the covariance between standard scores is the correlation between them and the means of standard score variables are zero.

To make certain distinctions clear, we will use b^* to stand for standard score regression coefficients. Although β is traditionally used in many areas of psychology, we reserve the Greek letter to refer to population values.

The example used above can be used here also. When the variables are converted to standard scores using the means and standard deviations on page 79, then

$$\mathbf{X}_+ = \begin{bmatrix} -.7559 & 1 \\ -.3779 & 1 \\ 1.1339 & 1 \end{bmatrix}$$

and

$$\mathbf{y} = \begin{bmatrix} -.8006 \\ -.3202 \\ 1.1209 \end{bmatrix}$$

Now the matrix Equation 6.19 has the following numerical form

$$\begin{bmatrix} 2.0000 & 0.0000 \\ 0.0000 & 3.0000 \end{bmatrix} \begin{bmatrix} 1.9971 \\ 0.0000 \end{bmatrix} = \begin{bmatrix} b_1^* \\ b_0^* \end{bmatrix}$$

Taking the inverse and multiplying out gives

$$\begin{bmatrix} 0.9986 \\ 0.0000 \end{bmatrix} \begin{bmatrix} b_1^* \\ b_0^* \end{bmatrix}$$

This example is consistent with the assertion that the standard score regression coefficient is the correlation. If we follow through with the matrix manipulations relevant to the variance of predicted scores and error scores, these will be found to be $s_{\hat{z}}^2 = r^2$ and $s_{e_z}^2 = 1 - r^2$, respectively.

MULTIPLE REGRESSION

Two Predictors: Raw Score Regression with Augmented X Matrix

In multiple regression, there are two or more predictors, x_j, that can be used simultaneously to predict the value of y. Suppose Sherlock Chan, our detective, knows not only the shoe size of the suspect but also the length of his trousers. (Perhaps the suspect had to make a hurried departure from a bedroom, leaving shoes and trousers behind.) Can one get a more accurate fix on the suspect's height by combining both variables rather than using either variable alone?

Probably so. Using the same criterion of minimum sum of squared deviations from predictions, it would seem that having another variable to work with should improve matters. At worst, trouser length could be ignored, leaving us no worse off than if we only had shoe size, and usually it should provide some additional accuracy. There are, of course, an infinite variety of ways to use these two variables to make the prediction. In multiple regression, we use one specific type of prediction called weighted linear combinations. Later we will see how this method can be used to encompass a wide variety of seemingly different prediction systems. First, we will begin with the simpler system.

The simplest case is the one in which there are only two predictors, x_1 and x_2; y is again the dependent variable. (The dependent variable is frequently referred to as the "criterion" in much of the psychological literature on multiple regression.) Then the problem is to find b_1 and b_2 to apply to x_1 and x_2 along with a constant b_0

in order to satisfy the least-squares criterion. That is,

$$\hat{y}_i = b_1 x_{i1} + b_2 x_{i2} + b_0$$

and it is desired that $\Sigma e_i^2 = \Sigma(y_i - \hat{y}_i)^2$ be as small as possible. More generally, with p predictors:

$$\hat{y}_i = \sum_{j=1}^{p} b_j x_{ip} + b_0 \tag{6.26}$$

the sum of squared errors of prediction should be as small as possible. How does one choose the b's so as to make them as small as possible? In matrices, the prediction is still formulated exactly as in Equation 6.15, $\hat{\mathbf{y}} = \mathrm{X}_+\mathbf{b}_+$. The only difference is that X_+ has three or more columns rather than two, and correspondingly, $\mathbf{b}_+$ has elements b_1, and $b_2, \ldots, b_p, b_0$. The errors $\mathbf{e}$ are also defined in the same way: $\mathbf{e} = \mathbf{y} - \hat{\mathbf{y}}$, and so is the criterion we wish to minimize, $\Sigma e_i^2 = \mathbf{e}'\mathbf{e}$. One can use the same process of stating that the mean error should be zero and the errors should be unrelated to the X's. That is, $\mathrm{X}_+'\mathbf{e} = 0$, as stated in Equation 6.13. Alternatively, Σe_i^2 can be differentiated with respect to each b_j, including b_0, and the resulting equations set equal to zero, and solved for the b_j. Either approach gives the solution expressed exactly as in Equation 6.19. To express the b's in terms of the original sums, sums of squares, and sums of products, results in very unattractive expressions. These will not be shown here, but will remain veiled behind the panels of the computer. They can also be found in standard references.

Deviation and Standard Score Solutions

For a number of reasons, it is useful to express the regression weights in terms of the variances and covariances rather than in terms of sums, sums of squares, and sums of products. For example, assume that X consists of a matrix of deviation scores and that $\mathbf{y}$ is likewise in deviation form. If this is the case, the intercept term b_0 equals 0. This allows us to omit the extra column of units that we had in X_+ and the intercept term b. Then, in matrices, we would still have

$$\hat{\mathbf{y}} = \mathrm{X}\mathbf{b} \tag{6.27}$$

and

$$\mathbf{e} = \mathbf{y} - \hat{\mathbf{y}} \tag{6.28}$$

In fact, the errors will be exactly the same as before. The requirement that the errors are uncorrelated with the x's, is

$$\mathrm{X}'\mathbf{e} = 0 \tag{6.29}$$

which, like Equation 6.19, leads to

$$\mathbf{b} = (\mathrm{X}'\mathrm{X})^{-1}\mathrm{X}'\mathbf{y} \tag{6.30}$$

by a similar route. Because X is, in deviation scores,

$$S_{xx} = (n - 1)^{-1}X'X \tag{6.31}$$

so

$$S_{xx}^{-1} = (n - 1)(X'X)^{-1} \tag{6.32}$$

and

$$\mathbf{s}_{xy} = (n - 1)^{-1}X'\mathbf{y} \tag{6.33}$$

Therefore,

$$S_{xx}\mathbf{s}_{xy} = (X'X)^{-1}X'\mathbf{y} \tag{6.34}$$

It is also true that, because the $n - 1$ terms cancel,

$$\mathbf{b} = S_{xx}^{-1}\mathbf{s}_{xy} \tag{6.35}$$

For conceptual purposes, Equation 6.35 is often a more useful form of expression for the slope coefficients than Equation 6.19. Often the intercept term is only of peripheral interest anyway, but if needed, it can be calculated from the b's and the means according to the generalization below:

$$b_0 = \bar{y} - \sum_j b_j\bar{x}_j \tag{6.36}$$

Because the units of the variables are often arbitrary, the original variables may be expressed in standard score form. Then the covariance matrix becomes the correlation matrix. The standard score regression coefficients may also provide some additional insight into the relative importance of the variables in the prediction, although this is not a foolproof process.

Defining b_j^* as the regressions coefficient for predicting z_y from the z_j, the special case of Equation 6.35 for standard scores gives

$$\mathbf{b}^* = R_{xx}^{-1}\mathbf{r}_{xy} \tag{6.37}$$

The general relation between raw score and standard score coefficients is

$$b_j^* = b_j(s_x/s_y) \tag{6.38}$$

This may be related back to the matrix expression for R, in Equation 2.38, and principles of inverses discussed in Chapter 5. We will not go into the details here, however.

The use of Equation 5.3 for the inverse of a 2 × 2 correlation matrix leads to the following formulas for the standard score regression weights for two predictors:

$$b_1^* = (r_{1y} - r_{12}r_{2y})/(1 - r_{12}^2)$$
$$b_2^* = (r_{2y} - r_{12}r_{1y})/(1 - r_{12}^2) \quad (6.39)$$

These formulas are simpler to compute from the statistics than from the raw scores. They also show how the coefficients are influenced by the various correlations.

The steps in computing raw and standard score weights by the different routes are shown in Exhibit 6-1, pages 88–89.

Variances and Correlations of $\hat{y}$

The total variance of y can be divided into two portions: an unpredictable portion, s_e^2, and a predictable portion, $s_{\hat{y}}^2$. The latter is actually the variance of the *predictions*, $s_{\hat{y}}^2$. Because of the definition of s_e^2, it equals $s_y^2 - s_{\hat{y}}^2$. Transposing $s_{\hat{y}}^2$ gives the equality

$$s_y^2 = s_{\hat{y}}^2 + s_e^2 \quad (6.40)$$

The $s_{\hat{y}}^2$ part may be thought of as variance due to regression and the s_e^2 part as variance due to deviations from regression, s_{reg}^2 and s_{dev}^2, respectively. At some later point, this mode of expression will be used when we want to emphasize these connotations. It is also worth observing some other connections. According to the methods described in Chapter 4, we can calculate the variance of predictions from the variances and covariances of the predictors:

$$s_{\hat{y}}^2 = \mathbf{b}'\mathrm{S}_{xx}\mathbf{b} \quad (6.41)$$

Substituting for one $\mathbf{b}$ the expression $\mathrm{S}_{xx}^{-1}\,\mathbf{s}_{xy}$ gives

$$s_{\hat{y}}^2 = \mathbf{b}'\mathrm{S}_{xx}(\mathrm{S}_{xx}^{-1})\mathbf{s}_{xy}$$

and the two inner terms give an identity, leaving

$$s_{\hat{y}}^2 = \mathbf{b}'\mathbf{s}_{xy} \quad (6.42)$$

Thus, this may be calculated quite simply as the sum of the products of each regression weight with its corresponding covariance with the dependent variable.

A similar method can be used to find the correlation of y with $\hat{y}$. Actually, it is more useful to work with the squared correlation. Chapter 4 showed that the

Raw Scores

$$\begin{array}{c} i \\ 1 \\ 2 \\ 3 \\ 4 \end{array} \quad \begin{array}{c} x_1 \; x_2 \; x_0 \\ \begin{bmatrix} 4 & 5 & 1 \\ 5 & 4 & 1 \\ 7 & 2 & 1 \\ 4 & 1 & 1 \end{bmatrix} \\ \mathbf{X}_+ \end{array} \quad \begin{array}{c} y \\ \begin{bmatrix} 6 \\ 3 \\ 5 \\ 2 \end{bmatrix} \\ y \end{array}$$

(a)

$$\begin{array}{c} \\ x_1 \\ x_2 \\ x_0 \end{array} \begin{array}{c} x_1 \quad x_2 \quad x_0 \\ \begin{bmatrix} 106 & 58 & 20 \\ 58 & 46 & 12 \\ 20 & 12 & 4 \end{bmatrix}^{-1} \\ (\mathbf{X}_+'\mathbf{X}_+)^{-1} \end{array} \begin{bmatrix} 82 \\ 54 \\ 16 \end{bmatrix} = \begin{bmatrix} b_1 \\ b_2 \\ b_0 \end{bmatrix} \quad \mathbf{X}'y = \mathbf{b}_+$$

(b)

$$\begin{bmatrix} .1786 & .0357 & -1.000 \\ .0357 & .1071 & -5.000 \\ -1.000 & -5.000 & 6.7500 \end{bmatrix} \begin{bmatrix} 82 \\ 54 \\ 16 \end{bmatrix} = \begin{bmatrix} .5714 \\ .7143 \\ -1.0000 \end{bmatrix}$$

$$(\mathbf{X}_+'\mathbf{X}_+)^{-1} \quad (\mathbf{X}_+'\mathbf{y}) = \quad \mathbf{b}_+$$

(c)

$$\begin{array}{c} \\ x_1 \\ x_2 \\ x_0 \end{array} \begin{array}{c} x_1 \quad x_2 \quad x_0 \\ \begin{bmatrix} 106 & 58 & 20 \\ 58 & 46 & 12 \\ 20 & 12 & 4 \end{bmatrix} \end{array} \begin{bmatrix} 82 \\ 54 \\ 16 \end{bmatrix} = \begin{bmatrix} b_1 \\ b_2 \\ b_0 \end{bmatrix}$$

$$(\mathbf{X}_+'\mathbf{X}_+)^{-1} \quad \mathbf{X}_+'\mathbf{y} \quad b_+$$

$$\bar{y} = 4.000$$

$$s_y^2 = 3.333$$

(d)

Exhibit 6-1 Raw score and standard score regression weights. Part (a) shows the raw data for two predictors and a dependent variable with an n of 4. The last column of the X_+ matrix is the constant vector of 1.0 whose inclusion will produce the value of b_0. Part (b) shows the symmetric $X_+'X_+$ matrix and the $X_+'\mathbf{y}$ vector. The lower-right element of $X_+'X_+$ is n, and the remaining elements of the last row and column are the sums of the x variables. The remaining diagonal elements are the sums of squares of x_1 and x_2, which are 106 and 46, respectively. The other, off-diagonal elements, are the sums of products of x_1 with $x_2 = 58$. The last element of $X_+'\mathbf{y}$ is the sum of y and the remaining two are its sums of products with x_1 and with x_2. Part (c) shows the inverse of $X_+'X_+$ and its product with $X_+'\mathbf{y}$, which yields

covariance of a linear combination (of which $\hat{y}$ is clearly an example) with a variable is the same as the linear combination of the covariances of the components with that variable. Here, the weights are the b's:

$$s_{y\hat{y}} = \mathbf{b}'\mathbf{s}_{xy} \tag{6.43}$$

This is exactly the same as Equation 6.42, so $s_{\hat{y}}^2 = s_{y\hat{y}}$.

Because the correlation is the covariance divided by the product of standard deviations, the squared correlation is

Exhibit 6-1 (*continued*)

Deviation Scores

$$\underset{\mathbf{X}}{\overset{x_1 \quad x_2}{\begin{bmatrix} -1 & 2 \\ 0 & 1 \\ 2 & -1 \\ -1 & -2 \end{bmatrix}}} \quad \underset{\mathbf{y}}{\overset{y}{\begin{bmatrix} 2 \\ -1 \\ 1 \\ -2 \end{bmatrix}}}$$

(e)

$$\begin{bmatrix} 2.000 & -.667 \\ -.667 & 3.333 \end{bmatrix}^{-1} \begin{bmatrix} .667 \\ 2.000 \end{bmatrix} = \begin{bmatrix} .5357 & .1071 \\ .1071 & .3214 \end{bmatrix} \begin{bmatrix} .667 \\ 2.000 \end{bmatrix} = \begin{bmatrix} .5714 \\ .7143 \end{bmatrix}$$

$$\mathbf{S}_{xx}^{-1}\,\mathbf{s}_{xy} \qquad = \qquad \mathbf{S}_{xx}^{-1}\,\mathbf{s}_{xy} \qquad = \mathbf{b}$$

(f) (g)

$$\begin{bmatrix} 1.000 & -.258 \\ -.258 & 1.000 \end{bmatrix}^{-1} \begin{bmatrix} .258 \\ .600 \end{bmatrix} = \begin{bmatrix} 1.071 & .277 \\ .277 & 1.071 \end{bmatrix} \begin{bmatrix} .258 \\ .600 \end{bmatrix} = \begin{bmatrix} .443 \\ .714 \end{bmatrix}$$

$$\mathbf{R}_{xx}^{-1}\,\mathbf{r}_{xy} \qquad = \qquad \mathbf{b}^*$$

(h)

$$\begin{bmatrix} 1.414 & 0 \\ 0 & 1.826 \end{bmatrix} \begin{bmatrix} .571 \\ .714 \end{bmatrix} [.548] = \begin{bmatrix} .443 \\ .714 \end{bmatrix}$$

$$\mathbf{D}_x \qquad \mathbf{b} \qquad s_y^{-1} = \mathbf{b}^*$$

(i)

b_+ (Equation 6.19), the column vector of regression coefficients. We see that $b_0 = -1.0000$, $b_1 = .5714$, and $b_2 = .7143$. This completes the calculation of the raw score regression weights. Part (d) gives the mean and variance of y for reference.

The next sections show the calculation of the regression weights from the covariance and from the correlation matrices. The deviation scores in the x and in the y are shown first in part (e), and S_{xx} and $\mathbf{s}_{xy}$ are in (f); (g) shows S_{xx}^{-1} and its product with $\mathbf{s}_{xy}$, which is $\mathbf{b}$ (Equation 6.35). Note that the entries in b are the same as those found earlier for b_1 and b_2 from $(X'X)^{-1}X'\mathbf{y}$.

In (h) the correlation matrices R_{xx} and $\mathbf{r}_{xy}$ are shown, followed by R_{xx}^{-1} and its product with $\mathbf{r}_{xy}$, which is $\mathbf{b}^*$ (Equation 6.37). The relationship of $\mathbf{b}$ to $\mathbf{b}^*$ (Equation 6.38) is shown in part (h), where D_{xx} contains the standard deviations of the x's and s_y^{-1} is the reciprocal of the standard deviation of y. The fact that this product gives $\mathbf{b}$ has been verified.

$$r_{y\hat{y}}^2 = s_{y\hat{y}}^2 / s_{\hat{y}}^2 s_y^2$$

Substituting Equations 6.42 and 6.43 gives

$$r_{y\hat{y}}^2 = (\mathbf{b}'\mathbf{s}_{xy})^2 / (\mathbf{b}'\mathbf{s}_{xy})\, s_y^2$$

Canceling like terms in numerator and denominator leaves

$$r_{y\hat{y}}^2 = \mathbf{b}'\mathbf{s}_{xy} / s_y^2$$

This provides a straightforward way of obtaining this correlation from the weights.

The correlation of y with $\hat{y}$ is called the *multiple correlation* of y with the x's. The uppercase R is traditionally used to denote it—this marks an exception to the rule of using uppercase letters for matrices only:

$$R^2 = \mathbf{b}'\mathbf{s}_{xy}/s_y^2 \tag{6.44}$$

If all the variables are in standard score form, then in place of $\mathbf{b}$ we have $\mathbf{b}^*$, $\mathbf{s}_{xy}$ becomes $\mathbf{r}_{xy}$, and s_y^2 becomes $s_{z_y}^2$, or 1.0. Therefore, in terms of correlations,

$$R^2 = \mathbf{b}^{*\prime}\mathbf{r}_{xy} \tag{6.45}$$

Curiously, we can show that R^2 can be gotten from just R_{xx}. The variance of the standard score predictions is the same as the squared multiple R. Because $\mathbf{b}^* = \mathrm{R}_{xx}^{-1}\,\mathbf{r}_{xy}$, $\mathbf{r}_{xy} = \mathrm{R}_{xx}\,\mathbf{b}^*$. Substituting Equation 6.45 gives

$$R^2 = \mathbf{b}^{*\prime}\mathrm{R}_{xx}\mathbf{b}^* \tag{6.46}$$

This equation should be familiar as the variance of weighted standard scores, where the weights are b^*. Also, we can show that

$$R^2 = s_{\hat{y}}^2/s_y^2 \tag{6.47}$$

Thus, one interpretation of R^2 is that it represents the proportion of the variance of y that is predictable. The expression, $1 - R^2$ is the ratio of error variance to total variance.

$$1 - R^2 = s_e^2/s_y^2 \tag{6.48}$$

This is the proportion of variance that is not explained by regression. The standard error of estimate, s_e, is

$$s_e = s_y\sqrt{1 - R^2} \tag{6.49}$$

Error Variance

The error variance is also directly related to the covariance matrix and the regression weights. Consider the column of deviations from regression: $\mathbf{e} = \mathbf{y} - \hat{\mathbf{y}}$. Its mean is fixed to be zero, so that

$$s_e^2 = \frac{1}{n-1}\Sigma e_i^2 = \mathbf{e}'\mathbf{e}(n-1)^{-1} \tag{6.50}$$

Consider just $\mathbf{e}'\mathbf{e}$ for the moment. By substituting the definition of $\mathbf{e}$, $\mathbf{y} - \hat{\mathbf{y}}$ in the equation, and applying some algebra, gives

$$s_e^2 = (\mathbf{y}'\mathbf{y} - \mathbf{b}'\mathrm{X}'\mathbf{y})(n-1)^{-1} \tag{6.51}$$

Although this is not necessary, it makes things a little easier to think about if we assume that we are dealing with deviation scores. The quantity $(n - 1)^{-1}\mathbf{X}'\mathbf{y}$ in the deviation score case is $\mathbf{s}_{xy}$, so

$$s_e^2 = s_y^2 - \mathbf{b}'\mathbf{s}_{xy} \tag{6.52}$$

Putting everything in scalar form gives

$$s_e^2 = s_y^2 - \Sigma b_j s_{jy} \tag{6.53}$$

This describes in general how to find the error variance without actually computing the deviations.

Having the variables in standard score form would lead directly to

$$1 - R^2 = 1 - \mathbf{b}^{*\prime}\mathbf{r}_{xy} \tag{6.54}$$

or

$$1 - R^2 = 1 - \Sigma b_j^* r_{jy} \tag{6.55}$$

In one form or another, we will make use of the expressions in this section in later chapters. Exhibit 6-2 illustrates these relations among the variances $s_{\hat{y}}^2$, s_e^2, s_y^2, R^2, and $s_{\hat{z}_y}^2$.

INTERPRETATION OF WEIGHTS

As humans, we are meaning-seeking creatures. To better understand the meanings of the regression weights, we might ask the following: What does it mean when one weight is larger than another? What does the sign mean? How much variance does one predictor account for? What are the causal relations implied by the weights?

In a sense, the weights are solutions of the "normal equations," like Equation 6.35, and that is all. They are merely the weights that allow us to minimize deviations from predictions. However, there are other popular ways of looking at weights which may be either invalid or valid, depending on the circumstances.

First, it may help to consider the equations geometrically and to visualize what the weights mean in that context. Visualizing what is happening in the two-predictor case is simplest. Imagine a motionless swarm of points in a transparent plastic cube, each point representing an observation. Each dimension of the cube represents a variable, and the vertical direction represents values on the dependent variable y. For convenience, we will call one face the "side", and the face attached to its right edge the "front." Viewed from the side, the left-right direction represents one predictor, x_1, and viewed from the front, the left-right direction represents x_2.

Viewed directly from the side, the points will represent the univariate regression on x_1; viewed from the front, they will represent the regression on x_2 by itself. A univariate regression line could be drawn on each of these two faces of the cube. Then distances from the regression lines in the vertical direction represent errors when a prediction is made using that variable alone. Suppose that deviations

$$\begin{bmatrix}6.000\\3.000\\5.000\\2.000\end{bmatrix} - \begin{bmatrix}4.857\\4.714\\4.429\\2.000\end{bmatrix} = \begin{bmatrix}1.143\\-1.714\\.571\\2.000\end{bmatrix}$$

$$\mathbf{y} \quad \hat{\mathbf{y}} \quad \mathbf{e}$$

(a)

$$[1.143 \quad -1.714 \quad .571 \quad .000] \begin{bmatrix}4 & 5 & 1\\5 & 4 & 1\\7 & 2 & 1\\4 & 1 & 1\end{bmatrix} = [0.0 \quad 0.0 \quad 0.0]$$

$$\mathbf{e'x} = \mathbf{0}$$

(b)

$$s_{\hat{y}}^2 = 1.809 \quad s_e^2 = 1.523 \quad s_{\hat{y}}^2 + s_e^2 = 3.332$$

(c)

$$[.5714 \quad .7143] \begin{bmatrix}2.0000 & -.6667\\-.6667 & 3.3333\end{bmatrix} \begin{bmatrix}.5714\\.7143\end{bmatrix} = 1.809$$

$$\mathbf{b'S}_{xx}b = s_y^2$$

$$[.6667 \quad 2.0000] \begin{bmatrix}.5714\\.7143\end{bmatrix} = 1.810$$

Exhibit 6-2 Variances, covariances, and correlations of predicted and error scores. The following matrices illustrate the relations among the variances and covariances of the predicted and error scores and their correlation, including the multiple correlation of the x's with y. Part (a) gives the predicted y's and the difference between predicted and actual, **e**. Part (b) verifies that X′**e** = 0. Part (c) gives the result of calculating $s_{\hat{y}}^2$ and s_e^2 from these score vectors; their sum is shown to equal the value of s_y^2 of part (d) within a small rounding error (see Equation 6.40). Part (d) calculates s_y^2 according to Equation 6.41. Note the similarity of

from the x_1 line are smaller than those from the x_2. Now imagine facing the cube from the side and then walking around the cube while looking at the swarm of points, mentally drawing a regression line through the swarm as viewed from that angle as you walk, and estimating the average degree of vertical departure from the line. If no viewpoint gives smaller errors than directly from the x_1 side, then only x_1 need be considered, and $b_2 = 0$.

The chances are great, however, that while making the circuit you will find some angle of view that gives a tighter-fitting regression line than was achieved from the x_1 side. If this spot is more toward the side than the front, then b_1 is greater than b_2, and vice versa. There is a gradual progression from $b_2 = 0$, through $b_1 = b_2$ (if the maximum is found viewing from the corner), to $b_1 = 0$ if the best fit is found while viewing directly from the front. (We will not work out the exact trigonometry of this, but the general idea can be imagined.) Thus, from a spatial point of view, finding the optimum values of the weights corresponds to finding the optimum angle for viewing the points.

What about negative weights? These simply mean that when the optimum view is found, in order for the slope to be from left to right, it is necessary to view it from the opposite side or from the back of the cube, depending on which weight is negative. Occasionally, the x_1 regression line slopes up from left to right when viewed from the side and the same for the x_2 line when viewed from the front, indicating positive x weights for each if used separately, but the optimum-fitting position is found when viewed from the back, meaning positive for b_1 and

$$(\mathbf{b}'\mathrm{S}_{xx})\mathbf{b} \quad = s_{\hat{y}}^2$$

$$s_{\hat{y}}^2/s_y^2 \quad = 1.8096/3.333$$

$$= .5429$$

(d)

$$[.443 \quad .714] \begin{bmatrix} .258 \\ .600 \end{bmatrix} = .545$$

$$\mathbf{b}^* \qquad \mathbf{r}_{xy} \quad = R^2$$

$$[.443 \quad .714] \begin{bmatrix} 1.000 & -2.58 \\ -.258 & 1.000 \end{bmatrix} \begin{bmatrix} .443 \\ .714 \end{bmatrix} =$$

$$\mathbf{b}^* \qquad \mathrm{R}_{xx} \qquad \mathbf{b}^* \quad =$$

$$= [.258 \quad .600] \begin{bmatrix} .443 \\ .714 \end{bmatrix} = .543$$

$$= \quad (\mathbf{b}^*\mathrm{R})\mathbf{b}^* \qquad = R^2$$

(e)

Equations 6.41 and 6.42 with the right-hand part of the equation. Finally, it shows the ratio for $s_{\hat{y}}^2/s_y^2 = R^2$ (Equation 6.47); the equivalence of this to Equation 6.51 is clear from the intermediate step.

Part (e) shows how to calculate R^2 in two other ways. First, it demonstrates the procedure used in Equation 6.45, followed by the procedure used in Equation 6.46. Both yield the same result, as shown in part (d).

negative for b_2! It is difficult to visualize the shape of the swarm of points in this case, but it does happen. Sometimes there are reasons to be suspicious of the results, but genuine cases do occur. This is one manifestation of what is called a "suppressor" effect, in which the variable with the surprising negative sign is called a suppressor variable.

Remember that correlations may be regarded as the cosines of the angles between the vectors that represent the variables and that in the standard score case the vectors are of equal length. This correspondence is useful here. If we think of the two vectors that represent the two predictors in this fashion, we may also think of there being a disk with the origin as the center and the two vectors in the plane of the disk—like two equal-length hands of a clock with a blank face. If the variables have a high positive correlation, the vectors are close together; if highly negative, they point nearly opposite; if near zero, they are at right angles. Now imagine the dependent variable y as a third vector, the same length as the other two vectors, and coming from the same origin, but pointing upward at some angle from the plane of the clock. Imagine the disk as horizontal, with a light shining from above.

The shadow of the y vector onto the disk is its projection onto the plane of the two predictor variables. The smaller the angle it makes with the disk, the longer its shadow. If this shadow stretches all the way out to the rim of the disk, the angle must be zero, and the y vector must be entirely in the plane of the disk. If there is no shadow, the y vector is orthogonal to it, and the angle is 90 degrees.

The length of the shadow is directly related to the cosine of the angle of the y vector with the disk. This cosine, or length, is the *multiple* correlation of y with the other two variables. Thus, the multiple correlation is the cosine of the angle of the y vector with the plane of x_1 and x_2.

The shadow or projection is a vector in the plane of the disk and the two predictor vectors are a *basis* for that plane. Every vector in that plane may be constructed as a linear combination of the basis vectors, and there is a specific pair of coefficients that do this. The coefficients that contruct the projection vector (the shadow) out of the predictor vectors are the *standard score regression coefficients.* In considering the various possibilities of the angle between the predictor vectors and where the projection vector falls with respect to them, we may realize that these relate to the relative size of the regression coefficients and their signs. These properties are illustrated in Figure 6-4.

So far this section has offered three interpretations of the weights b_j or b^*. The first is simply algebraic; the weights are the values that solve the normal

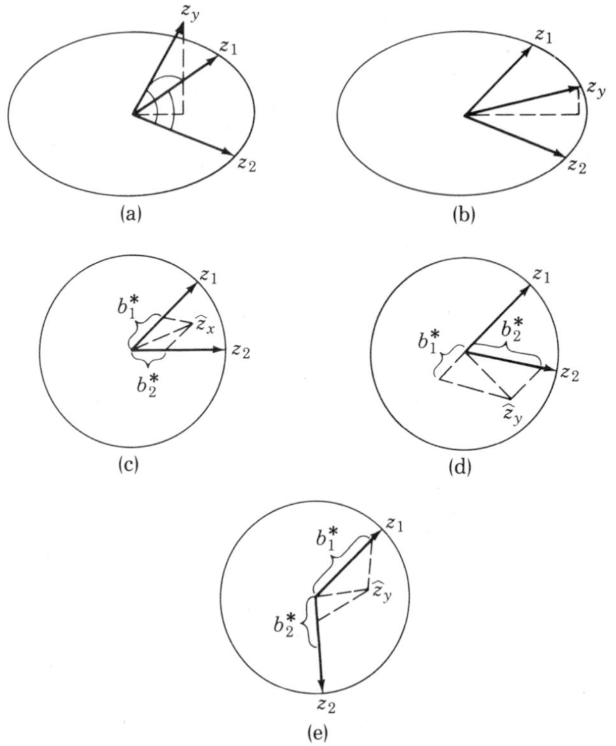

Figure 6-4 Vectors $\mathbf{z}_1$ and $\mathbf{z}_2$ define a plane, represented by a disk. The multiple correlation of $\mathbf{z}_y$ with $\mathbf{z}_1$ and $\mathbf{z}_2$ is the cosine of the angle of its vector and the plane. The predictable part of y is its projection onto the plane. Parts (a) and (b) represent low and high multiple correlations, respectively. The three circular figures are disks viewed from above to show the $\mathbf{z}_y$ vector being constructed out of $\mathbf{z}_1$ and $\mathbf{z}_2$. Parts (c) and (d) show the effect on the weights of where the projection falls with respect to the predictor vectors. Part (e) shows the effect of a change in the angle between the two predictors; the relation between $\mathbf{z}_y$ and $\mathbf{z}_1$ is the same as in (c), although the weights change due to the position of $\mathbf{z}_2$.

equations given in matrix form in Equation 6.30 and its variation, Equation 6.35. The second interpretation is a visual-heuristic one, based on the angle of view from which the smallest predictive error is achieved. The third is a vector interpretation in which the weights are applied to the predictor vectors, and the result is used to reconstruct the projection of z_y onto the plane defined by the other two vectors. These various concepts can aid in interpreting many cases, but fundamentally, the weights represent parameters for the data in a mathematical model, or simply empirical constants that provide optimal predictions. Any deeper interpretation is fraught with ambiguity.

If the predictors are orthogonal, then they have properties that are quite clear. In standard score form, uncorrelated predictors give b_j^* that are equal to the corresponding r_{jy}, as just noted. Thus, the weight for x in forming the most predictive composite is the same as its slope. Also, since $\mathrm{R}_{xx} = \mathrm{I}$, $\boldsymbol{b}^* = \boldsymbol{r}_{xy}$, and $R^2 = \Sigma_j r_{jy}^1$, each predictor is "contributing variance" in proportion to both its direct correlation and its standardized regression weight. It would be nice if this or a similar relation could be found in the correlated predictor case as well, but it cannot.

Predictors can hardly be uncorrelated unless they are variables under the control of the investigator, as they are in common experimental designs. This is one advantage of manipulated, as opposed to observed, variables. More will be said on this in chapters 10 and 11.

If the predictors are correlated, then the picture becomes more complex, except perhaps if R is greater than .99, for example. If that happy state of affairs does exist, then one may be able to interpret the raw score weights in quite a literal, functional fashion. The regression equation is expressing what amounts to a functional dependence of y on the x's. If $b_1 = 1.0$, $b_2 = 2.0$, and $b_0 = 0$, for example, and R is very high, then the data may be interpreted as $y = x_1 + 2x_2$. Thus, every change in x_1 will lead to an equal change in y and every change in x_2 will lead to twice as great a change in y. These may have direct, substantive interpretation within the context of the problem being studied.

In the majority of cases encountered, however, the multiple R is not very high. As a result, one cannot rely on such direct interpretations of the coefficients because they will change if more predictors are added to the system. One cannot overemphasize the dangers of overlooking this fact. If it is found that $b_1 = 1.0$, $b_2 = 2.0$, and R is only moderately high, one might be tempted to conclude that a change in x_1 leads to an equal change in y. This is not necessarily true. Real-world relationships may depend on a network of variables, some of which are unobserved. If these variables *were* observed, b_1 might turn out to be zero, or even a negative number.

Because the variances of the variables commonly reflect measurement units, one usually focuses on the values of the standardized weights. These depend on the variable's correlation with the y and its intercorrelations with the other predictors, because these intercorrelations determine the elements of R_{xx}^{-1}. Moreover, the effect of these may be hard to predict from examination of the intercorrelations, unless the latter are all very small compared to the validities. There are principles, however, that help in determining what to expect.

Generally speaking, one expects that the relative sizes of the b_j^* will correspond fairly closely to the relative sizes of the r_{jy}, but that their differences may be more pronounced. In the following example, there are two moderately highly correlated predictors, one of which has a somewhat higher correlation with the dependent variable than the other although both are of the same sign. The result is a pair of weights that are (a) more different in size than the validities are, and (b) both lower than the validities.

$$\begin{bmatrix} 1.00 & .60 \\ .60 & 1.00 \end{bmatrix}^{-1} \begin{bmatrix} .50 \\ .35 \end{bmatrix} = \begin{bmatrix} .453 \\ .078 \end{bmatrix}$$
$$\mathrm{R}_{xx}^{-1} \quad \mathbf{r}_{xy} = \mathbf{b}^*$$

This effect will be less pronounced if the intercorrelation is lower, and become exaggerated if it is higher. The weight may even become negative for the second variable if the intercorrelation is high enough.

The addition of a third variable can have a pronounced effect on the weights. Take the following example, which is like the preceding one, except that the intercorrelation of the predictors is lower:

$$\begin{matrix} 1 & 2 \end{matrix} \qquad y$$
$$\begin{bmatrix} 1.00 & .30 \\ .30 & 1.00 \end{bmatrix}^{-1} \begin{bmatrix} .50 \\ .35 \end{bmatrix} = \begin{bmatrix} .434 \\ .220 \end{bmatrix}$$
$$\mathrm{R}_{xx}^{-1} \quad \boldsymbol{r}_{xy} = \boldsymbol{b}^*$$

Suppose a variable very closely related to x_1 were added to the system, one which correlated .80 with it and had the same correlations with the other two variables. For example, suppose y is the grade given in a statistics course, x_1 is a numerical reasoning test, and x_2 is a vocabulary test. Now suppose we bring in x_3, which is a *second* numerical reasoning test similar to the first. Then the correlations might look like this:

$$\begin{matrix} 1 & 2 & 3 \end{matrix} \qquad y$$
$$\begin{bmatrix} 1.00 & .30 & .80 \\ .30 & 1.00 & .30 \\ .80 & .30 & 1.00 \end{bmatrix}^{-1} \begin{bmatrix} .50 \\ .35 \\ .50 \end{bmatrix} = \begin{bmatrix} .224 \\ .204 \\ .220 \end{bmatrix}$$
$$\mathrm{R}_{xx}^{-1} \quad \boldsymbol{r}_{xy} = \boldsymbol{b}^*$$

Note that the introduction of a variable similar to x_1 has greatly reduced the size of its weight. From these data, we might have concluded that the three variables contribute about equally. In a sense this may be true, but to conclude that vocabulary and numerical reasoning are about equally important to statistics performance is quite incorrect. What has happened is that the part of y's variance that is predictable from numerical reasoning has been divided up between two observed measures of it.

This last example illustrates a general, although far from universal principle: The size of the b^* weight for a predictor tends to increase with the validity correlation and to decrease with the absolute size of its correlations with other predictors. The squared validity correlations indicate how much variance is shared between each predictor and the dependent variable. In general, when predictors correlate, some of the variance each shares with y is also shared between the two predictors. This is reflected in a reduction of the size of the regression weight for each predictor. Again, this is an *approximate* principle. It is not always followed very closely, and may even be contradicted, particularly where there are a number of predictors with complex interrelations. This principle is perhaps most useful for composing predictor batteries: one tries to include predictors that correlate as much as possible with y, and as little as possible with all other predictors.

A heuristic device that is useful here—provided it is not taken too literally—is to represent the variance of two predictors and dependent variables in a "Venn" diagram (see Figure 6-5). The variance of each variable in the diagram is represented by a circle and the circles overlap. The degree of overlap shows the amount of shared variance, represented by the squared simple correlations among the variables. The multiple R^2 is the variance that y shares with either x_1 or x_2, analogous to an intersection of y with the union of x_1 and x_2. Figure 6-5 illustrates two cases, one in which x_1 and x_2 do not overlap (correlate zero), and one in which they overlap appreciably. In both cases the overlap of x_1 and x_2 with y is the same. The intersection of y with the union of the other two is smaller when x_1 and x_2 overlap appreciably.

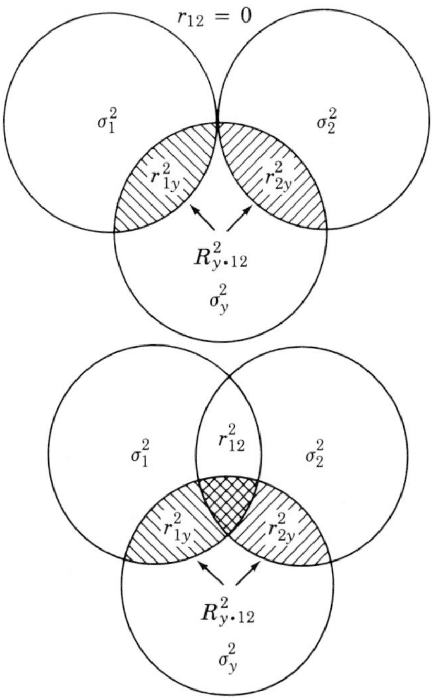

Figure 6-5 "Venn" diagrams show the effect on multiple correlation of correlated predictors. The multiple R^2 is the proportion of variance of y, which is shared with either x_1 or x_2. In the upper figure, the predictors are uncorrelated, and this is the sum of variance shared with x_1 and variance shared with x_2. The presence of correlation between the predictors means that the latter two overlap, as in the lower figure, so that the variance shared with x_1 or x_2 is less than their sum.

Rather than interpreting importance from the size of the b_j^* weights in the correlated case, a different principle is suggested here. The value of a variable as a predictor depends on how much variance that variable is *sufficient* in itself to account for and how much it is *necessary* for. The former is given by its squared validity. That is, if all we have is this one variable, we can account for that much variance. It is *sufficient* to account for that much, r_{2y}^2. To see how much is *necessary* to account for, we must compute R^2 from a battery including it and then again from a battery containing all other predictors except it. The difference between these two R^2's is how much we lose by excluding that variable. So, the variable is *necessary* only for that much of the variance in the context of the current battery of predictors. But remember that the amount of variance that a variable is necessary for predicting will depend on which other variables are present.

The heuristics described for the interpretation of multiple regression work well for most of the situations that one is likely to encounter. However, the following example contradicts most of them:

$$\underset{\mathbf{R}_{xx}}{\begin{bmatrix} 1.00 & -.80 \\ -.80 & 1.00 \end{bmatrix}^{-1}} \underset{\boldsymbol{r}_{xy}}{\begin{bmatrix} .30 \\ .30 \end{bmatrix}} = \underset{\boldsymbol{b}^*}{\begin{bmatrix} b_1^* \\ b_2^* \end{bmatrix}}$$

Note that each predictor in this example seems to account for a small proportion (9 percent) of the variance of y, so we would expect R^2 to be less than .18. On computing it, we would find that R is .90!

Here is a case where the proportion of variance accounted for by two combined variables is much greater than the sum of the variances attributable to each. The Venn diagram idea can no longer be applied, and each variable is "necessary" for more variance than is "sufficient." There is nothing illegal or self-contradictory about these data, although they might spoil our faith in the general principles. However, it is better to know about such things from the beginning than to be surprised by them in the future.

The only interpretation that works well here is the vector one. The correlation of −.80 between the two predictors means that there is an angle of nearly 180° between them—which is actually 159°. This means that even though there is a large angle between each of the predictors and y, y must be close to the plane of the two predictors. The situation is illustrated by the vectors in Figure 6-6.

This example is curious in another respect. That is, the standard score regression weights are both 1.5. While we think of standard score weights as being in the same metric as the correlation coefficient, they are not really correlations in the multipredictor case.

Situations like this are empirically rare, but not unheard of—particularly in less extreme forms. Consider a job where success is dependent on both speed and accuracy. We wish to select candidates by giving a job sample test, and we tell candidates that both speed and accuracy are important. Some go fast and make many errors; others go carefully and get little done—the correlation between the speed and accuracy aspects of the test is negative. The most successful candidates are those who are both fast and accurate, even though such persons are quite rare. Note here that the coefficients are modeling the task. Their size tells us that it is

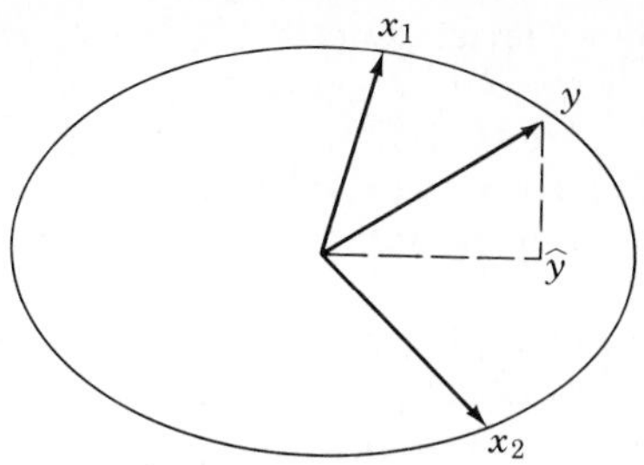

Figure 6-6 A three-dimensional view of an unusual configuration of variables. The two predictors are separated by 159° [$\cos^{-1}(-.80)$], and y is separated from each by 72.5° [$\cos^{-1}(.30)$].

very important to be fast and very important to be accurate, even though neither is very important on its own.

It should also be noted that this example is not curvillinear, as there is no "interaction" between the two predictors. This is a straight linear model, albeit an unusual and surprising one.

The preceding discussion of regression weights has been from a purely descriptive point of view. The problems of statistical inferences made from sample weights to the corresponding population weights have not been considered, and are often tricky and subtle. These will be dealt with at some length in Chapter 9 after the general concepts of statistical inference have been reviewed in Chapter 8.

Dependence among Variables

One technical problem that was alluded to in Chapter 3 is the problem of linear dependence among variables. Here, the total system has $p + 2$ vectors, that is, the p predictors plus y and the constant vector. This means that unless n is greater than $p + 1$ the system is bound to be linearly dependent. In particular, this means that there *will be* a set of weights such that a linear combination of the other p + 1 will equal y exactly. Thus, the multiple R *will be* 1.0 if $n = p + 1$. Such a finding is not empirically meaningful, however.

If n is less than $p + 1$, then X_+, which includes the constant vector, will have fewer rows than columns, and $X_+'X_+$ will not have an inverse. The computer will indicate this, and valuable time—yours and the computer's—will have been wasted. Remember that $X_+'X_+$ may fail to have an inverse for other reasons such as the way the variables are defined, if this makes them linearly dependent. Several examples were given in Chapter 3, including gain scores and their components, ipsative measures, and total scores and their components. In such cases it is usually only necessary to eliminate one of the set of linearly dependent variables—either the composite, where there is one, or the least interesting variable.

The preceding paragraph referred to *linear* dependence among variables. If the predictors are *non*linear functions of each other, then linear dependence is unlikely to occur. Users of multiple regression, however, should be warned that a condition approaching linear dependence may happen in instances where there are nonlinear functional relations among predictors, as, for example, when both x and x^2 are included as predictors and the mean is large compared to the standard deviation, or when both x and $1/x$ are included. Except in cases of very exact predictions, functional dependence among predictors can lead to puzzling values for the weights. Therefore, one should not include several different functions of the same observed variable in a multiple regression except under carefully considered circumstances. Some of these circumstances are noted in Chapter 9.

We may encounter some terms that are related in meaning to linear dependence of vectors and singularity of matrices. An "ill-conditioned" matrix is one that is singular or nearly so. In the latter case, calculations that employ insufficient accuracy can lead to erroneous solutions for equations. This is a kind of rounding error, and should not be a problem with widely used computer packages unless a warning message has been given. *Colinearity* refers to two vectors that are colinear; that is, they fall in a single dimension. *Multicolinearity* refers rather generally to a set of linearly dependent vectors. These two terms are sometimes used loosely to refer to variables that correlate fairly highly without being linearly dependent. As will be seen in Chapter 9, multicolinearity of predictors leads to increased sampling uncertainty concerning regression weights.

APPENDIX

Derivation of Estimates by Minimizing Σe_i^2

In this section, we will use elementary calculus to derive the least squares estimates of the regression parameters. Those with no experience with differential calculus may prefer to skip this section, although the calculus used is both limited and elementary. We make use of the principle such that when a function is at a maximum its derivative is zero and uses the formula for the derivative of a^2 and of the derivative of the sum of functions of a variable. The method used for the latter two conditions may be unfamiliar to some readers.

The criterion of goodness of fit is Σe_i^2. Our goal is to make this quantity as small as possible through judicious choice of b_1 and b_0. Once the data are in hand, the x and y values are not variables: they are fixed constants. The only thing that affects the errors is the choice of b_1 and b_0, and the process of taking the derivative of the error function may be thought of as symbolically varying b_1 and b_0 and observing the effect on Σe_i^2. Starting from the definition of e_i, we have

$$\Sigma e_i^2 = \Sigma(y_i - b_1 x_i - b_0)^2$$

We wish to take the partial derivative of this function, first with respect to b_1 and then with respect to b_0. There is a theorem which states that the derivative of the sum of a set of functions of a variable is the sum of the derivatives of the separate functions. The sum of squared errors is an example of this; there are n functions of b_1 and b_0 that are summed. If $[u(b)]^2$ is a function of b, its derivative, $d[u(b)]^2/db$, is $2u(b)d[u(b)]/db$. This is $2u(b)$ times its derivative. Consider the first term in the sum above,

$$e_1^2 = (y_1 - b_1x_1 - b_0)^2$$

Taking the partial derivative of e_1^2 with respect to b yields

$$de_1^2/db_1 = 2(y_1 - b_1x_1 - b_0)(-x_1)$$

because the derivative of $u(b)$, the function in parentheses, with respect to b_1, is $-x_1$. A similar expression will result as each term in the sum is differentiated; all that changes is the subscript on x and y. These are then summed to give:

$$d(\Sigma e_i^2)/db_1 = 2\Sigma(y_i - bx_i - b_0)(-x_i)$$

Multiplying out the right-hand side gives $(-x_iy_i + b_1x_i^2 + b_0x_i)$. Summing each term and factoring b_1 and b_0 out of the sums in which they appear (since they are constants with respect to the summation) gives

$$d(\Sigma e_i^2)/db_1 = -2\Sigma x_iy_i + b_1\Sigma x_i^2 + b_0\Sigma x_i$$

If the function is a minimum, this will equal zero.

The same process is carried out with respect to b_0. The derivative of $(y_i - bx_i - b_0)^2$ with respect to b_0 is $2(y - b_1x_i - b_0)(-1)$, since the latter term is the coefficient of b_0. Therefore,

$$d(\Sigma e_i^2)/db_0 = \Sigma 2(y_i - b_1x_i - b_0)(-1)$$

Multiplying this out and summing each term separately results in

$$d(\Sigma e_i^2)/db_1 = -2\Sigma y_i + b2\Sigma x_i + 2nb_0$$

This too is equal to zero when the minimum is reached. Putting the two equations together, setting them equal to zero, and dividing by two results gives

$$0 = -\Sigma y_i + b_1\Sigma x_i + nb_0$$

$$0 = -\Sigma x_iy_i + b_1\Sigma x_i^2 + b_0\Sigma x_i$$

If the term in each equation not involving b or b_0 is moved to the other side of the equal sign, the following results:

$$b_1\Sigma x_i + nb_0 = \Sigma y_i$$

$$b_1\Sigma x_i^2 + b_0\Sigma x_i = \Sigma x_iy_i$$

When the order of these two equations is interchanged, they yield

$$\begin{bmatrix} \Sigma x_i^2 & \Sigma x_i \\ \Sigma x_i & n \end{bmatrix} \begin{bmatrix} b_1 \\ b_0 \end{bmatrix} = \begin{bmatrix} \Sigma x_i y_i \\ \Sigma y_i \end{bmatrix}$$

which is exactly like the scalar form of Equation 6.18.

SUMMARY

This chapter covers a good deal of ground. Because multiple regression serves as a foundation for the remaining chapters, the reader should be familiar with it before moving on. First, there is the basic concept of linear regression in which one variable, y, is predicted from another, x, by means of a linear regression equation. This is appropriate when the data indicates that a linear relationship between x and y is a reasonable description, at least as a first approximation, or when theoretical considerations suggest it. Visual inspection is important in deciding whether linear regression is reasonable, although methods for evaluating this question statistically will be given in chapters 9 and 10.

The fundamental equation in linear relationships between empirical as opposed to mathematical variables is

$$y_i = b_1 x_i + b_0 + e_i$$

where b_1 represents the slope of the line relating y to x, b_0 is the intercept, and e_i is the amount, positive or negative, by which the ith value of y deviates from the line.

The presence of error means that there is some choice in deciding just which line one will use in predicting y—that is, the values of b_1 and b_0. The traditional way of making this choice is to select those values for the slope and intercept that minimize Σe_i^2. This is called the *least squares* principle. It leads to the following values for b_1 and b_0:

$$b_1 = s_{xy}/s_x^2$$

$$b_0 = \bar{y} - b_1 \bar{x}$$

If X_+ is a two-column matrix in which the second column consists entirely of ones and the first contains the values of x, and y is a column matrix of the corresponding y values, then a matrix formulation of the solution for the parameters is

$$\mathbf{b}_+ = (\mathbf{X}_+'\mathbf{X}_+)^{-1}\,\mathbf{X}'\mathbf{y}$$

The least squares solution guarantees that the errors are uncorrelated with x and that the mean error is zero. If there is more than one predictor, and these principles continue to hold, Equation 6.19 is a general solution, requiring the addition of columns to X_+. In this case, we have *multiple regression,* and the column matrix $\mathbf{b}_+$ contains $p + 1$ elements $b_1, b_2, \ldots, b_p, b_0$. The slope may also be found from the covariance matrices as follows:

$$\mathbf{b} = \mathrm{S}_{xx}^{-1}\,\mathbf{s}_{xy}$$

If the variables are in standard score form, then the intercept is zero, and in the single predictor case the slope of the regression is r_{xy}. In general, the standard score regression weights b_j may be obtained from

$$\mathbf{b}^* = \mathrm{R}_{xx}^{-1}\,\mathbf{r}_{xy}$$

or from the raw score regression weights where

$$b_j^* = b_j\,(s_j/s_y)$$

There are a number of useful relationships involving the weights and the variances, covariances, and correlation of the predictions and the errors. The total variance in y is the sum of the variance of the predictors and the error variance:

$$s_y^2 = s_{\hat{y}}^2 + s_e^2$$

These two components are sometimes referred to as variance due to prediction and variance due to deviations, respectively.

Solution b, and the principles involving the variance of weighted sums, yield the following two expressions for $s_{\hat{y}}^2$:

$$s_{\hat{y}}^2 = \mathbf{b}'\mathrm{S}_{xx}\mathbf{b}$$

and

$$s_{\hat{y}}^2 = \mathbf{b}'\mathbf{s}_{xy}$$

where $\mathbf{b}'\mathbf{s}_{xy}$ is also an expression for the covariance of y with $\hat{y}$.

The squared multiple correlation R^2 of y with the best combination of x's is the ratio $s_{\hat{y}}^2/s_y^2$ (see Equation 6.47). It may be found in the following ways:

$$R^2 = \mathbf{b}'\mathbf{s}_{xy}/s_y^2$$

or

$$R^2 = \mathbf{b}^{*\prime}\mathbf{r}_{xy}$$

or

$$R^2 = \mathbf{b}^{*\prime}\mathrm{R}_{xx}\mathbf{b}^*$$

The standard error of estimate s_e depends on R^2 and s_y:

$$s_e = s_y \sqrt{1 - R^2}$$

In the case of multiple predictors, the interpretation of the weights may not be simple. One way to interpret them is simply as the constant terms in a linear functional equation that relates y to the x's. A second way to interpret the weights is as defining the direction in the multivariate space of the predictors which gives maximum correlation with y. A third interpretation may be based on the angular separation among vectors.

To some extent the size of the standard score regression coefficient reflects the importance of the corresponding variable in prediction, especially when the predictors are uncorrelated (orthogonal). In general, however, the relationship between the importance of a variable and its regression weight is more complex and involves consideration of the simple correlation of the variable with the dependent variable and of the decrease in R^2 which occurs when that single variable is left out of the equation.

PROBLEMS

Below is a small X_+ matrix, including the constant column of ones, and the y vector.

$$\underset{\mathrm{X}_+}{\begin{bmatrix} 5 & 10 & 1 \\ 3 & 0 & 1 \\ 1 & 4 & 1 \\ 1 & 3 & 1 \\ 0 & 8 & 1 \end{bmatrix}} \quad \underset{y}{\begin{bmatrix} 5 \\ 5 \\ 3 \\ 1 \\ 1 \end{bmatrix}}$$

1. Compute $\mathrm{X}'_+\mathrm{X}_+$ and X'_+y.

2. Select the appropriate parts of $\mathrm{X}'_+\mathrm{X}_+$ and X'_+y to compute the separate *univariate* regression equations for predicting y from x_1, and then from x_2.

3. Using the inverse of $\mathrm{X}'_+\mathrm{X}_+$ given below compute the bivariate regression equation. Then, verify that it is the inverse.

$$\underset{(\mathrm{X}'_+\mathrm{X}_+)^{-1}}{\begin{bmatrix} .06564 & -.00718 & -.09538 \\ & .01641 & -0.6769 \\ & & .72923 \end{bmatrix}}$$

4. Find the covariance matrices S_{xx} and $\mathbf{s}_{xy}$ and use them to compute the regression coefficients. Verify that they are the same as in Problem 3.

5. Use the appropriate equation to compute the intercept from the variables' means and the regression coefficients. Compare your answer to b_0 from Problem 3.

6. Compute $\hat{y}$ and e from the results of Problem 3.

7. Compute $s_{\hat{y}}^2$ and verify that it equals $\mathbf{b}'\mathbf{s}_{xy}$ and $\mathbf{b}'\mathrm{S}_{xx}\mathbf{b}$.

8. Verify that $\mathbf{e}'\mathrm{X} = 0$ and that $s_e^2 = s_y^2 - s_{\hat{y}}^2$.

9. Convert the variables to standard score form. Obtain R_{xx} and $\mathbf{r}_{xy}$ using any method.

10. Find $\mathbf{b}^*$ from R_{xx} and $\mathbf{r}_{xy}$ and use it to get $\hat{z}\hat{y}$.

11. Verify from the results of previous problems that $\mathrm{R}_{y.12}^2 = (\mathbf{b}'\mathrm{S}_{xx}\mathbf{b})/s_y^2 = \hat{\mathbf{z}}'\hat{\mathbf{z}}(n-1)^{-1}$. Then verify that $(\mathbf{y} - \hat{\mathbf{y}})'(\mathbf{y} - \hat{\mathbf{y}})(n-1)^{-1} = s_y^2(1 - \mathrm{R}_{y.12}^2)$, and that $\hat{\mathbf{y}}'\hat{\mathbf{y}}(n-1)^{-1} = \mathrm{R}_{y.12}^2\, s_y^2$.

12. Use the following R_{xx} and the three values of $\mathbf{r}_{xy}$ for three different y's to get the appropriate b^*'s. Note the effect of the different values for $\mathbf{r}_{2y}$ on b^*, $\mathrm{R}_{y.12}^2$, and $\mathrm{R}_{y.12}$.

$$
\begin{array}{c}
\\ x_1 \\ x_2 \\ \\
\end{array}
\begin{array}{c}
\begin{array}{cc} x_1 & x_2 \end{array} \\
\begin{bmatrix} 1.00 & .60 \\ .60 & 1.00 \end{bmatrix} \\
\mathrm{R}_{xx}
\end{array}
\quad
\begin{array}{c}
y_1 \\
\begin{bmatrix} .56 \\ .40 \end{bmatrix} \\
\mathbf{r}_{xy}
\end{array}
\quad
\begin{array}{c}
y_2 \\
\begin{bmatrix} .56 \\ .24 \end{bmatrix} \\
\mathbf{r}_{xy}
\end{array}
\quad
\begin{array}{c}
y_3 \\
\begin{bmatrix} .56 \\ .56 \end{bmatrix} \\
\mathbf{r}_{xy}
\end{array}
$$

13. Use the following R_{xx} with the $\mathbf{r}_{xy}$ in Problem 12 to obtain three new sets of $\mathbf{b}^*$. Note the differences between these and the results from Problem 12, and the differences in the multiple R's.

$$
\begin{array}{c}
\\ x_1 \\ x_2 \\ \\
\end{array}
\begin{array}{c}
\begin{array}{cc} x_1 & x_2 \end{array} \\
\begin{bmatrix} 1.00 & .28 \\ .28 & 1.00 \end{bmatrix} \\
\mathrm{R}_{xx}
\end{array}
$$

14. Below are R_{xx}, $\mathbf{r}_{xy}$, and R_{xx}^{-1} for three predictors. Use them to find $\mathbf{b}^*$, R, and R^2.

$$
\underset{\mathrm{R}_{xx}}{\begin{bmatrix} 1.00 & .60 & .28 \\ & 1.00 & .28 \\ & & 1.00 \end{bmatrix}}
\quad
\underset{\mathbf{r}_{xy}}{\begin{bmatrix} .56 \\ .56 \\ .40 \end{bmatrix}}
\quad
\underset{\mathrm{R}_{xx}^{-1}}{\begin{bmatrix} 1.5965 & -.9035 & -.1940 \\ -.9035 & 1.5965 & -.1940 \\ -.1940 & -.1940 & 1.1086 \end{bmatrix}}
$$

15. Find the proportions of variance uniquely contributed by each variable. (Note that you can use the results of problems 13 and 14 to aid in this.)

16. Suppose variables 2 and 3 in Problem 14 are the same as those in Problem 13. Comment on the effects of the inclusion of variable 1 on the regression coefficients for variables 2 and 3.

17. If equal standard score weights were used in Problem 14, what correlation with y would result?

ANSWERS

1. $\begin{bmatrix} 36 & 57 & 10 \\ 57 & 189 & 25 \\ 10 & 25 & 5 \end{bmatrix}$ $\begin{bmatrix} 44 \\ 73 \\ 15 \end{bmatrix}$

X'_+X_+ $\quad$ X'_+y

2. $\begin{bmatrix} .0625 & -.125 \\ -.125 & .450 \end{bmatrix}$ $\begin{bmatrix} .875 \\ 1.250 \end{bmatrix}$ $\begin{bmatrix} .0156 & -.0781 \\ -.0781 & .5906 \end{bmatrix}$ $\begin{bmatrix} -.03125 \\ 3.156 \end{bmatrix}$

$(X_1'X_1)^{-1}$ $\quad$ $\mathbf{b}_1$ $\quad$ $(X_2'X_2)^{-1}$ $\quad$ $(X_2'X_2)^{-1}X_2'\mathbf{y} = \mathbf{b}_2$

3. $b' = [.9333 \quad -.1333 \quad 1.800]$

4. $\begin{bmatrix} 4.00 & 1.75 \\ 1.75 & 16.00 \end{bmatrix}$ $\begin{bmatrix} 3.5 \\ -.5 \end{bmatrix}$ $\begin{bmatrix} .2626 & -.0287 \\ -.0287 & .06564 \end{bmatrix}$ $\begin{bmatrix} .9333 \\ -.1333 \end{bmatrix}$

S_{xx} $\quad$ $\mathbf{s}_{xy}$ $\quad$ S_{xx}^{-1} $\quad$ $\mathbf{b}$ (Same as from Problem 3)

5. $b_0 = 3.0 - (.9333)(2.0) + (.1333)(5.0) = 1.80$. Same as Problem 3.

6. $\hat{\mathbf{y}}' = [5.133 \quad 4.600 \quad .200 \quad 2.333 \quad .733]$
$\mathbf{e}' = [-.133 \quad .400 \quad .800 \quad -1.333 \quad .267]$

7. $s_{\hat{y}}^2 = 3.333$ from all three sources.

8. $X'\mathbf{e} = [.002 \quad .002 \quad .001]$ (Note rounding error.)
$s_e^2 = .6667$; $s_{\hat{y}}^2 = 4.000 = 3.333 + .667$.

9. $R_{xx} = \begin{bmatrix} 1.000 & .219 \\ .219 & 1.000 \end{bmatrix}$ $\quad \mathbf{r}_{xy} = \begin{bmatrix} .8750 \\ -.0625 \end{bmatrix}$

10. $b^{*\prime} = [.9333 \quad -.2667]$; $\hat{\mathbf{z}}_y = [1.067 \quad .800 \quad .400 \quad -.333 \quad -1.133]$

11. $R_{y.12}^2 = \mathbf{b}^{*\prime}\mathbf{r}_{xy} = \mathbf{b}^{*\prime}R_{xx}b^* = .8333$; $3.333/4.0 = .8333$ $s_{\hat{z}_y}^2 = 3.333/4 = .8333$; $s_e^2 = (1 - R^2)$ $s_y^2 = .667$. From Problem 7, $s_{\hat{y}}^2 = 3.333$ and $R_{y.12}^2 s_y^2 = (.8333)(4.0)$.

12. Criterion variable

	y_1	y_2	y_3
b_a^*	.500	.650	.350
b_b^*	.100	−.150	.350
R^2	.320	.328	.392
R	.566	.573	.626

Note that $b_1 - b_2$ is consistent with $r_{x1} - r_{y1}$. It is largest where both validities are larger in set d, predictor b is a suppressor.

13. Criterion variable

	y_1	y_2	y_3
b_1^*	.486	.535	.438
b_2^*	.264	.090	.438
R^2	.378	.321	.490
R	.615	.567	.700

Multiple R's tend to be higher, except for x_2, where there is now no suppressor variable.

14. $\mathbf{b}^{*\prime} = [.310 \quad .310 \quad .226]$; $R^2 = .438$; $R = .662$.

15. Variables 1 and 2 are both necessary for $.438 - .378 = .060$. Variable 3 is necessary for $.438 - .392 = .046$.

16. Regression weight for x_2 is substantially reduced.

17. $r^2 = .434$

READINGS

Green, P. E., and J. D. Carroll (1976). *Mathematical tools for applied multivariate analysis.* New York: Academic Press, pp. 259–270.

Kerlinger, F. N., and E. J. Pedahzur (1973). *Multiple regression in behavioral research.* New York: Holt, Rinehart, and Winston, pp. 11–80.

Lindeman, R. H., P. F. Merenda, and R. Z. Gold (1980). *Introduction to bivariate and multivariate analysis.* Glenview, IL: Scott, Foresman, pp. 93–110.

Tatsuoka, M. M. (1970). *Discriminant analysis: The study of group differences.* Champaign, IL: Institute for Ability and Personality Testing, pp. 1–32.

van de Geer, J. F. (1971). *Introduction to multivariate analysis for the social sciences.* San Francisco: Freeman, pp. 93–106.

7

Partial Correlation and Stepwise Regression

PARTIAL CORRELATION

Statistical Control of Variables

When two variables are correlated, the interpretation is notoriously ambiguous. What accounts for the correlation? Is the relationship between the variables a direct one, or is it due to the influence of another variable? Partial correlation analysis attempts to resolve some of the ambiguities in interpretation. For example, suppose we are interpreting educational data of a nonexperimental kind. We have data from 100 high schools consisting of the average mathematics achievement score in each high school, and the average per-pupil expenditure in the school. (Note that the school, not the individual student, is the observational unit.) A significant correlation is found between the two variables, tending to support the position that increased expenditures lead to higher achievement.

It could be argued that the observed relationship is due to the dependence of both variables on another, one having to do with the characteristics of the families from which the students come. That is, better educated families tend to live in more affluent districts and to have higher achieving children. By this reasoning, the relationship between achievement and expenditure is spurious, or at best, derivative. It is a result of both variables being correlated with another, the parents' educational level. The accepted method of testing this alternative explanation is to find the average parents' educational level for each school and "partial it out"—that is, compute the correlation between achievement and expenditure while removing the influence of parents' education on each. The question is, do expenditures and achievements still correlate when this is done?

Using the same three variables, we might want to analyze the correlations differently. Observing the correlation between parents' education and achievement, we might wonder whether *this* can be accounted for in terms of the relationship of both variables to expenditure levels. We would partial out the latter from the first two to test this possibility. We see that which variable is partialed out depends on which variable is considered to be the primary relationship and which is the alternative explanation.

So, one major use for partial correlation analysis is to test whether an observed correlation between two variables disappears if the influence of a third is taken into account. When one variable is partialed out, the resulting coefficient is called a *first-order* partial. Any number of variables could be partialed out, in which case the result is a second-, third-, or pth-order partial correlation. In fact, sometimes the observed, direct correlation between two variables is called a *zero-order* correlation. This does not mean the correlation is close to zero; it simply means that no variables have been partialed out.

In the preceding types of cases, we would expect that the partialed variable—hereafter called a control variable—would reduce or even eliminate the apparent relationship between two variables. In a second type of situation, it may be expected to *enhance* the relationship. Suppose depressive patients are treated with different dosages (0, 1, 2, or 3 milligrams) of a drug to improve their moods. Observers rate the cheerfulness of the patients during the period after administration. Then the regression of cheerfulness on dosage is determined, along with the correlation between the two variables, according to the methods used in Chapter 6.

Suppose a moderate correlation is found. This is encouraging, but in all likelihood there is also some relationship between a patient's mood before drug administration and subsequent to it. To some degree, this may mask the effectiveness of the drug. Partialing out pretreatment cheerfulness will reveal this, in the sense that the partial correlation between post-treatment cheerfulness and drug dosage, using pretreatment cheerfulness as the control variable, will be higher than the zero-order correlation between them. This latter example is also called an analysis of covariance, which is discussed in Chapter 12, but the ideas are the same.

Thus partial correlation analysis may be used to determine the relationship between two variables after the influence of one or more other variables has been taken into account. This may be either for the purpose of evaluating alternative explanations for a correlation involving a third variable, or to enhance the correlation between two variables by removing a source of variance that is extraneous to the primary question.

Procedures in Partial Correlation Analysis

Having learned what partial correlation analysis is supposed to do, we will now consider how it is done. The effect of partial correlation analysis is somewhat more limited than might be assumed from verbal descriptions of it, although it is indeed a useful procedure. Moreover, it is related to a number of aspects of multiple regression analysis, factor analysis, and other multivariate methods.

The description of partial correlation can be simplified if we assume that all the variables have been converted to standard scores; this will not affect any correlations. A partial correlation is nothing more than a correlation between two error- or deviation-from-regression scores. That is, if we want the partial correlation between z_2 and z_y with z_1 partialed out, we can obtain it in the following way. First, find the regression coefficient for estimating z_y from z_1 (the control variable), and calculate the error scores $e_y = z_y - b^*_{1y}z_1$. Second, find the regression of z_2 and z_1. Now calculate the deviations from the following regression: $e_2 = z_2 - b^*_{12}z_1$. Note that subscripts have been added to the slope coefficients in order to indicate which regression is being considered. Finally, compute the correlation between the deviations from the two regression lines, e_y and e_2. This is the partial correlation between y and x_2:

> A partial correlation between two variables is the correlation between those parts of the variables that are not predictable from the control variable.

Under special circumstances it may have some additional special properties, but in general, this is all a partial correlation represents.

More than one variable can be partialed out. Suppose we want to know the relationship between sons' IQs and sons' incomes by partialing out sons' educations and fathers' incomes. This is just an extension of the same process. Either variable may be removed first. When sons' educations are partialed out, we obtain the deviations from regression of sons' incomes on sons' educations and the deviations from regression of sons' IQs on sons' educations. The correlation between these two deviation scores would be the first-order partial. Next, we find the regression of each of these *error* scores on the second control variable, fathers' incomes. This gives a second-order deviation score: those parts of sons' incomes and sons' IQs that are not predictable from either sons' educations or fathers' incomes. The correlation between these two error scores is the second-order partial correlation between the two variables. Third order partials can be constructed by regressing the second-order errors on a third control variable, and so on, for as many control variables as there are. The order in which the control variables are removed does not matter; the final correlation is the same. This is because of the properties of linear combinations in which the coefficients have been derived by the least-squares principle.

If the preceding discussion has been assimilated, one further level of complication can be introduced: multiple predictors of y and multiple control variables. For example, suppose we want the *multiple* correlation of y with x_3, x_4, and x_5 after x_1 and x_2 are partialed out of all four; this would show how well y can be predicted by x_3, x_4, and x_5 once the relationship with x_1 and x_2 is removed. All that needs to be done is to find the deviations from multiple regression of the dependent variable on the two control variables and then the deviation from regression of each of the three predictors estimated from the two control variables. This gives e_y, e_3, e_4, and e_5. Then we find the multiple correlation of e_y with the other variables. The important concept is that we have a dependent variable, a set of independent variables, and a set of control variables. We transform the former in such a way as to make the variables all uncorrelated with the control variables. Then the deviations on the dependent variable may be predicted using the deviations on the independent variable.

The notation used to specify the nature of a partial correlation involves inserting a dot in a bevy of subscripts on a correlation. For example,

$$r_{12\cdot3},\ r_{34\cdot12},\ r_{y5\cdot1234}$$

The subscripts to the right of the dot indicate the variables partialed out, whereas those to the left indicate the two variables to be correlated. Thus, the first example is the partial correlation of variables 1 and 2, controlling for 3; the second is the (second-order) partial of variables 3 and 4, controlling for 1 and 2;

and the third is the fourth-order partial of y and predictor 5, controlling for variables 1 through 4. Complex multiple correlations are expressed in this way also. Here, the first subscript denotes the dependent variable. The ones that follow it are the predictors, and the partialed variables are again indicated to the right of the dot. So the expression $R_{y12 \cdot 345}$ is the multiple correlation of y with x_1 and x_2 after the association with x_3, x_4, and x_5 is partialed out.

Occasionally, partial correlations may be misinterpretated. This confusion arises over the use of the term "controlling for," and "control" variable, or even over the expression "correlation of y with x_2, holding x_1 constant." Except under special circumstances, the control variable is neither controlled nor held constant. Only deviations from regression are computed, and the correlation between the deviations is determined. This is the same as "holding the control variable constant" only when *all* the regressions among the variables are *exactly linear.* Because this rarely occurs for most data, partialing is not the same as holding constant. In the real world, the relationship between y and x_2 is somewhat different at different values of x_1. Imagine, for example, the swarm of points in a clear plastic box viewed from the side. Now imagine slicing the box vertically, so that the slices are removable. Holding x_2 constant and observing the relationship between x_1 and y means pulling out one of these slices, laying it down on the table, and examining the swarm of points trapped in the slice. Do the same thing with another slice, and so on. If the distribution of points is purely linear, then the regression of y on x_2 revealed in each slice will be the same, and the within-slice correlation will always be the partial correlation $r_{y2 \cdot 1}$. For real data, this will not exactly be the case; the regressions and correlations in different slices will be different, although perhaps only slightly. The computed partial correlation represents a weighted average of these correlations across the slices.

Partial Correlations and Vectors

It may be useful to visualize partial correlations in terms of vectors. For example, suppose there are three variables, x_1, x_2, and y in standard score form so that the lengths of the corresponding vectors are equal. Suppose all three correlations among them are .50. The correlations represent the cosines of the angle between the corresponding vectors. A trigonometric table will tell us that the angle whose cosine is .50 is 60°. We may therefore visualize the vectors as iron rods welded together at their ends. Viewed at right angles to the plane they define any two vectors that meet in a "*v*," with the two arms at 60°, and the third rod foreshortened, as shown below.

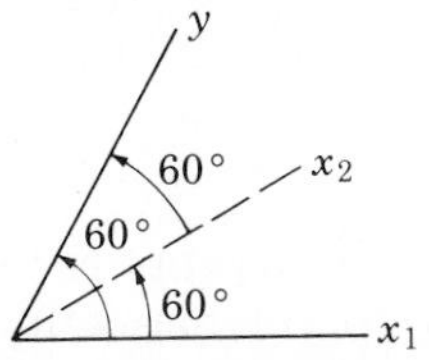

In partial correlation, we look at the angle between two vectors *orthogonal to a third.* It is as if we observe the angle between two vectors as seen from the direction in which the third points. The partial correlation is the cosine of this angle. Imagine the three rods, and the angle between two of them seen while sighting along the third angle. The angle will undoubtedly be different from the preceding angle. In fact, it will look like the figure below, in which the circle at the bottom left represents the foreshortened view of the control variable when we sight down it.

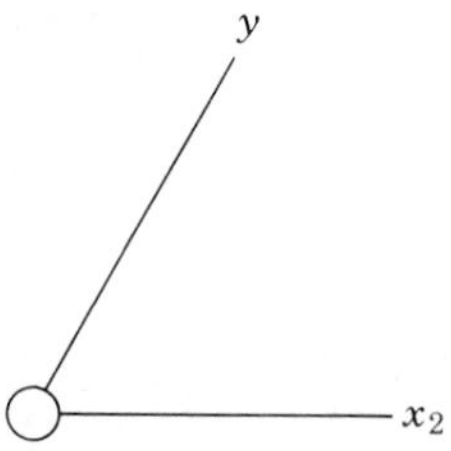

The primary effect is the change in angle; how much it changes depends on the three original angles. There is also an effect on the lengths of the two legs: they are shortened only a little if their angle with the control vector is close to 90°, otherwise they are shortened more. (Remember that negative correlations correspond to angles of more than 90°.) The smaller the angle of y and x_2 with the control variable, the greater the effect. For example, if these two angles were both 45° instead of 60, and the angle between x_1 and x_2 were still 60°, the following would result:

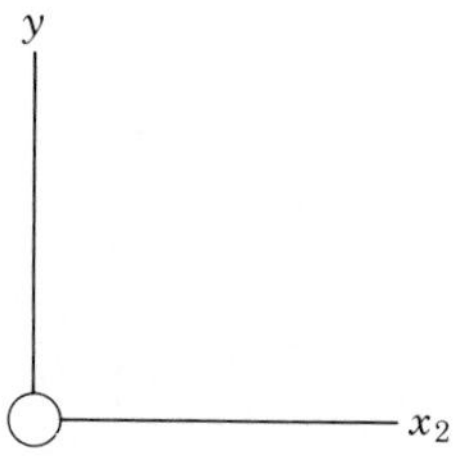

Viewed from the x_1 direction, y and x_2 are orthogonal.

Equations for Partial Correlation

The geometrical and numerical concepts used above may be translated into equations. For example, the correlation between y and x_2 with x_1 partialed out is

$$r_{y2\cdot 1} = \frac{r_{y2} - r_{12}r_{1y}}{\sqrt{(1 - r_{12}^2)(1 - r_{1y}^2)}} \tag{7.1}$$

The origin of this equation is based on the fact that $r_{2y \cdot 1}$ is the correlation between e_y and e_2 and that the correlation equation expresses r as the ratio of the covariance to the product of the two standard deviations. For simplicity, we assume the variables are in standard score form. Then, $b^*_{1y} = r_{y1}$ and $b^*_{12} = r_{21}$, and the covariance between e_y and e_2 is

$$\frac{\Sigma e_y e_2}{n-1} = \frac{\Sigma(z_y - r_{1y}z_1)(z_2 - r_{12}z_1)}{n-1}$$

Multiplying the equation out gives

$$\frac{\Sigma e_y e_2}{n-1} = \frac{\Sigma z_y z_2 - \Sigma r_{1y} z_1 z_2 - \Sigma r_{12} z_y z_1 + \Sigma r_{1y} r_{12} z_1^2}{n-1}$$

Because the correlations are constants, they may be put to the left of the summation signs. We can put each term over the denominator separately:

$$\frac{\Sigma e_y e_2}{n-1} = \frac{\Sigma z_y z_2}{n-1} - \frac{r_{1y}\Sigma z_1 z_2}{n-1} - \frac{r_{12}\Sigma z_y z_1}{n-1} + \frac{r_{1y} r_{12} \Sigma z_1^2}{n-1}$$

Because the product terms are all correlations, and $\Sigma z_1^2/(n-1) = 1.0$, then

$$\frac{\Sigma e_y e_2}{n-1} = r_{2y} - r_{1y}r_{12} - r_{12}r_{1y} + r_{1y}r_{12} = r_{2y} - r_{1y}r_{12}$$

The standard deviations of e_y and e_2 are $\sqrt{1 - r_{1y}^2}$ and $\sqrt{1 - r_{12}^2}$, respectively. Dividing by these deviations gives Equation 7.1, the correlation between the deviation scores.

Note that the partial correlation depends first on the original zero-order correlation between the two variables. Subtracted from this is a correction for their correlations with the control variables. (Note what happens if one of these correlations is opposite in sign to r_{y2}.) The denominator terms are corrections for the reduced variance of x_2 and y, that is, the reduced vector lengths.

This equation can be translated into a matrix formulation involving linear combinations. Remember that the correlation is the regression weight if the variables are in standard score form. Suppose Z is an $n \times 3$ matrix of the standard scores z_1, z_2, and z_y; $\mathbf{e}_2$ is a column vector of deviations of scores on z_2 from their regression on z_1; and $\mathbf{e}_y$ is a vector of deviations of scores on z_y from its regression on z_1. To formulate the latter two in matrix terminology, we define two column vectors of weights: $\mathbf{w}_2' = (-r_{12}, 1.0, 0)$ and $\mathbf{w}_y' = (-r_{1y}, 0, 1.0)$. Then

$$\mathbf{e}_2 = \mathbf{Z}\mathbf{w}_2$$

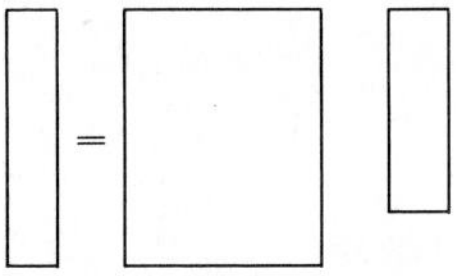

and

$$\mathbf{e}_y = \mathbf{Z}\mathbf{w}_y$$

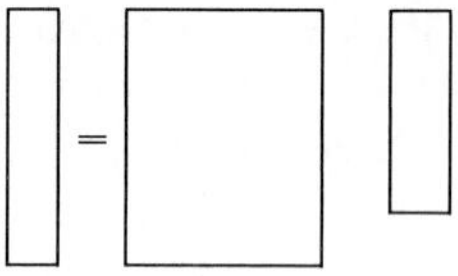

The covariance between e_2 and e_y will be $\mathbf{e}_y'\mathbf{e}_2/(n-1)$. Substitution shows that this will be $\mathbf{w}_y'\mathrm{R}\mathbf{w}_2$. Similarly, the variances of the two error scores will be $\mathbf{w}_2'\mathrm{R}\mathbf{w}_2$ and $\mathbf{w}_y'\mathrm{R}\mathbf{w}_y$, respectively. Because the correlation is the covariance divided by the product of the two standard deviations, we have the matrix formulation

$$r_{2y\cdot1} = \frac{\mathbf{w}_2'\mathrm{R}\mathbf{w}_y}{\sqrt{(\mathbf{w}_2'\mathrm{R}\mathbf{w}_2)\,(\mathbf{w}_y'\mathrm{R}\mathbf{w}_y)}} \tag{7.2}$$

Equation 7.2 holds for any number of predictors provided the appropriate definitions are made of $\mathbf{w}_2$ and $\mathbf{w}_y$. If there are c control variables, $\mathbf{w}_2$ consists of the c standard score regression coefficients for predicting z_2 from the control variables (with reversed signs), a 1.0 weight for z_2, and a zero for z_y. Correspondingly, $\mathbf{w}_y$ consists of the c standard score regression coefficients for predicting z_y from the control variables, a zero weight for z_2, and a 1.0 for z_y. In the single control variable case, the regression weights are the simple correlations of the variables with x_2 or y.

More generally, c control variables are partialed out of a predictor variable x_2 and the dependent variable y, $\mathbf{w}_j = (-\mathbf{b}_j^*, 1.0, 0)$ and $\mathbf{w}_y = (-\mathbf{b}_y^*, 0, 1.0)$. Here, $\mathbf{b}_j^* = \mathrm{R}_{cc}^{-1} r_{cj}$ and $\mathbf{b}_y^* = \mathrm{R}_{cc}^{-1}\mathbf{r}_{cy}$, where R_{cc} is the matrix of intercorrelations among the control variables, and $\mathbf{r}_{cj}$ and $\mathbf{r}_{cy}$ are the vectors of correlations of the control variables with x_j and y, respectively. In more general cases, the partial correlation represents the correlations between deviations from multiple regressions. We can go further and think of partialing the control variables out of a whole collection of other variables. The weight vectors that define the residual variables all have essentially this same form, even though their entries differ. All that is needed is to include zeros in the vector for each variable that is not partialed out.

Sometimes we see the terms "part" correlations or "semi-partial" correlations. These are closely related to partial correlations, but they are not the same. They refer to the situation in which the control variables are partialed out of only one variable, ordinarily the independent variable. That is, it is the correlation between e_2 and y, rather than between e_2 and e_y. Note that if the partial correlation is zero, so is the part correlation. In fact, the only difference between part and partial is in the denominator, where only the variance of x_2 is reduced. Therefore the part correlation is always lower in absolute value than the partial correlation. The equation for the three-variable case is

$$p_{2y\cdot1} = \frac{r_{2y} - r_{1y}r_{12}}{\sqrt{1 - r_{12}^2}}$$

We have explained the partialing process as if we had quite literally created new variables whose regressions were then examined. It is not necessary to carry out these operations in practice, since partial correlations can be computed from the variances and covariances among the variables involved.

It may be useful to think in terms of squared part and partial correlations as proportions of shared variance. A squared *part* correlation is the proportion of the variance of y that is shared with x_2, but not x_1. The squared partial correlation, on the other hand, takes into consideration the variance lost by both variables in the overlap with x_1. It is the proportion of the *remaining* variance of y that is shared with x_2. In the diagram below, the partial correlation $r_{y2 \cdot 1}$ is the ratio of the shaded area to the total remaining area of the circle, whereas the part correlation is the ratio of the shaded area to the original area of the whole circle, indicated by the dotted lines.

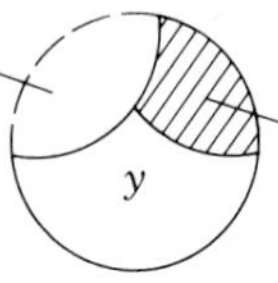

These principles continue to hold as more variables are partialed out. Indeed, one can think of "x_1" as a composite variable made up of all the variables partialed out so far and "x_2" as any of the variables not yet partialed.

Calculations of First-Order Partials

Suppose we have the following intercorrelations among three variables and want to find the correlation of x_2 with y, partialing out x_1:

	1	*2*	*y*
1	1.00	.60	.80
2		1.00	.50
y			1.00

Then,

$$r_{2y \cdot 1} = \frac{.50 - (.60)(.80)}{\sqrt{(1 - .60^2)(1 - .80^2)}} = \frac{.50 - .48}{\sqrt{(.64)(.36)}} = \frac{.02}{.48} = .0417$$

Therefore, in this case a substantial correlation is almost reduced to zero. In the following example, the correlation is actually increased by partialing:

	1	*2*	*y*
1	1.00	.20	.40
2		1.00	.90
y			1.00

$$r_{1y \cdot 2} = \frac{.40 - (.20)(.90)}{\sqrt{(1 - .20^2)(1 - .90^2)}} = \frac{.40 - .18}{\sqrt{(.96)(.19)}} = \frac{.22}{.427} = .515$$

The difference between the two cases can be understood in terms of the shared variance concept. In the first, most of the variance shared between x_2 and y is also shared with x_1. In the second, y shares so much variance with x_1 that the small overall proportion shared with x_2 is a good part of the remainder.

PARTIAL COVARIANCE ANALYSIS

The partialing process can be carried out on the correlation or covariance matrix. This procedure is employed in advanced multiple regression and in factor analysis, so it helps to have some familiarity with it.

Suppose we have four variables in standard score form; their means are zero, their variances 1.0, and their covariances are their intercorrelations. As already noted, the regression coefficient for predicting one variable from another is, in this case, the correlation between the two. The correlation matrix is shown in the upper left panel of Table 7-1.

If z_1 is to be partialed out first, the regression coefficients of the other three variables would be .6, .5, and .4, respectively. Suppose we were to compute the covariances among the resulting deviation scores. For example, the partial covariance of z_2 and z_3 would be

$$\Sigma(z_2 - r_{12}z_1)(z_3 - r_{13}z_1)/(n - 1) = r_{23} - r_{13}r_{12}$$

which is $.700 - .600 \times .500 = .400$. In a similar way, we can compute the partial covariances of z_2 with z_4 and z_3 to be .240 and .500, respectively. The variances of the residual variables, that is, of the deviations from regression on z_1, will be $1 - r_{1j}^2$, or .64, .75, and .84, respectively.

We could also partial z_1 out of itself, although this may seem rather useless. We could calculate the regression of z_1 on z_1, find the slope to be 1.0 and the intercept 0, painfully calculate all the deviations from the regression line, myopically not notice that they are all zero, doggedly compute their variance and their covariances with the other deviations, which are all zero too. The only immediate

Table 7-1 Steps in Partial Covariance Analysis

	Data 1	2	3	4		*Step 1* 1	2	3	4
1	1.00	.60	.50	.40	1	0	0	0	0
2	.60	1.00	.70	.48	2	0	.64	.40	.24
3	.50	.70	1.00	.70	3	0	.40	.75	.50
4	.40	.48	.70	1.00	4	0	.24	.50	.84
	Step 2 1	2	3	4		*Step 3* 1	2	3	4
1	0	0	0	0	1	0	0	0	0
2	0	0	0	0	2	0	0	0	0
3	0	0	.50	.35	3	0	0	0	0
4	0	0	.35	.75	4	0	0	0	.505

benefit of this is that it allows us to keep a complete matrix of partial variances and covariances, as in the upper right panel of Table 7-1.

Table 7-1 is the variance-covariance matrix for deviations from regression on z_1, including those for z_1 on itself. These are the variances and covariances among the variables when the influence of z_1 is removed. Note that these deviation scores are no longer standard scores. The zeros in the first row and column remind us that there is now nothing left of z_1. We will call the entries in this section first-order partial variances and covariances, and symbolize the covariances as $s_{jk\cdot 1}$ and the variances as $s^2_{j\cdot 1}$, adding more subscripts to the right of the dot later on to denote the partialing of other variables.

Now we will partial out the residual portion of z_2 from the other two variables and the covariance and variances of these as well. The variance of the predictor is no longer 1.0, so we must remember that a regression coefficient is the covariance divided by the predictor's variance. Here, the coefficients are $s_{23\cdot 1}/s^2_{2\cdot 1}$ and $s_{24\cdot 1}/s^2_{2\cdot 1}$, respectively, and the deviations from the predictions are $z_{3\cdot 1} - (s_{23\cdot 1}/s^2_{2\cdot 1})z_{2\cdot 1}$ and $z_{4\cdot 1} - (s_{24\cdot 1}/s^2_{2\cdot 1})z_{2\cdot 1}$, respectively. These deviations will be denoted $z_{3\cdot 12}$ and $z_{4\cdot 12}$. After some algebraic simplifciation, we find that their covariance, $s_{34\cdot 12}$, is

$$s_{34\cdot 1} - (s_{23\cdot 1}) \times (s_{24\cdot 1})/s^2_{2\cdot 1}$$

The variances of these second-order deviations are

$$s^2_{3\cdot 1} - (s_{23\cdot 1})^2/s^2_{2\cdot 1}$$

and

$$s^2_{4\cdot 1} - (s_{23\cdot 1})^2/s^2_{2\cdot 1}$$

Numerically, these are

$$s_{34\cdot 12} = .500 - .400 \times .240/.640 = .500 - .150 = .350$$

$$s^2_{3\cdot 12} = .750 - (.400)^2/.640 = .750 - .250 = .500$$

and

$$s^2_{4\cdot 12} = .840 - (.240)^2/.640 = .840 - .090 = .7500$$

Partialing $z_{2\cdot 1}$ out of itself leaves nothing, making its variance and covariances reduce to zero. The second-order partial variance-covariance matrix is shown in the lower left panel of Table 7-1. The nonzero entries that remain are the variances of the deviations of the two remaining variables from their respective multiple regressions on z_1 *and* z_2 and the covariance between the two.

The next to last step is the partialing of $z_{3\cdot 12}$ out of $z_{4\cdot 12}$. Again, the regression coefficient is the partial covariance over the partial variance of the predictor—in this case, .350/.500 = .700. The final deviation variable is

$$z_{4\cdot 123} = z_{4\cdot 12} - (s_{34\cdot 12}/s^2_{3\cdot 12})\, z_{3\cdot 12}$$

Because there are no other variables, there are no covariances, but the variance, $s^2_{4\cdot 123}$, is

$$s^2_{4\cdot 12} - (s_{34\cdot 12})^2/s^2_{3\cdot 12} = .750 - .245 = .505$$

The complete matrix $S_{\cdot 123}$ is shown in the lower right panel of Table 7-1.

This is called the semifinal step, even though there are no further partial covariances to compute. For reasons of completeness and symmetry, we can partial $z_{4\cdot 123}$ out of itself to give a null matrix here when there are only four variables.

At each step in this procedure, the regression coefficients of each variable are, as always, the covariances divided by the variance or residual variance of the variable partialed out at that step. The next step's covariances are always their covariance at this step, minus the product of their covariances with the partialed variable divided by the variance of the variable being partialed. For example,

$$s_{jk\cdot 123} = s_{jk\cdot 12} - (s^2_{j3\cdot 12}) \times (s_{k3\cdot 12})/s^2_{3\cdot 12}$$

if variable 3 is the one partialed out after x_1 and x_2. The variances at the next step are always the variances at this step minus the squared covariance with the partialed variable divided by its variance at this step. For example,

$$s^2_{j\cdot 123} = s^2_{j\cdot 12} - (s_{j3\cdot 12})^2/s^2_{3\cdot 12}$$

Perhaps it is clear how a computer algorithm could be written to do this. To keep the subscripts straight, it helps to denote the variances as s_{jj} instead of s^2_j. Several computer programs for regression refer to these residual proportions of variance as "tolerances."

The meaning of the terms in the partial covariance matrices and their relationship to partial correlations should also be explained further. Consider the expression for $s_{23\cdot 1}$, which is also $r_{23} - r_{12}r_{13}$. Dividing this expression by the square roots of the corresponding diagonal entries, $\sqrt{1 - r^2_{12}}$ by $\sqrt{1 - r^2_{13}}$, gives the partial correlation $r_{23\cdot 1}$. This will be true of higher order partials as well, giving a general formula for higher-order partial correlations in terms of the preceding-order partial variances and covariances:

$$r_{jk\cdot 12\cdots p} = \frac{s_{jk\cdot 12\cdots p-1} - (s_{jp\cdot 12\cdots p-1})(s_{kp\cdot 12\cdots p-1})}{(s^2_{j\cdot 12\cdots p-1})^{1/2}(s^2_{k\cdot 12\cdots p-1})^{1/2}} \tag{7.3}$$

The pth-order partial correlation, $r_{jk\cdot 12\cdots p}$, is thus expressed in terms of the p-minus first-order partial covariances of j and k with p and the partial variances of j and k. This holds for all values of p.

A second aspect of the partial variance-covariance matrix involves multiple correlations. Consider $s^2_{3\cdot 12}$, which is .500. This is one minus the squared multiple correlation of z_3 with z_1 and z_2. This can be verified by going back to the correlation matrix. The expression $1 - s^2_{4\cdot 12}$ is also the squared multiple correlation of z_4 with z_1 and z_2. Finally, the lone entry in the last section, $s^2_{4\cdot 123}$ is $1 - R^2_{4\cdot 123}$, the squared multiple correlation of z_4 with the other three variables. These examples demonstrate a general principle. When we begin with the correlation matrix, the diagonal of the pth order partial covariance matrix is the complement of (that is, one minus) the squared multiple correlation of each variable with the p variables that have been partialed out. If we start from the covariance matrix instead, then these are the residual variances of the variables—in other words, $s_j^2\,(1 - R^2_{j.12\ldots p})$—but then R^2 can be computed after division by s_j^2.

HIERARCHICAL AND VARIABLE-SELECTION MULTIPLE REGRESSION

Strategies in Selecting Predictors

For a variety of reasons, the investigator may prefer not to include all the variables in the final regression equation unless the data indicate that it is necessary. For example, a priori considerations of a theoretical, a practical, or even a political nature may warrant giving precedence to some predictors over others. This leads to a *hierarchical* multiple regression. On the other hand, we may select, on a purely empirical basis, those variables which are most important in the regression equation, either for the practical purpose of reducing the amount of measurement or computation that will be necessary in the future or in order to identify which variables are most important, thus guiding future research. This leads to a *variable-selection* analysis. Both of these strategies are varieties of *stepwise multiple regression,* in which variables are entered into the regression one step at a time. The multiple R and the regression equation are examined after each step, and the computing process may in fact terminate after any one rather than continuing to get the regression equation for the complete set of predictors. References to stepwise multiple regression in the literature may refer to either variable selection or heirarchical multiple regression, although usually the former.

Suppose an employment office routinely gives a typing test to applicants for clerical positions yielding two scores: speed in words per minute (WPM) and number of errors (NE). Applicants are also interviewed by a professional personnel technician and given a numerical rating on poise (PO), appearance (AP), and general suitability (GS). It would save time for both the office and the applicant, and money for the organization if there were no interview. However, if the interview furnishes relevant information that predicts the quality of an employee's performance, it should be retained. The typing test, on the other hand, is simple to give. (Who would hire someone to do typing without giving him or her a typing test, anyway?) The question here is whether the ratings from the interview provide information over and above that provided by the typing test and whether this results in selecting better employees than the typing test alone.

A hierarchical regression analysis may provide the answer to this question, assuming that a suitable variable reflecting job performance for a group of employees can be found. If so—and this is the most difficult problem in such studies—then the job performance score or rating (JP) is the dependent variable in the multiple regression. The multiple R (and standard error of estimate) is obtained on the basis of a multiple regression of JP on WPM and NE *alone.* This could be done by listing JP, WPM, and NE as the first two variables in the matrix and examining $s^2_{y.12}$, then finding $R^2_{y.12}$.

Alternatively, we can just obtain the multiple correlation from the inverse of the 2 × 2 correlation matrix involving these two predictors, then compute the multiple R on *all five* variables and compare it with the first one. It will be larger, of course, because the use of more variables necessarily reduces the amount of error in prediction, but the increase may be small. If the increase is small, then the personnel office may conclude that it is not worth the trouble of giving the interview. If the improvement in prediction is appreciable, then the interview is retained. The statistical significance of the improvement in R will be discussed in Chapter 9.

The results of this analysis can then be used to give weighted scores for the applicants, and the decision on hiring would be based on the weighted score. The weights can be the regression weights (the *raw* score weights) themselves, but it is more convenient to convert them to simple integers which are approximately proportional to the weights, for example, WPM − NE + P + AP + 2GS. The correlation of a weighted score based on these simple weights with the criterion can be rather simply calculated using the methods outlined in Chapter 4. This accounts for the amount of loss in predictive accuracy through their use.

Suppose, on the other hand, the personnel office is simply interested in reducing the number of variables that need to be considered, with no particular preferences as to which these should be. Here the process of variable-selection multiple regression can be used to decide which variables are best and how many need to be used. This is done by selecting the variable with the highest zero-order correlation with the dependent variable, and then a second variable that, when included with the first, will give the highest multiple R, and then a third variable that will give the greatest increase in R when included with the first two, and so on, continuing this process by adding the variable that will give the greatest increment to R, until further increments are considered negligible. Then the multiple regression will include just those variables that have been selected. At each stage, the variable that gives the largest increase in R has the highest current partial correlation. The variables are said to "enter" the equation, and at any given step the variables are divided into two sets, one that has already entered the regression equation, and one that has not. The most promising candidate from the latter set is the one selected to enter the equation next.

The entry of a new variable or set of variables into the equation alters the regression weights of the variables already included because the inverse of the matrix used in the computation is a new one. The only exception to this rule is the case where the predictors are uncorrelated. However, if the process is carried to completion—that is, if all variables are entered into the equation—the final regression weights are *exactly the same as if the complete solution had been calculated in the first place.* This is because the inverse of the covariance or correlation matrix is still being computed. When all of the variables are in the equation, the matrix is always the same, no matter what has taken place in intermediate steps. In fact, as we saw earlier, the partial correlation is defined in almost exactly the same way as the regression coefficient. Calculating the regression weights by the stepwise procedure looks different from calculating the inverse, but it has exactly the same effect. In practice, very small differences may be found. These are attributable to the rounding errors, which are different because the steps are carried out in a different order.

Variable-selection MR is highly susceptible to errors of inference—that is, mistaking chance effects for real ones. In fact, a large proportion of the published results using this method probably present conclusions that are not supported by the data. Consequently, analyses of this type should not be carried out until after reading Chapters 8 and 9.

Forced and Free Variables

Suppose we wanted to combine the hierarchical and variable-selection strategies in the same analysis. For example, suppose the personnel office wanted to include in the regression equation only those interview variables that raised R over and above its value attributed to the two typing variables, not all three. This can be done by entering WPM and NE, and then seeing which of the interview variables, if any, has a substantial partial correlation once WPM and NE have been entered. The variable with the largest second-order partial is entered if its contribution is large enough to make it worth including, otherwise the process stops. If the additional variable is entered, then we would calculate and examine the third-order partials of the two variables remaining to see if either is appreciable. If so, we would include that variable and then calculate and examine the fourth-order partial of the remaining variable. It too may then be included in the equation. If a computer is used, it can carry out all the steps. Then the investigator works backwards, deleting the variables included in the reverse order of their entry into the equation.

We may think of this process as *forcing* the two typing variables into the equation while the interview variables are *free* to enter if their contribution is large enough, given that they add to the prediction made by the forced ones. These are two general concepts for which all types of analysis are special cases—including the conventional ones in which all variables are included in the equation. In conventional analysis, all variables are forced. In a pure variable-selection case, all variables are free, entering only if they contribute. In the hierarchical analysis, there are two levels of forcing, involving one set of variables and then a second. In the hierarchical analysis mentioned here, the typing variables are forced, whereas the interview variables are free.

There can be many levels of forcing, up to a complete specification of the order in which the variables will enter the equation. For example, for the convenience of the personnel office, it would be simpler to count the number of words typed in the fixed time and not to proofread for errors. Furthermore, it would be simpler to use only GS from the interview, followed in plausibility by PO and AP. In this case, the R for WPM is found first, then the one for WPM and NE, then for WPM and NE plus GS, then WPM, NE, and GS plus PO, and finally these four variables plus AP. The increment in R^2, with the addition of each variable, is used to decide whether to include that variable and all those variables specified earlier in the equation. This decision can be based on the absolute size of the increment or questions of statistical significance (to be discussed in Chapter 9). It should be emphasized that this procedure depends on having a clear a priori order for the variables, specified on practical or theoretical grounds, before the data are gathered. If the order is decided post hoc, the inferential validity of these decisions is suspect.

There can be different levels of freedom, also, perhaps reflecting a milder degree of preference of one variable over the other. For example, we may want the typing variables to enter before the interview variables, but only if they have appreciable power. This would be a kind of variable-selection method that is biased toward the typing variables but does not require them. Perhaps it can now be seen that the possible strategies are endless, and that each may be appropriate, depending on the goals of the investigator and the characteristics of the variables. As always, the first question is "What is the question?" and the second question is "What are the variables?"

Modern computer programs for multiple regression, which will be examined in Chapter 9, use the concept of an *inclusion level* to specify the exact type of analysis desired. An inclusion level is a number assigned to each variable that specifies the level the variable has in the hierarchy and whether it is forced or free.

Because there are so many different correlations, it may be useful to sort them out in terms of the concept of variance-accounted-for. We know that the squared multiple correlation is the proportion of variance accounted for by the predictors. If there are two sets of predictors, p in one and q in the other, then the difference between the correlation using both sets, R^2_{p+q}, and the correlation using just the p in one set, $R^2_{p+q} - R^2_p$, is the proportion of variance that is accounted for by the q additional predictors, that is, the proportion of variance for which these are necessary.

The squared partial correlation, $r^2_{jy \cdot p}$, between the dependent variable and some variable x_j when p-other variables have been partialed out, is the proportion of *residual* variance that is accounted for by x_j. It is $(R^2_{p+j} - R^2_p)/(1 - R^2_p)$, where R^2_{p+j} means that x_j has been added to the set of p others. On the other hand, it is the squared *part* correlation that gives the proportion of the total y variance that x_j accounts for.

Example of Hierarchical and Variable-Selection MR

The correlation matrix used earlier in this chapter will provide the data for an example of the two major types. Variable 4 will be the dependent variable. We will first do a hierarchical analysis and assume for the sake of simplicity that the hierarchy is x_1, x_2, and x_3, and that these variables will be forced into the regression equation in that order. We will also assume that the sample size is very large, so that the effects of sampling are negligible. To emphasize certain distinctions, the predictors will have variances of 1.0 while the variance of y is assumed to be 4.0.

The hierarchical analysis is shown in Exhibit 7-1. The correlation matrix, R, is given first, including the correlations of the predictors with y. Then the results of entering each variable are shown in a series of three steps.

If we had a hierarchical conception of the a priori importance of these variables in the example, we would presumably conclude that x_1 accounts for a moderate amount of variance, x_2 makes a modest additional contribution, and x_3 makes a substantial additional contribution. Therefore, we would include all of the variables in the ultimate regression equation. Note, however, that the final R is .7036, minutely higher than the correlation of x_3 by itself, and that the other two end up with weights near zero. This may call into question our preconceptions about the variables. However, as we will see in Chapter 9, regression equations are notoriously unstable from a sampling point of view. We should act on these results quite cautiously, sticking to the original hypothesis, unless our data is based on a sample size of several hundred.

Exhibit 7-2 shows the course of a variable-selection analysis of the same data. Here, the variable selected at each step for inclusion in the regression equation is the one with the highest correlation with y, partialing out all variables previously entered into the equation.

Data

	1	2	3	y
1	1.00	.60	.50	.40
2	.60	1.00	.70	.48
3	.50	.70	1.00	.70
4	.40	.48	.70	1.00

$$s_y^2 = 4.00 \quad R^2 = 0 \quad R = 0 \quad \sqrt{1 - R^2} = 1.00 \quad s_{\text{est}} = 2.00$$

Step 1 and **Partial**

1	2	3	y	b^*	$r_{jy\cdot 1}$
0	0	0	0	.400	———
0	.64	.40	.24	———	.327
0	.40	.75	.50	———	.630
0	.24	.50	.84		

$$s_{y\cdot 1}^2 = 3.36 \quad R_{y\cdot 1}^2 = .160 \quad R_{y\cdot 1} = .400 \quad \sqrt{1 - R_{y\cdot 1}^2} = .917 \quad s_{\text{est}} = 1.833$$

Step 2 and **Partial**

1	2	3	y	b^*	$r_{jy\cdot 12}$
0	0	0	0	.175	———
0	0	0	0	.375	———
0	0	.50	.35	———	.572
0	0	.35	.75		

$$s_{y\cdot 12}^2 = 3.000 \quad R_{y\cdot 12}^2 = .250 \quad R_{y\cdot 12} = .500 \quad \sqrt{1 - R_{y\cdot 12}^2} = .866 \quad s_{\text{est}} = 1.732$$

Step 3 and **Partial**

1	2	3	y	b^*	$r_{jy\cdot 123}$
0	0	0	0	.0875	———
0	0	0	0	−.0625	———
0	0	0	0	.7000	———
0	0	0	.505		

$$s_{y\cdot 123} = 2.020 \quad R_{y\cdot 123}^2 = .495 \quad R_{y\cdot 123} = .7036 \quad \sqrt{1 - R_{y\cdot 123}^2} = .711 \quad s_{\text{est}} = 1.421$$

Exhibit 7-1 Example of a hierarchical stepwise analysis in which the order of the variables is 1, 2, and 3. The correlation matrix for the three predictors and y is shown first. At step 1, x_1 is entered, with a weight of .400 (equal to its correlation), giving a "multiple" R of .16 and a s_{est} of 1.833. Also shown is the partial covariance matrix $s_{\cdot 1}$ and the partial correlations $r_{2y\cdot 1}$ and $r_{3y\cdot 1}$. At step 2, x_2 is entered resulting in $b_1^* = .175$, $b_2^* = .375$, $R_{y\cdot 12}^2 = .250$, and $s_{\text{est}} = 1.732$. The partial covariance matrix and $r_{3y\cdot 12}$ is also shown. Finally, at step 3, x_3 is entered, giving $R_{y\cdot 123}^2 = .495$ and $s_{\text{est}} = 1.421$. The weights are now: .0875, −.0625, and .7000. The only nonzero element of the partial covariance matrix is $s_{y\cdot 123}^2$, which is .505, $1 - R_{y\cdot 123}^2$. Note that at step 1 the part correlation $p_{y2\cdot 1}$ equals $.24/(.64)^{1/2}$, or .30. After step 2, the part correlation, $p_{y3\cdot 12}$, equals $.35/(.50)^{1/2}$. Then $R_{y\cdot 123}^2$ equals $r_{y1}^2 + p_{y\cdot 2.1}^2 + p_{y\cdot 12}^2 = .495$.

By this definition, the first variable entered is the one with the highest zero-order r, in this case x_3, which has an r of .700, accounting for .490 of the variance and giving an s_{est} of 1.428. Then x_1 has the highest partial r, so it is entered next. The second-order partial of x_2 is very low, but we enter it anyway to complete the example.

Data				*Step 1*					*Partial*
1	2	3	y	1	2	3	y	b^*	$r_{jy\cdot 3}$
1.00	.60	.50	.40	.75	.25	0	.05		.081
.60	1.00	.70	.48	.25	.51	0	−.01		−.020
.50	.70	1.00	.70	0	0	0	0	.700	
.40	.48	.70	1.00	.05	−.01	0	.51		

$s_y^2 = 4.000$ | $s_{y\cdot 3}^2 = 2.040$

$R^2 = 0$ | $R^2 = .490$

$R = 0$ | $R = .700$

$\sqrt{1 - R^2} = 1.000$ | $\sqrt{1 - R_{y\cdot 3}^2} = .714$

$s_{est} = 2.000$ | $s_{est} = 1.428$

Step 2					*Partial*	*Step 3*					*Partial*
1	2	3	y	b^*	$r_{2y\cdot 13}$	1	2	3	y	b^*	$r_{2y\cdot 123}$
0	0	0	0	.0667	−.058	0	0		0	.0875	——
0	.4267	0	−.0267			0	0	0	0	−.0625	——
0	0	0	0	.6667		0	0	0	0	.7000	——
0	−.0267	0	.5067			0	0	0	.505		

$s_{y\cdot 13}^2 = 2.028$ | $s_{y\cdot 123}^2 = 2.020$

$R_{y\cdot 13}^2 = .4933$ | $R_{y\cdot 123}^2 = .495$

$R_{y\cdot 13} = .702$ | $R_{y\cdot 123} = .7036$

$\sqrt{1 - R_{y\cdot 13}^2} = .712$ | $\sqrt{1 - R_{y\cdot 123}^2} = .711$

$s_{est} = 1.424$ | $s_{est} = 1.421$

Exhibit 7-2 Example of a variable selection analysis of data from Exhibit 7-1. Variable x_3 enters first because it has the highest zero-order validity. It has a regression weight of .700 and therefore gives a "multiple" R of .700, an $R_{y\cdot 1}^2$ of .490, and a s_{est} of 1.428. The partials are .081 and −.020 for x_1 and x_2, so x_1 is entered next. This gives $b_1^* = .067$, $b_3^* = .667$, and $R_{y\cdot 13}^2 = .493$ with a s_{est} of 1.424. The $r_{2y\cdot 13} = -.036$. When it is entered, the regression weights are .0875, −.0625, and .7000, respectively, and $R_{y\cdot 123}^2 = .495$ with a s_{est} of 1.421. The final results are the same as for the previous analysis in all respects. The part correlations here are $p_{y1\cdot 3} = .05/(.75)^{1/2}$ and $p_{y2\cdot 13} = .0267/(.4246)^{1/2}$. Then $R_{y\cdot 123}^2 = r_{y3}^2 + p_{y\cdot 3}^2 + p_{y\cdot 13}^2 = .495$, as before.

This is an issue on which there is widespread misunderstanding. Many users of multiple regression mistakenly think that the regression weights are affected by the order in which variables were added to the equation. Not so! However, a weight depends on which variables are *included at that stage*, not on the order in which they are entered. All variable-selection procedures are just computing

$R_{xx}^{-1}\mathbf{r}_{xy}$. That is, the final regression weight for x_1 in the example is the same whether the variables were entered 1, 2, and then 3 or 3, 2, and then 1. The final weight is .0875 after Step 3 of Exhibit 7-2 as well as after Step 3 of Exhibit 7-1. However, its value at Step 2 is different because in the first case the other predictor is x_2, whereas in the second case it is x_3. Anthropomorphically, we may state this principle as "The variables know who else is at the party, but they don't care when they arrived."

If we were carrying out this analysis with real data, we would probably conclude that x_3 was sufficient as a predictor. The other variables raise R by only .0036 and reduce the s_{est} from 1.428 to 1.421. These are negligible improvements. The cautions stated earlier concerning the sampling fluctuations of regression weights should be kept in mind, however.

The course of a stepwise multiple regression involving a number of predictors can be somewhat surprising in that variables that initially look unimportant can end up contributing a large amount of variance, and those that enter the equation first may end up doing little. The reason is that the patterns of correlations among the predictors, that is, the way in which they share their variance, interact with the way they share variance with the dependent variable.

NONLINEAR REGRESSION

Recognizing Nonlinearity

Earlier in this chapter, it was stated that the most important step in any analysis is the plotting of variables against each other—especially the plotting of y against the predictors. This allows the investigator to see whether the univariate regressions are reasonably linear. If they are not linear, the model used to analyze the data is inappropriate. The low correlations do not reflect a lack of relationship between the variables: they reflect that the relationship is not of the form $y = bx + b_0$.

Curvilinearity can take on a limitless variety of forms. What to do about it, and how to interpret it, is a topic that could occupy a ponderous tome. All that can be attempted in this brief section is to point out the major types of curvilinearity that are likely to be encountered and to suggest some ways to handle them. The primary goal should be to transform the variables in such a way as to remove the curvilinearity, particularly if there is some substantive theoretical justification for the transformations made. A less desirable but still useful approach is to incorporate additional terms in the regression model that describe or capture the nature of the curvilinearity in the equation. That is, one may include x_1^2, x_2^2, $x_1 \times x_2$, x_1^3, and so on, as predictors.

To do this, we could mentally or physically draw a curve to fit the points on the plot. If the trend is the same direction throughout the range of the curve, rather than reaching a maximum or minimum and reversing itself, the curve is said to be *monotonic*. The curves in Figure 7-1(a) are all monotonic with a positive overall trend, whereas those in 7-1(b) are monotonic with a negative trend. Several nonmonotonic curves are shown in 7-1(c). These curves tend to increase to a maximum and then begin decreasing, or vice versa. Or, there may be no overall trend at all in these curves.

Monotonic curves in which y increases at a faster and faster rate [see curves 1 and 2 in Figure 7-1(a)] are called *positively accelerated,* or *concave-upward,* whereas those that tend to level off are called *negatively accelerated,* or *concave downward.* Common functions that are positively accelerated are "power functions," where $y = bx^a$ with a greater than 1.0, and the exponential functions, where $y = bc^x$. If $y = bx^a$ is indeed the true relationship, then a plot of log y versus log x (log-log coordinates) will be linear with a slope of a and an intercept of log b. The other function is an *exponential* function, and, if it is appropriate, the plot of log y and x (semi-log coordinates) will be linear, with a slope equal to log c and intercept log b. Both log-log and semi-log approaches may be used in plotting the data, although where there is a theoretical reason for expecting one rather than the other, the decision may be made on theoretical grounds. Both are likely to be characterized by heteroskedasticity—unequal variance around the regression line—in the original data, since there are larger deviations from the regression line at larger values for x. Using the transformation may smooth this out.

Functions having an overall positive trend that increase at a decreasing rate may be examples of power functions $y = bx^a$ in which a is positive but less than

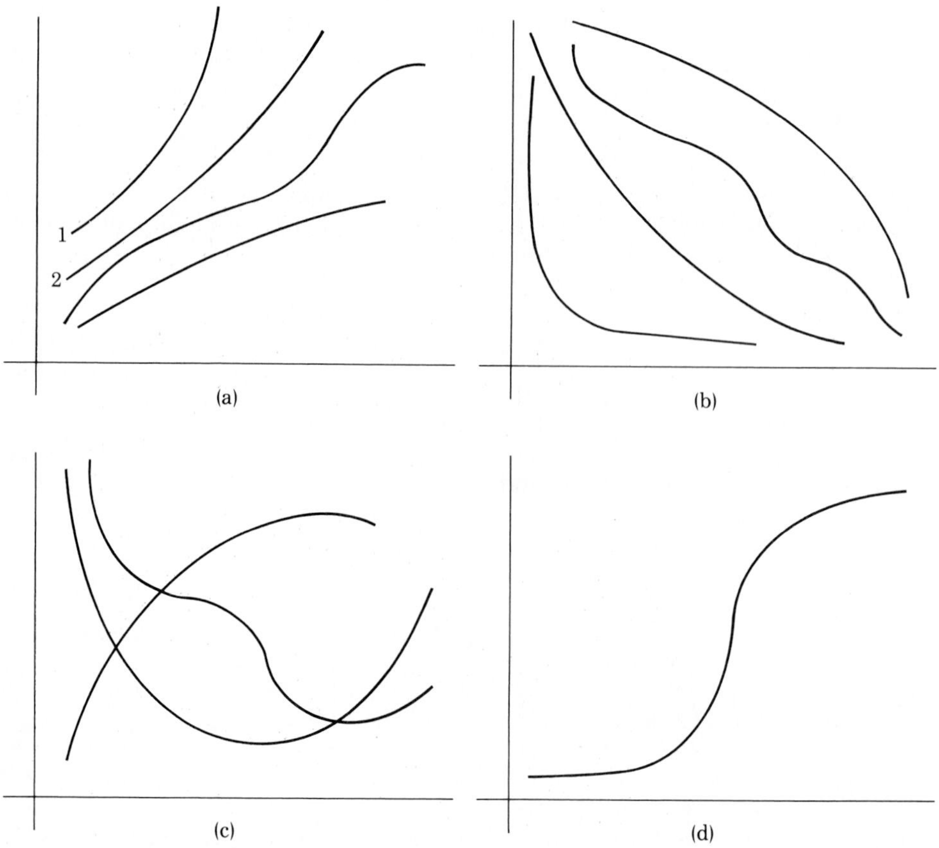

Figure 7-1 Varieties of curves. (a) Monotonic with positive trend; (b) monotonic with negative trend; (c) nonmonotonic curves; (d) an ogival curve.

one—for example, $y = \sqrt{x}$ for $a = 1/2$, or of logarithmic form, $y = b \log x$. The function $y = \sqrt{x}$ will be linearized in log-log coordinates but the slope will be less than 1.0, whereas the $y = b \log x$ requires an exponential (antilog) transformation of x, e^x, in order to become linear. An experienced investigator may be able to discriminate which approach will be appropriate by examining the curves, but trial and error is likely to work best for most of us.

Occasionally, the transformations suggested above may not seem to work in cases where the plots of the original suggest that they should. There are a variety of reasons for this. One that may be superficial is that the transformations are all based on the assumption that the given origin for x is the correct one, and that there is no constant in the function. That is, it might be that $\log (x - k)$ should be taken instead of $\log x$. Methods for finding an optimum k are too complex for this book, but a simple trial and error process may be successful.

Curves with a negative overall slope that are concave upward [see Figure 7-1(b)] have two possible simple theoretical bases. One is $y = b/x$, or more generally, $y = b/x^a = bx^{-a}$, where a and b are positive. Log x versus log y should be linear with a negative slope here, since this is also a power function, but one with a negative exponent. An alternative with a similar shape is the negative exponential bc^{-x} which should be linear in log y versus x. It could be that here, too, one should use $\log (x - k)$ instead of $\log x$ in a given case.

Sigmoid curves like 7.1(d) are commonly encountered when y is a probability of some event—for example, the probability of detecting a faint light plotted against the intensity of the light. Here, the trend might be made linear by the use of a probability function such as the integral of the normal curve (called "probit analysis"). If it is a probability, then the transformation is $y' = z(y)$, where $z(y)$ is the normal deviate corresponding to y. For example, if $y = .84$, $z(y) = 1.0$, according to Table 1 in the Appendix.

It should be noted that the commonly available statistical packages have built-in procedures for making the transformations noted above. In SPSS-X this can be accomplished by the COMPUTE procedure; in BMD, the same can be done through the use of TRANSFORMATION. SAS has ample procedures for creating new variables in the DATA step. These procedures can be used to replace variables with transformed versions. Note that this transformation must be specified completely by the user; no complex curve-fitting is carried out (to determine b_0, for example), although some packages have procedures for doing this. Investigators should examine those available to them to get an idea of just how far the computer can go with respect to selecting the particular kinds of functions and transformations needed.

Polynomial Curve-Fitting

When there seems to be no simple transformation that will make a relationship linear, it is common practice to use polynomial functions of the variables as a way of capturing the curvilinearity. A general polynomial in x is $a_0 + a_1x + a_2x^2 + \cdots + a_mx^m$, but using more than two powers is rarely justified or interpretable. The quadratic function

$$y = b_0 + b_1x + b_2x^2 \tag{7.4}$$

is usually sufficient for purely empirical work, at least where there is considerable scatter around the curve. If b_1 and b_2 are both positive, then the function is monotonically increasing, and in fact, concave upward. If they are both negative, the reverse is true. If b_1 is positive and b_2 is negative, then y increases first until the influence of x^2 overpowers that of x, at which time it reaches a maximum and then decreases with increasing rapidity. The reverse happens if b_1 is negative and b_2 is positive.

An attractive aspect of this function is that if b_2 is zero, it reverts to the linear form. So, $y = b_0 + b_1 x$ is a special case of the more general quadratic Equation 7.4. If $b_1 = 0$ instead of b_2, then we have the special case of the power function with $c = 2$. Equation 7.4 represents quite a variety of curves relating y to x.

The quadratic, a function involving both x and x^2, can also be used as a predictive equation. In fact, we can do this in a multivariate case; for each x_j there is an x_j^2, and we could do a multiple regression of y on the $2p$ predictors thus created. This could be done, for example, if we wanted to relate preference for auto designs to a variety of quantitative factors (for instance, length, weight, horsepower, trunk capacity, and so on) if we suspected that there would be curvilinear dependence of a fairly simple type—such as an optimum level on each factor. For example, most people prefer a car of intermediate length—one that is neither too small to sit in nor too long to fit in the garage.

In cases like these it is often best to use a hierarchical approach, forcing in the linear forms of the variables first, and adding squared terms only if they are necessary. In most cases, if the values of x_j are all positive, the correlation of x_j with x_j^2 may be very high, with resulting near-linear-dependence. It is then advisable to use not just x_j^2 but $(x_j - c_j)^2$, where c_j is some constant near the mean of x_j.

We can also "crossbreed" different x's to form product terms like x_1x_2, x_3x_8, and so on. Although we can form multiple products such as $x_1x_2x_3x_4$, there is ordinarily little justification for using more than the simple products like x_jx_k. Again, it is wise to subtract constants: in other words, use $(x_j - c_j)(x_k - c_k)$ to avoid dependencies. The product terms are sometimes referred to as "interactions" because if they are predictive there is an effect of being high on one variable *and* low on the other and vice versa. The concept of interaction is actually more general than this, but the term has a certain degree of appropriateness.

Having introduced polynomial regression, a number of cautionary remarks are worth noting. First, it often represents blind empiricism of the most dreary, mindless sort. Picture yourself trying to explain to a colleague just why the product of age times vocabulary plus the square of the number of older siblings should predict reaction time in a two-choice paradigm—particularly a colleague who is skeptical about multiple regression in the first place. Also, note the rapidity with which variables proliferate. Just using the linear and square terms gives $2p$ predictors. Including all the products gives $p(p - 1)/2$ more, making a total of 65 terms if we start with 10 predictors.

The purpose of this section has been to suggest the elementary approaches that can be taken if it appears that the regression of y on x is curvilinear. The first approach is one of transforming x or y or both in such a way as to make the relationship linear. (Remember that if there are several x's and a multiple regression after transformation is planned, there can be only a single transformation of y, not a different one for each x. Otherwise, we do not have the same variable in each case.) It is also desirable to have a theoretical explanation for why the transformation is necessary. In certain cases it may reduce the degree of heteroskedasticity.

A general family of transformations of independent variables is the polynomial, where from each x_j successively higher powers are generated and used as predictors in addition to x_j. It is rare that a theoretical basis can be found for terms beyond x^2, so the analysis is often limited to this case. Polynomial regression must be used with some restraint, however. Not only is the result often difficult to interpret; there are other pitfalls as well. Because the powers of a variable are very highly correlated, some degree of orthogonalization is important. We should also be aware of the fact that additional terms "count" as additional variables as far as the final multiple regression is concerned, and that we can rapidly come close to having a linearly dependent set of predictors.

A CAUTIONARY NOTE

In general, partial correlation analysis is affected by any lack of reliability or validity in the variables. In many ways these effects resemble tuberculosis as it occurred a generation or two ago: They are widespread, the consequences are serious, the symptoms are easily overlooked, and most people are unaware of their etiology or treatment.

If a control variable is less than perfectly reliable, partialing does not completely remove its effect. Realistically, any variable must be thought of as being composed of an underlying "true" variable plus a random error. Correlations between variables must reflect covariances between their respective true scores since, by definition, the random component does not correlate with anything.

On the other hand, the variance of a variable consists of the sum of true score variance plus error variance. When these representations are included in the formulas for partial correlations, some surprising conclusions result. In general, partialing out an observed variable incompletely partials out the underlying true score—the deviations from the regression will still correlate with the true score of the partialed variable, although to a lesser degree. For example, when we partial out the score on an intelligence test, we are not partialing out intelligence. If we find that a partial correlation is not zero, indicating that an observed correlation between school achievement and per-pupil expenditures is not completely accounted for by parents' education, it could be because we have used a fallible measure of parents' education (all measures are fallible), such as their report of their own educational level. If we could measure the variable infallibly, the partial correlation might be different.

Unreliability of measurement is not the only problem. Often the control variable is not precisely the variable we would like to measure, but merely related to it. If we measure "education" by reported years of schooling, for example, the error of measurement is the error of determining the true number of years of schooling. However, the amount of schooling is only a rough index of amount of education, even if it can be determined accurately. People with the same number of years of school vary widely in the amount of education they have received. That is, the amount of schooling is less than completely *valid* as a measure of education. Therefore, even if we partial out a precise measure of years of schooling, we are still not partialing out *education.*

The extent to which imperfect reliability and/or validity causes difficulties depends on the degree to which it occurs. If reliability correlations are in the upper nineties, then the problem is negligible, unless we are trying to interpret very small correlations. If they are in the sixties, then the problem is likely to be severe. The same is true of the effect of the validity.

Does this mean that we should not bother with partial correlations and related methods such as stepwise multiple regression and path analysis? Certainly it is preferable to investigate the effect of partialing—even imperfectly—rather than not do so at all. Still, one has to take results with a grain of salt rather than to assume that conclusions are ironclad.

Where measurements are fallible, there is a tendency to simplify them, dividing subjects into groups such as high versus low. Such divisions are not infallible; think of the persons near the borders. Some who should be highs are actually classified as lows, and vice versa. In addition, the "barely highs" are classified the same as the "very highs," even though they are different. Therefore, reducing an unreliable variable to a dichotomy makes the variable more unreliable, not less.

The effects of unreliability, invalidity, and coarse classification can be quantified, but the complexities are considerable. A discussion of the unreliability problems can be found in Linn and Werts (1982). It is also possible to use programs such as LISREL (Joreskog and Sorbom, 1982) to compensate for these effects under some circumstances. Such methods are not complete panaceas, however, since they may require estimates of quantities that are difficult to come by and/or questionable assumptions about variables. Where available, however, these methods can be very useful.

SUMMARY

Partial correlation analysis is frequently used to clarify the nature of the relationships among variables. It may be used to show the extent to which a simple, zero-order correlation of one variable with another can be explained by the relationship of both to a third variable. It is also used to bring the relationship of two variables into sharper focus by removing the effect of extraneous variance shared by two variables. Thus partial correlation analysis may be used either to decrease or to increase a superficial correlation. It may be used if we suspect that the relationship between two variables can be explained by the dependence of both on a third variable. It is also used when we suspect that a strong relationship between variables may be obscured by a third.

A variable whose influence on the others is removed by the partialing process is referred to as a *control* variable. In general, however, it is incorrect to infer that the control variable is being controlled or held constant in any literal sense. Rather, a partial correlation is a correlation between residual scores—the deviations of the two variables of interest from the regression of each on the control variable. Because of the properties of correlations between weighted sums, the correlation of x_2 with y when x_1 (denoted $r_{y2\cdot1}$) is partialed out is

$$r_{y2\cdot1} = \frac{r_{y2} - r_{12}r_{1y}}{\sqrt{(1 - r_{12}^2)(1 - r_{1y}^2)}}$$

If there are a number of control variables, then scalar expressions for partial correlations may become unwieldy if one formulates them in terms of the original correlations. The matrix expression presented in Equation 7.2 can be used for any number of variables. Let $r_{2y \cdot c}$ be the partial correlation of x_2 with y for a whole set of control variables, designated as c. Then determine the *multiple* regression of x_2 on that set and also the multiple regression of y on it. A $c + 2$ weight vector, w_2, consists of the negatives of the standard score regression weights for predicting x_2, a weight of 1 for x_2, and a weight of zero for y. Similarly, w_y contains the regression weights for predicting y and its unit weight, and a zero for x_2. Then the partial correlation is a slightly generalized form of Equation 7.2:

$$r_{2y \cdot c} = \frac{\mathbf{w}_2'\mathrm{R}\mathbf{w}_y}{\sqrt{(\mathbf{w}_2'\mathrm{R}\mathbf{w}_2)\,(\mathbf{w}_y'\mathrm{R}\mathbf{w}_y)}}$$

This partial correlation could also be computed from the covariance matrix S instead of R; then the weight vectors contain the raw score regression weights.

Sometimes it is helpful to use partial covariances instead of partial correlations. In some applications, it is useful to think of partial correlation or partial covariance analysis as proceeding in a sequential manner. First one variable is partialed out of the other p, then another is partialed out of the remaining $p - 1$, and so on. The removal of each variable is referred to as "sweeping it out." Then an alternative general formula for $r_{2y \cdot c}$ can be expressed, using the partial covariances left after all but the last control variable has been removed. This is summarized in Equation 7.3.

An important topic related to partial correlation analysis is stepwise multiple regression, which is concerned with how the multiple correlation increases as additional variables are added to the regression equation. There are two main forms of stepwise multiple regression: variable selection and hierarchical. In variable selection multiple regression, the computer chooses the variable to be added next to the regression equation, adding the one which seems likely to increase the multiple correlation the most. In hierarchical multiple regression, the order of entry of the variables is determined by the investigator on the basis of a priori considerations. Hybrid types of analysis are also possible. Both of these are discussed in some detail in Chapter 9.

All of the analyses described here are based on the assumption that the relationship between dependent and independent variables is essentially linear. It is an empirical fact that many are not. Scatterplots to determine the nature of the empirical relationships are an essential part of any regression analysis. For relationships that appear to be curvilinear, or that are expected to be, on theoretical grounds, there are two types of responses to cuvilinearity. One is to transform the original variables by some mathematical function that removes curvilinearity. A second is to introduce additional terms into the regression equation that attempt to account for this curvilinearity. Most commonly, this takes the form of introducing derived variables that are the squares, cubes, and so on, of the observed predictors, and even the products of different predictors or powers of predictors. Users of regression should be wary of the presence of curvilinearity and cautious in adjusting to it.

When control variables are less than perfectly reliable, as they generally are, partial correlation analysis should be treated with caution. The problem is that partialing out the observed variable does not completely remove all the variance that is due to the underlying true score. The lack of complete validity in the control variables also increases the ambiguity of a partial correlation analysis, but in a less predictable way.

Computer Applications

A computer is clearly necessary for all but the simplest partial correlation and stepwise multiple regression analyses, and the reader is probably aware that common statistical packages contain such programs. Chapter 9 describes how these programs relate to the concepts and procedures considered in this chapter.

PROBLEMS

Below is a matrix of scores for six persons on four variables, and the corresponding covariance and correlation matrices variables. Also given is the inverse of the covariance matrix for the first two predictors.

X

	1	2	3	4
1	3	2	2	2
2	−2	0	−1	−2
3	1	6	1	0
4	−1	−2	0	6
5	0	−4	−3	−2
6	−1	−2	1	−4

S_{xx}

	1	2	3	4
1	3.2	3.2	1.6	1.6
2	3.2	12.8	4.0	1.6
3	1.6	4.0	3.2	1.6
4	1.6	1.6	1.6	12.8

R_{xx}

	1	2	3	4
1	1.000	.500	.500	.250
2	.500	1.000	.625	.125
3	.500	.625	1.000	.250
4	.250	.125	.250	1.000

$$S_{cc}^{-1} = \begin{bmatrix} .4167 & -.1042 \\ -.1042 & .1042 \end{bmatrix}$$

1. Compute $r_{24\cdot 1}$ by the correlation formula.

2. Predict x_2 and x_4 from x_1, obtain the deviations from the predictions, and the correlation between the two sets of deviations. Compare your result to your answer from Problem 1.

3. S_{cc}^{-1} is the inverse of the covariance matrix for the first two variables. Assume they are control variables and compute the regression coefficients for predicting x_3 and x_4 from them. Calculate and then correlate the two corresponding residual variables.

4. Obtain the corresponding correlation between $x_{3\cdot 12}$ and $x_{4\cdot 12}$ directly from the covariance matrix using the formulas for variances and covariances of weights combinations.

5. Prove algebraically the equivalence of the correlations from Problems 3 and 4.

6. How can we tell from the correlation matrix that $R^2_{4\cdot 12}$ will be larger than $R^2_{4\cdot 23}$?

7. Below is the vector of regression weights for predicting x_4 from x_1, x_2 and x_3. Use these to compute $R^2_{4\cdot 123}$. Compare the results to $R^2_{4\cdot 12}$ and $R^2_{4\cdot 1}$. (*Hint:* The latter two are obtainable from information we already have.)

$$\mathbf{b}' = [.3889 \quad -.1111 \quad .444]$$

8. Can you relate $R^2_{4\cdot 123}$ to $R^2_{\cdot 12}$ and $r_{34\cdot 12}$?

ANSWERS

1. $r_{24 \cdot 1} = 0$

2. $\hat{\mathbf{x}}'_{2 \cdot 1} = [3 \ -2 \ 1 \ -1 \ 0 \ -1]$; $\hat{\mathbf{x}}'_{4 \cdot 1} = [1.5 \ -1.0 \ .5 \ -.5 \ 0 \ -.5]$; $\hat{\mathbf{e}}'_{2 \cdot 1} = [-1 \ 2 \ 5 \ -1 \ -4 \ -1]$; $\hat{\mathbf{e}}'_{4 \cdot 1} = [.5 \ -1.0 \ -.5 \ 6.5 \ -2.0 \ -3.5]$

$$r_{ee} = \frac{0}{(9.6)^{1/2}\,(12.0)^{1/2}} = 0$$

3. $\mathbf{b}'_{3 \cdot 12} = [.25, .25]$; $\mathbf{b}'_{4 \cdot 12} = [.50, 0]$; $\mathbf{e}'_{3 \cdot 12} = [-.75 \ -.50 \ .75 \ -.75 \ 2.00 \ -1.75]$; $\mathbf{e}'_{4 \cdot 12} = [-.5 \ 1.0 \ .5 \ -6.5 \ 2.0 \ 3.5]$ Same as $e_{4 \cdot 1}$ because $b_{42} = 0$.

$$r_{ee} = \frac{4.0}{(9.0)^{1/2}\,(60.0)^{1/2}} = .1721$$

4. $\mathbf{w}'_{e3} = [-.25 \ -.25 \ 1.0 \ 0]$; $\mathbf{w}'_{e4} = [-.5 \ 0 \ 0 \ 1.0]$; $\mathbf{w}'_{e3}\,\mathrm{S}_{xx}\mathbf{w}_{e4} = .8$; $\mathbf{w}'_{e3}\,\mathrm{S}_{xx}\mathbf{w}_{e3} = 1.8$; $\mathbf{w}'_{e4}\,\mathrm{S}_{xx}\mathbf{w}_{e4} = 12.0$

$$r_{ee} = \frac{.8}{(1.8)^{1/2}\,(12.0)^{1/2}} = .1721$$

5. $\mathbf{e}_{3 \cdot 12} = \mathrm{X}\mathbf{w}_{e3}$; $\mathbf{e}_{4 \cdot 12} = \mathrm{X}\mathbf{w}_{e4}$; $s_{e3e4} = 1/(n-1)\ \mathbf{e}'_{3 \cdot 12}\mathbf{e}_{4 \cdot 12} = 1/(n-1)\ (\mathrm{X}\mathbf{w}_{e3})'\mathrm{X}\mathbf{w}_{e4} = 1/(n-1)\ \mathbf{w}'_{e3}\mathrm{X}'\mathrm{X}\mathbf{w}_{e4} = \mathbf{w}'_{e3}\,(1/(n-1)\ \mathrm{X}'\mathrm{X})\,\mathbf{w}_{e4} = s'_{e3}\mathrm{S}_{xx}\mathbf{w}_{e4}$
Since s_{c3c4} is the covariance between errors, s_{e3e4} is the same by either route. It can be shown in the same way for s^2_{e3} and s^2_{e4}.

6. Because while r_{14} equals r_{34}, r_{12} is less than r_{23}.

7. $R^2_{4 \cdot 123} = 1 - (s^2_{e4 \cdot 123}/s^2_4)$; $w'_{e4 \cdot 123} = [.3889 \ -.1111 \ .4444 \ -1.0]$
$s^2_{e4 \cdot 123} = 11.644$, so $R^2_{4 \cdot 123} = .0903$; from Problem 5, $s^2_{e4 \cdot 12} = 12$, so $R^2_{4 \cdot 12} = .0625$; $R^2_{4 \cdot 1} = .0625$

8. $R^2_{4 \cdot 123} = R^2_{4 \cdot 12} + r^2_{34 \cdot 12}\,(1 - R^2_{4 \cdot 12}) = .0903$

READINGS

Linn, R. L., and C. E. Werts (1982). Measurement error in regression. In *Statistical and methodological issues in psychology and social science research,* edited by G. Keren. Hillsdale, NJ: Erlbaum, pp. 131–154.

Mulaik, S. A. (1972). *The foundations of factor analysis.* New York: McGraw-Hill, pp. 77–94.

8

Statistical Inference in Multivariate Statistics

BASIC PRINCIPLES OF INFERENCE*

The Sampling Distribution

It may be surprising to have completed seven chapters of a statistics book and found no mention of statistical tests, significance, or probabilities of any kind. This is not because these matters are unimportant in multivariate statistics. Indeed, statistical inference is perhaps even more important in multivariate statistics than in the simpler cases encountered in an elementary course. The fact that inferences are being made about more than one parameter makes statistical inferences more complicated practically and more difficult theoretically. The basic ideas of regression have been presented in a purely descriptive way first to provide a general understanding of what is involved before introducing the complications of inference.

All modern statistical inference centers on the problem of inferring the values of population parameters from sample data. That is, the data analyzed—in our case, a set of scores by a set of persons or other observational units on a set of variables—is assumed to be only a portion of the potential data that could have been gathered. We gather a set of data so that we may generalize from it to a universe of potential data that is in some sense comparable. The data at hand come from a sample of units and we wish to generalize from the sample to the population. This is true at least in principle, even when no serious attempt has been made to define a population of data or observations and sample from it according to some plan such as random sampling.

In analyzing data, we compute sample *statistics*—means, variances, correla-

*Before reading this chapter, it may be helpful to review the concept of inference in an elementary statistics book.

tions, regression coefficients, multiple correlations, and so on. These describe the *sample,* even though the main purpose in computing them is to *estimate* the corresponding *population parameters.* That is, we are using the sample statistics as *estimators* of the population parameters. We know that the estimators are not exactly equal to the parameters, yet we have to use them in order to come to conclusions about the parameters. The process we use in coming to these conclusions is statistical inference.

The central concept in statistical inference is that of the *sampling distribution,* which might better be called the "sample statistic probability distribution." It defines the probabilities that various values of the sample statistic will occur, given that a random sample of a certain size has been drawn from the relevant population.

In order to make inferences about the parameters, we need to know the properties of the sampling distributions of the statistics we are using as estimators. Certain terms and concepts are used to describe the characteristics that are most relevant to the inferential process. The most important ones are described in the following section.

Characteristics of Sampling Distributions

Like other distributions, sampling distributions have means, variances, and shapes. We need to know what these are and what they depend on. The terminology for these characteristics is defined below.

If an estimator is *unbiased,* this means that the mean of its sampling distribution is equal to the parameter. The sample mean is always as unbiased estimate of the population mean. We use $n - 1$ in the denominator of the sample variance to avoid a bias that would result from using the more natural n. On the other hand, there may be a substantial bias in the sample R as an estimator of the population squared multiple correlation P. It is possible to correct R to make it nearly unbiased.

Note that Greek letters represent population parameters whereas Roman letters represent the statistics that we use as their estimators: r and s are estimates of ρ and σ. P is used as an uppercase Greek rho. This notation extends to matrices, for example, S is the estimator of the population covariance matrix Σ.

The standard deviation of the sampling distribution is called the *standard error* of the estimator and the variance is called the *sampling variance.* These tell us the extent to which the estimator is likely to deviate from its average value; the latter is also called the *expected value* of the estimator. The standard errors depend on the population parameters in systematic ways, so we need to pay attention to what affects them as well as their particular numerical values. It is possible to have different estimators of the same parameter. For example, we can use the sample median to estimate the population mean. Obviously, we prefer to use the estimators that have the smallest standard errors. The ratio of the standard errors of two estimators of the same parameter is called their relative *efficiency.* Under most circumstances, the sample mean is more efficient than the median as an estimate of the population mean, but there are times when the reverse is true.

Another desirable property of an estimator is *consistency.* This means that the discrepancy between estimator and parameter can be expected to diminish as the sample size increases, approaching zero as the sample becomes infinite. This property implies that we can treat the estimator as if it were the parameter if the sample is large enough. For most of the estimators that we deal with, consistency operates in a particularly simple and nice way. The estimators are unbiased and the standard errors of the estimators are inversely proportional to the square root of the sample size—or something very closely related to it. In some types of multivariate analysis, we sometimes use an estimator that is relatively inefficient

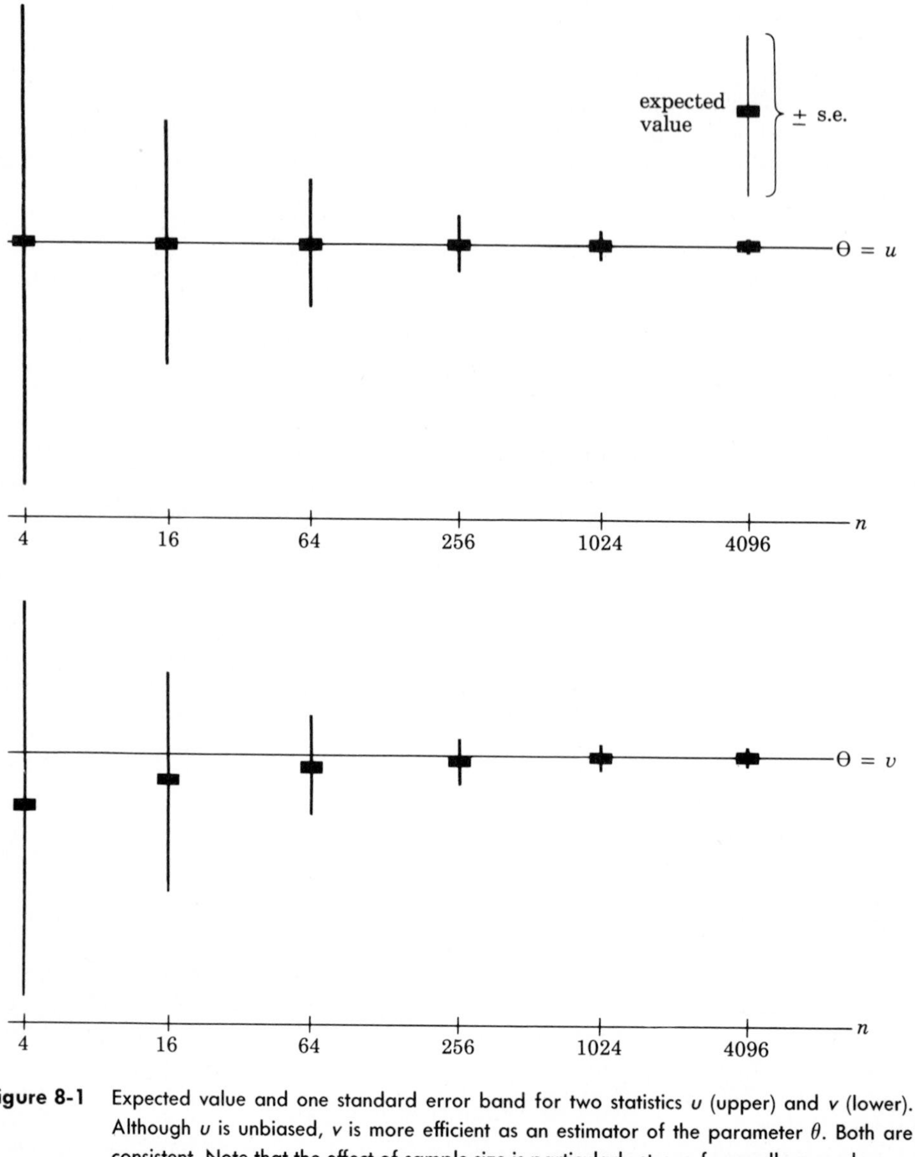

Figure 8-1 Expected value and one standard error band for two statistics u (upper) and v (lower). Although u is unbiased, v is more efficient as an estimator of the parameter θ. Both are consistent. Note that the effect of sample size is particularly strong for smaller samples.

(although consistent), perhaps in order to avoid some extreme computational difficulty.

Figure 8-1 illustrates the properties of estimators. The behavior of two different estimators, u and v of the same hypothetical parameter θ, is shown as a function of sample size. The horizontal line represents the true value of θ, and sample size increases from left to right. The short horizontal bars represent the expected values of the estimators, and the thin vertical lines represent the standard errors. In the upper part we can see that u is unbiased, whereas the lower portion shows that v is biased. However, the lengths of the vertical lines show that v is slightly more efficient than u. The effect of increasing sample size shows that both estimators are consistent.

Inferences about parameters, such as testing hypotheses and the construction of confidence intervals, depend on the proportions of the sampling distributions that are cut off by certain values for the statistic. Such proportions depend on the expected value and standard error of the estimator, that is, where the distribution is located, how spread out it is, and its exact shape. This shape depends on the form of the parent distribution, the estimator that is being considered, and the sample size. For normal parent population distributions, the sampling distribution of the mean is also normal, whereas that for variance has the chi-square distribution (see p. 141), and that for the correlation coefficient is a complex function too frightening to specify.

Normality is a useful property for sampling distributions. Normal distributions are completely specified by their means and variances, needing no other parameters. They are widely tabled and very regular in form. We sometimes transform an estimator so that it has a normal sampling distribution. The Fisher z' transformation for the correlation coefficient is the most common example, but there are others. Many sampling distributions have another property called *asymtotic normality*. This means that their sampling distributions become increasingly like the normal distribution as their sample sizes increase and that their distributions are normal for infinite samples. For practical purposes, the

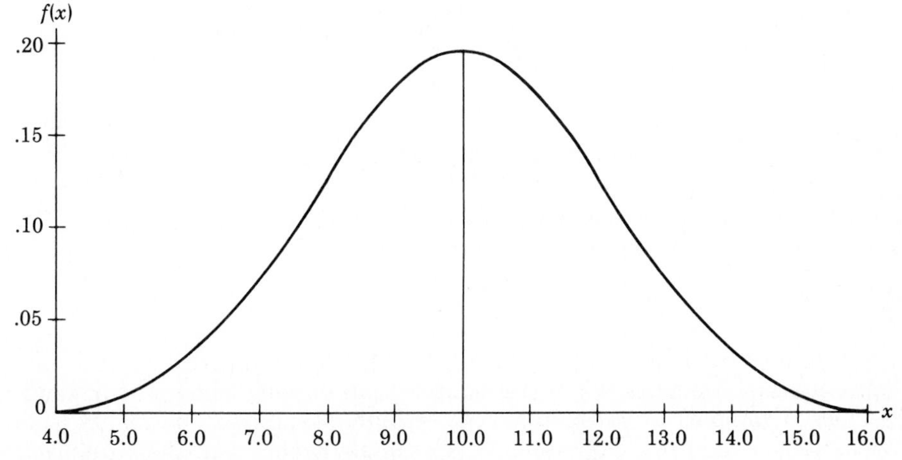

Figure 8-2 Normal population distribution with $\mu_x = 10.0$ and $\sigma_x^2 = 4.0$.

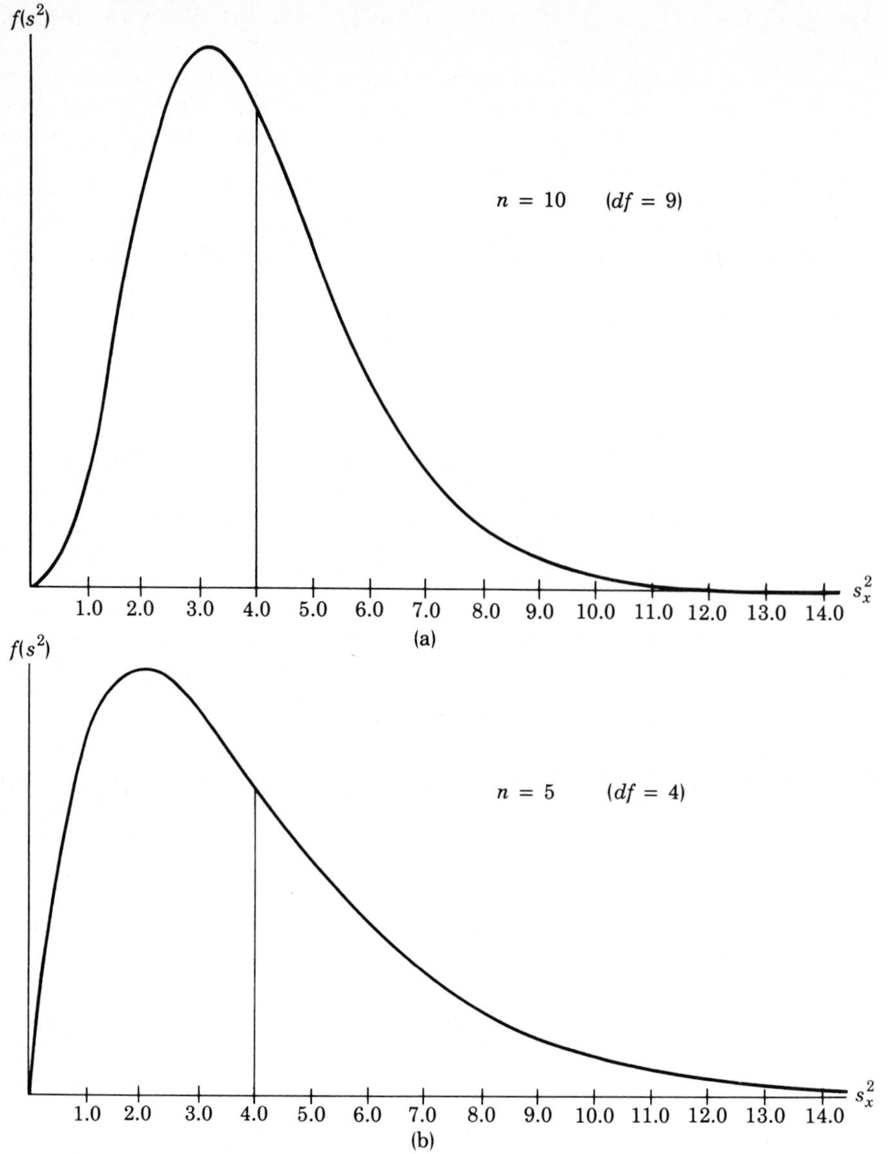

Figure 8-3 Sampling distribution of the sample variance of a normal distribution where $\sigma^2 = 4.0$. Sample size = 10 in (a) and 5 in (b).

sampling distributions can be treated as if they were normal if the sample is "large enough." Deciding whether a sample is large enough, however, depends on specifics. Examples of estimators that exhibit asymptotic normality include means from nonnormal populations, the sample proportion, the variance, and even the correlation coefficient.

A few sampling distributions are illustrated in Figures 8-2 and 8-3. Figure 8-2 is a parent population that is normally distributed with a mean of 10.0 and a variance of 4.0. Figure 8-3 shows the sampling distribution of the sample variance when $n = 10$ (upper) and $n = 5$ (lower). The mean of both distributions is 4.0, showing that the sample variance is an unbiased estimator. Note, however, that in both cases the mode is not a 4.0; somewhat lower values are more likely to occur than is the true variance. With the larger sample, the standard error is smaller and the distribution is also closer to normal.

In multivariate statistics, we are frequently concerned about making inferences regarding more than one parameter of a population at a time. After all, we are dealing with more than one variable, and each variable will have its own distribution and therefore its own parameters. If multiple variables are observed, there are multivariate distributions and parameters that describe the multivariate distributions—the correlations between variables, the regression coefficients for predicting one variable from several others, and so on.

We therefore have to consider the *joint* sampling distribution of several estimators. Estimators may not be independent of each other in the sense that an accidental overestimate of one parameter may tend to be systematically accompanied by an over- or underestimate of another parameter. On the other hand, some estimators are independent of each other. So, we need to think not only of the expected value and sampling variance of an estimator but of the sampling covariance between two estimators. This is important in understanding the sampling behavior of regression coefficients.

The sampling distribution is never observed directly; it is a completely theoretical concept, constructed on the basis of assumptions about the population. It can be approximated where the population is known by what is called "Monte Carlo simulation," a method whereby random samples are repeatedly drawn from the population and the value of one or more estimators is calculated. The value's of the estimator are tabulated each time, and as the number of samples increases, the resulting distribution approximates the sampling distribution for the statistic. A related technique is called "bootstrapping," in which the sample itself is treated as a population in order to find out the nature of an otherwise unknown sampling distribution.

HYPOTHESIS TESTING

The General Process

By far, the most common process in statistical inference is *hypothesis testing.* Here we compare the sample estimator to some hypothesized population value for a parameter, and reach a conclusion as to whether the sample value might reasonably have occurred if the hypothesized value for the parameter were true. When we say that some statistic is "significant," we mean that it has been compared to a hypothesized value, typically zero, for the corresponding parameter, and found to be different enough from zero to conclude that zero is not the population value.

Hypothesis testing is based on the sampling distribution of the statistic. In hypothesis testing we typically have a *research* hypothesis that we wish to test with the data. For example, we believe that there is a nonzero correlation between two variables, or we think that it is possible to predict some variable using some weighted combination of others, or we think that the difference between two means is not zero, and so on. We wish to test these research hypotheses with our data. To do so, we use a kind of reverse reasoning process. We say, "suppose the correlation *is* zero," or that "the dependent variable is *not* predictable," or that "the difference between the means *is* zero," or whatever. The latter are *null* hypotheses about population parameters. We then construct the sampling distribution for the sample statistic under the assumption that the null hypothesis is true. Then the actual value of the sample statistic is compared to the sampling distribution. If the obtained sample statistic is one that is unlikely—that is, if it falls in an extreme part of the sampling distribution—we reject the null hypothesis. We are saying in effect, "If the null hypothesis were true, it is very unlikely that a sample statistic like the one we found would occur. Therefore, the null hypothesis is most likely false." When we say that some statistic is "significantly different from zero" or simply "significant," we mean that the null hypothesis has been rejected because the obtained statistic falls into the extreme part of the sampling distribution constructed around the hypothesized value of zero.

We cannot directly test the research hypotheses because they are not specific; they cover a whole range of possible values for the parameter. It is not possible to set up a simple way of testing such diffuse hypotheses, so we fall back on formulating a specific converse to the experimental hypothesis that we can test formally. Rejecting it supports the plausibility of the idea that we are really interested in. If it happens that our experimental hypothesis *is* specific, then we can test it directly, and it becomes the "null hypothesis." This is done in some model-fitting procedures in multivariate statistics, particularly confirmatory factor analysis.

The Specifics

In any given application, these general ideas have to be translated into a specific procedure. First, we have to be more exact about what is meant by "unlikely to occur if the null hypothesis is true." This is defined as a probability that is called the *alpha level* of the test, where alpha refers to the probability that we will reject the null hypothesis when it is true. Ordinarily this is set at some low value such as .05 or .01 because we wish to make such false rejection unlikely; it is considered very undesirable to reject the null hypothesis incorrectly. The alpha level used defines cut-off values near one or both extremes in the sampling distribution. These "critical values" define two regions for the sample statistic: an acceptance and a rejection region. If the statistic falls into the acceptance region, the null hypothesis is accepted; if it falls into the rejection region, the null hypothesis is rejected. The regions are defined so that the rejection region contains a proportion of the sampling distribution equal to alpha when the null hypothesis is true.

There can be either one or two rejection regions, depending on the nature of the research hypothesis. If deviations from the null hypothesis only make sense if they occur in one direction, not both, then there is only one rejection region. This situation is called a one-tailed test. If the null hypothesis could be false in either direction, that is, if the true value of the parameter could be greater or less than the null hypothesis, then there are two rejection regions, one at each end of the acceptance region. This is a two-tailed test.

In either case, alpha is the proportion of the sampling distribution that is beyond the critical values. In the one-tailed case the area is all at one end of the distribution; in the two-tailed case, half of it is at each end. In fact, it is possible to have most of alpha at one end and only a little at the other, but this is not common in research.

Example

We can apply these ideas to the chi-square distribution shown in Figure 8-3, and use this distribution to test a specific hypothesis about a variance. In this case, we have a null hypothesis that is not zero. Suppose an experimenter wishes to use a small computer to present stimuli. She wants to present them at random intervals after a ready signal, so that the intervals are normally distributed with a mean of 10 seconds and a variance of 4 seconds, just like the population distribution shown in Figure 8-2.

The computer manual contains instructions for generating random times with a normal distribution with any specified mean and variance. She programs the instructions, and tests to see if the computer is operating correctly by examining a sample of 10 generated times:

4.79 8.83 8.22 13.66 11.37 11.20 2.25 8.01 14.46 10.59

Many characteristics of the numbers could be examined, but she decides to test their variance. It is 14.25, when it should be 4.0. Could this have occurred by chance? We use the chi-square sampling distribution to see if this is plausible. Theoretical statistics tells us that $(n - 1)s^2/\sigma^2$ has the chi-square distribution with $n - 1$ degrees of freedom (df), so we find $(n - 1)s^2/\sigma^2 = 32.0625$. This should be compared to the theoretical sampling distribution. Since there is no reason to expect the variance to be high rather than low, a two-tailed test is appropriate.

The experimenter decides on an alpha of .05, which means .025 in each tail. According to Table 2 of the Appendix, which contains percentage points of the chi-square distribution, we see that for 9 df the critical points are 2.70 for the lower and 19.0 for the upper. The obtained value of 32.0625 is far beyond the upper limit, so the null hypothesis is rejected; we conclude that the program is not operating correctly.

This process is shown graphically in Figure 8-4, which is the same as the Figure 8-3, except that the abscissa has been relabeled to reflect the chi-square scale rather than the scale of the variance itself. Note that if the chi-square was between 2.70 and 19.00, corresponding to sample variances of 1.2 to 8.44, the null hypothesis would have been accepted because the statistic fell in the acceptance region.

Therefore, in almost all applications, hypothesis testing consists of formulating a research hypothesis that reflects a parameter of a distribution, turning the formulation around to make a null hypothesis about the parameter, and using a sampling distribution to test the hypothesis. This is similar to the more intuitive process we sometimes use. Suppose all the numbers generated by the computer had been whole numbers, or that they had all been between zero and one, or all

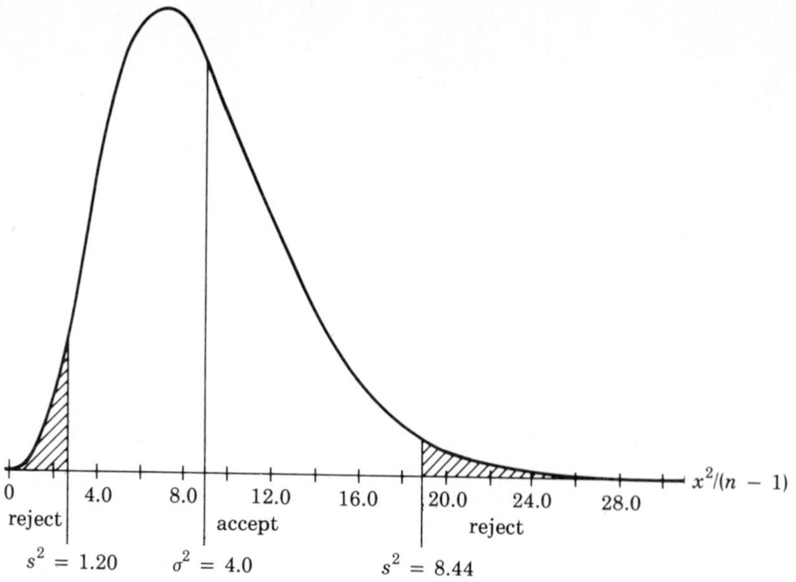

Figure 8-4 Chi-square distribution with 9 degrees of freedom. Acceptance and rejection regions are labeled for $\alpha = .05$, which is two-tailed. Note that this is the same distribution as Figure 8-3, except that the scale is different.

negative. We could immediately have concluded that something was wrong because all of these sequences are extremely unlikely to occur if the computer is sampling from a normally distributed population with a mean of 10 and a variance of 4. Hypothesis testing formalizes this kind of inferential process and makes it efficient, allowing us to reach valid conclusions more often.

Power

Hypothesis testing procedures guarantee that, when true, the null hypothesis will be rejected just alpha of the time, provided the subsidiary assumptions we are making—such as normality of populations, independent random sampling, and so on—are also true. Another side to the issue, though, is that we want to reject the null hypothesis when it is false. If the true variance of the generated numbers in our example was not exactly 4.0, but was not too different (say 2.0, 3.0, 6.0, or even 10.0) then there might be a fairly good chance that the sample variance would fall between 1.22 and 8.44, leading us to accept the null hypothesis, although incorrectly. The ability of a hypothesis-testing procedure to reject the null hypothesis when it is false is called its *power,* and our test should have power sufficient for our purpose in doing the study and the analysis.

To understand statistical power we have to suppose again. We suppose that σ^2 was really 8.0 instead of 4.0. If a random sample of 10 is drawn from a normally distributed population with a variance of 8.0, there is a sampling distribution for the variances the experimenter might get. It too follows the chi-square form, although it centers around 8.0 instead of 4.0. Sometimes she would get lower values than 8.0, and sometimes higher. This sampling distribution is shown in

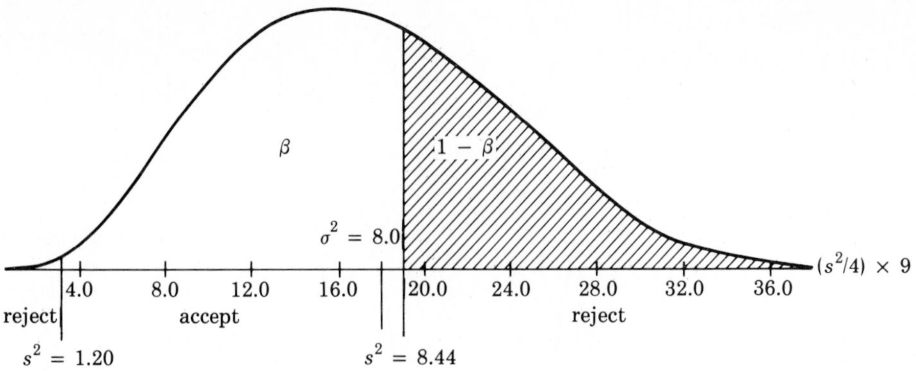

Figure 8-5 Sampling distribution for $(s^2/4)/(n-1)$ when the true variance $\sigma^2 = 8.0$, and $n = 10$. Vertical lines indicate positions of boundaries of the acceptance region of Figure 8-4.

Figure 8-5. Remember that in testing the hypothesis $\sigma^2 = 4.0$, sample variances of 1.20 to 8.44 would have led to accepting the null hypothesis. Over half the time, the sample variance will be between these limits when the true variance is 8.0, as illustrated in Figure 8-5. Thus, the power is relatively poor here. In fact, it is only when the departures from 4.0 are fairly extreme that there is a very good chance of rejecting the null hypothesis.

The power of a statistical test is measured by the probability of rejecting the null hypothesis when it is false. This is typically denoted $1 - \beta$, where β is the probability of wrongly accepting a false null hypothesis.* Wrongly accepting is called a Type II error, whereas wrongly rejecting is referred to as a Type I error. Statisticians sometimes refer whimsically to Type III errors (making an arithmetical mistake), Type IV errors (doing the wrong analysis), and so on, although Types I and II are the only errors that have formal status.

The probability of making a Type I error is completely controlled by the investigator; it is equal to the alpha level set. The Type II error probability is influenced by a number of factors. For example, it may be affected by the amount of departure from the null hypothesis. If departure is extreme, then power will tend to be high. When samples are large, the sample value can be expected to be close to the true value, so the acceptance region is narrow. Furthermore, the sample statistic is likely to be close to its true non-null value and therefore the sample statistic is unlikely to fall into the acceptance region. Finally, the power is affected by the alpha level used. With a small alpha level, the acceptance region is relatively wide, that is, a relatively large deviation is needed for significance. Therefore, there is a conflict between avoiding Type I and Type II errors. Similarly, the use of a one-tailed test increases power if the appropriate tail is being used because the boundary of the acceptance region is being moved away from the true value. The effects of true parameter value, sample size, and alpha are illustrated in Figure 8-6.

Other considerations also influence the power of an analysis. For example, research designs may differ in their power through reducing extraneous variance.

---

*β is also used to stand for the population regression weight. The intended meaning for β should be clear from the context, however.

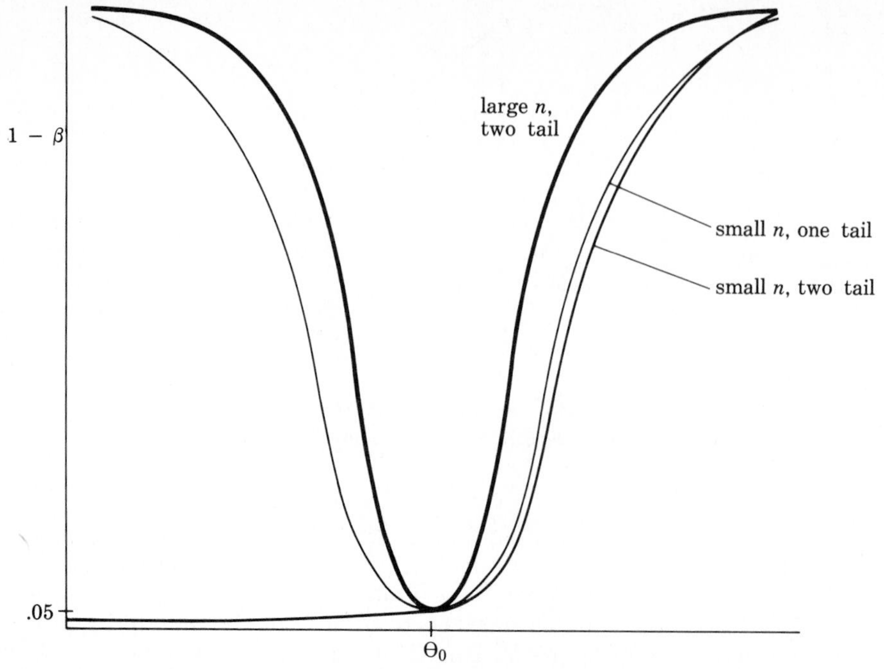

Figure 8-6 Power functions for a hypothetical statistic as a function of the true value of the parameter θ_0. The different curves represent large and small samples and one- and two-tailed tests.

Also, there may be more than one way to analyze the same data, and more than one statistical test that will shed light on the same research hypothesis. These need to be kept in mind also since they affect power.

CONFIDENCE INTERVALS

The General Idea

Hypothesis testing is used to rule out some particular value for a parameter as being unlikely in the light of the data. But what is the true value for the parameter? The sample estimator is a best guess as to its value, but we know that this value cannot be exactly correct. Nearby values will be almost as reasonable. The *confidence interval* (c.i.) is a way of defining the range of values for a parameter that is still plausible given the sample data.

Suppose a sample correlation is .40, based on an n of 28. This correlation is significant; the null hypothesis, $\rho = 0$, can be rejected at $\alpha = .05$, although just barely. Our best guess is that ρ is .40, but what other values are plausible? We define this range by calculating a confidence interval. For these data, we can be pretty sure (that is, 95-percent sure) that ρ is somewhere between .03 and .67. How we arrive at exactly those limits is a technical question that will be addressed later. The point is that the sample value, together with the confidence interval, provides the best guess as to the true value and the range of other values the parameter could reasonably have. In this case, the sample correlation is moderately positive, but we see the true value could be close to zero or as large as .67.

It is important to be clear on just what a confidence interval represents. If the 95-percent confidence interval for the correlation is .03 to .67, it means that if we say that the correlation *is* between .03 and .67, this statement has a 95-percent chance of being correct, and a 5 percent chance of being wrong, that is, of the true correlation either being less than .03 or greater than .67. It is called a "confidence" interval, not a certainty interval, because we are confident that it contains the parameter, but not certain. One of the things that the study of statistics teaches is that the only thing that is certain is uncertainty.

Greater than 95-percent confidence can be achieved, but only at the cost of a wider interval. For the present example, a 99-percent c.i. would run from −.09 to .73 and a 99.9-percent c.i. would run from about −.23 to .79. Even greater confidence would require even wider limits. The only interval that we can be absolutely certain contains the true correlation is the interval from −1.00 to 1.00. We know that interval contains the parameter—even without any data! There is a tendency for the concepts of an acceptance region to be confused with the confidence interval, and indeed they are related. The main difference is that the acceptance region gives the limits within which the sample statistic will fall with .95 probability when the null hypothesis about the parameter is true, whereas the confidence region is constructed around the sample statistic and is the range within which the true value of the parameter lies with .95 probability. Ninety-five percent of all such statements are true (the correlation is in the interval) and five percent are false (it is outside the interval). If we want to reduce the percentage that are outside, we must make the interval wider.

Intuitively, the sample estimate, such as the r_{xy} of .60, represents our best guess as to the true population correlation. The c.i. represents the range of alternatives that are not unreasonable in the light of the data.

It is time to formalize our description of a confidence interval. This is given in the following generalization:

> The $1 - \alpha$ confidence interval for a parameter θ is that set of values for θ, $\theta_l \leq \theta \leq \theta_u$, such that
>
> $$Pr(\theta_l \leq \theta \leq \theta_u) = 1 - \alpha \tag{8.1}$$

The boundaries θ_l and θ_u are constructed from sample data, and depend on the sampling distribution for the estimate of θ, on the sample size, and on how small we wish to make α. The confidence interval is formed around the sample estimate and can be thought of as including all the reasonably plausible alternative values for the parameter. Where sampling distributions are symmetric, the c.i. is symmetric around the estimate. For example, a c.i. for a regression coefficient β_j is $b_j \pm \hat{\sigma}_{b_j} t_\alpha$, where $\hat{\sigma}_{b_j}$ is the estimated standard error of b_j and t_α is the value of student's t for the appropriate alpha level and degrees of freedom.

There is a subtle point of statistical inference in the probability expressed in

Equation 8.1. What does it mean to say that the true value of the parameter is within the interval with probability $1 - \alpha$? In any given instance, it either is in the interval or is not; we never know which is true.

We can understand the probability in the following way. Suppose we make many assertions about parameters being within their intervals, each statement being based on its own data. The c.i. means that $1 - \alpha$ of those statements will be true, and that the parameter will be within the interval we say it is in. The remaining α of the statements will be false; in those cases the parameter will be outside the interval. It is in this sense that we can say, for example, that with $n = 28$, and $r_{xy} = .40$, the .95 c.i. for ρ_{xy} is .03 to .67.

Calculating Confidence Intervals

Calculating specific confidence intervals involves the use of whichever sampling distribution is appropriate for the estimator from which it is calculated. First, we find the point on the continuum of the sampling distribution that cuts off the upper $\alpha/2$ and add it to the estimator. Then we find the point that cuts off the lower $\alpha/2$ and subtract it from the estimator. These form the limits for the c.i. This procedure is simplest in a symmetric sampling distribution in which the points in the sampling distribution are the same except for the sign.

For the nonsymmetric distributions, two approaches can be taken. One is to transform the scale of the statistic so that it has a symmetric distribution, and the other is to work with the nonsymmetric distribution directly.

The correlation coefficient has a nonsymmetric sampling distribution. In fact, the form of the distribution depends also on the value of ρ. However, the Fisher z' transformation of r has a nearly normal sampling distribution with a standard error that depends only on n: $(n - 3)^{-1/2}$. What we do here is transform the obtained correlation by the transformation, presented in Appendix Table 3, and define the limits of the c.i. on the z' scale. Then we transform these limits back to the r scale by using the table in reverse.

This is how we found the limits of .03 to .67 for our correlation of .40 on a sample of 28. For z', the relevant sampling distribution is normal, and a standard score of 1.96 cuts off the upper .025 of the distribution, according to Appendix Table 1. Since the standard error is $(n - 3)^{-1/2}$, multiplying this by $(28 - 3)^{-1/2}$ gives $(1.96)(.2) = .392$. Adding this to the z' transformation of .40, which is .424, gives .816. Subtracting the same amount to get the lower limit gives .032. So, the .95 c.i. on the z' scale is .032 to .816. Translating these limits back to r, using some interpolation, gives .03 to .67, as expected. Actually, the use of Appendix Table 3 is not crucial since $z' = \frac{1}{2} \log_e [(1 + r)/(1 - r)]$. This can be calculated with a moderately sophisticated pocket calculator.

The second approach, working with the sampling distribution directly, can be illustrated with the variance. For example, as we saw earlier, the sampling distribution for the variance is the $\chi^2/(n - 1)$ distribution. Given our sample variance of 32.0625, obtained with $n = 10$, we found from Appendix Table 2 the points that cut off the lower 2.5 percent and the upper 2.5 percent. They are 2.70 and 19.0, respectively. Dividing these by 9, and multiplying by s^2 yields 9.61875 to 67.6875 as the c.i. for σ^2.

There is a direct connection between hypothesis-testing and confidence intervals. Suppose we ask, "Given this sample statistic, what null hypotheses about the population parameter could be rejected?" If we systematically tried to find null hypotheses that could be rejected, we would conclude that those parameters outside the c.i. are the ones from which the sample value is significantly different, whereas those within it are ones that would not be rejected if they were the null hypothesis. Thus we may think of the c.i. as containing those parameter values from which the sample statistic is not significantly different.

MULTIVARIATE CONSIDERATIONS

Multiple Parameters and Statistical Inference

In multivariate methods, the investigator attempts to make inferences about more than one parameter, and this introduces a new class of problems in deciding how to go about making those inferences. As a result, outright errors in inference—that is, conclusions drawn that ought not to be drawn—are more frequent than in univariate analysis. In fact, some of the most widely used computer packages almost encourage inferential errors because of the way they present results.

A two-predictor multiple regression actually represents a rather complex situation. As always, the first question is, what is the question? What are we trying to come to conclusions about? Exactly what conclusions are we trying to reach? Are we trying to find out what P is in the population? Are we most concerned about the standard error of estimate? Do we want to know whether or not β_1 is zero, or whether both β_1 and β_2 are zero? What is the c.i. for β_1? For β_2? For both simultaneously? The list could go on. Not all the possible questions are appropriate except in very special cases, but there usually are several possibilities, and each one may correspond to a type of inferential analysis.

The questions the investigator is primarily concerned about can usually be translated into statements about the regression coefficients, including tests of hypotheses about them or confidence intervals for them. For example, the hypothesis $P = 0$—P is an uppercase Greek rho—can be expressed as $\beta = 0$, where β is the vector of population regression weights. In the same way, an analysis of variance tests the hypothesis that $\mu_j - \mu_k = 0$ for all pairs of means. On the other hand, we would also like to know which weights are not zero, or which pairs of means are in fact different. In addition, we would like to know the confidence intervals for the regression coefficients, or for the mean differences. That is, we would like to make inferences about individual parameters, as well as omnibus conclusions about whether or not they are all zero.

Principles of Probability

In order to understand the problems and strategies of multiple inferences, we need to review a few of the principles of probability and introduce some notation. Probability means that numbers between 0 and 1.0 are attached to events. The events can be anything that can be clearly specified—for example, that it will rain next Tuesday, that you will get an A in this course, that ρ is between .60 and .80, or that b falls between .35 and .65. The expression $\Pr(A)$ is used to stand for the probability of some event A.

Corresponding to any event's occurrence is its opposite, or nonoccurrence: that it does not rain on Tuesday, that you get some grade other than A, that ρ is not between .60 and .80, or that b is not between .35 and .65. The nonoccurrence of an event is called its *complement*, whose probability is denoted $\Pr(\bar{A})$. An important principle is

$$\Pr(\bar{A}) = 1 - \Pr(A) \tag{8.2}$$

We also need to consider *joint* events, the co-occurrence of two or more events. For example, suppose you receive an A (event A) *and* that it also rains on Tuesday (event B). That is, both events occur, or the statements specifying each event are both true. Such compound or joint events and the associated probabilities are referred to as $\Pr(A \text{ and } B)$—that is, the probability that both A and B occur or are true.

A *union* of events means that one or the other or both occur or are true. The probability that one or the other or both occurs is referred to here as $\Pr(A \text{ or } B)$. The "or both" part of the definition is important.

Joint events and unions are connected by the formula

$$\Pr(A \text{ or } B) = \Pr(A) + \Pr(B) - \Pr(A \text{ and } B) \tag{8.3}$$

This states that the probability of the union of two events is the sum of the probabilities of the separate events minus the probability of their joint event. The probability of the joint event needs to be subtracted because otherwise the instances of co-occurrence are being counted twice, once in $\Pr(A)$ and once in $\Pr(B)$. Both unions and joint events are illustrated in Figure 8-7. Here, the area of the enclosing square is 1.0. We can think of the probabilities of various events, unions, and joint events as being reflected by the areas of the corresponding circles or segments.

In some situations, A and B are *mutually exclusive*. This means they cannot both occur or be true. For example, a subject cannot be both blue-eyed (A) and brown-eyed (B): $\Pr(A \text{ and } B) = 0$. In this case, $\Pr(A \text{ or } B)$ reduces to $\Pr(A) + \Pr(B)$. Such events simplify many applications of probability because we do not have to find each $\Pr(A \text{ and } B)$.

Conditional probabilities are also important. A conditional probability, denoted $\Pr(A|B)$, is the probability that an event occurs given that another event has occurred or must occur. For example, we may express in this way the probability that a subject has brown hair (A) given that he has brown eyes (B). An important relationship is

$$\Pr(A|B) = \Pr(A \text{ and } B)/\Pr(B) \tag{8.4}$$

where Pr(brown-haired given brown-eyed) is the ratio of brown-eyed brunettes to the total number of people having brown-eyes.

A special situation occurs when $\Pr(A|B) = \Pr(A)$. This means that the occurrence of B does not affect the probability of A. Such events are called *independent*. Because of Equation 8.4, an equivalent statement is that if A and B are independent, then

$$\Pr(A \text{ and } B) = \Pr(A)\Pr(B) \tag{8.5}$$

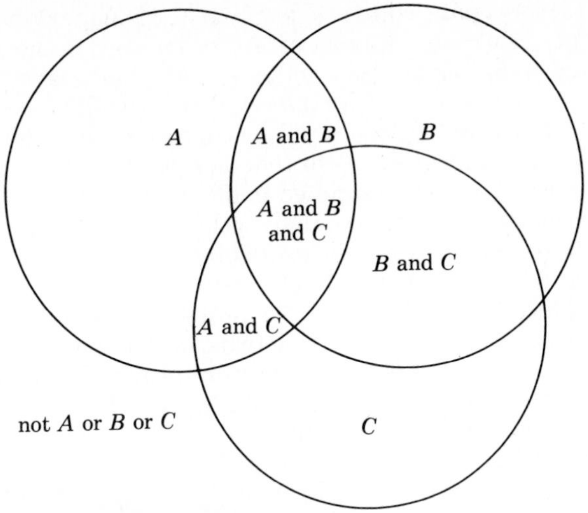

Figure 8-7 Venn diagram of events A, B, and C, and their intersections and unions.

If x is a variable with possible values $x_1, x_2, \ldots, x_j \ldots$, and y is another variable with possible values $y_1, y_2, \ldots, y_k \ldots$, then the probability that a variable takes on a certain value is referred to as $\Pr(x = x_1)$, or $\Pr(y = y_3)$, and so on. These can be abbreviated as $\Pr(x_1)$, $\Pr(y_3)$, and so on. The probability of the joint occurrence of x_1 and y_3 is $\Pr(x_1 \text{ and } y_3)$ and the probabilities of unions are $\Pr(x_1 \text{ or } y_3)$, and so on. Conditional probabilities are $\Pr(x_1 | y_3)$. Different values of x are mutually exclusive, because x cannot take on different values at the same time.

Variables (as opposed to just pairs of events) are independent if $\Pr(x_j \text{ and } y_k) = \Pr(x_j) \Pr(y_k)$ for all j and k. Independence of variables is important because it allows us to determine the joint probabilities from just the simple or *marginal* probabilities.

The principles of probability for complex events described in this section are important for multivariate statistical inference because such inference often involves multiple statements about parameters. The conclusions we wish to draw from our data may revolve around the joint truth of two or more statements about parameters or around the joint occurrence of two or more events or the occurrence of their union.

STRATEGIES IN MULTIVARIATE INFERENCE

Complex Null Hypotheses

Suppose we have two predictors and are trying to reach a conclusion about the regression weights for predicting a dependent variable—for example, the length of time it takes to read a paragraph is to be predicted from the number of words in the paragraph and the average number of letters per word in the paragraph. Thus there are a number of paragraphs and for every paragraph there are three numbers: the time taken to read it; the number of words, x_1; and the average number of letters per word, x_2. We could be concerned with the null hypothesis that both regression weights are zero, or that $\beta_1 = 0$ or that $\beta_2 = 0$. The methods of testing each hypothesis will be examined in detail in the next chapter. For now, suppose that in a given instance the null hypothesis is true for both weights. If we test *both weights simultaneously*, the probability of a Type I error is α, because the null hypothesis is true. Alternatively, we can test each weight separately, using the same α level, say .05. What is the probability of making a Type I error now? It is no longer .05 now that we have two chances to make an error, one for each weight. Define event A as a Type I error on the first weight and event B as a Type I error on the second weight. "Making a Type I error" could now mean making one on β_1, the event (A), making one on β_2, the event (B), or making one on both, the event (A and B). So, the probability of making at least one Type I error if we test b_1 and b_2 separately is Pr(A or B).

Equation 8.3 relates the probability of a union to the probability of the separate events. In this instance, Pr(A) = Pr(B) = α, so Pr(A or B) would be found if we knew Pr(A and B). Unfortunately, this probability is hard to find, because it depends on specific properties of the data.

All is not lost, however, for we can place some boundaries on Pr(A or B), and even make a good guess as to which boundary it is closest to. The lower limit for Pr(A and B) is zero, because it is a probability. This would make Pr(A or B) equal to 2α. This would only happen if A and B were mutually exclusive, that is, if they could not both happen with the same data. This does not seem reasonable, as there is no reason to expect that a Type I error for one parameter precludes a Type I error about the other. The world usually gives us the opportunity to make as many errors as we want. Theoretical statistics beyond the scope of this book shows that this is the case here, also. Indeed, we could be wrong about both parameters. The formula shows that this has its advantages, because it reduces the probability of being wrong about one or both. What is most important is that we have found one boundary, Pr(A or B) $\leq 2\alpha$.

The other boundary is α. If it is true that *whenever* a Type I error is made about β_1 one is also made about β_2, then Pr(A and B) = Pr(B) = Pr(A). Then Equation 8.3 reduces to just Pr(A or B) = Pr(A) = α. Thus, under any circumstances, the probability of at least one Type I error given two tests using α is somewhere between α and 2α; For k tests, the limits would be α and $k\alpha$.

It is hard to decide with certainty which boundary is likely to be closer, although usually we expect it to be closer to 2α. The reason for this goes back to the concept of independence, Pr (A and B) = Pr(A)Pr(B) for independent events. Sometimes it is easier to find the probability of an event by subtracting the probability of its complement from one. The probability of making at least one Type I error is one minus the probability of not making any. If A and B are independent, so are their complements: Pr($\overline{A}$ and $\overline{B}$) = Pr($\overline{A}$)Pr($\overline{B}$). The probability of not making a Type I error is $1 - \alpha$, so the probability of not making an error on either of two independent tests is $(1 - \alpha)(1 - \alpha)$ or $(1 - \alpha)^2$. If α is .05, this is .95 × .95, or .9025. Subtracting this from unity leaves .0975, which is pretty close to .1000, or 2α. Thus, testing two *independent* true null hypotheses means that the probability of rejecting at least one of them is a little less than 2α.

If we are testing b_1 and b_2 for significance in the same sample, however, they are clearly not independent. Theoretical analyses show that the degree of nonindependence is not very great unless the correlation between the variables is substantial, so we may not be far off by treating the tests as if they were independent. The nonindependence may make the probability either closer to α or to 2α, depending on its source, but generally it will be close to 2α. Assuming that the probability is 2α is conservative, but probably only mildly so. Now we have a useful principle: If k hypotheses are tested at a given alpha level, the probability of at least one Type I error occurring when all null hypotheses are true is between α and $k\alpha$, although probably closer to the latter. If the tests are independent, the probability is $1 - (1 - \alpha)^k$; this limit is probably not far off in the majority of common cases since then there is only a moderate lack of independence.

In multivariate analyses, we want to know what our alpha level actually is and attempt to control it while keeping power at a reasonable level. The elementary probability principles described above are basic to these goals.

HYPOTHESISWISE AND EXPERIMENTWISE ALPHA LEVELS

When testing multiple hypotheses or forming multiple confidence intervals, we often want to keep the probability of making *any* Type I errors at α. This is called controlling the *experimentwise* or *analysiswise* alpha level. There are a number of procedures for doing this. We will encounter several of them, including Bonferroni tests, Scheffé intervals and their extension to Roy intervals, the Tukey tests, and omnibus testing.

Controlling analysiswise alpha levels is important when the research is exploratory and the investigator is looking for anything in the data that might be of value. Because there are so many opportunities for false positives, it is important to have conservative methods for controlling Type I errors.

As a field of research advances, and an investigator becomes more sure of his ground, the research hypotheses that lead to statistical tests are likely to have genuine status as hypotheses. In that case, the investigator should be willing to risk a probability equal to α of making a Type I error on each test. Then it is legitimate to treat each test as if it were the only one being made, and to use the ordinary statistical tests of hypotheses. In such cases, we are said to be controlling the alpha level hypothesiswise. In some contexts it is also called a comparisonwise or parameterwise alpha.

When making post hoc tests or comparisons, the analysiswise alpha level is almost always primary, except in cases where an omnibus test has been made. The hypothesiswise approach, on the other hand, is almost always appropriate when making a priori tests or comparisons.

Controlling Analysiswise Alpha Levels

Most multivariate analyses are exploratory in that we do not know just what correlations or mean differences might be present. Therefore, we need to control the analysiswise alpha level. In the majority of cases this is done by having a vector null hypothesis, such as one stating that the vector of deviations of group means from the overall mean is zero. This is a null hypothesis about a vector, and the acceptance region is correspondingly multidimensional, centered around the null hypothesis.

The boundaries of the acceptance region are constructed such that when the null hypothesis is true, the probability is at most α that a point representing sample estimates will fall outside it. The tests described in the following sections correspond to different sizes and shapes for their acceptance regions.

Bonferroni Tests

The most appropriate approach to hypothesis testing in the multivariate case depends on the purpose of the analysis. (What is the question?) Often, the question is, "Is prediction possible at all, using x_1 and/or x_2?" The most appropriate method here is the test of significance for the multiple correlation (see Chapter 9), which in effect tests the hypothesis that all the β_j are zero, also called an omnibus test.

Occasionally, it is desirable to test this overall hypothesis by testing each regression coefficient, adjusting the alpha level to take account of the fact that several tests are being made. If b_1 is tested using $\alpha/2$ instead of α to define the critical value, and the same is done for b_2, the reasoning of the section on probability leads to the conclusion that the probability of rejecting *at least* one null hypothesis when both are true is then no greater than the sum of the two separate probabilities, or α. Thus, if we tested b_1 and then b_2 at the .025 level, the probability that we find one or the other to be significant when both population values are zero is no greater than .05. This is the *Bonferroni* approach to testing hypotheses for several parameters. In general, the principle states that:

> If k hypotheses are tested at the α/k level, the probability that at least one Type I error is made is no greater than α.

Unfortunately, the tables of test statistics usually give only a limited number of values that are often not exactly the ones that are needed. This rarely leads to ambiguity, however. For example, if we are is testing three hypotheses, and seek an overall α of .05, the relevant probability is .05/3, or .0167, which is not in the table. However, .05 and .01 are, and the computed value of the statistic presents no problem if it is beyond the .01 level (reject) or does not reach the .05 (accept), but only if it falls somewhere between. Interpolation can then be used, although it is likely to be crude, it may well be accurate enough. Many modern computer programs provide exact significance levels along with the test statistics.

Omnibus Test Followed by Individual Tests

The omnibus test of significance is a test of a hypothesis about a vector. For example, in multiple regression we test H_o: $\boldsymbol{\beta} = \mathbf{0}$. That is, we hypothesize that all the regression coefficients are zero. As with a univariate test, there are acceptance and rejection regions, and the test should be such that when H_o is true, the probability is at most α that the vector of sample regression coefficients falls in the rejection region. The acceptance region is an area for two predictors, a volume for three, and in general, a p-dimensional hypervolume. The term "hypervolume" sounds impressive, but there is no need to hyperventilate; it is just a term that describes the idea of a volume extended to any dimensionality.

Depending on the strategy adopted, there is considerable choice in the shape of the acceptance region. For example, when we use a Bonferroni test on the individual regression coefficients, it is a p-dimensional box; we are rejecting H_o: $\beta = \mathbf{0}$ when any one of the coefficients exceeds its adjusted acceptance region. For the omnibus F test described in Chapter 9, the acceptance region is a p-dimensional ellipsoid centered at the point defined by the null hypothesis. This ellipsoid has a considerably smaller volume than does the Bonferroni box, so power is higher.

Null hypotheses other than $\beta = \mathbf{0}$ are also possible. Some theory may dictate a non-null H_o, or common sense may suggest H_o: $\beta = (1,1, \ldots, 1)'$. In these cases the acceptance region has the same shape but is centered on the null hypothesis, assuming normal sampling distributions.

The omnibus test keeps the Type I error probability at α. When H_o is rejected, it is natural to want to know which of the weights are non-zero. It seems reasonable to take a hypothesiswise approach to testing individual weights, once the overall null hypothesis has been rejected. The reasoning here is that we are now generally willing to accept the probability of a Type I error about each coefficient separately, rather than taking the stance that we want to keep the probability at α of wrongly rejecting null hypotheses about *any* of the coefficients. If we hold the latter point of view, then we would need a procedure that gives an analysiswise alpha control such as making a Bonferroni adjustment of the alpha level for each test. This seems excessively conservative under most circumstances, and is likely to lead to puzzling results. For example, we may reject the overall null hypothesis $\beta = \mathbf{0}$, but when we test individual coefficients with an analysiswise alpha level it may turn out that no hypotheses about individual coefficients *can* be rejected. This is because the analysiswise adjustment makes the acceptance region very wide for the individual coefficients.

The concept of a confidence interval also extends to more than one parameter, but since it too becomes multidimensional, we should call it a region. Such a region requires a mathematically defined boundary so that statements such as "β is within the following boundary" would be true with probability $1 - \alpha$. In the case of regression coefficients, the region is defined by the equation of a p-dimensional ellipsoid. The dimensions of the ellipsoid will depend on the observed variance-covariance matrix, and the sample size. This is the same shape as the acceptance region for a null hypothesis, except that the acceptance region is centered on the point representing the null hypothesis, and the confidence region (c.r.) is centered on the sample coefficients.

Thus the confidence region may be rather difficult to define or understand intuitively. Still, we can make use of the idea that it contains all those values for β that cannot be rejected by the data at the given alpha level. This allows us to explore the boundaries of the confidence region by suggesting a number of different values for the vector—or whatever set of parameters is relevant—and testing to see if these "hypotheses" can be rejected. In this way, a general idea of the size and shape of the confidence region could be gained. This idea is developed further in Chapter 9.

Open-Minded Significance Testing in Multivariate Analysis

Significance testing is particularly crucial in multivariate analysis and we must be particularly alert to the dangers of post hoc hypotheses and misleading alpha levels. The presence of multiple variables and the multitude of statistics concerning their relationships provides myriad opportunities for "seeing" something in the data. We want to find things in our data, and our eyes and brains are well-adapted to detecting patterns. Unfortunately, the patterns may be the result of purely random events, and we end up chasing a will-o'-the-wisp.

For example, if we collect data on 20 predictors and examine their correlations with the dependent variable, throwing out the 18 that have the lowest values, we cannot test the significance of regression on the remaining two as if the 18 did not exist. Yet this is just what widely available computer programs are likely to encourage. The same problem comes up in a variety of forms. We must be alert to the possibility and ask ourselves what the range of alternatives really is in any analysis.

A nonstatistical anecdote may bring the problem into better focus. The first spring my family became houseowners, we dug a flower bed in the former cow-pasture which was our backyard and optimistically planted rows of several varieties of flower seeds. Most sprouted within the times specified on the seed packages, except for one row of flowers—the larkspur. This section was anxiously scanned every day. Eventually, however, some small shoots appeared, which seemed to be more or less in the rows where the larkspur had been planted. Moreover, these shoots had pointed leaves, as the seed package indicated, unlike the common weeds that had also sprung up throughout the garden. The main thing, though, was that they seemed to be in rows, as they would if they had been planted. There were somewhat fewer shoots than one might have hoped for, but with cultivation and care they were likely to produce a brave showing, even after removing the ones that were most out of line. After all, some seeds do stray, fall accidentally from hand or package, or are disturbed by bird or foot.

Time passed, and the flowers grew, including the larkspur. They now stood in three sturdy rows whose mild irregularity had almost disappeared. The trouble was, they were looking less and less like the picture on the now-frayed seed package. Doubt set in, but was pushed aside. Look how straight the rows are now! Would that happen by chance? Besides, the plants are so neat and green. Buds showed, and eventually opened, after a fashion, revealing not the bright violets, pinks, and whites expected, but the most nondescript, dullest, greenish-grey blooms imaginable. Doubt was soon replaced by certainty: these were not larkspur but one of the less common, and less colorful variety of meadow weeds of the area. By then it was too late in the year to plant again.

In data as well as in flower beds, the eye is anxious to see flowers, not weeds, and to perceive a crooked line as fairly straight. Tests of hypotheses that are uninfluenced by peeks at the data by either the investigator or the computer are one of the few safeguards that the conclusions arrived at will be worth looking at.

SUMMARY

With few exceptions, the investigator analyzes data from a sample of observations for the purpose of generalizing to a population. The numbers that characterize the data—statistics—are used to estimate numerical characteristics of the population, called parameters. The central tool in this generalization process is the sampling distribution. This is derived by assuming specific characteristics for population parameters, and thereby deducing the likelihood that particular values for the sample statistics would occur if the assumption were true. The sampling distribution is thus a probability distribution for possible values of the statistic. Important characteristics of the sampling distribution are its mean—the expected value of the statistic—and its standard deviation, which is called the standard error of the statistic.

A statistic is used as an estimator of a population parameter. Important qualities of estimators are that they be unbiased, *consistent*, and have minimum variance. The sampling distribution of a statistic depends on the exact form of the population distribution as well as on the nature of the statistic under consideration. Important sampling distributions included the normal, the chi-square, the t, and the F distributions. A characteristic of many commonly used estimators is asymptotic normality.

An important aspect of inference is hypothesis testing. In hypothesis testing, the investigator proposes that a population parameter has a certain specific value. Then, a statistic is computed from the data and the hypothesis is accepted or rejected, depending on where the obtained value falls in the sampling distribution for the statistic. In effect, cut-off values are set up such that the sample statistic, or a test-statistic derived from it, will fall outside those values only a specified proportion of the time, α. This guarantees the hypothesis will be rejected α proportion of the time (Type I error) when it is true. If H_o is false but the test statistic falls within the acceptance region, leading to an erroneous acceptance of H_o, this is called a Type II error; its probability is denoted β. β probability depends on a number of factors, whereas α is determined simply by the investigator's choice.

A second important inferential tool is the confidence interval. This is determined on the basis of the sample data, which is stated to contain the population parameter with a certain degree of confidence $1 - \alpha$, such as .95. The sample estimate is our best guess as to the value of the population parameter and the c.i. provides a band around the best guess that we are pretty sure does include the true value. Confidence intervals are constructed using the sampling distribution of the parameter in much the same way that hypotheses are tested.

Multivariate statistical inference introduces new complications because we wish to form inferences about many parameters and because many statistics are available from the data. In order to clarify the nature of these problems, it is desirable to review the basic principles of probability theory, defining the probability of an event A, $\Pr(A)$; the probability of the complement of this event, $\Pr(\overline{A})$; and the probability of the union, $\Pr(A \text{ or } B)$; and intersection of events, $\Pr(A \text{ and } B)$. Basic to the specification of probabilities is a sample space. An important kind of probability is conditional probability, the probability that an event occurs given that it is known that another event does occur. The independence of events is defined in terms of a special characteristic for their conditional probability, and the independence of variables—that is, all possible combinations of events—is done in a similar way.

The probability of unions and intersections of events is very relevant to multivariate inference. The null hypotheses and the confidence intervals are complex because they are concerned not with just one parameter but with many. If one tests k hypotheses, each at α, and all of the null hypotheses are true, the probability of making at least one Type I error is at least α and at most $k\alpha$, and most likely fairly close to the latter. Therefore, if one tests each at α/k instead of α, the probability that any Type I error is made is at most α. This approach to hypothesis testing in the multiple hypothesis case is called the Bonferroni approach. It seeks to control the experimentwise alpha level by making the hypothesiswise alpha level smaller.

There are procedures that test hypotheses about a number of parameters simultaneously, such as a test that all the regression weights are zero. We refer to these tests as omnibus tests. An effective strategy in multivariate inference is to first make an omnibus test at a specific alpha level. If the hypothesis is rejected, then individual parameters are tested using the nominal alpha level.

PROBLEMS

1. (a) Using the z' transformation, construct a two-tailed acceptance region for H_o: $\rho = 0$ based on an n of 28 and $\alpha = .05$.
   (b) Do the same for $n = 103$.
   (c) In which case is a Type I error more likely?
   (d) Suppose $r = .35$. Is it significant in either case?

2. Repeat Problem 1, but based on H_o: $\rho = .40$

3. (a) If $r = .40$, construct a .95 c.i. for ρ when $n = 28$ and when $n = 103$.
   (b) Do the same using .99 c.i.
   (c) What do these c.i. say about the significance of $r = .4$?

4. (a) If $s^2 = 12.0$ based on $n = 16$, is H_o: $\sigma^2 = 10$ rejected at $\alpha = .02$?
   (b) What other values for s^2 based on this sample size and α level would not lead to rejection of H_o: $\sigma^2 = 10.0$?
   (c) What is the range of values for σ^2 that would not be rejected as null hypotheses, based on these data?

5. (a) With n's of 28 and 103, and an .05 α level, what is the power of H_o: $\rho = 0$ when $\rho = .20$?
   (b) When $\rho = .40$?

6. An experimenter computes correlations between longevity and 10 dietary variables, and finds that the correlation with parsnip consumption is significant at the .05 level. He reports this correlation as significant, not mentioning the other variables investigated. (a) What is the true α level? (b) What is the probability that he has made a Type I error?

7. If the experimenter in Problem 6 wanted to make a Bonferroni correction, what α level would he use?

ANSWERS

1. (a) $-.392 \leq z_r \leq .392$ or $-.373 \leq r \leq .373$ (b) $-.196 \leq z_r \leq .196$ or $-.194 \leq r \leq .194$ (c) The probability is the same since it is defined by α. (d) $r = .35$ is significant in (b) but falls in the acceptance region in (a).

2. (a) $.0316 \leq z_r \leq .8156$ or $.0316 \leq r \leq .683$ (b) $.2276 \leq z_r \leq .6196$ or $.224 \leq r \leq .550$ (c) Same as (b). (d) Not significantly different from .4 in both cases.

3. (a) $.032 \leq \rho \leq .673$; $.224 \leq \rho \leq .551$ (b) $-.0924 \leq \rho \leq .735$; $.164 \leq \rho \leq .593$ (c) r of .4 is significant at .05 but not .01 for $n = 28$, and significant at both for $n = 103$.

4. (a) $\chi^2 = 18.0$, n.s. (b) $3.486 \leq s^2 \leq 20.385$ (c) $5.886 \leq \sigma^2 \leq 34.42$, that is, .98 c.i.

5. (a) $\beta = .77$ for $n = 28$; $\beta = .47$ for $n = 103$ (b) .44; .011

6. (a) $.05 \le \alpha_{true} \le .50$; $\alpha_{true} = .401$ if independence is assumed. (b) The true probability is unknown.

7. .005

READINGS

Hays, W. B. (1982). *Statistics,* 3rd ed. New York: Holt, Rinehart, and Winston, pp. 176–270.

Myers, J. L. (1979). *Fundamentals of experimental design.* Boston: Allyn and Bacon, pp. 290–312.

9

Inference in Multiple Regression

SAMPLING MODELS

In multiple regression, we obtain the values of the *sample* statistics, multiple R, s_e, the b_j, and the b_j^*, but what we would really like to know are the values of the corresponding parameters. Like all sample statistics, these parameters are subject to sampling fluctuations. As everywhere else in statistics, the procedures of estimation and hypothesis testing are necessary. The basic population model used in multiple regression is

$$y_i = \beta_1 x_1 + \beta_2 x_2 + \cdots + \beta_p x_p + \beta_0 + e_i$$

Unlike the equations in Chapter 6, this one is expressed in terms of the population coefficients β instead of the sample b's. The inferential process follows from a series of assumptions. The first is that e_i is normally distributed with mean zero and constant variance, independent of the values of the x's. For example, suppose $p = 2$ and $\beta_1 = 3$, $\beta_2 = 1$, and $\beta_0 = 3$, and that a particular set of values for each of the x's is selected in the population. Now suppose that $x_1 = 2$, and $x_2 = -1$ and that the distribution of y for that particular subpopulation is examined. When this is done, the assumption is that the conditional distribution of y will be normal, that its mean is $3(2) + 1(-1) + 3 = 8$, and that its variance is the same as for any other combination of x values. This normality and equality of variance is called *homoskedastic conditional normality.* It is the analogue of the homoskedasticity assumption that we make for descriptive purposes in multiple regression. Note that the assumption is made about the population, not the sample. The sampling errors in multiple regression are based on this assumption.

There are two situations in which the assumption of homoskedastic conditional normality holds true. One is a *fixed constants* situation. Here, the investigator *chooses* a set of combinations of x values; they are under his control, as in a controlled experiment. This means that there is no population distribution of x's sampled. The only distribution involved is that for y, which is assumed to be normal for any combination of x values. A second situation in which this assumption holds true is *multivariate normality*, a special case of a random variables model or random effects model. Here, we have a population of persons or other elements, and attached to each element is a vector of $p + 1$ values representing the scores on the p predictors and the dependent variable. The combinations of x values that occur in the data depend on sampling, and we have to take what the sample offers. Homoskedastic conditional normality will hold in this latter case if the distribution of all these variates is *multivariate normal*. Not only does each variable have a normal distribution, but for a specified value of $p - 1$ of the variables, the conditional distribution for the remaining variable is normal. If the assumption of multivariate normality holds, it also guarantees that all the regressions are linear, not curvilinear.

To illustrate the difference between these two situations, imagine two studies of the same topic, one in which one model is appropriate and one in which the other is appropriate. Suppose we are interested in the effect of word length (WL) and the size of type (ST) upon reading speed. The latter is defined in terms of the time it takes to recognize and speak a word after it is flashed on a screen, (RT). The investigator could say to himself, "Words come in various lengths, so I will pick some which will adequately cover the range to construct my stimuli." He proceeds to select words with 2, 4, and 6 letters, up to 12 letters per word. (Words as long as heteroskedasticity—18 letters—are too unfamiliar to subjects.) He types out the words and uses a photographic enlargement process to vary the size of the type by ½, 1, 1½, and 2—up to 4 times the size of the original letters. Typically, he would represent each word using each type size (See *crossed designs* in Chapter 11), although it is not necessary to do so. Therefore, only certain predetermined combinations of values of WL and ST appear in the experiment. This is often called a *fixed* effects model.

In contrast, he could take a naturalistic approach by randomly sampling books from the library and words from the books. Some of the words sampled are long and some are short, and some typeface would be large and some small. Each of the sampled words is then reproduced by a standard process that duplicates the size of the word as printed. Here, the investigator has sampled from a population of words, and each word has its three characteristics of length, type size, and difficulty (average RT). The statistics of the sample depend on the population parameters and the random effects of sampling. The investigator's choices are not the determining factor. This is often called a *random* effects model.

Both of these approaches have their place, and fortunately, the inferential procedures of multiple regression as well as the descriptive ones are equally appropriate to each. Although most of the theory used in multiple regression is derived in the fixed concepts context, it is also usually applicable to the multivariate normality context. (Occasional exceptions will be noted.) What is important in each case is that the respective assumptions concerning homoskedasticity hold to a reasonable degree. But even this becomes less important for inferences about parameters as the sample size increases. The regressions also must be linear in both the fixed and the random effects model if the inferential process is to be strictly valid.

Estimating P^2, σ_e^2, and β_i

Sampling distributions for estimators depend on both sample size and population parameters, sometimes on more than one parameter. The sample data provides sample statistics that estimate the population parameters. We would like these estimates to be good ones—that is, unbiased, having minimum variance, efficient, and consistent. Finally, we would like to know how good the estimates are and what form their sampling distributions take. If it turns out that the ordinary least-squares estimates derived earlier are not optimum in these respects, we need to know what alternatives to use.

Under either the multivariate normal or fixed constants models, the assumption about the model in the population is the same as the one we have been making about the data anyway—that is, linear, homoskedastic regression. Therefore, these estimates are consistent, coming arbitrarily close to the parameters in arbitrarily large samples. Some of the least-squares estimates are biased, although this decreases as sample size increases and is easy to correct except for negligible quantities. Whether these estimates are minimum variance is a complex question, so various aspects of it will be taken up separately.

Because it involves the definition or sampling properties of several other estimators, we begin with the problem of estimating σ_e^2. The total variance in y, σ_y^2, can be partitioned into two parts (Equation 6.40)—one that is predictable from the x's and one that is not. The former is designated $\sigma_{\hat{y}}^2$ and the latter, σ_e^2. Here is one area where the two sampling models differ. In the case of the fixed model, the variance of y depends on which values of the x's are included in the experiment. Assuming that WL and ST do affect RT, RT's variance will be large if a wide range of WL and ST is used and small if a narrow range is used. Thus σ_y^2 has no constant definition, it depends on the conditions of the experiment. This is not the case for the multivariate normal model.

In either case, however, σ_e^2 can be estimated from the data: $ss_{dev}/(n - p - 1)$ has σ_e^2 as expected value, where ss_{dev} equals $\Sigma(y - \hat{y})^2$, and, in the case of normal homoskedasticity, is distributed as $\chi^2\,\sigma_e^2/(n - p - 1)$. Thus, we can use

$$\hat{\sigma}_e^2 = ss_{dev}/(n - p - 1) = s_e^2\,\frac{n - 1}{n - p - 1} \tag{9.1}$$

as our estimate of σ_e^2 and be quite confident about it. Note, however, that this is *not* s_e^2, which uses $n - 1$ in the denominator. It is larger, although the amount clearly depends on the relationship of p to n. If $n = p + 1$ or less, there will be no deviations from the regression line. Equation 9.1 and common sense tell us that then we also have no way of estimating σ_e^2.

A related problem occurs within R^2, which can be a substantial overestimate of P^2. (Remember that P is an uppercase Greek rho.) Since P^2 is defined analogously to R^2 as

$$P^2 = 1 - \frac{\sigma_e^2}{\sigma_y^2} \tag{9.2}$$

this may lead to estimation problems: s_e^2 is an underestimate of σ_e^2, whereas s_y^2 is an unbiased estimate of σ_y^2. This makes R^2 an overestimate of P^2. To correct the bias, we use the unbiased estimates σ_e^2 and σ_y^2:

$$\hat{P}^2 = 1 - \frac{\hat{\sigma}_e^2}{s_y^2} \tag{9.3}$$

This has a direct relationship with R^2, which shows the amount of bias in R^2:

$$\hat{P}^2 = R^2 - (1 - R^2)\frac{p}{n - p - 1} \tag{9.4}$$

The square root of Equation 9.4 is used as the estimate of P. If $\hat{P}^2 < 0$, we assume $P^2 = 0$, since a square cannot be negative.

In computer packages, the estimates $\hat{P}^2$ and $\hat{P}$ are referred to as "unbiased estimates" of P^2 and P. Some bias remains, but the amount is small unless $n - p$ is below 10.

The reader can calculate the effects of Equations 9.1 and 9.4 in any given instance, but some general principles can be observed. The same fraction, $p/(n - p - 1)$, occurs in both. Therefore, the amount of bias and correction depends on the ratio of p to $n - p - 1$. Thus, no matter how large the sample, the amount of correction needed is the same, as long as the ratio of predictors to subjects stays the same. On the other hand, the amount of correction is small when there are many more subjects than variables. If $p = 10$ and $n = 21$, it can be calculated easily that $\hat{\sigma}_e^2 = 2s_e^2$. In the same case, $\hat{P}^2 = 2R^2 - 1$ and $R^2 = .80$ drops to $\hat{P}^2 = .60$. If R^2 is less than .50, the correction makes $\hat{P}^2$ negative, and P^2 is assumed to be zero. Note that the relation of $\hat{P}^2$ to R is exactly the same if $n = 201$ but $p = 100$. On the other hand, if $p = 2$ and n is still 21, the correction is mild: $\hat{\sigma}_e^2 = 1.1s_e^2$ and $\hat{P}^2 = 1.1R^2 - .1$.

The biases in s_e^2 and R^2 as estimates of the corresponding population parameters can be thought of as the result of a kind of "over-fitting" of the sample data, or a "boot-strap" effect. In the sample, we capitalize on chance relationships and make a closer fit than is possible in the population. The extreme of this is when $p = n - 1$, whereupon the sample deviations can be reduced to zero, even if $P = 0$.

The sum of squared deviations from the regression line is divided by the *degrees of freedom*—that is, by the number of deviations $y - \hat{y}$ that are free to vary—to provide an estimate of σ_e^2. The degrees of freedom concept for deviations is widely useful because it allows us to keep track of what is going on in multivariate statistics. As a general rule, a degree of freedom is used for each parameter that is estimated in a linear system. Thus, in a single variable one loses a degree of freedom by estimating the mean, which leads to the necessity of using $n - 1$ in the denominator to estimate σ^2. In univariate regression there are two parameters to estimate, b_0 and b_1, which means two degrees of freedom are lost. In multiple regression, the estimation of $p + 1$ parameters leads to the use of $p + 1$ degrees of freedom, which makes $n - p - 1$ the denominator of the estimate of σ_e^2.

In terms of error variance, it is as if one is spending degrees of freedom to purchase error variance. The expenditure of each degree of freedom results in getting one nth of the error variance. One can inflate R^2 by involving more and more variables, estimating more and more parameters, and thereby reducing s_e^2. Unfortunately, error variance is like the gold in fairy tales, which tends to disappear from the buyer's purse when he goes to look for it. The unbiasing effects of Equations 9.1 and 9.4 attempt to "uninflate" our currency and to tell us how much prediction can really be obtained by fitting the parameters of the regression equation. Later, we will find that even the unbiased estimates are optimistic in some respects.

The b_j are unbiased estimates of the β_j; which are also consistent and therefore, they are generally used as the estimates. Later in this chapter, though, we will see that it may be preferable in some cases to use a different estimate of β_j, one that is biased but has less sampling variance. These estimates are called "ridge" estimates.

Sampling Errors and Confidence Intervals in Regression Parameters

If we observe the methods of investigators who employ multiple regression, it becomes apparent that they focus on point estimates of P^2, σ_e^2 and the b_j. However, to find out the stability of the estimates, the primary basis for making such inferences is the sampling distribution of the statistic. The standard deviation of the sampling distribution, the standard error of the statistic, is useful in measuring how far off we can expect the estimate to be, particularly if the sampling distribution is more or less normal or at least symmetric. It was noted above that $\hat{\sigma}_e^2$ has $\chi^2\sigma_e^2/(n - p - 1)$ as a sampling distribution. Thus, the methods reviewed for inferences about variance in Chapter 8 are appropriate for constructing confidence intervals and testing hypotheses about it.

The sampling distribution of R^2 or $\hat{P}$ does not have a simple form, so our degree of uncertainty about P^2 may be rather uncertain. Perhaps the best procedure for making confidence intervals is to treat P like r and use the Fisher z' transformation, with $1/(n - p - 2)$ as the estimate of the sampling variance of z'. Combining the unbiasing processes used in Equations 9.1 and 9.4 with these procedures helps to provide realistic estimates of how accurate a prediction is possible *in the population.*

A phenomenon observed throughout application of multiple regression is the instability of the regression coefficients. The expression "bouncing betas" was current for many years when the Greek letter was used to refer to the standardized regression coefficient. The sample raw score regression coefficient is, as was noted, an unbiased estimate of its population counterpart. Moreover, its sampling distribution is normal when the population is normal, as we are assuming here. The problem is its standard error:

$$\sigma_{b_j} = \frac{\sigma_e}{\sigma_j(1 - P_j^2)^{1/2}n^{1/2}} \tag{9.5}$$

According to this equation, the standard error of the regression coefficient for x_j is directly proportional to σ_e—that is, it decreases as prediction is more accurate—and is inversely proportional to the sample size. (The standard deviation of x_j in the denominator simply reflects a scaling effect that is also in the denominator of β_j.) The other term, $(1 - P_j^2)$, is the proportion of the variance of *predicator j,* which *is not itself predictable from the other predictors.* Because it is in the denominator, as the amount of variance that a *predictor* shares with *other predictors* goes up, the instability of its regression weight increases. As a consequence, adding a variable to the predictive battery is likely to *increase* the uncertainty about the regression coefficients of those that were originally in the equation. Thus, including a lot of variables in the regression equation may actually make our conclusions *less* certain than if only a judiciously selected few are included for study. Variables with the most stable regression coefficients will be the ones that are uncorrelated with the other predictors. The terms in this equation do much to explain "bouncing betas."

A theoretical standard error is of limited practical use unless it can be estimated from sample data. This can be done rather straightforwardly by substituting the sample estimates for the population quantities:

$$s_{b_j} = \frac{\hat{\sigma}_e}{(n-1)^{1/2} s_j (1 - R_j^2)^{1/2}} \tag{9.6}$$

This equation may look cumbersome because it requires the computation of each R_j^2. This is not a severe problem, however, because the sampling variances of the b_j are a gift; they are $\hat{\sigma}_e$ times the diagonal terms of $(X'_+X_+)^{-1}$ or of $(n-1)^{-1} S_{xx}^{-1}$. That is, $s^{jj} = 1/s_j^2(1 - R_j^2)$. In fact, the off-diagonal elements of these matrices are the sampling covariances of the b_j, a fact that also helps to explain bouncing betas, particularly the fluctuations in the relative sizes of the coefficients for the different variables.

Strictly speaking, Equation 9.6 is only correct for fixed predictors. However, random variable predictors are usually treated in the same way.

Confidence intervals for single regression coefficients may be constructed through the use of the t distribution. This distribution should be familiar as the one used for a variety of purposes in elementary statistics—especially the t test for the significance of the difference between means. The theoretical definition of this distribution is that it is the ratio of a normal distribution and the square root of a chi-square divided by its *df*. Consequently, the t distribution is frequently used in inferences about an estimator that is normally distributed, and divided by its *esimated* standard error such as b_j/s_{b_j}.

This is the case when forming a c.i. for b_j. Specifically, the $1 - \alpha$ c.i. is given by

$$b_j - t_{1/2\alpha}\, s_{b_j} \le \beta_j \le b_j + t_{1/2\alpha}\, s_{b_j} \tag{9.7}$$

where $t_{1/2\alpha}$ is for $n - p - 1$ *df*. Appendix Table 4 gives the relevant points for the t distribution.

Note that the preceding confidence interval is "parameterwise" (see Chapter 8), which is appropriate when considering each coefficient individually. Remember that the probability that all p regression coefficients are within their respective c.i. is somewhere between α and $p\alpha$, and probably closer to the $p\alpha$. A method for defining a c.i. for all the coefficients jointly will be discussed later in this chapter.

For some applications, we may want to compare the standard score regression weights to their standard errors. We may alter Equation 9.6 to make it appropriate for a standard score weight just by deleting the standard deviation terms:

$$s_{b_{j^*}} = \frac{(1 - R^2)^{1/2}}{[(1 - R_j^2)(n - p - 1)]^{1/2}} = \frac{(1 - R^2)^{1/2}}{[r^{jj}(n - p - 1)]^{1/2}} \tag{9.8}$$

This is not strictly correct, however. The sampling behavior of standard score regression coefficients is more complicated, but is probably accurate enough.

Examples

Exhibit 9-1 demonstrates the basic calculations for the methods described so far in this chapter. The first section shows the covariance and correlation matrices for these predictors, two of which are rather highly correlated, and one, x_3, which has a lower correlation with these two. All three predictors have low to

$$n = 20 \qquad s_y^2 = 25{,}000$$

$$\mathbf{S}_{xx} = \begin{bmatrix} 4.000 & 4.800 & 3.200 \\ 4.800 & 9.000 & 3.600 \\ 3.200 & 3.600 & 16.000 \end{bmatrix} \qquad \mathbf{s}_{xy} = \begin{bmatrix} 6.000 \\ 7.500 \\ 8.000 \end{bmatrix}$$

$$R_{xx} = \begin{bmatrix} 1.000 & .800 & .400 \\ .800 & 1.000 & .300 \\ .400 & .300 & 1.000 \end{bmatrix} \qquad \mathbf{r}_{xy} = \begin{bmatrix} .600 \\ .500 \\ .400 \end{bmatrix}$$

$$\mathbf{S}_{xx}^{-1} = \begin{bmatrix} .7533 & -.3753 & -.06622 \\ -.3753 & .3091 & .00552 \\ -.06622 & .00552 & .07450 \end{bmatrix} \qquad \mathbf{S}_{xx}^{-1}\mathbf{s}_{xy} = \mathbf{b} = \begin{bmatrix} 1.175 \\ .1104 \\ .2401 \end{bmatrix}$$

$$\mathbf{R}_{xx}^{-1} = \begin{bmatrix} 3.013 & -2.252 & -.530 \\ -2.252 & 2.781 & .066 \\ -.530 & .066 & 1.192 \end{bmatrix} \qquad \mathbf{R}_{xx}^{-1}\mathbf{r}_{xy} = \mathbf{b}^* = \begin{bmatrix} .470 \\ .066 \\ .192 \end{bmatrix}$$

$$1.175 \times 6.0 + .110 \times 7.5 + .240 \times 8.0 = 9.795$$
$$\Sigma b_j s_{jy} = s_{\hat{y}}^2$$

$$25.000 - 9.795 = 15.205 \qquad 9.795/25.0 = .3918$$
$$s_y^2 - s_{\hat{y}}^2 = s_e^2 \qquad s_{\hat{y}}^2/s_y^2 = R^2$$

$$19 \times 15.205/16 = 18.056; \qquad 1.0 - 18.056/25.0 = .2778$$
$$(n-1)s_e^2/(n-p-1) = \hat{\sigma}_e^2; \qquad 1 - \hat{\sigma}_e^2/s_y^2 = \hat{\mathrm{P}}^2 \quad \sqrt{\hat{\mathrm{P}}^2} = \hat{\mathrm{P}};$$

$$.470 \times .600 + .066 \times .500 + .192 \times .400 = .3918$$
$$\Sigma b_j^* r_{jy} = s_{\hat{z}_y}^2 = R^2$$

$$(.3918)^{1/2} = .6259 \qquad 1 - .3918 = .6082$$
$$(R^2)^{1/2} = R \qquad 1 - R^2$$

$$(.2778)^{1/2} = .527; \quad .3918 - .6082 \times 3/16 = .2778$$
$$R^2 - (1 - R^2)p/(n-p-1) = \hat{\mathrm{P}}^2$$

Exhibit 9-1 Basic computations for inferences in multiple regression.

moderate correlations with y. The next section shows the inverses of the two matrices and the two sets of regression weights. The b_j^* was calculated from $\mathbf{R}_{xx}^{-1}$ $\mathbf{r}_{xy}$, but is also $b_j^* = s_j b_j / s_y$.

The next section shows the calculation of $s_{\hat{y}}^2$ according to Equation 6.41 and the corresponding figure for standard scores (Equation 6.46). Then s_e^2 is found from Equation 6.52, and it is demonstrated that $R^2 = s_{\hat{y}}^2/s_y^2$ (Equation 6.47). We can also find R from R^2 or from the standard score error variance, $1 - R^2$. The last line of Exhibit 9-1 uses Equation 9.1 to get the unbiased estimate $\hat{\sigma}_e^2$, and on the far right, the unbiased estimate $\hat{P}^2$ from Equation 9.3. The latter is also calculated from $1 - \hat{\sigma}_e^2/s_y^2$. We also calculate $\hat{P}$.

The correction for bias increases the estimate of the population variance to 18.1, a modest increase which reflects the fact that this n is fairly large relative to p. If one finds the two standard errors of estimate by taking the square roots, the results are $\hat{\sigma}_e\ \sqrt{15.2} = 3.90$ and $s_e = \sqrt{18.1} = 4.25$, so the effect on these standard deviations is less than on the variances.

There is a similar effect on R^2 when it is corrected for the degrees of freedom ratio; .392 becomes .278, and on R, .626 dropping to .527.

Exhibit 9-2 shows how to get c.i.'s for the parameters whose unbiased estimates

(a) Confidence Interval for σ_e^2

$$\hat{\sigma}_e^2 \div (\chi^2_{1-1/2\alpha,df}/df) \le \sigma_e^2 \le \hat{\sigma}_e^2 \div (\chi^2_{1/2\alpha,df}/df)$$

$$18.056 \div (28.80/16) \le \sigma_e^2 \le 18.056 \div (6.91/16)$$

$$10.031 \le \sigma_e^2 \le 41.81$$

(b) Standard Errors and Confidence Intervals for Weights

Raw score weights: β

b_j	$1/s_j^2(1 - R_j^2)$	$\hat{\sigma}_e^2$	s_{bj}^2	s_{b_j} (Eq. 9.6)	$t_{.025}$	.95 c.i. (Eq. 9.7)
1.175	.7533	18.056	.7159	.8461	−2.120	−.619 to 2.969
.110	.3091	18.056	.2937	.5420	−2.120	−1.039 to 1.259
.240	.0745	18.056	.0708	.2661	−2.120	−.324 to .804

(c) Standard Score Weights: β^*

b_j^*	$1/(1 - R_j^2)$	$1 - R^2$	$s_{b_j^*}$ (Eq. 9.8)	$t_{.025}$	.95 c.i.
.470	3.013	.6082	.3384	2.120	−.247 to 1.187
.066	2.781	.6082	.3251	2.120	−.623 to .755
.192	1.192	.6082	.2129	2.120	−.259 to .643

Exhibit 9-2 Confidence intervals for σ_e^2, β_j, and β_j^*. The c.i. for σ_e^2 is given according to the method outlined in Chapter 8. The c.i. for the weights are found by using the standard error Equations 9.6 or 9.8 where $1/s_j^2(1 - R_j^2)$ is found from $\mathbf{S}_{xx}^{-1}$. These are used in conjunction with the t value for .025 and 16 df to give the c.i., according to Equation 9.7 or its standard score counterpart.

were found in Exhibit 9-1. For σ_e^2, the .025 and .975 χ^2 values for 16 *df* are 6.91 and 28.80, respectively (see Appendix Table 3). Dividing each into 16 df and multiplying by $\hat{\sigma}_e^2$ gives the upper and lower boundaries of 41.81 and 10.03, respectively. Thus, we are uncertain about the error variation by a factor of about four because of the small df in this example. The c.i. for the standard error of estimate can be found by taking the corresponding square roots. The result is a .95 confidence that it lies between 3.17 and 6.47.

The rest of Exhibit 9-2 estimates the standard errors and c.i.'s for the regression weights. Parts (a) and (b) deal with the raw score weights and part (c) with the standard score weights. The first columns show the weights, followed by the reciprocals of s_j^2 $(1 - R_j^2)$ from the diagonals of the inverse. The standard errors for x_1 and x_2 reflect the fact that these two variables share almost two-thirds of their variance with other predictors. The third columns give $\hat{\sigma}_e^2$ and the fourth show the squared standard errors, column three multiplied by column 2 (since it is a reciprocal) and then divided by n, which is 20. The fifth column gives the standard errors themselves. The lower section repeats these calculations in standard score form. The standard errors for the standard score weights show the effects of the correlations of the predictors with each other. Note that x_1 and x_2 are highly predictable, primarily because of their high correlations with each other, so their standard errors are higher than the one for x_3.

These standard errors, and the appropriate t for $n - p - 1$ *df* (that is, 16, which is found in Appendix Table 4 to be 2.120), are used to find the .95 c.i. for the raw score and the standard score weights. The two sets of regression coefficients, standard errors, and c.i.'s, differ only by the ratio of the standard deviation of the variable to the standard deviationof y. All the c.i.'s include 0 within their limits and thus are not significantly different from zero. Furthermore, the indication is that b_2 *could* be substantial and negative. On the other hand, considerably higher values than the one obtained cannot be ruled out. The true standard score regression weight for x_1 could even exceed 1.0.

There is a wide range of uncertainty about each of the regression coefficients. However, these are parameterwise c.i.'s, so the likelihood that *all three* parameters are within their respective individual intervals is less than .95—somewhere between .95 and $1 - 3(.05) = .85$. Also, the substantial correlation between x_1 and x_2 makes their sampling distributions nonindependent. The effects of this will be discussed later in this chapter.

The calculation of the regression weights, the multiple correlation, and the standard error of estimate are relatively straightforward once one understands an inverse and musters the patience or computer facilities to compute it. However, computed statistics are of relatively limited value. They describe only the sample, whereas our interest is usually in the population. If we wish to estimate the corresponding population quantities, then we should remove any bias in the estimates. This process can be painful because the unbiased estimates look worse than their biased counterparts. Nevertheless, they are the realistic estimates of how strong the relationships are and how accurate the prediction is in the population. Unbiasing is employed by all but the most unsophisticated users and computer programs.

Forming c.i.'s around the unbiased estimates is highly desirable. Knowing the c.i. means that we have not only the unbiased estimate as a best guess of the parameter value but also the limits within which the parameter should lie. This seems to be important information. However, c.i.'s are currently reported in only a minority of studies using multiple regression.

HYPOTHESIS TESTING IN MULTIPLE REGRESSION

Testing R^2 for Significance

The null hypothesis of initial interest in a multiple regression is that the predictors are collectively useless. This may be stated formally in either of two equivalent ways:

$$H_0\colon P = 0$$

or

$$H_0\colon \beta_1 = \beta_2 = \cdot\cdot\cdot\ \beta_p = 0$$

The two equations are equivalent because the only way P can be zero is if all the β's are zero and because all the β's are zero when P is zero. The statement could be made in terms of P^2 or the β^* instead.

Even if P is zero, R is very unlikely to be zero. Some position for the regression line can always be found that will account for some of the sample variance. That leads to bias in R.

In multiple regression, we separate the variance into predictable variance and error variance. When the null hypothesis is true, "predictable" variance really just represents capitalization on chance, that is, error variance. In fact, recalling that $ss_{reg} = \Sigma(\hat{y}_i - \bar{y})^2$, we expect under H_0 that on the average $ss_{reg} = p\sigma_e^2$. The sum of squares for regression is in proportion to the number of fitted parameters. Dividing by p gives $ms_{reg} = ss_{reg}/p$, so this mean square just provides an estimate of error variance. Now we have a way of testing the null hypothesis because when H_0 is true we simply have two estimates of error variance, one from ms_{reg} and one from the unbiased $\hat{\sigma}_e^2$. Comparing these two estimates of error variance leads to the standard test of H_0 in multiple regression.

Using terminology parallel to ms_{reg}, we define

$$ms_{dev} = ss_{dev}/(n - p - 1) = \hat{\sigma}_e^2$$

The standard test of H_0 is

$$F_{p,n-p-1} = \frac{ss_{reg}/p}{ss_{dev}/(n - p - 1)} = \frac{ms_{reg}}{ms_{dev}} \quad (9.9)$$

That is, the ratio of ms_{reg} to ms_{dev} has the F distribution with p and $n - p - 1df$. An equivalent F can be expressed in terms of R^2

$$F = \frac{R^2/p}{(1 - R^2)/(n - p - 1)} \quad (9.10)$$

because $R^2 = ss_{reg}/ss_{tot}$.

The ratio of two estimates of the same variance has the F distribution, and under H_0, both ms_{reg} and ms_{dev} estimate σ_e^2. Strictly speaking, the distribution has this form only when the errors are normally distributed and homoskedastic. Some care should be taken to ensure these conditions hold to a reasonable degree. The null hypothesis is rejected when the F ratio is "too large" according to the critical values given for .05 and .01 given in Appendix Table 5, or according to the probability levels provided by a helpful multiple regression program.

This is an example of an omnibus test because it is testing all the regression coefficients at once. It can be followed by t (or $F = t^2$) tests of individual coefficients, following the protection levels strategy of the previous chapter. Computer programs provide the F ratios, often under the heading "F-to-delete." These and other computer applications are discussed later in this chapter.

Exhibit 9-3 shows the application of Equations 9.9 and 9.10 using data from Exhibit 9-1. In part (a), the sums of squares are calculated from the variances given in the earlier exhibit. (With the original data, they could be calculated directly.) Both sums of squares are divided by their respective degrees of freedom, 3 and 16, to form the mean squares. Then the ratio of these is found, which is F. This is compared to the critical value for $\alpha = .05$ in Appendix Table 5, which is 3.24. Thus H_0 is *rejected* at this alpha level, indicating that ms_{reg} is not just another estimate of σ_e^2 and that it is not reasonable to infer that all the regression coefficients are zero—even though zero was within the c.i.'s for all the coefficients. We will return later to this apparent paradox.

Part (b) of the exhibit shows the equivalent calculations based on R^2, again from Exhibit 9-1. These give exactly the same results, as they should.

Other Hypotheses about P^2 or σ_e^2

Occasionally, situations occur where we want to test something other than a simple null hypothesis about P. For example, we may want to know whether the multiple correlation involving the current set of variables is the same in two or more groups in order to find out whether a prediction equation is equally effective in each group.

There are several possible hypotheses here, involving different analyses or data, so it is important to be clear on which hypothesis is relevant. We could have the same or different predictors, the same or different criteria, and the same or different groups, and several of the eight combinations could be relevant. Some are more frequently encountered than others.

(a) Analysis of Sums of Squares

Source	Sum of Squares	*df*	Mean Square	*F*
Regression	186.105	3	62.035	3.44
Deviations	288.895	16	18.056	
Total	475.000	19		

(b) Analysis of Squared Multiple Correlation

R^2	.3918	3	.1306	3.44
$1 - R^2$	.6082	16	.03801	

Exhibit 9-3 *F* tests for significance of regression. Part (a) shows the analysis based on sums of squares and part (b) is based on squared multiple correlations. Data are taken from Exhibit 9-1.

In a series of papers, Steiger and Browne (1984) present methods for dealing with several of these situations. The methods are not simple, and are not implemented in any routinely available computer packages.

The most straightforward question is whether prediction is equally accurate when the same predictors and same criteria are used in different groups. While often described as if the question is defined in terms of multiple correlations, this is not the preferred method since P^2 can be influenced by such things as the relative heterogeneity of groups on the predictors. Instead, the question should be framed and tested by comparing the error variances. This can be done via Bartlett's test (see Chapter 11) of the hypothesis H_0: $\sigma^2_{e_1} = \sigma^2_{e_2} = \cdots \sigma^2_{e}$. The related question of whether the regression coefficients are the same in all groups can be tested by the extended form of analysis of covariance described in Chapter 12.

If the criteria are different as well as the groups, then it is not clear what should be done. The most obvious procedure is to perform Bartlett's test on the various $1 - R^2$ for different groups, although the rationale for the procedure is not well-established. Besides, testing whether one criterion in one group can be predicted as well as another criterion in another group amounts to a comparison of apples and oranges—or even beef stew and fruit salad.

A third type of comparison arises when the question is whether one battery of predictors is more effective than the other in the same group. If one battery is a subset of the other, then the methods of hierarchical multiple regression described later in this chapter would apply. But if the question is whether one battery provides more effective prediction than another, separate one, then an alternative procedure is used. Again, this procedure compares error variances, but takes into consideration their lack of independence. The following F test is used:

$$F = \left(\frac{\hat{\sigma}^2_{e_1}}{\hat{\sigma}^2_{e_2}} + \frac{\hat{\sigma}^2_{e_2}}{\hat{\sigma}^2_{e_1}} - 2\right)\left(\frac{n - 2}{1 - r^2_{e_1e_2}}\right) \tag{9.11}$$

In this equation, $\hat{\sigma}^2_{e_1}$ and $\hat{\sigma}^2_{e_2}$ are the estimated error variances using predictor batteries 1 and 2, respectively. The correlation between the deviations using battery 1 and those using battery 2 is $r_{e_1e_2}$. This correlation can be found using the methods for correlation of weighted sums described in Chapter 4.

Having performed the two multiple regressions—one using battery 1 to get $\mathbf{b}_1$ and one using battery 2 to get $\mathbf{b}_2$—Equation 4.17 may be used to get $r_{e_1e_2}$. In the equation, $\mathbf{w}$ would consist of the elements of $\mathbf{b}_1$, followed by a zero for each variable in battery 2, and a -1 for y. Similarly, $\mathbf{v}$ would have a zero for each member of battery 1, followed by the elements of $\mathbf{b}_2$, and again a -1 for y.

TESTING HYPOTHESES ABOUT REGRESSION WEIGHTS

Testing Single Regression Coefficients

Determining whether the null hypothesis value for a regression coefficient is within the c.i. is the same as testing that hypothesis, but having a procedure for testing coefficients more directly is also useful. Assuming normal homoskedasticity, if β_{0j} is the true value of the parameter, then

$$t = \frac{b_j - \beta_{0j}}{s_{b_j}} \tag{9.12}$$

has the t distribution with $n - p - 1$ df. We can use this equation to test particular H_0 about a single regression coefficient. Alternatively, we could use the fact that $F_{1,df} = t^2_{df}$ and avoid taking the square root to find s_{b_j}.

On rare occasions, we may have a specific nonzero hypothesis about β_j. Ordinarily, the null hypothesis is H_0: $\beta_j = 0$. When β_{0j} is set equal to zero, the square of Equation 9.12 equals the ratio of the amount that the error sum of squares is reduced when b_j is included in the regression equation to the estimated error variance. If ss_p is the sum of squared deviations from the regression line when all the predictors are included and ss_{p-1} is the sum of squares when all but variable j is used, the squared version of Equation 9.12 becomes

$$F = \frac{ss_p - ss_{p-1}}{ss_p/(n - p - 1)} \tag{9.13}$$

This test is an example of the common form noted earlier in which we test whether estimating this one b_j parameter "buys" more than its proportional share of the error variance. An equivalent form based on R^2 is

$$F = \frac{(R^2_p - R^2_{p-1})}{(1 - R^2_p)/(n - p - 1)} \tag{9.14}$$

This equation relates more closely to Equation 9.13 if we think of the numerator as being $(1 - R^2_{p-1}) - (1 - R^2_p)$. Slightly generalized forms of Equations 9.13 and 9.14 will be introduced later. The first column of Appendix Table 5 shows that, for moderate df, H_0 is not rejected unless F is at least 4.0. Thus estimating this parameter buys at least four times its proportional share of variance. If the df are small, F must be substantially larger than this.

The preceding formulas provide two alternative ways of testing hypotheses about individual regression coefficients. Their use is appropriate only if a parameterwise stance is taken toward inferential errors. That is, if the null hypothesis about a given β_j is true, the probability is α that it will be rejected. If there are p predictors, and the regression coefficient for each one is tested, remember that the probability of rejecting at least one of the null hypotheses about individual coefficients when all of them are true is between $p\alpha$ and α, and probably closer to $p\alpha$. Testing the hypothesis that all the regression coefficients are zero is best accomplished by the method described in the previous section. If the hypothesis is rejected, then using the protection levels idea, individual hypotheses might be tested by one of the formulas described in this section.

Exhibit 9-3 can be used as a basis for calculating the examples shown in Exhibit 9-4. Employing Equation 9.12 is simple here since the s_{b_j} have already been calculated. The results are shown in the column labeled t of Exhibit 9-4. Comparing these values to the value for 16 df in Appendix Table 4, it becomes apparent that H_0 could not be rejected for any of these values, at any reasonable

Predictor	b_j	s_{b_j}	t	$R^2_{x \cdot j}$	$R^2_{y \cdot 123} - R^2_{y \cdot jk}$	F
1	1.175	.8462	1.39	.3187	.0731	1.92
2	.110	.5420	.20	.3905	.0013	.03
3	.240	.2661	.90	.3611	.0307	.81

Exhibit 9-4 Testing of significance of individual regression coefficients in two ways. The b_j and s_{b_j} are from the previous exhibit. The t is calculated according to Equation 9.12 with $\beta_{0j} = 0$. $t_{.025, 16} = 2.12$. $R^2_{y \cdot \bar{j}}$ refers to the multiple R with all but variable j. The F values for $R^2_{y \cdot 123} - R^2_{y \cdot jk}$ agree with the corresponding t values.

level of significance. This is entirely consistent with the fact that zero was within the c.i. for each.

This result is rather surprising since we rejected the hypothesis that all weights are zero. Thus although we are reasonably confident that not all the weights are zero, we cannot be as confident that any individual weight is not zero. Such seemingly contradictory results are not uncommon in multivariate statistics. Some of the reasons for this are discussed in the next section.

Hypotheses about β

Testing a hypothesis about several parameters simultaneously means that the acceptance region is multidimensional. Testing whether the vector of regression weights is 0, 0, . . . , 0 is the same as testing whether the point defined as 0, 0, . . . , 0 is within the multiparameter *confidence region,* the one that contains the vector of weights β with probability $1 - \alpha$. To understand this better, it helps to understand what influences the confidence region, and to see what one looks like.

In cases where the sampling distributions are symmetric, the confidence region is the same shape as the multiparameter acceptance region. The difference is that the confidence region is centered at β, whereas the acceptance region is centered at β_0. Thus, descriptions of one are the same as descriptions of the other except for location. The theoretical sampling distribution depends on n and the inverse of Σ_{xx}. We use S_{xx}^{-1} as a way of estimating the latter. As noted earlier, the off-diagonal elements of S_{xx}^{-1} indicate the degree of nonindependence of the sample regression coefficients. In fact, s^{jk}/n estimates the "sample covariance" of b_j and b_k, the degree to which overestimates of one tend to go with overestimates of the other, and vice versa.

As always, the sampling characteristics of an estimator depend on the population parameters σ^{jk}, not the sample estimates s^{jk}. However, under the multivariate normal model the s^{jk} can be used to estimate the σ^{jk}, and under the fixed constants model, $s^{jk} = \sigma^{jk}$, since the x's are known, not sampled. Thus s^{jk} can be used to help see what happens to the b_j under sampling.

If the elements of a variance-covariance matrix tend to have the same sign—all positive in this case—then the elements of the inverse tend to have the opposite sign; thus two of the three elements here are negative, and the exception, s^{23}, is quite small. These negative sampling covariances mean that overestimates of one regression weight tend to occur with underestimates of others. In cases where all the sample weights are positive, testing the hypothesis that all are zero is comparable to asking whether all three are overestimates of the true weights. The fact that the overall H_0 is rejected indicates that the data are such that this is unlikely to be the case.

If the null hypothesis were true, then the sampling distribution for the three weights simultaneously would be three-dimensional. It would be dense around the point (0, 0, 0) and thin out as we moved away from that point. However, unless all the predictors have equal variances and are uncorrelated, the density does not thin out uniformly; rather, it thins out more rapidly in some directions than others. The inverse of the variance-covariance matrix describes the way it thins out. In testing a multivariate hypothesis, we construct acceptance and rejection regions just as for a single parameter. If we want to divide the three-dimensional space into an acceptance region and a rejection region such that the former has probability $1 - \alpha$, it is natural to want the acceptance region to be as small as possible. In the case of multivariate normality, in which the regression coefficients have this distribution, the boundary of the region is an ellipsoid, which is elliptical at any cross section. (Imagine a burly football player holding the sides of a football between his palms and pushing in.) If the sample values of all three coefficients are such that they define a point within the ellipsoid, H_0 is accepted; outside, it is rejected. There is not simply a shape to the region. It also has a location and an orientation. Its center is at (0, 0, 0) (the null hypothesis), and the degree of tilt of the axes depends on the off-diagonal items of S_{xx}^{-1}. Its relative thinness in any direction (how hard the player is squeezing) depends on a combination of the sizes of the diagonal elements of this matrix and of the off-diagonal ones. The overall volume of the ellipsoid depends on $\sigma_e^2/(n)$. In this particular case, the matrix shows that the ellipsoid is thickest in the b_1 direction and thinnest in the b_3. Furthermore, it is substantially flattened due to the correlation between b_1 and b_2.

By testing the hypothesis that all the weights are zero we are seeing whether the point in the three-dimensional space corresponding to (b_1, b_2, b_3) is inside or outside of the ellipsoid. In this example, $H_0{:}\beta = 0$ was rejected, although barely at the .05 level. Thus the point corresponding to (1.175, .110, .240) is just outside the acceptance region, as illustrated in Figure 9-1. It would look like a spacewalker floating just above a giant ellipsoidal space station.

The acceptance region in this figure is actually three-dimensional, and the upper part of the figure is a slice through it along the plane where $b_3 = 0$. Here it appears that the sample point (1.176, .110, .240) is just *within* the acceptance region. This is a misleading result of the perspective. If the slice is taken instead through the plane where $b_3 = .240$, as in Figure 9-1(b), then it can be seen that the sample point is just outside the acceptance region.

Two additional slices through the ellipsoid are shown in 9-1(c) and (d), corresponding to sections at $b_3 = .48$ and $b_3 = .72$.

Thus *any* point outside the ellipsoid leads to rejection of H_0 and corresponds to a significant multiple correlation. The shape and orientation of the figure show the directions in which there is the greatest uncertainty. For any given values of b_2 and b_3, the acceptance interval for b_1 is large compared to the corresponding intervals for the other coefficients, especially b_3 (the effect must be visualized). This corresponds to the relatively large standard error for this coefficient.

The marked downward tilt of the ellipsoid in the b_1, b_2 plane is essentially a reflection of the large negative entry in the inverse, s^{12}. The smaller effects of s^{23} and s^{13} could be seen if slices were taken through these planes. The fact that the tilt of the axis is negative implies that the uncertainty about $\beta_1 - \beta_2$ is much greater than the uncertainty about $\beta_1 + \beta_2$. These are among the things that can

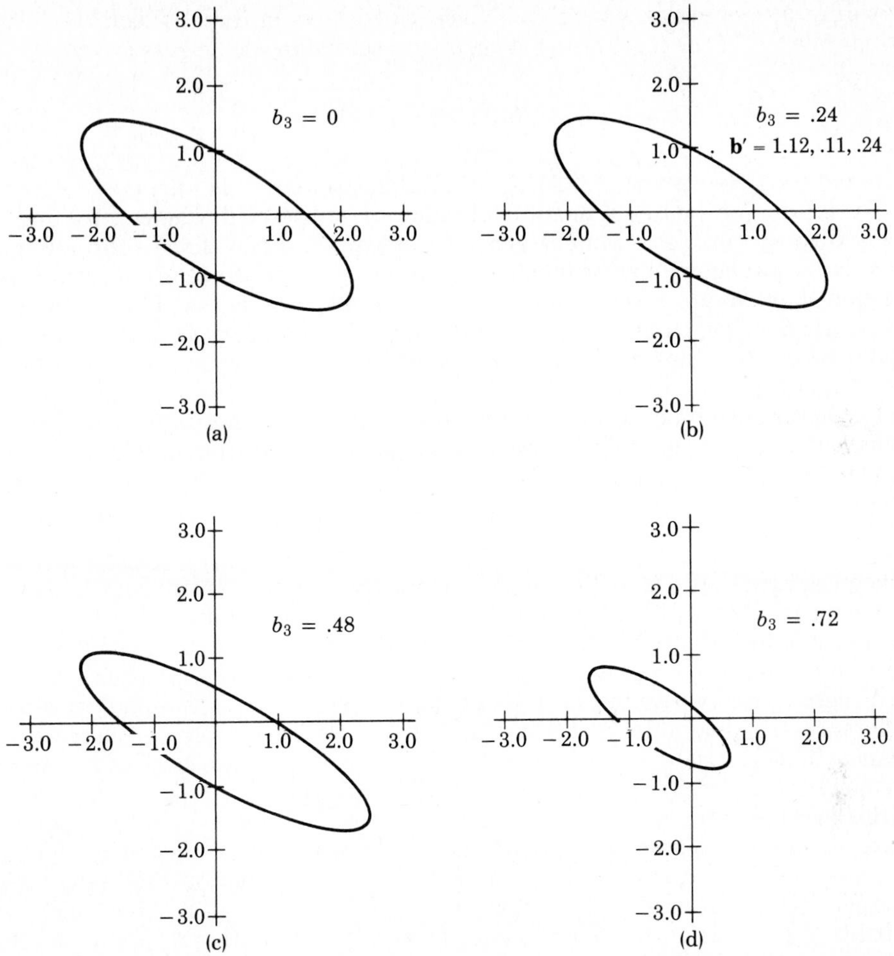

Figure 9-1 Slices through the acceptance region for **b** parallel to the b_1, b_2 plane. Combinations of values for the sample regression coefficients that are outside the figure would lead to rejection, at $\alpha = .05$. In panel (b) we see that the point (1.12, .11, .24) lies just outside the figure.

usually be discerned by study of S_{xx}^{-1} and construction of the ellipsoid, although the latter is a laborious process.

Inferences about Differences

Suppose we wanted to make an inference about the relative sizes of two population regression weights—that is, test the hypothesis H_0: $\beta_1 = \beta_2$. This hypothesis can be restated as H_0: $\beta_1 - \beta_2 = 0$. Now, inferences about $\beta_1 - \beta_2$ involve the sampling variances of the individual b's and the sampling covariance between them. As noted earlier, these quantities are estimated from the corresponding entries in S_{xx}^{-1}, which are to be divided by n, or directly from T_{xx}^{-1}. The standard error of the difference between two regression coefficients is

$$s_{b_j - b_k} = (t^{jj} + t^{kk} - 2t^{jk})^{1/2}\,\sigma_e \tag{9.15}$$

where t^{jk} is an element of $(X'X)^{-1}$, or equivalently of $[(n - 1)^{-1}S_{xx}^{-1}]$.

The effect of t^{jk} bears comment. Note that it is subtracted, whereas if x_j and x_k are positively correlated, ts^{jk} is generally *negative*. Thus, since subtracting a negative is adding, we may be much more uncertain about the *relative* sizes of two weights than we are about the individual weights themselves. This is another factor that helps explain the phenomenon of "bouncing betas," since the observation often has to do with the relative sizes of regression weights in different samples.

The formulas can be used to test hypotheses about the relative sizes of the two coefficients in the usual way, provided it is a genuine a priori hypothesis:

$$t = (b_j - b_k)/s_{b_j - b_k} \tag{9.16}$$

When the hypothesis is more like an attempt to examine the trustworthiness of an empirical observation about the relative sizes of two coefficients, we resort to the much more conservative Scheffè process. Here we use the same ratio, but the critical value for rejection is $(pF_{p,df})^{1/2}$ instead of t, where df equals $n - p - 1$. This highly conservative test is needed to make up for the fact that many possible differences are available for testing, and that large ones could be observed by chance. This test, along with the others in this section, are part of a general system for inferences about regression coefficients.

It is perhaps more common for interest to focus on differences in the standard score regression coefficients, although their relative sizes have only an ambiguous interpretation. With standard score coefficients the inferential process is more complex, but where n is large (perhaps $n - p \geq 100$ or more), it is reasonable to substitute terms from R^{-1} for the ones from S_{xx}^{-1} in Equation 9.15 and use the result as in Equation 9.16.

The direction of the long axis of the ellipses in Figure 9-1 supports the observation that there is great uncertainty about the differences in regression coefficients in cases like the present one where correlations are positive. We can slide up and down the axis a long way in this direction without getting outside the ellipsoid.

Generalized Inference about Weights

So far we have tested hypotheses about individual weights and about differences between weights. Both are examples of weighted combinations of weights, and the methods used are members of a system for making inferences about all possible combinations of weights. If $\mathbf{c}$ is a vector that defines a weighted combination of weights, then the sampling variance of the weighted combination is

$$\sigma_c^2 = \sigma_e^2 n^{-1}(\mathbf{c}'\Sigma_{xx}{}^{-1}\mathbf{c}) \tag{9.17}$$

Substituting the sample estimates for the population quantities gives

$$\hat{\sigma}_c^2 = \hat{\sigma}_e^2 n^{-1}(\mathbf{c}'\mathrm{S}_{xx}^{-1}\mathbf{c}) \tag{9.18}$$

The various formulas for standard errors that have been used here to test hypotheses about individual regression coefficients, or differences between them, are special cases of this one, where **c** has a very simple form. For example, $\mathbf{c}' = (1, 0, 0, \ldots)$ is for finding the standard error of b_1 itself; $c' = (1, -1, 0, 0, 0, \ldots)$ is for finding the standard error of the difference $b_1 - b_2$, and so on.

These formulas are used to test hypotheses about combinations of coefficients or to form c.i.'s for these combinations. In cases where sampling distributions are symmetric, the multiparameter confidence region has the same shape and size as the acceptance region. The only difference is that it is centered at **b** instead of at β_0, as shown in Figure 9-2.

The estimated standard error Equation 9.18 is used in two different ways, depending on whether the **c** vector defines a genuine a priori hypothesis or is suggested by the data. In the former case, the c.i. for the combination is

$$\mathbf{c}'\mathbf{b} \pm (F_1\hat{\sigma}_e^2\mathbf{c}'\mathrm{S}_{xx}^{-1}\mathbf{c})^{1/2}n^{-1/2} \tag{9.19}$$

Correspondingly, the acceptance region for a null hypothesis H_0: $\mathbf{c}'\beta = \beta_0$ is

$$\mathbf{c}'\beta_0 \pm (\mathrm{F}_1\,\hat{\sigma}_e^2\mathbf{c}'\mathrm{S}_{xx}^{-1}\mathbf{c})^{1/2}n^{-1/2} \tag{9.20}$$

Note that F_1 is the critical value of F for 1 and for $n - p - 1$ df at the chosen Type I error rate.

For combinations or hypotheses that are suggested by the data, F_1 must be replaced by pF_p, where F_p is the F for p and $n - p - 1$ df, to ensure that the error rate does not exceed α. For the c.i.,

$$\mathbf{c}'\mathbf{b} \pm (pF_p\,\hat{\sigma}_e^2\mathbf{c}'\mathrm{S}_{xx}^{-1}\mathbf{c})^{1/2}n^{-1/2} \tag{9.21}$$

will always contain $\mathbf{c}'\beta$ with probability $1 - \alpha$ or more for all **c**. Correspondingly,

$$\mathbf{c}'\beta_0 \pm (pF_p\hat{\sigma}_e^2\mathbf{c}'\mathrm{S}_{xx}^{-1}\mathbf{c})^{1/2}n^{-1/2} \tag{9.22}$$

is an acceptance region that leads to rejection of H_0 only α percent of the time when it is true for all **c**. This allows us to try any conceivable combination, whether suggested by the data or not. The latter two equations are the Scheffé intervals.

In effect, the last two equations draw a sphere around the ellipses in Figures 9-1 or 9-2. Everything within the sphere is acceptable. If the sphere contains much more volume than the ellipse, that is the price paid for drawing the sphere that contains this ellipse and any other ellipses that might have occurred. In effect, it allows the investigator to say, after seeing the data, "I think I'll use $\mathbf{c}' = (1.176, .1104, .2401)$," thereby getting maximum results by peeking at the data, and maintaining the true alpha level.

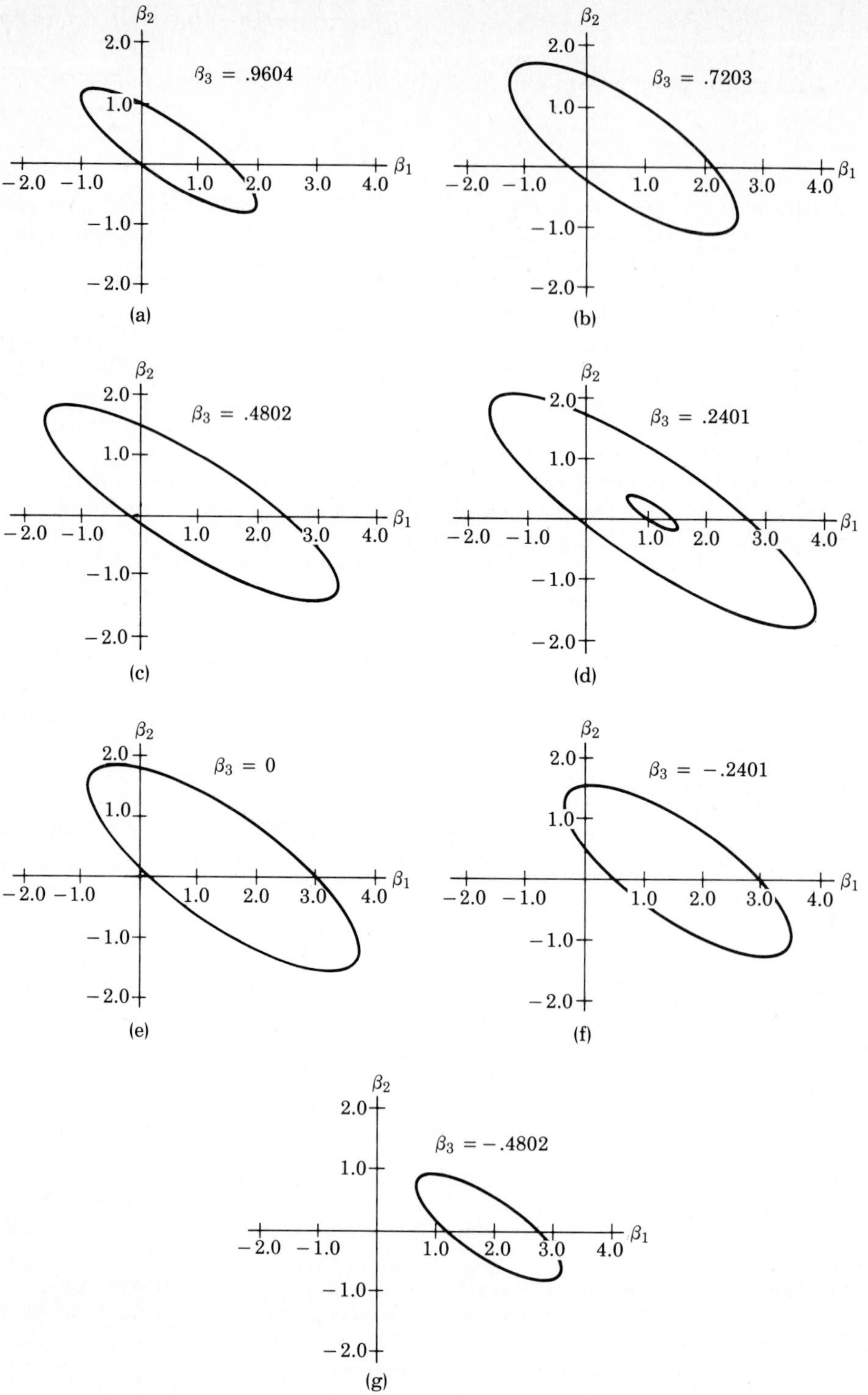

Figure 9-2 Sections through the .95 confidence region of the data from Exhibits 9-1 to 9-5. Panel (d) shows that (0, 0, 0)′ is just outside the confidence region. The small ellipse in panel (d) is a section through the ellipsoid that would result if $n = 104$.

OTHER INFERENCES ABOUT WEIGHTS

Sensitivity

As we have observed, the accuracy of prediction may be relatively insensitive to departures from the optimum least squares weights. Thus one may wish to explore whether other sets of weights are about as good as **b**. One may have an explicit, a priori hypothesis about the other set or an alternative set may be suggested by the data, or there may be something in between. One may be interested in absolute prediction with these weights, or only be concerned with relative prediction—the correlation between a composite based on these weights and y. One may wish to explore in a descriptive way the effects of using other weights, and wish to decide on the basis of the data whether these weights are plausible alternatives to the obtained weights and/or to the null hypothesis.

Recall the methods used in Chapter 4 for assessing the effects of using a weight vector. Let $\mathbf{w} = (w_1, w_2, \ldots, w_p)$ stand for some alternative to **b**, so that

$$\hat{y}_{iw} = \Sigma w_j x_j \tag{9.23}$$

and

$$e_{iw} = \hat{y}_{iw} - y_i \tag{9.24}$$

Using the sum of squared deviations that results from using **w**, $ss_{dev:w}$ can be readily calculated by attaching -1.0 as a tail to **w** and by pre- and postmultiplying the complete matrix of variables by this augmented vector:

$$ss_{dev:w} = [\mathbf{w}' \quad -1.0] \begin{bmatrix} T_{xx} & \mathbf{t}_{xy} \\ \mathbf{t}'_{xy} & t_y^2 \end{bmatrix} \begin{bmatrix} \mathbf{w} \\ -1.0 \end{bmatrix}$$

Alternatively,

$$s^2_{dev:w} = [\mathbf{w}' \quad -1.0] \begin{bmatrix} \mathrm{S}_{xx} & \mathbf{s}_{xy} \\ \mathbf{s}'_{xy} & s_y^2 \end{bmatrix} \begin{bmatrix} \mathbf{w}' \\ -1.0 \end{bmatrix} \tag{9.25}$$

The effect of using **w** instead of **b** can be assessed by comparing $ss_{dev:w}$ to ss_{dev} or $s^2_{dev:w}$ to s^2_e.

The above equations show the effect of varying the weights on *absolute* prediction. Suppose we are interested in relative prediction reflected in the correlation between the criterion and a combination that has alternative weights **w** instead of **b**. Then the correlation is

$$r_{y\hat{y}_w} = \frac{\mathbf{w}'\mathbf{s}_{xy}}{(\mathbf{w}'\mathrm{S}_{xx}\mathbf{w})^{1/2} s_y} \tag{9.26}$$

according to Chapter 4.

Exhibit 9-5(a) shows that using $\mathbf{w}' = (.25, .25, .25)$ instead of $\mathbf{b}' = (.470, .066, .192)$ leads to $s^2_{dev:w} = .625$, which is barely higher than $s^2_{dev} = .608$. Also, $r_{y\hat{y}_w} = .612$, is barely lower than R = .626. Even using (0, .5, .2) does not do badly ($s^2_{dev:w} = .690$ and $r_{y\hat{y}_w} = .558$), in spite of the fact that the best predictor, x_1, is not used at all and that x_2, which has the lowest weight in $\mathbf{b}_1$, has the highest in $\mathbf{w}$.

Confidence Regions for β

These ideas can be extended to define a confidence region for β. Recalling the interchangeability of hypothesis testing and confidence regions: $1 - \alpha$ confidence region contains all the hypotheses that cannot be rejected by the data. One can therefore test whether a given trial vector $\mathbf{w}$ of weights is within the confidence region by testing H_0: $\beta = \mathbf{w}$. Explicitly, $\mathbf{w}$ will be within the confidence region unless

$$F = \frac{(ss_{dev:w}\ ss_{dev})/p}{ss_{dev}/(n - p - 1)} \tag{9.27}$$

exceeds the tabled value for α and $p, n - p - 1\ df$. This can be expressed as a ratio

$$S = \begin{bmatrix} 1.0 & .8 & .4 & .6 \\ .8 & 1.0 & .3 & .5 \\ .4 & .3 & 1.0 & .4 \\ .6 & .5 & .4 & 1.0 \end{bmatrix}$$

b	$\mathbf{w}_1$	$\mathbf{w}_2$
.470	.250	0
.066	.250	.500
.192	.250	.200
−1.000	−1.000	−1.000

(a)

$s^2_{dev:w}$	.608	.625	.690
$r_{y\hat{y}_w}$	.626	.612	.558

(b)

$F_{3,16}$	.15	.72
$F_{2,16}$	.23	1.06
$F_{3,96}$	.89	4.32
$F_{2,96}$	1.37	6.35

(c)

Exhibit 9-5 Effect of using $\mathbf{w}_1$ (equal weights) and $\mathbf{w}_2$ (0 weight for x_1). For simplicity, the variances of all four variables are assumed to equal 1.0. The σ^2_e is .608, and R is .626. Equation 9.26 is used to find the $s^2_{dev:w}$, and Equation 9.27 to find $r_{y\hat{y}_w}$. Equations 9.28 and 9.30 test whether the weights are outside the confidence region. First assume $n = 20$, and then $n = 100$. Slightly different results occur if more significant figures are used in the weights, due to the effects of rounding.

of $ss_{dev:w}$ to ss_{dev} and F_α:

$$\frac{ss_{dev:w}}{ss_{dev}} \geq \frac{F_\alpha}{n - p - 1} + 1 \tag{9.28}$$

If the weights are for standard scores, then $ss_{dev} = (n - 1)(1 - R^2)$ and the $ss_{dev:w^*}$ is found from the correlation matrix.

Where relative prediction is of concern—where correlations are of more interests than error variances—Equation 9.28 should be adjusted since a scale adjustment would be allowed in the predictions. In that case,

$$F = \frac{(R^2_{y \cdot p} - r^2_{y\hat{y}_w})/(p - 1)}{(1 - R^2_{y \cdot p})/n - p - 1} \tag{9.29}$$

seems more appropriate.

Analyses such as those shown here often lead to the conclusion that the data permit considerable latitude in the weights that can be used. This is especially true if $n - p$ is under 100 and if there are at least moderate correlations among the predictors. The latter, it should be recalled, increases the uncertainty regarding the true weights.

These concepts and equations are applied in Exhibit 9-5. Assuming first that $n = 20$, the F to determine whether (.25, .25, .25) is within the confidence region (Equation 9.28) is only .15. Allowing scale adjustment, that is, using the weights for relative prediction, yields an F of .22. The second set of weights, (0, .5, .2), yields F values of .72, and 1.06, respectively. Thus, either set of weights is clearly within the confidence region.

If n were 100, equal weights would yield F's of .92 and 1.38. So even if the sample is this large, the equal weights are within the confidence region. On the other hand, the (0, .5, .2) weights would lead to F's of 4.48 and 6.63, clearly significant in both cases. For the larger sample, we would then conclude that this vector is outside the confidence region.

Applying Equation 9.28, we compute that $ss_{dev:w}$ would have to be as low as .506 in order for a weight vector to be outside the .95 c.r. when $n = 20$. But for $n = 100$, any weights that lead to an $s^2_{dev:w}$ of less than .591 indicate a weight vector is outside the confidence region.

Figure 9-2 illustrates this principle. The figure shows seven slices through an ellipsoid that represents the confidence region for β. Any point within the ellipsoid is within the .95 c.r.; any point outside is not. From the slice through the plane where $\beta_3 = 0$, we see that $(0, 0, 0)'$ is just outside the confidence region. The confidence region is the same shape and orientation as the .05 acceptance region shown in Figure 9-1.

Note how the weights are employed in this discussion. They can be used in a literal fashion to form a composite score, which is a prediction of the value of y for each individual, denoted $\hat{y}_{icw}$. Then $ss_{dev:w}$ is the sum of squared discrepancies between y_i and $\hat{y}_{iw}$, $\Sigma(\hat{y}_{iw} - y_i)^2$, and $s^2_{dev:w}$ is this quantity divided by $n - 1$. We could also apply the trial weights either to deviation scores or standard scores of the predictors and compare the prediction to the comparable form of y.

There are two approaches to computing these sums of squares or variances. One method actually computes the discrepancies, whereas the other is based on applying the methods of Chapter 4 to the covariance matrix. Let $c_1, c_2, \ldots, c_p$ refer to the values of the weights to be used in a particular trial set in order to distinguish them from the b's. Then a $p + 1$ element vector is formed in which the first p elements are the trial weights and the last element is -1. Applying this vector of weights to the complete set of scores, including y, is one way to find the differences between the predictions and the actual y_is. The differences are squared and summed to form $ss_{dev:c}$. Dividing by $n - 1$ gives the variance. This can be accomplished using the procedures available in the common statistical computer packages. In SPSS^x, we would form a new variable which is the composite of the original $p + 1$ variables by applying the weights to them in the COMPUTE procedure. Then $s^2_{dev:c}$ would be the variance of this new variable. BMDP offers the same process using TRANSFORMATION cards, and in SAS, it is accomplished in the DATA step.

The meekness of the example data in accepting such widely different sets of weights deserves comment. Note that in $\mathbf{w}_1$ and $\mathbf{w}_2$ the sum of the weights for x_1 and x_2 is about .5, which is more or less the same as for $b_1 + b_2$. We can increase one weight as long as we decrease the other algebraically to compensate. On the other hand, the data are much less supportive of patterns where *both* x_2 and x_1 get high weights or both get low ones. Figure 9-2 should show that the long axis of the confidence region has a negative slope with respect to β_1 versus β_2. Correspondingly, the sampling distribution would show a negative or compensatory relationship between b_1 and b_2. This is reflected in the negative entry in s^{12}, indicating, as noted earlier, that a sampling deviation in one direction for b_1 is likely to be accompanied by a deviation in the other direction for b_2, and vice versa. From the point of view of confidence regions for the coefficients, a large value for one of the weights may be assumed, provided the other is assumed to have a small value, or vice versa.

The overall conclusion to be reached from this exploration may be that variation in the sets of weights makes surprisingly little difference in the overall accuracy of prediction. Indeed, with a sample as small as 20, only the most general conclusions can be reached concerning the relative weights unless the multiple R is considerably higher than it was here. On the other hand, a moderately large sample of 100 permits fairly distinct inferences, at least if there are only a few predictors. However, an investigator may be willing to work with weights that do nearly as well as the obtained ones, even though they may be outside the confidence region in a large sample. These are only vague generalizations, however. The investigator is better off exploring her own data in this manner than she is in acting on such principles. This is an important and neglected practice among users of multiple regression, although it is easily implemented through the use of statistical computer packages.

INFERENCE IN STEPWISE REGRESSION

Capitalization on Chance

Chapter 7 describes stepwise multiple regression and its two main variations, hierarchical and variable-selection MR. These need to be considered from an inferential point of view. The latter, in particular, provides enticing pitfalls for the unsophisticated investigator. Moreover, multiple regression typically requires a considerable investment in time, money, ego, or status and so the investigator wants to find positive results. Therefore, it is particularly important that the investigator, or at least those who read or review the results, be protected against seeing positive results where there are only chance effects.

Hierarchical Analyses

Consider the investigator who arranges the predictors into an a priori order for entry into the regression equation. Suppose there are two subsets of variables, one of which has priority over the other. The priority may be based on past experience. For example, set A may be known to predict y and the question is whether or not set B will increase the predictive accuracy. Or the priority may be based on theory. For example, there may be theoretical reasons for expecting set A to be effective, whereas set B may have no good theoretical basis, or it may be based on practicality. For example, set A may be cheaper or more convenient to work with, whereas set B may be expensive or difficult. In all these cases, it is assumed that the second set will not be of any use if the first is not, so the second will not be tested unless the first is found to provide significant predictive power. Therefore set B will not be tested unless set A is significant. If this seems a rather stiff requirement, and that really all the investigator has in mind is a preference for set A over set B, then he is in the variable selection mode and must use some variant of the methods of the next section. Although the hierarchical procedures generalize to more than two sets, and only a single variable may be in each, the strictness of the hierarchical structure must be maintained for the methods of this section to apply. That is, sets of variables are tested according to a strictly defined a priori order, and as soon as a set is found to be nonsignificant, *no further tests are made.*

This may be recast in more formal terms. Let there be p_1 predictors in the first set and p_2 in the second so that $p_1 + p_2 = p$. Then a hypothesis, H_{01}: $\beta_1, \beta_2, \ldots \beta_{p1} = 0$, can be tested at a given alpha level. If the null hypothesis is true, the probability of its rejection is α. A second hypothesis concerns the second set of variables: H_{02}: $\beta_{p_1+1}, \beta_{p_1+2}, \ldots \beta_{p2} = 0$. If the first null hypothesis is true, then there is no reason to expect the second one to be false, so if the first null hypothesis is accepted, the second will not be tested. That is, either both null hypotheses are true, both are false, or the first is false and the second is true; the reverse possibility is considered impossible.

This is an example where the hypotheses are nested. More importantly, from the point of view of controlling Type I error probabilities, the probability of *any* Type I error here is α. Suppose both nulls are true. We will either reject H_{01} and not H_{02} or we will reject both. The probability of either of these things happening is α because the second outcome is a subset of the first and Pr(A or B) = Pr(A) if B is a subset of A. The other possibility is that H_{01} is false and H_{02} is true. In this case, the probability of a Type I error is no more than α. The only way a Type I error could occur is by rejecting H_{02} after rejecting H_{01}, which happens with a probability less than α if H_{01} is false and H_{02} is true, since sometimes we accept H_{01}. There is not even an opportunity to falsely reject H_{02} if H_{01} is accepted—wrongly in this case. Thus, no matter what the true state of affairs, the probability of a Type I error is at most α when the hierarchical procedure is followed.

If it should happen that H_{01} is true and H_{02} false, there is very little chance of finding out about it. We are in a situation analogous to that of making a one-tailed test when the null hypothesis is false in the opposite direction. Thus, unless we are dead certain that this is not the case—or if we would rather not know about it if it is—we should adapt the methods described in the next section.

The tests of the hierarchical hypotheses take what should by now be a familiar form: Is the reduction in error sum of squares more than would be expected by chance on the basis of the number of degrees of freedom that are expended to achieve it? For example, H_{01} is tested as if the other variables in the battery were not present. Then each hypothesis lower in the hierarchy is tested by subtracting its ss_{dev} from the ss_{dev} resulting from the previous step, and dividing by the number of new parameters—or new predictors—estimated in the new step. This is divided by the estimate of error variance based on all the predictors included so far. Thus, in considering the sth null hypothesis in the hierarchy, involving p_s predictors, we refer to the deviations from regression based on all the variables up to and including the sth set as $ss_{dev:s}$. Then the relevant test is

$$F = \frac{(ss_{dev:s-1} - ss_{dev:s})/p_s}{ss_{dev:s}\Big/\left(n - \sum_{v=1}^{s} p_v - 1\right)} \tag{9.30}$$

In the denominator, p_v represents the number of variables in each set v, up to and including the set s.

In terms of the multiple correlation, Equation 9.30 can be expressed as

$$F = \frac{(R_s^2 - R_{s-1}^2)/p_s}{(1 - R_s^2)\Big/\left(n - \sum_{v=1}^{s} p_v - 1\right)} \tag{9.31}$$

Thus we have formulas for testing the sth null hypothesis in the hierarchy for which the regression weights for all the variables in set s are zero. Note that the equations hold even for the first set because the numerator in Equation 9.30 is ms_{reg} and the denominator is ms_{dev} and in Equation 9.31, R_{s-1}^2 is zero. These procedures are illustrated in Exhibit 9-6.

Equal Weights as a Hypothesis

In many cases, weighting the predictors equally works nearly as well as the sample regression weights and using equal standard-score weights works even better. Moreover, if we apply the sample weights to new data as a form of validation (a procedure to be discussed on pages 191–192 under "cross validation") the equal weights may be found to do even better than the least-squares ones. Suppose we tested the complete weights b against the effectiveness of the equal weights and used b only if it significantly improved over the equal weights. This has considerable merit, particularly when all predictors correlate positively with each other and all validities have the same sign.

Including the p predictors in the battery reflects some conviction on the part of the investigator that they will be of value. Since we are often unsure as to just which predictors are likely to be more important, a personal hypothesis may be roughly characterized as $H_1{:}\beta = k/s_j$. This means that, except for differences in standard deviation, which reflect the different units variables are measured in, their regression weights are equal.

$$\begin{array}{cccc} 1 & 2 & 3 & y \end{array}$$
$$\begin{bmatrix} 1.00 & .60 & .50 & .40 \\ .60 & 1.00 & .70 & .48 \\ .50 & .70 & 1.00 & .70 \\ .40 & .48 & .70 & 1.00 \end{bmatrix}$$

	Step 1		*Step 2*	
Variable	b^*	r		r
1	—	.40		—
2	—	.48	b^*	.327
3	—	.70	.40	.630

Variable Entered	1	2
R_s^2	.160	.250
R_{s-1}	0	.160
Increase	.160	.090
p_s	1	1
F	5.33	3.24

Exhibit 9-6 Hierarchical stepwise multiple regression using data from Exhibit 7-1. Assuming that $n = 30$, and that the hierarchy is 1,2,3, the F may be calculated from Equation 9.34. Alpha is set at .05. At step 1 the F required for significance with 1,28 df is exceeded, and H_{01} is rejected. At step 2, the F for 1,27 df is not exceeded, and H_{02} is accepted, ending the analysis. The b^* column shows the weights for variables already entered and the r column shows the partial correlation of predictor j with y, partialing out the variables already entered.

This hypothesis can be tested by making a composite variable, which is a weighted combination of all p predictors using the weights of the reciprocals of the standard deviations:

$$x_c = x_1/s_1 + x_2/s_2 + \ldots x_p/s_p.$$

Using the methods described in the previous section, the hypothesis can be tested by seeing if this weight vector is in the confidence region. If maintaining accuracy of relative prediction is the concern, then Equation 9.29 is the relevant one. If absolute prediction is the goal, then Equation 9.27 is appropriate. Note that rejecting the equal weights hypothesis in its various forms is the same as concluding that differential weights are necessary.

Note that if the composite variable is included along with all p original variables and is used as set 1 and other p variables as set 2, there will be a linear dependence problem, and the inverse for all $p + 1$ will not be computable. This can be avoided by including only $p - 1$ of the original variables in set 2. The multiple R based on these p predictors will be the same as that based on the original p. Note, however, that if the hypothesis of equal weights is rejected, it is simplest to do an analysis employing all p observed variables in order to determine those weights.

The procedure, then, is to see how much of the variance is accounted for by a simple composite variable. Then a test is made to see if the optimum weights perform significantly better. This can be done with any other composite as well, of course. If not, then one would accept H_1. If the increase is significant, then one would accept H_2 to the effect that not only are the weights nonzero, they are also unequal.

If equal standard score weights are used, it will be necessary to find out the standard deviations of the predictors before regression analysis is performed.

When the standard deviations are expected to be approximately equal, however, raw score weights—that is, a simple sum of the predictors—may be used instead.

Polynomial Regression and Hierarchical Testing

An obvious place to apply these methods is polynomial regression—that is, a situation where we are investigating the possibility of curvilinear regression by predicting with x_j^2, x_j^3, and so on, as well as x_j itself. The powers of x form a natural hierarchy, one in which it is often reasonable to apply the stringent decision-rules such as not looking at the contribution of x^3 if x^2 and x do not predict. Where there are several predictors, natural hierarchical sets are formed by the powers of the predictors. So, we test all the linear x_j as a group, then the contribution of all the x_j^2, and so on. If the product terms are used then they too will have a natural place in the hierarchy. In the preference for car designs example (p. 128), this might take the form of using first all the x_j, then all the x_j^2, then all the $x_j x_k$, and so on, if higher powers are to be allowed, inserting the simple powers before the product terms. Thus polynomial regression falls quite naturally into the hierarchical framework, although just how the hierarchy is formed depends on the nature of the expectations about the variables.

General Nonhierarchical Testing

There is a confusing multiplicity of procedures for testing hypotheses in multiple regression, and it is now necessary to add still another. In the preceding section, the investigator had a nested hierarchy of hypotheses, but suppose this is not the case. Suppose that the investigator has groups of variables and wishes to test each group and that there are k sets of variables, with p_q in Set q, with $p = \Sigma_{q=1}^{k} p_q$. Then $ss_{dev:\tilde{q}}$ represents deviations resulting from using *all but* Set q, and $ss_{dev:p}$ still represents all p predictors, and correspondingly, $R_{\tilde{q}}^2$ and R_p^2. A generalization of Equation 9.13 is

$$\mathrm{F} = \frac{(ss_{dev:\tilde{q}} - ss_{dev:p})/p_q}{(ss_{dev:p})/(n - p - 1)} \tag{9.32}$$

The equivalent generalization of Equation 9.14 is

$$F = \frac{(R_p^2 - R_{\tilde{q}}^2)/p_q}{(1 - R_p^2)/(n - p - 1)} \tag{9.33}$$

Note that if $p_q = 1$, this test is the same as the one for testing a single regression coefficient.

Here, there are k null hypotheses, about one for each set of coefficients. Each one is tested using the same denominator, assuming that all the other null hypotheses are false. If this is the case, then only this denominator is an estimate of σ_e^2. Thus, this procedure is generally more powerful than the hierarchical one if nominal F levels are used. It loses some power when some hypotheses are true and some are false, which is the price paid for not having an a priori order.

Note, however, that in this case we are back in the multiple hypothesis-testing business. Either we take a hypothesiswise stance toward the vector of coefficients in each set or we must control the analysiswise alpha level. In the latter case, we must make a Bonferroni-type correction of the alpha levels used for the test of each set, or use a protection-levels approach, first testing the overall significance of R_p^2 and making the other tests only if R_p^2 is rejected.

Unfortunately, most current computer programs require multiple analyses for this approach. The example used in Chapter 7 regarding the prediction of job performance of typists (p. 119) can be used to illustrate the difference between the current method and the hierarchical one. Here, $k = 2$, $p_1 = 2$ (the two typing scores), and $p_2 = 3$ (the three interview ratings). In the hierarchical analysis, we first test the multiple R for the two typing measures as if they were the only variables in the analysis. If this is significant, we go on to test the improvement provided by the interview variables. This would be appropriate if we had good reason to expect that the interview would not be valid unless typing skill was. But if our expectations were less definite, either or both sets might be valid, and the tests noted here would be used.

If so, we need to decide whether to take a hypothesiswise or analysiswise stance with respect to alpha levels. The former may be appropriate from a purely research point of view. Suppose there are two questions: (1) Does typing predict performance? and (2) Does the interview predict performance? We wish to answer each, as a separate entity, and so we take the hypothesiswise approach, bearing in mind that there is an enhanced probability of at least one false positive answer since we tested two hypotheses. But if the question is, Does either group predict, and if so, which one? Then we need to take the more conservative route. The latter is likely to be a case in which the interest is purely practical or in which the research is exploratory.

Not all statiticians agree with these recommendations. Some reasons are as follows. Suppose one null hypothesis is true and the other not. Then the sum of squares for the set with the null coefficients should really be made part of the ss_{dev}. Some investigators consider the hierarchical analysis to be the only valid one. The recommendations here are based on the principle that, unless one knows which set has true null hypothesis, one is likely to inflate the error by including predictable variance in ss_{dev}. Thus the recommended procedure has the greatest power, whereas other tactics can be used to control Type I error frequency.

VARIABLE-SELECTION METHODS

Significance of Prediction

It is in the variable-selection multiple regression strategy that the inferences are most vulnerable unless care is taken to control Type I probabilities. Indeed, most computer programs for multiple regression are positively satanic in their temptation toward Type I errors in this context. This can be avoided by acknowledging that several predictors have an opportunity to enter the regression equation at any step. This adjustment can be accomplished in a variety of ways. The most convenient methods, given the nature of the modern computer programs, are the Bonferroni approach and the protection levels strategy.

Recall that in variable-selection procedures, the predictor entering the regression equation is the one with the highest correlation with the dependent variable after all previously entered variables are partialed out, and that no predictor is entered unless a specified improvement is achieved. In the inferential context, this decision-rule involves testing the regression weight of each of the variables not yet entered to see if it "significantly" improves R^2, using Equation 9.14 or one of its equivalents, 9.12 or 9.13. For example suppose two variables have already been entered in the regression equation and there are five others. For each of these five variables, Equation 9.14 is calculated, where $p = 3$. The variable yielding the highest F ratio is added to the predictor set, provided its F ratio is significant.

The problem is how to define "significant." If we use the nominal tabled value with 1 and $n - p - 1$ df for a given alpha level, this ignores the fact that there were five predictors which had the opportunity to exceed this level. If the null hypothesis is true and all their weights are zero, then the probability that at least one manages to exceed the required level is greater than α, probably quite close to 5α in this case. Thus, if we really wish to use the .05 level here, we should compare the largest F to the value tabled for *.01* and include the variable only if its F exceeded this. If this is not true for any variable, the analysis ends, and only the first two variables are in the regression equation.

This approach is used at every step, even when there are no variables in the equation. So, if q variables have entered the equation and there are r candidates remaining, with $q + r = p$, then the F level that is required for significance in Equation 9.14—or its equivalents when one wants to use α as the Type I probability—is the F corresponding to α/r. Alternatively, if the computer calculates the Fs for all the candidate variables and also provides their nominal alpha levels, these levels should be multiplied by r to give a realistic criterion for accepting a new variable. Obviously, if there are many candidate variables at any stage and none of them has a very high F value, then the resulting product can exceed 1.0, an impossible value for a probability. This is not a serious problem, however, as we are merely trying to decide whether the largest of the obtained values could have been obtained by chance when it is actually zero. If the nominal alpha levels multiplied by the number of candidates is less than some predetermined alpha level such as .05, then it is reasonable to conclude that it is unlikely the particular variable has a zero regression weight, and allow it to enter the regression equation.

There is another practice that should be avoided if serious inferential errors are not to be made. This has to do with testing the significance of R when a subset of variables has been selected. Almost all major computer packages print the F ratios of Equations 9.9 or 9.10 for the *selected* set of variables. Taking this F ratio literally will inflate the Type I error rate because of the extent to which chance relations will be capitalized on in selecting the variables. Suppose the null hypothesis is true for all p variables. None of their sample correlations with y are likely to be exactly zero, and the more variables we have the more likely it is that some of the r_{jy} will be relatively large. The program will select those variables to be in the regression equation. The R^2 will look large, and the F ratio is likely to look significant. For the q variables that by chance are the most predictive, the F ratio

$$\frac{(R^2/q)/(1 - R_q^2)}{1 - R_q^2/(n - q - 1)}$$

is very likely to exceed the tables F and q and $n - q - 1$ df, even when the null hypothesis is true for all variables. This F ratio is therefore misleading. Another practice that should be avoided if the analysis has been a variable-selection one is using the F ratio to judge the "significance" of the R^2 resulting from including q variables. Using Equation 9.9, and taking the numerator degrees of freedom as q is misleading since it was designed for use after all the p predictors have been included, but here we have selected the q best ones instead. These q will obviously do better than the remainder, so an F ratio based on these predictors will be much larger than one based on all predictors. Thus when the null hypothesis is true and we select the q predictors that happen by chance to have the largest regression weights in the sample, we can get an F ratio that looks significant a substantial proportion of the time, particularly if q is only a small fraction of p. *We should not compute it and should not pay any attention to an* F *ratio that has been computed, even if it is done by a highly respectable computer program.*

In addition, it is probably better to use p, not the number of variables selected, in Equation 9.1 to estimate σ_e^2 and in Equation 9.3 to estimate P^2. This may be viewed as slightly conservative, but in a sense, *all* the variables are in the equation, even though some of them have been given zero weights. Because this is done on the basis of sample data, in a sense, they are in the equation. These procedures are illustrated in Exhibit 9-7.

Omnibus Test Followed by Tests of Single Parameters

What F ratio is interpretable in the ordinary sense if a variable-selection method is used? The answer is only one, the one based on all the predictors. If

	Step 1				Step 2			
Variable	b^*	r	R_{in}^2	F	b^*	r	R_{in}^2	F
1	—	.40	.160	5.33	—	.081	.493	.18
2	—	.48	.230	8.38	—	−.020	.490	.01
3	—	.70	.490	26.90	.700			

	Step 1	Step 2
Candidates	3	2
p_q	1	1
$n - p - 1$	28	27
F required	6.48	5.63
Variable entered	3	None
R_3^2	.490	—
R_{s-1}^2	0	—
Increase	.490	—
F	26.90	—
$\hat{\sigma}_e$	1.447	—

Exhibit 9-7 Variable selection stepwise multiple regression with Bonferroni corrections, using $\alpha = .05$. The column labeled r gives the partial correlation for each variable not yet in the equation, controlling for those that already are. Column R_{in}^2 gives the value R^2 will have if the variable of that row is entered at this step and F is the ratio of Equation 9-14. At step 1 there are three candidates so that the F required for significance corresponding to a nominal alpha level of $.05/3 = .0167$. Calculations show that for 1 and 28 *df* this should be 6.48. Variables 2 and 3 exceed this value, so the higher one is selected. At step 2 there are two candidates, so the F required for significance is that for alpha level $.05/.2 = .025$ and 1,27 *df*. Clearly, neither 1 nor 2 reaches this value.

this is significant, then the alpha level provides the usual protection against Type I errors. If the null hypothesis is true, then a significant F will be found here α of the time. Thus we may go on to perform a variable-selection analysis of the same predictors without increasing the Type I error probability when H_0 is rejected here. This provides an alternative to the Bonferonni approach proposed in the previous section.

There are two ways this can be done: forward and backward. The forward approach is the ordinary variable-selection method in which the process starts with no predictors and adds, one at a time, the variable that most reduces the error of estimate until no variable contributes enough to earn its keep or achieve significance.

In the backward approach, as might be expected, the process takes place in the other direction. We start with the complete multiple regression system and *drop* the variable that contributes least to the prediction, provided its regression weight is not significant or that it is contributing too small a percentage of variance to be worth keeping. If a variable is dropped, then the regression weights for the remaining $p - 1$ are recomputed. (Remember that this deletion will change the weights of the others unless the predictors are orthogonal.) These are examined in the same way as was done previously for all p variables, and again a variable may be dropped. The process continues until no variable is dropped from the equation. The remaining variables are the ones selected for use in the regression equation. The forward and backward solutions do not have to arrive at the same final subset, although they are likely to, and usually will only differ from each other only by a variable or two.

Excluding variables that have the "least significant" regression weights one at a time is similar to seeing if zero values for the corresponding regression weights give a point that is within the confidence region for the whole set. For example, if there are four predictors, and x_1 and x_4 are selected by the backward solution and $R^2_{y.14}$ is not significantly less than $R^2_{y.1234}$ by Equations 9.28 or 9.29, then point (b_1, 0, 0, b_4) is within the confidence region of the complete set of regression weights. Either forward or backward solutions can be used to find those variables for which zero is a plausible value for the regression weight. Although both solutions work, they will not always succeed in finding the largest possible set of zero weights.

Some computer packages offer the option of trying all possible subsets of variables. This is guaranteed to find the best possible solution employing a given number of variables. In view of the amount of sampling uncertainty involved, selecting the "best" subset in a sample is almost always an unreliable process. Monte Carlo studies have demonstrated that if several samples are taken from the same population, a variable selection method rarely selects exactly the same subset in different samples.

A certain amount can be said against variable-selection methods, in spite of the fondness that some investigators and program-writers have for them. First, these methods are highly subject to sampling instability due to capitalizing on chance effects in the current sample. After all, the regression weights for the complete set are the best estimates of the population weights that we have—at least by the methods discussed so far. Selecting only some variables indicates that the weights for the others are zero. Concluding that something which is not significant is in fact zero is a widespread fallacy. Using zeros for some variables and the obtained weights for the others when these are set to zero may be appreciably farther from the true weights than the ordinary least-squares weights, even though it is within the confidence region. Thus, we should be cautious in using variable-selection methods, adopting them only when there is a real need to reduce the number of predictors and/or a clear distinction between predictive and nonpredictive variables, preferably one that can be bolstered by some theoretical considerations.

PREDICTION AND ACCURACY OF PREDICTION

Predictive Accuracy and Error of Estimate

We have already encountered the distinction between variance around the sample regression line and variance around the regression line in the population, s_e^2 and σ_e^2. Adjusting s_e^2 to $\hat{\sigma}_e^2$ takes account of the fact that fitting parameters to the data at hand will most likely result in a closer fit than is possible in the population. In this section, we consider still a third type of deviation from a regression line. That is, if we use the regression coefficients from a *sample* to make predictions in a second sample or in a population, what size can we expect for the deviations?

As usual, the measure of accuracy is a squared deviation. The expected squared deviation from the regression line in the population, $E(y - \hat{y}_\beta)^2$ is σ_e^2. The subscript β indicates that y_β is a prediction based on the population regression equation. It is this quantity of which $\hat{\sigma}_e^2$ is an unbiased estimate. If the coefficients derived from sample data are to be used in actual prediction, we would be concerned with a somewhat different quantity

$$d^2 = E(y - \hat{y}_b)^2 \tag{9.34}$$

where d^2 denotes the expected squared deviation of a y from a prediction based on the coefficients from the sample.

How accurate would predictions be if we took the coefficients derived from the sample and applied them in the population? Although we would not do this literally, we may intend to use the regression equation to predict the value of y for individuals not in the present sample. For example, we may wish to perform a validity study to select potential college students. First, we would gather data on predictors for a sample of student applicants and then determine some dependent variable such as grade point average after they are admitted. Then we would perform a regression analysis, estimate the coefficients, and apply them to the scores of new applicants with a view toward predicting their GPAs. We will use d^2 as the measure of how accurate *these* predictions, based on the sample regression line, are expected to be in the population (or a new sample), assuming that there are no changes in the applicant population, grading standards, and various other unrealistic assumptions.

What we hope to do is estimate d^2 on the basis of the data in the original sample. If this can be done, then we would expect the estimate $\hat{d}^2$ to be greater than $\hat{\sigma}_e^2$ since the true d^2 is bound to be greater than σ_e^2. The latter represents deviations from the regression line that are optimum in the population, whereas d^2 represents deviations from a different regression line, one resulting from the sample coefficients. Note that d^2 is the accuracy that is to be expected, on the average, in a second or "cross-validation" sample (see following section). Deviations in any given cross-validation sample may be somewhat larger or smaller than this, but they represent what is to be expected across all possible cross-validation samples.

In order to estimate d^2 from sample data we would use the following equation (Browne, 1975b):

$$\hat{d}^2 = \hat{\sigma}_e^2 \frac{(n+1)(n-2)}{n(n-p-2)} = ss_{dev} \frac{(n+1)(n-2)}{n(n-p-1)(n-p-2)} \quad (9.35)$$

A couple of things should be noted about this equation. First, to a close approximation, it reduces to $\hat{\sigma}_e^2[1 + (p+1)/(n-p-2)]$. This is similar to the correction made in Equation 9.1 going from s_e^2 to $\hat{\sigma}_e^2$. However, Equation 9.35 is *applied* to $\hat{\sigma}_e^2$ and is more extreme. In an extreme case, where $n = 10$ and $p = 5$, this $d^2 = 3\hat{\sigma}_e^2$, which itself was $2.25s_e^2$. So, d^2 will inflate rapidly if p increases beyond $1/2n$. It further places $n - 2$ as the absolute limit for p if we intend to make predictions from the sample regression equation. Thus the variance around the sample regression line, s_e^2, is not only an underestimate of the variance around the population line but also σ_e^2 is an underestimate of the deviations to be expected if the sample line is used in the population of a second sample. These corrections become substantial if p is more than a small fraction of n. They should motivate the investigator toward using only a few variables in prediction equations when actual predictions are to be made from them.

A slightly different question is what the *correlation* will be between the predictions and the actual scores in the population. This is still a third quantity to go with R and P. The P assumes that the population weights are to be used to make predictions in the population, whereas the present concern is with the use of the *sample* weights in the population or a second sample.

When we correlate predictions with actual dependent variable scores, it does not matter whether the predictions are based on weights of 10 for x_1 and 5 for x_2 or 1.0 for x_1 and .5 for x_2, as long as the signs and ratios among the weights stay the same. However, which pair of weights are used *does* affect the squared deviations of predictions from actual scores. The multiple correlation and the error variances are directly tied together in the case of the least squares by the regression weights. If some other weights are used, however, the connection breaks down.

If we are concerned with the accuracy of relative prediction in the population or another sample, then we need an estimate of this accuracy. Such a correlation may be designated

$$w = \rho_{y\hat{y}_b} \quad (9.36)$$

which is the correlation between y scores in the population and the predictions of them based on sample regression coefficients. The population multiple P, on the other hand, is $\rho_{y\hat{y}}$, the correlation of y with predictions of y based on the true population weights. The procedures for estimating this correlation w are too cumbersome to include here, except to observe that w will be less than $\hat{P}$ for the same reasons that d^2 is greater than σ_e^2. Thus, $\hat{P}$ is an overly optimistic estimate of how well the b_j will work if they are actually used in another sample.

Cross-Validation

As users of multiple regression became aware of its inferential pitfalls, they developed a commonsense way to estimate how accurate the sample weights are as estimates: they try them out on new data. As time passed, this practice became formalized as a set of *cross-validation* techniques.

In its simplest form, cross-validation consists of determining the weights in one sample and applying them to the predictor scores of a new sample. The correlation of these scores with y in the new sample—which is a simple correlation, since only one slope parameter is estimated in the new sample—is the *cross-validation correlation,* whose expected value is w of Equation 9.36. One can also calculate the average squared deviation of the actual y's from the predictions in the new sample—no parameters are estimated in the new sample—which is an estimate of d^2.

Since there are usually two samples—obtained by splitting a large sample *randomly* into two smaller ones—a set of regression weights could be determined on the second sample also and applied to the first. These two estimates of the b_j, w, and d^2 could be averaged to make one overall best guess as to their values. This is called *double cross-validation.* Assuming the cross-validation correlations are significant, we could combine both samples into one to arrive at an overall estimate of the regression parameters.

Still another procedure that can be used if a lot of computer time is available is the *jackknife.* In the case of multiple regression, the jackknife works as follows. First, calculate a set of regression coefficients using all but one of the observations. Use these to predict the nth observation's y value, recording the value of the prediction. Now put that observation back in the sample, remove a different one, and recalculate the regression coefficients using these new values to predict the y value of the new holdout, again recording the prediction. Continue this process until all n members of the sample have served as the holdout, then calculate $r_{y\hat{y}}$ and $s^2_{y-\hat{y}}$. The former is an estimate of w and the latter of d^2.

The reason why this method is valid is that each prediction is based on data that is independent of the data being predicted. The advantage of this method over splitting the sample into halves and cross-validating is that the weights are estimated each time with $n - 1$ observations instead of $n/2$ and the accuracy statistics are based on all n. In cross-validation, they are based on only $n/2$. Whether this is superior to double cross-validation followed by averaging is debatable, however.

In certain situations cross-validation may be the only possible means of evaluation. For example, when using variable selection, the estimation equation for d^2 (Equation 9.35) does not apply, and neither does the jackknife. In fact, single cross-validation may be the only procedure that makes sense, since often it turns out that different sets of predictors are selected in the two samples.

Cross-validation also may be used to test the hypothesis $P = 0$. This is a good method to employ when variable selection has been used. We assume that the total sample has been divided randomly into two halves, a validation and a cross-validation (c-v) sample. The weights are estimated in the validation sample and tested in the other. The test is quite simple. First, the weights estimated in the validation sample are used to estimate the y's in the c-v sample, and then to find $r_{y\hat{y}}$ in the c-v sample. The $r_{y\hat{y}}$ is tested for significance as an ordinary correlation because the weights were determined independently.

Naturally, the methods described in Chapter 4 for computing correlations of weighted composites could be used instead of directly computing the $\hat{y}$. Also, if double cross-validation is used, we determine two $r_{y\hat{y}}$'s and test each, applying a Bonferroni correction to the significance levels.

Cross-validation has the additional advantage of not being based on any assumptions other than that both samples are from the same population—the one to which generalization of results is desired. The same is true of the jackknife. Statistical formulas on the other hand, are of necessity based on assumptions concerning the nature of the variables, their distributions, and so on. These assumptions may be violated, perhaps in ways obvious from the data, so the appropriateness of the formulas used in inference may be called into question. We may hope that the inferences are fairly robust with respect to violation of assumptions, but we never know. So, although it is less elegant in its reliance on a commonsense rationale, cross-validation has much to recommend it.

Predicting New *y* Values

As we have seen, there are a variety of reasons for conducting a multiple regression analysis. Sometimes interest focuses on the regression equation itself. In other cases, the regression equation is used to make predictions. The regression weights are used on individuals with known x scores but unknown y scores to predict what their y scores should be. This is presumably done for some practical purpose, such as selecting from a pool of candidates.

Let $\mathbf{x}_i$ be the vector of scores of a new individual on the predictors, expressed as deviations from means in the sample used to estimate the weights. It should come as no surprise that our prediction of y is

$$\hat{y}_i = \bar{y}_i + \mathbf{b}'\mathbf{x}_i \tag{9.37}$$

Alternatively, if $\mathbf{x}_i$ is in raw scores,

$$\hat{y}_i = b_0 + \mathbf{b}'\mathbf{x}_i \tag{9.38}$$

These are equivalent ways of making "best guess" predictions of y for a new observation.

If we want not only a best guess but also to estimate an interval within which we can be pretty sure the y value does in fact lie, then things become a bit more complicated. This interval is analogous to a confidence interval. We want to be $1 - \alpha$ confident the y value for this vector of x scores will lie within this interval. This is called a *prediction* interval, because c.i. refers to an interval for a parameter.

For this estimate, we cannot simply use $\hat{\sigma}_e$, which represents our estimate of the standard deviation of the y scores around the regression line. Instead, we need to define a separate "standard error" for each vector of x values, so that the width of the prediction interval depends on the $\mathbf{x}_i$.

We call this standard deviation $s_{y|x_i}$. Given $\mathbf{x}_i$ in *deviation score* form, it is estimated as

$$s_{y|\mathbf{x}_i} = \hat{\sigma}_e(1 + \mathbf{x}_i{}'S_{xx}^{-1}\mathbf{x}_i\, n^{-1})^{1/2} \tag{9.39}$$

where n is the sample size of the group used to find $\mathbf{b}$. The $1 - \alpha$ prediction interval for the y score of individual i is

$$\hat{y}_i - s_{y|x_i} t_{1/2\alpha} \leq y_i \leq \hat{y}_i + s_{y|x_i} t_{1/2\alpha} \tag{9.40}$$

This formula makes the prediction interval larger for x scores that are more deviant from their means. This is understandable in terms of the uncertainty about the regression coefficients. Since the **b** values in the prediction are only estimates, these cause part of the error in prediction. Insofar as $\mathbf{b} \neq \beta$, the effect of errors in the coefficients is larger the more deviant the x value. After all, if x_i is close to the mean, then prediction based on the sample b_j is not much affected by the sampling errors: predictions will be close to the mean also. But if x_i is far from the mean, then the errors in the b_j will have a much greater effect. This is analogous to the error made in aiming at a target. Standing close to a target, a small aiming error has a small effect, whereas standing at a distance, the same error will result in missing the target by the proverbial mile. The $\boldsymbol{x}_i \mathrm{S}_{xx}^{-1} \boldsymbol{x}_i$ term takes account of this increased uncertainty about "distant" $\mathbf{x}_i$ by making the boundaries of the interval wider.

Sometimes we see references to an interval for the *average* or *expected* y value given $\mathbf{x}_i$. This is a much smaller interval since it refers only to the average y for the $\mathbf{x}_i$. But here our interest is in the accuracy of predicting y_i, which, on the average, is no better than $\hat{\sigma}_e$ and becomes worse as we deviate from the mean.

RIDGE REGRESSION

Colinear Predictors and Regression Instability

After decades of complaining about the sampling instability of regression coefficients, something to do about it has been found. This is called *ridge regression.*

Recall that one of the factors that makes a sample regression weight a poor estimator of the population weight is a high correlation with the other variables. Ridge regression is particularly applicable in this regard.

The basic idea of ridge regression is quite simple, and the ideas are simplest to describe using standard score weights $\mathbf{b}^* = \mathrm{R}_{xx}^{-1} \mathbf{r}_{xy}$. It can be proven that there is a diagonal matrix K such that $\mathbf{b}^*_{ridge} = (\mathrm{R}_{xx} + \mathrm{K})^{-1} \mathbf{r}_{xy}$ will give a smaller squared error in the population. That is, if

$$\hat{z}_{y\,ridge} = \sum_j z_j\, b^*_{ridge}$$

then

$$\mathrm{E}(z_y - \hat{z}_{y\,ridge})^2 < \mathrm{E}(z_y - \hat{z}_{yb^*})^2$$

provided the right K is chosen.

The effect of ridge regression is greatest when the predictors are highly correlated. In fact, ridge regression was developed to deal with the instability that develops when the predictors are nearly linearly dependent. However, there is an effect even when the predictors are uncorrelated.

The problem, then, is to find K. The effect on errors of prediction is not very sensitive to the exact values of K. The elements of K are typically small positive quantities such as .01 to .10 in the standard score case. Using K has the effect of pretending the variances are a bit greater than they really are, while keeping the covariances at the observed values.

Ridge coefficients tend to be a little smaller in absolute value and a little less different from each other than are the least squares coefficients. It is as if the point representing the vector of coefficients were moved toward the origin. Ridge corrections seem most useful when the linear dependence is greatest and when the validity correlations of the variables are in about the same order as the predictors' average correlations with each other.

If the weights themselves are to be used for prediction, we might well consider using ridge regression, particularly when there is near linear dependence. Note that the least squares weights are unbiased estimates of β. The ridge weights are biased, but in this case the biased estimates will have smaller squared errors.

SUMMARY

Sampling Models and Estimation

A lot of ground has been covered in this chapter, beginning with a description of the sampling models that underlie inference in multiple regression. Here, we considered two separate concepts, the multivariate normal model and the fixed constants model. The multivariate normal model states that the vector of scores on the p predictors and the dependent variable is an element from a population whose density is the multivariate normal distribution. That is, not only are all the univariate distributions normal, but for any specified value for p of the variates, the distribution on the other variate is also normal. A consequence of multivariate normality is that all of the regressions of any variable on any one or any combination of the other variables are linear and homoskedastic. Exactly normal distributions are impossible in practice, but are often a reasonable approximation to reality. Statistical inference seems to be fairly robust under moderate violations of the assumptions. The normality assumption greatly simplifies much of the mathematics that underlies statistics.

In the fixed constants model the investigator's interest is confined to those values of the independent variables used in the analysis. The independent variables are not sampled at all, but deliberately determined. It is assumed that the distribution of y is normal with constant variance when all the values of the x's are specified. The first model applies to situations in which the independent variables represent characteristics of individuals, whereas the second applies to independent variables that are experimentally manipulated. Both models lead to the same inferential procedures in the examples considered in this chapter.

Corresponding to the sample least squares estimates b_j of the regression coefficients are the population regression coefficients, β_j, the former are unbiased estimates of the latter. However, the sample s_e^2 is an underestimate of σ_e^2 and R is an overestimate of the population correlation P. In order to get unbiased estimates the following equations are used:

$$\hat{\sigma}_e^2 = s_e^2 \frac{n-1}{n-p-1}$$

$$\hat{P}^2 = R^2 - (1 - R^2)\frac{p}{n-p-1}$$

Like any other statistic, the regression coefficients, error variance, and multiple correlation are all subject to sampling fluctuations. Their sampling variances are all inversely affected by the df, $n - p - 1$. The estimated error variance has, like other sample variances, the chi-square distribution. The distribution of the multiple R is complex, but the z' transformation can be used for inferential purposes such as setting confidence intervals. The sampling distribution of the regression coefficients is normal with a standard error that depends on the multiple correlation of that predictor with all the other predictors, R_j^2, according to the equation

$$s_{b_j} = \frac{\hat{\sigma}_e}{(n-1)^{1/2}\, s_j\, (1 - R_j^2)^{1/2}}$$

The sampling characteristics of regression statistics are used to estimate the population parameters, and to test hypotheses and make confidence intervals for these parameters.

Testing Hypotheses

The basic null hypothesis in multiple regression states that the population multiple correlation is zero, H_0: $P = 0$. Equivalent to this is the hypothesis that all the regression weights are zero: H_0: $\beta = 0$. The test of this hypothesis is straightforward; it is an F test that can take a variety of equivalent forms. One is

$$F = \frac{ss_{reg}/p}{ss_{dev}/(n-p-1)}$$

An equivalent ratio can be obtained from R^2:

$$F = \frac{R^2/p}{(1-R^2)/(n-p-1)}$$

There are a variety of other hypotheses that can be tested. Some questions can be answered using methods relating to previous ones. One such question is whether the error variance is the same in two populations. Another is whether the predictive accuracy for one set of predictors is equal to that for a second set, assuming data on both are available from the same sample.

A variety of different hypotheses can be tested about the regression coefficients, in addition to the hypothesis that they are all zero. The simplest test determines whether a given coefficient is zero. If the overall null hypotheses $\beta = 0$ is rejected, then individual coefficients may be tested using parameterwise alpha levels as described in Chapter 8. The distinction between one- and two-tailed significant tests was also noted. In a one-tailed test, only deviations from the null value in one direction can lead to rejection. This increases power at the cost of requiring that all deviations in the other direction, no matter how large, lead to acceptance.

Corresponding to the confidence interval for a single parameter is the p-dimensional confidence region for a set of p parameters. One states with confidence $1 - \alpha$ that the point defined by *all* the parameters is within the region. In the case of regression weights, this region is ellipsoidal. A trial set of regression coefficients can be substituted for the least squares estimates and, using the methods described in Chapter 4, we can find the error variance around these predictions. Wide variation in the coefficients often has relatively little effect on the amount of error. Thus, we may find alternative weights that do not significantly contradict the data, even though the least squares weights give the smallest discrepancies.

Inference in Stepwise Regression

In stepwise multiple regression, the predictors enter the regression individually or in sets, rather than all at once. There are two main varieties—hierarchical and variable selection—and a variety of hybrid types. In a hierarchical analysis, the predictors are arranged in subsets on the basis of a priori considerations concerning their priority in the regression equation. The sets are tested in this order to see if their addition to the regression equation significantly reduces the error sum of squares over the variables previously entered. The test is made using Equation 9.19. This is an example of the type of test which uses the ratio of reduction in error to the number of degrees of freedom that were used to estimate parameters. The Type I error probability is kept at α by not testing any new set once any set does not significantly improve prediction.

Two common applications of the hierarchical strategy are polynomial regression and the equal weights hypothesis. In polynomial regression, powers of the predictors as well as the predictors themselves are used. The equal weights hypothesis formalizes the common observation that equal standard score regression weights often do nearly as well as the least squares regressions weights in the sample, and may do even better upon cross-validation. This chapter described procedures for using hierarchical multiple regression to test whether regression is linear or to test the equal weights hypothesis.

Another type of stepwise regression is variable-selection multiple regression. In the forward variety, variables enter the equation one at a time. The variable entered at each step is the one that increases the multiple R the most. This method can be used to find an effective subset of predictors, but it must be used with care if population inferences are to be made since the nominal printed alpha levels are distortions of the true ones. Here, a Bonferroni adjustment should be made.

An alternative is to use the variable-selection methods only after the overall significance is established by fitting the complete regression equation and using the test of Equations 9.8 or 9.9. This can be followed by a second analysis in the variable-selection mode in which variables are entered in order of effectiveness. An alternative to the forward method of selection, after the omnibus test has led to rejection, is a backward solution in which the variable contributing the *least* to the regression equation is dropped at each step until only variables having a significant weight remain. In these ways, a variable-selection method may be used without gross distortion of significance levels while retaining a reasonable degree of power.

Prediction and Accuracy of Prediction

If the sample regression weights are used to make predictions using the x scores of new cases to make predictions of their y values, we need a way of estimating how accurate those predictions will be. The expected squared difference between actual y values and predictions for new data, d^2, is expected to be larger than σ_e^2 because the latter assumes use of the population regression equation, whereas in practice we are using that derived from the sample. Equation 9.35 provides an estimate of d^2, which is about as much larger than $\hat{\sigma}_e^2$ as $\hat{\sigma}_e^2$ is larger than s_e^2. Thus, if the number of predictors is an appreciable fraction of the sample size, s_e^2 and R^2 may well be grossly inflated as indicators of how accurate predictions based on the sample regression equation will be.

There are two additional aspects of predictive accuracy. One is the expected *correlation* between predictions and actual y values in a new sample or in the population. This correlation is complicated to estimate, but it must be less than P probably by about the same proportion that d^2 is greater than $\hat{\sigma}_e^2$. The second type of predictive accuracy has to do with prediction of each individual score. One can find a prediction interval for the score of a new individual by making use of the regression equation and σ_e, according to Equations 9.39 and 9.40. The former shows that predictions are expected to be most accurate for individuals whose x scores are near the mean.

A general method for evaluating the accuracy of a set of estimated regression weights is *cross-validation* in which weights developed in one sample are applied to a new one. The correlation between predicted and actual y scores and/or the error variance is found in the new sample, and this correlation shows how well the regression equation can be expected to work in practice. There are a number of varieties of cross-validation, including the jackknife, which can be used on a single sample. Cross-validation of the regression equation is important, particularly in cases where the assumptions may not hold very well or where a variable-selection method was used, since then the statistical estimation formulas do not hold exactly.

Ridge Regression

An alternative to ordinary least squares regression is *ridge regression.* Ridge estimates of the weights are slightly biased, although they sometimes provide significant reduction in sampling variability. They are particularly important as the predictors approach linear dependence, but are somewhat more accurate as estimates in all cases. In ridge regression, instead of using the inverse of R_{xx} to calculate the standard score weights, $(R_{xx} + \mathrm{K})^{-1}$ is used, where K is a diagonal matrix with small entries of about .05.

COMPUTER APPLICATIONS

Packages of programs to perform complex statistical analyses were an early fruit of the first computer revolution that took place in the 1950s. The "Biomed" package, BMD—or BMDP in later versions—compiled under the editorship of W. J. Dixon, was an early example which continues to find wide acceptance. But there are many others, and new ones are still being marketed. This section will describe how to implement the methods presented in this text using three of the most widely used packages: SPSS[x] (SPSS Institute, 1983), BMDP (Dixon, 1981), and SAS (SAS Institute, 1979).* This is not an endorsement of these packages over others; indeed, it is necessary to work actively with the packages, rather than passively accepting default options and output, if we are to carry out the analyses according to the principles proposed in this book.

The sections that follow describe the implementation of the various techniques using these three packages. Practical considerations limit the amount of detail that can be presented; our purpose is simply to tie the concepts and procedures of this book to the commands and output of the systems. The reader will need to refer to the systems' manuals for many details. Also, for reasons of space, the sections on one system will refer to some details that are given in the description of another system. Terms that are in boldface, capital letters refer to commands in a particular system.

COMPUTER IMPLEMENTATION OF MULTIPLE REGRESSION METHODS

BMDP

The general-purpose regression program in BMDP is called BMDP2R, Stepwise Regression. There are five other programs that can be used for this purpose, but this one fits most directly into the methods and concepts that have been described in this chapter.

BMDP is an integrated system for data analysis, and the user must expect to spend some time becoming familiar with its general features, file handling, input and output control, and the like. Here, we are concerned only with the specification of the regression itself and the interpretation of the output.

Central to the operation of P2R—and several other multiple regression equations—are statistics called F-to-enter and F-to-remove. F-to-enter reflects the contribution that a variable not in the equation would make to R^2 if it were added to the regression equation, whereas F-to-remove reflects the decrement to R^2 that would result from the removal of the variable that is in it. This occurs in the context of a particular subset of the total set of predictor variables being in the equation at a particular step, with the remainder not being in the equation. Suppose that there is a total of p predictors and that at some step there are s of them in the regression equation. Then F-to-remove is the F ratio of Equation

---

**SAS User's Guide: Statistics*. Cary, NC: SAS Institute, 1982.
BMDP Statistical Software, edited by W. J. Dixon. Berkeley, CA: University of California Press, 1981.

9.13, except that p is replaced by s, the number of variables that are currently in the equation. F-to-enter is the same equation, except that p is now replaced by $s + 1$ since there would be one more variable if any one of the variables not currently included were entered. Remember that Equation 9.13 is equivalent to the square of Equation 9.12, and can also be interpreted as F in Equation 9.14. Also, these F ratios are proportional to the corresponding squared partial correlation of the variable.

Variable selection in P2R and other similar programs is carried out in a foward mode by adding at each step the variable with the highest F-to-enter; in backward mode it drops the variable with the lowest F-to-remove. An important point to remember is that the true alpha levels of these F statistics are not the same as the nominal ones that one would read from a table due to the variable-selection process. If several variables can enter the regression equation, the odds are much higher than 1 in 20 that the highest F ratio will exceed the nominal 5 percent level.

Program BMDP2R permits the user to do a complete multiple regression, simple forms of hierarchical analysis, or any of several varieties of variable selection multiple regression. The user specifies the desired type primarily through the use of the FORCE=, LEVELS, and ENTER commands. Backward variable selection also involves use of REMOVE. These can all be carried out by setting METHOD equal to "F." The other options under METHOD are quite specialized in purpose.

The LEVELS command assigns numbers to the variables that reflect the user's preferences concerning the inclusion of the variables in the regression equation. The program considers variables for inclusion in order of their LEVELS, beginning with the lowest level. Whether a variable enters or not is determined by its contribution to the regression equation, as determined by its F-to-enter. All variables whose LEVELS are below the maximum set by FORCE= are forced into the regression equation (except that variables whose LEVELS are 0 will not be allowed to enter the equation).

It is possible to manipulate these parameters to perform a hierarchical or a complete multiple regression. A hierarchical analysis can be carried out by giving all the variables LEVELS that are higher than FORCE= while setting the value of F-to-enter very low (the default value is 4.0, approximately the nominal 5 percent value). We would assign the same LEVELS value to the variables that are in a given subset of the hierarchy but different LEVELS to different subsets. Then the program will enter the variables into the regression equation in the desired order since all variables will pass the F-to-enter test when this criterion has been set very low, and will be entered in the intended order. A complete multiple regression can be carried out rather simply by setting FORCE= greater than the highest value assigned to LEVELS.

Variable-selection multiple regression makes use of the ENTER and REMOVE commands which specify the nominal F value for the contribution that variables must have in order to enter or remain in the regression equation. If forward variable-selection is desired, only the ENTER command is used, but the general mode of operation is first to enter variables and then to remove them. The ENTER command is followed by two numbers that control the entry and removal of variables. The first number is the minimum F-to-enter that a variable must have to be allowed in the regression equation. The second number is the lowest F-to-remove value a variable may have to allow its removal once it is entered. Setting the first number very low, say .001, and the second very high, say

1,000, will lead to all the variables entering and none being removed (until REMOVE comes into play). The higher the F-to-enter, the more selective the program will be. Lowering the F-to-remove means that entered variables can be removed if their contribution is reduced by the addition of later variables. The defaults here are 4.0 for both F-to-enter and F-to-remove.

After a set of variables has been settled on through the use of ENTER, REMOVE comes into operation. This command has F-to-enter and F-to-remove values, also. If backward selection is desired, we should set these values lower than the ones for ENTER and it will remove the variables in reverse order of apparent contribution. All variables can be removed in order. Although the P2R set up contributes to the possible confusion if one is trying to perform a pure backward stepping process, the system described here should work.

Note that P2R and other variable selection programs print what they call an *analysis of variance table.* It shows ss_{reg}, ss_{dev}, df, and F ratios for the variables in the regression equation as if no variable selection had taken place. It is best to simply ignore this table since the F ratios are spuriously high when variables have been selected.

SAS

SAS offers several programs to perform multiple regression analyses, although some of the operations described in this chapter are difficult to perform using this system. The most complete possibility is offered by PROC GLM. Unfortunately, the complexities of this program require a more in-depth description than can be offered here regarding the various types of analyses. This section provides a brief description of its simplest application, followed by a description of an additional program, STEPWISE.

In the GLM multiple regression, the dependent and independent variables are specified with a MODEL statement such as MODEL Y = X1 X2 X3. The program proceeds in a hierarchial fashion, entering the variables one at a time in the order listed.

The output consists of several tables which show the course of the analysis. First is the anova table for the overall analysis when all the variables are in the equation, and for R^2. No adjusted R^2 is presented. Following this are three tables that assess the proportions of variance contributed by variables and offer information about the significance of each predictor. The descriptions in these tables are rather cryptic, employing terms such as TYPE I or TYPE IV SS. The first of the three tables presents the TYPE I SS, associated F value, and significance. The TYPE I SS represents the increment that occurs in ss_{reg} when the listed variable is added to the regression equation *in that position.* This may be used as a basis for a hierarchial analysis. If there is only one variable at each level of the hierarchy, then the test information is provided directly. When there are several variables at one or more levels, the user can sum the associated ss and df to form the $ss_{dev:s}$ and $ss_{dev:s-1}$ of Equation 9.30 by subtraction from the CORRECTED TOTAL, which gives ss_t.

The TYPE IV SS numbers of the next SAS table are the numerator terms of Equation 9.13. These represent the amounts of variance uniquely contributed by each variable when all the others are included in the analysis. These values do not depend on the order of insertion of the variables, unlike those on the previous table. The third table lists the b's, including b_0 as the INTERCEPT, on the column labeled ESTIMATE. T FOR HØ is the t of Equation 9.12, which is the square root of the F in the table above. The STD ERROR OF ESTIMATE is, rather confusingly, the standard error of the regression coefficient, that is, the denominator of Equation 9.12.

If we want to use a variable-selection mode, then we must employ the procedure STEPWISE. It operates in several modes; FORWARD, which enters variables one at a time in order of apparent contribution; BACKWARD, which is a backward elimination procedure; and STEPWISE, which is similar to the forward and backward mode used by BMDP2R. There are two other modes also.

The STEPWISE output first prints the analysis of variance table, computing not only the misleading F but also printing an erroneous probability for it. The remainder of the information printed at each step is like the output of P2R. The column labeled TYPE II SS is the amount of variance produced by that variable, considering the other variables that are currently in the regression equation. It is as if the only variables in the analysis were the ones currently in the regression equation. The F column is the same as the F-to-remove of P2R. A Bonferroni correction is easily made here since the exact probabilities of the F's are given as PROB F. We multiply the probability corresponding to the variable entered at that step by the number of variables not in the equation, plus one.

Still a third regression program is offered by SAS, RSQUARE. This program performs all possible regressions for those who are anxious to find the very best sets of predictors for each size of predictor set. Think carefully before using it, particularly if p is more than 5 or 6, because the potential number of regressions is 2^p.

SPSS^x

The modernized version of SPSS^x has a different type of syntax for specifying programs and their options. Many users find it more difficult, so one should consider expert consultation in implementing an analysis. Procedure REGRESSION is quite different from the other packaged programs in many respects. It does not have the inclusion levels feature of BMDP2R. Instead, the analysis is controlled by a series of keywords.

The METHOD subcommands of SPSS^x specify the type of analysis. ENTER means all variables will be forced; FORWARD means forward variable selection will be used, and BACKWARD means the reverse selection process; STEPWISE means that a backward and forward procedure similar to BMDP2R will be used.

Some ingenuity must be employed for a hierarchical analysis, but this can be done in reverse, by making use of the REMOVE subcommand. Specify ENTER, thereby entering all variables, then use REMOVE to eliminate the last group of variables; a second REMOVE takes out the next set, and so on. Then the various R's or s_e^2 will be available.

Forward and backward selection can be controlled using FIN or PIN to specify the entrance criteria, and/or FOUT or POUT to specify F-to-remove. Note that PIN and POUT cannot be taken literally because of the variable selection aspect.

The SPSS^x program allows one to assess the contribution of sets of variables, as

---

SPSS^X User's Guide. New York: McGraw-Hill, 1983.

in Equations 9.32 or 9.33. This is done by specifying sets of variables in the TEST keyword. A second useful feature is that the sampling variance-covariance matrix of the b's can be obtained by requesting BCOV in the STATISTICS keyword.

The output provides information similar to that found in the BMDP version, but is even more detailed. The same kinds of cautions and interpretations apply.

PROBLEMS

1. In a four-predictor multiple regression where $n = 30$, it is found that $ss_{reg} = 120$; $ss_{dev} = 280$. Compute R^2 and s_e^2 and estimate σ_e^2 and P^2. Now compute these statistics for $n = 13$ and $n = 205$.

2. The following is a covariance matrix and the inverse of $\mathbf{S}_{xx}$ where $n = 44$, $ss_{dev} = 148.05$, and $\hat{\sigma}_e^2 = 3.701$. Find s_{b_j} for each x and the t's for significance.

$$\begin{array}{c} \\ x_1 \\ x_2 \\ x_3 \\ y \end{array} \begin{array}{c} \begin{array}{cccc} x_1 & x_2 & x_3 & y \end{array} \\ \begin{bmatrix} 8.0 & 2.5 & 1.0 & 4.0 \\ 2.5 & 8.0 & 6.0 & 4.0 \\ 1.0 & 6.0 & 10.0 & 5.0 \\ 4.0 & 4.0 & 5.0 & 7.5 \end{bmatrix} \\ \mathbf{S} \end{array} \quad \begin{array}{c} \begin{bmatrix} .1412 & -.0610 & .0225 \\ -.0610 & .2536 & -.1461 \\ .0225 & -.1461 & .1854 \end{bmatrix} \\ \mathbf{S}_{xx}^{-1} \end{array} \quad \begin{array}{c} \begin{bmatrix} .4334 \\ .0401 \\ .4326 \end{bmatrix} \\ \mathbf{b} \end{array}$$

3. For the data in Problem 2, $R^2_{y.123} = .54094$; $R^2_{y.12} = .40635$; $R^2_{y.23} = .36364$; $R^2_{y.13} = .54009$. Use these to test the significance of b_1, b_2, and b_3. Compare the resulting F's to the t^2 from the previous problem.

4. Test the significance of $R^2_{y.123}$ from Problem 3.

5. Use the results of Problem 2 to form .95 confidence intervals for the three coefficients.

6. Find $s_{b_1-b_2}$. Test the hypotheses $\beta_1 = \beta_2$ and $\beta_2 = \beta_3$ as a priori hypotheses. Also test them as post hoc conclusions.

7. Use the methods of Equations 9.19 through 9.22 to test a hypothesis about some linear combination of weights that you decide on.

8. Using $\mathbf{w}_1 = (.3, .3, .3)$ and $\mathbf{w}_2 = (.3, .5, -.3)$, find $ss_{dev:w}$ and $r_{y\hat{y}}$.

9. Test to see whether the sets of weights in Problem 8 are within the confidence region and whether they produce significant reductions in correlation.

10. In a prediction system, the variables are classified into sets. Set 1 has 3 predictors, set 2 has 2, and set 3 has 4. $R^2_{y.1} = .25$; $R^2_{y.12}$ (for variables 1 to 5) $= .40$ and $R^2_{y.123}$ (for all 9 predictors) $= .45$. Perform the hierarchical analysis, testing the contribution of each set using $n = .85$.

11. If in the previous problem $R^2_{y.13} = .32$ and $R^2_{y.23} = .41$, (remember that $R^2_{y.12} = .40$), what is the significance of the contributions of each set of variables?

12. Find a paper in the literature where multiple regression was used. Identify the

dependent variable and the predictors. What was the unit of observation—that is, what defined the rows of the data matrix? What was the purpose of the regression analysis? What mode of analysis was used? How did it relate to the purposes? Would you have done the analyses the same way?

13. Using the data in Problem 1, find out the expected squared discrepancy if the sample weights are applied to new data.

14. Treat the data of Problem 2 as a cross-validation sample. In the validation sample, $\mathbf{b}'$ was found to be (.30, .20, .15), based on $n = 50$. Find $r_{y\hat{y}}$ in the cross-validation sample and test it for significance.

15. The results of a regression analysis are shown below. Three new cases are observed on the predictors, and we wish to make predictions of their y values:

$$\underset{\mathbf{S}_{xx}}{\begin{bmatrix} 4.0 & 1.0 \\ 1.0 & 2.0 \end{bmatrix}} \quad \underset{\mathbf{b}}{\begin{bmatrix} .125 \\ .615 \end{bmatrix}} \quad \underset{\overline{\mathbf{x}}}{\begin{bmatrix} 3.0 \\ 5.0 \end{bmatrix}} \qquad \bar{y} = 2.0;\ \hat{\sigma}_e = 1.70;\ n = 25.$$

The new raw score $\mathbf{x}$ vectors are $\mathbf{x}_1' = 4,6$; $\mathbf{x}_2' = 3, 5$; and $\mathbf{x}_3' = 6,8$. What are the .95 prediction intervals?

ANSWERS

1. $R^2 = .300$; $s_e^2 = 9.66$; $\hat{\sigma}_e^2 = 11.2$; $\hat{\mathrm{P}}^2 = .188$; $n = 13$; $\hat{\sigma}_e^2 = 35.0$; $\hat{\mathrm{P}}^2 = -.05$ (should be set to zero); $n = 205$; $\hat{\sigma}_e^2 = 1.40$; $\hat{\mathrm{P}}^2 = .286$

2. $s_{b_1} = .1103$; $s_{b_2} = .1478$; $s_{b_3} = .1263$; $t_1 = 3.93$; $t_2 = .27$; $t_3 = 3.42$

3. $F = 15.45, .0738, 11.27$; $df = 1{,}40$ (Same as the t^2 of Problem 2)

4. $F = 15.71$; $df = 3, 40$

5. $.4334 \pm .2226$; $.0401 \pm .2986$; $.4326 \pm .2551$

6. $s_{b_2-b_3} = .2109$; $s_{b_2-b_3} = .2509$; $t = 1.86, -1.56$ (Hint: see Equation 9.16) Post hoc acceptance regions are differences falling between $\pm.616$ and $\pm.732$, respectively (Hint: see Equation 9.22) with $\mathbf{c}' = (1, -1, 0)$ and $c' = (0, 1, -1)$

8. $ss_{dev:w_1} = 161.25$; $ss_{dev:w_2} = 279.07$; $r_{y\hat{y}_{w_1}} = .708$; $r_{y\hat{y}_{w_2}} = .402$

9. $F_{3,40} = 1.188$; $F_{3,40} = 11.80$; $F_{2,40} = 1.75$; $F_{2,40} = 16.54$; $\mathbf{w}_1$ is within the .95 confidence region; $\mathbf{w}_2$ is not by both criteria

10. $F_{3,81} = 9.00$; $F_{2,79} = 9.875$; $F_{4,75} = 1.70$

11. Contribution of set 1: $F_{3,75} = 1.82$; set 2: $F_{2,75} = 8.86$; set 3: $F_{4,75} = 1.70$

13. $\hat{d}^2 = 13.50$

14. $r = .707$; $t = 4.584$, $df = 42$, $p < .001$

15. $\hat{y}_1 = 2.75 \pm 3.57$; $\hat{y}_2 = 2.0 \pm 3.53$; $\hat{y}_3 = 4.25 \pm 3.64$

READINGS

Bock, R. D. (1975). *Multivariate statistical methods for behavioral sciences.* New York: McGraw-Hill.

Browne, M. W. (1975). Predictive validity of a linear regression equation. *British Journal of Mathematical and Statistical Psychology, 28,* 79–87.

Browne, M. W. (1975b). A comparison of single sample and cross-validation methods for estimating the mean squared error of prediction in a multiple linear regression. *British Journal of Mathematical and Statistical Psychology, 28,* 112–120.

Cattin, P. (1980). Note on the estimation of the squared cross-validated multiple correlation of a regression model. *Psychological Bulletin 87:*63–65.

Darlington, R. B. (1968). Multiple regression in psychological research and practice. *Psychological Bulletin 69:*161–82.

Darlington. R. B. (1978). Reduced variance regression. *Psychological Bulletin 85:*1238–55.

Darlington, R. B., and C. M. Boyce. (1982). Ridge and other new varieties of regression. In *Statistical and methodological issues in psychology and social sciences research,* edited by G. Keren. Hillsdale, NJ: Erlbaum.

Freedman, D. A. (1983). A note on screening regression equations. *American Statistician, 37,* 152–55.

Green, B. F. (1977). Parameter sensitivity in multivariate methods. *Multivariate Behavioral Research, 12,* 263–288.

Hocking, R. R. (1976). The analysis and selection of variables in linear regression. *Biometrics, 32,* 1–44.

Kerlinger, F. N., and E. J. Pedhazur (1973). *Multiple regression in behavioral research.* New York: Holt, Rinehart, and Winston, pp. 53–80.

Miller, A. J. (1984). Selection of subsets of regression variables. *Journal of the Royal Statistical Association,* Series A, *147,* 389–425.

Morris, J. D. (1982). Ridge regression and some alternative weighting techniques. *Psychological Bulletin 91:*203–10.

Morrison, D. F. (1976). *Multivariate statistical methods,* 2nd. ed. New York: McGraw-Hill. pp. 102–115, 203–310.

Rozeboom, W. W. (1978). Estimation of cross-validated multiple correlation: A clarification. *Psychological Bulletin 85:*1348–51.

Steiger, J. H., and M. W. Browne (1984).The comparison of interdependent correlations between optimal linear composites. *Psychometrika, 49,* 11–24.

Tatsuoka, M. M. (1970). *Discriminant analysis: The study of group differences.* Champaign, IL: Institute for Ability and Personality Testing. pp. 1–32.

Wilkinson, L., and G. E. Dallal (1981). Tests of significance in forward selection regression with an F-to-enter stopping rule. *Technometrics, 23,* 377–80.

10

One-Way Analysis of Variance

USING QUALITATIVE PREDICTORS

Imagine a professor of statistics sitting at his desk, leafing through the *Annals of Statistics* and chuckling from time to time, when there is a tap on his door. He mumbles in reply and the door opens to reveal a student—in fact, a hesitant, hangdog student. The student accepts an invitation to sit down—albeit somewhat uncomfortably, because he has to perch on the pile of printouts on the chair. The student has come to make a confession. He had taken the professor's course the previous semester and had been applying multiple regression to various problems, but recently, the computer refused to accept his last job.

"Umm-hmm. Tell me more about it," said the professor.

It seems that the student had been supplementing his scholarship funds by doing statistical work for various people. He had recently accepted a commission from the Dean of Students to analyze some data and then predict the amount of beer consumed by each student from various variables which were available. Nothing had predicted—not SAT scores, GPA, parents' income, height, distance traveled to school, nor anything else had passed the criteria he had learned in the course—criteria similar to those described in the previous chapter.

Finally, in his desperation, it had occurred to the student that this variable might be predicted from which residence hall or fraternity house a student lived in. There were 13 possible residences. He tried numbering the residences from 1 to 13, and using the individual's residence number as a variable, but that hadn't done anything as a predictor either. (He had already expected this, since it was only a nominal scale variable.) Then he had thought of treating each residence as a *separate variable,* giving a score of 1 on the variable to all those students who lived in the corresponding place and a 0 to everyone else. He ended up with 13 predictors, each of which was dichotomous. He knew that regression analyses should only be done with continuous predictors, but he had gone ahead and done it anyway, hoping the computer wouldn't notice.

No sooner had he reached the front of the queue at the computer center and submitted his job when the computer—one of the earliest talking models—announced for all to hear, "Your covariance matrix is singular. Please correct." The student retrieved his input and scurried over to the professor's office to confess his use of dichotomous predictors.

"Never mind," the professor said. "Many students at your stage of development have the same impulses. It's nothing to be ashamed of. I did the same thing myself when I was young. In fact, there is nothing really wrong with doing it. All you have to do to fix up a singular covariance matrix is remove one or more of the predictors. Then go back and try again. But I would advise you to take the second semester of my course before trying this sort of thing in the future, so you will understand more about it, particularly since you were a B-minus student in the first course."

His load lifted, the student thanked the professor and strode toward the door, but the professor went on. "One more thing. In this case you should call it an analysis of variance instead of multiple regression."

The student hurried out, puzzled over this last remark. He knew a bit about analysis of variance, but there didn't seem to be any connection. There were F tests, though. Perhaps he should take the second semester course. In the meantime, there was just enough time to resubmit the job. If all went well, he would have the study wrapped up in a day or two, and the dean would pay him off soon enough so he would not miss another payment on his Ferrari.

The Connection

Suppose there are three groups of subjects in an experiment, two experimental groups, $E1$ and $E2$, a control, C, and a dependent variable, y. For simplicity, let us make the groups small—three subjects in $E1$ and two each in $E2$ and C. Then we'll define three predictor variables. The first is constructed by giving a score of 1 to those in group $E1$ and 0 to everyone else; the second by giving a score of 1 to those in group $E2$ and 0 to everyone else; and the third by giving a score of 1 to those in C and 0 to everyone else. This is called *dummy* coding. The resulting score matrix which we will call X_d, will look like Table 10-1. For now, we will not include the extra column of unities that is used to produce b_0 in the score matrix. This will circumvent the singularity problem, and it will be shown later that it does not affect the point we are trying to make.

Table 10-1 Dummy Coding of Qualitative Variables

Group	x_1	x_2	x_3
	1	0	0
First Experimental	1	0	0
	1	0	0
Second Experimental	0	1	0
	0	1	0
Control	0	0	1
	0	0	1

Suppose that we ignore the rather peculiar nature of X_d and that we want to use it to predict y as accurately as possible in the least squares sense. That is, we want to find a vector of weights, which we will call **v** rather than **b**, since it is not exactly a vector of regression weights, so that $\hat{\mathbf{y}} = X_d\mathbf{v}$. The least squares solution for **v** has the usual form:

$$\mathbf{v} = (X_d'X_d)^{-1}X_d'\mathbf{y}$$

Because of the way X_d is constructed, that is, because each subject gets a score of 1 on exactly one of the variables, $X_d'X_d$ will have a special form. *It will be a diagonal matrix whose diagonal entries are simply the number in each group.* $X_d'y$ is special also. As each row of X_d' multiplies the single column of y, all it does is *sum the scores of the persons in that group,* so the entries in $X_d'y$ are the group sums. When these are multiplied by $(X_d'X_d)^{-1}$ the latter simply divides the sums by the number in the group, so **v** *is the group means,* $\bar{y}_{\cdot h}$. When we multiply $X_d\mathbf{v} = \hat{\mathbf{y}}$, the special nature of X_d means that $\hat{y}_i$ *is the mean of the group to which it belongs:* Repeating the notation of Chapter 6, $\mathbf{e} = \mathbf{y} - \hat{\mathbf{y}}$ and $\mathbf{e}'\mathbf{e}$ is the sum of squared deviations from regression, called the residual. In this case, $e_i = y_i - \bar{y}_{\cdot h}$, so $\mathbf{e}'\mathbf{e}$ is therefore *the sum of squared deviations from the group means;* in analysis of variance this is called the within-groups sum of squares. If this is subtracted from $\Sigma_i(y_i - \bar{y}_{\cdot\cdot})^2$, the total sum of squares for y, the result would be the sum of squares due to "regression," but here it represents the sum of squares between groups. We could do the usual F test for significance of regression as in the previous chapter, but here we end up comparing a within-group mean square to a between-groups mean square (ms). These calculations are presented in Exhibit 10-1.

Before going further with the correspondences, the next section will review briefly one-way analysis of variance concepts and return to the question of why the unit vector was not included in X_d.

$$\underset{X_d}{\begin{matrix} 1 & 0 & 0 \\ 1 & 0 & 0 \\ 0 & 1 & 0 \\ 0 & 1 & 0 \\ 0 & 0 & 1 \\ 0 & 0 & 1 \\ 0 & 0 & 1 \end{matrix}} \quad \underset{\mathbf{y}}{\begin{matrix} E1 \begin{cases} 2 \\ 4 \end{cases} \\ E2 \begin{cases} 5 \\ 7 \end{cases} \\ C \begin{cases} 7 \\ 9 \\ 8 \end{cases} \end{matrix}} \quad \underset{X_d'X_d}{\begin{bmatrix} 2 & 0 & 0 \\ 0 & 2 & 0 \\ 0 & 0 & 3 \end{bmatrix}} \quad \underset{X_d'\mathbf{y}}{\begin{bmatrix} 6 \\ 12 \\ 24 \end{bmatrix}} \quad \underset{(X_d'X_d)^{-1}}{\begin{bmatrix} \tfrac{1}{2} & 0 & 0 \\ 0 & \tfrac{1}{2} & 0 \\ 0 & 0 & \tfrac{1}{3} \end{bmatrix}} \quad \underset{\mathbf{v}}{\begin{bmatrix} 3.0 \\ 5.0 \\ 8.0 \end{bmatrix}} \quad \underset{\hat{\mathbf{y}} = X\mathbf{v}}{\begin{bmatrix} 3.0 \\ 3.0 \\ 6.0 \\ 6.0 \\ 8.0 \\ 8.0 \\ 8.0 \end{bmatrix}} \quad \underset{\mathbf{e} = \mathbf{y} - \hat{\mathbf{y}}}{\begin{bmatrix} -1.0 \\ 1.0 \\ -1.0 \\ 1.0 \\ -1.0 \\ 1.0 \\ 0.0 \end{bmatrix}}$$

Exhibit 10-1 Dummy code matrix X_d and calculations leading to predicted y scores. $X_d'X_d$ is a diagonal matrix consisting of the group sizes and $X_d'\mathbf{y}$ is the sum of scores in each group; therefore, **v** is a vector of group means. Then $\hat{\mathbf{y}}$ consists of the mean for the corresponding group, and **e** consists of the deviations from the group means.

ONE-WAY ANALYSIS OF VARIANCE

Sums of Squares

The majority of readers may have already studied analysis of variance, (*anova*), perhaps in great detail. At this point, it may be useful for them to review a familiar textbook on anova, provided that differences in notation do not create more confusion than clarification. In any event, one-way anova will be reviewed briefly in this section.

One-way analysis of variance is the simplest anova case. Here, the observations are presumed to come from different experimental *conditions,* for example, different experimental treatment groups or different natural classes such as which residence hall a student lives in. For purposes of the analysis, it does not matter whether the groups are experimental or natural, although there may be different interpretations of the results depending on which is the case. A "one-way" design assumes that there is only one way or mode of classification of the groups, as opposed to a cross-classification (factorial) design or a hierarchical one. (Even these can be treated initially as one-way designs by ignoring the hierarchical or factorial structure.)

Like other statistical methods, anova has a descriptive and an inferential aspect. We will examine the descriptive aspect first, once we have set forth some notational conventions. First, a double subscript y_{ih} will be used, where h refers to the group and i refers to the individual observation within the group. There will be n_h observations within group h, g different groups, and $\Sigma\, n_h = n$. The mean of a particular group h will be denoted $\bar{y}_{.h}$ and the overall or grand mean, $\Sigma_i \Sigma_h\, y_{ih}/n$ will be $\bar{y}_{..}$. Then there is the "total sum of squares," which is the sum of squared deviations from the grand mean,

$$ss_t = \sum_i \sum_h (y_{ih} - \bar{y}_{..})^2 \tag{10.1}$$

The total sum of squares may be broken down into two parts. One part is attributable to differences between the means of the different groups and the other to the deviations of the individual observations from the mean of its group. That is, the deviations from the grand mean can be divided into two segments:

$$y_{ih} - \bar{y}_{..} = (y_{ih} - \bar{y}_{.h}) + (\bar{y}_{.h} - \bar{y}_{..}) \tag{10.2}$$

Substituting this in Equation 10.1, followed by some algebra requiring familiarity with the mode of operation of summation signs, gives the following result:

$$ss_t = \underbrace{\sum_i \sum_h (y_{ih} - \bar{y}_{.h})^2}_{\text{within}} + \underbrace{\sum_h n_h (\bar{y}_{.h} - \bar{y}_{..})^2}_{\text{between}} \tag{10.3}$$

The "within" part represents the sum of squared deviations from the group means (ss_w) and the "between" part represents the deviations of the group means from the grand mean (ss_b). Note that there are n terms in each of the sums of squares, ss_b and ss_w, but in ss_b, n_h of these are the same in each group h. The fact of n deviations in each ss needs to be considered in order to see the connections with regression. Equation 10.3 shows how the total sum of squares is broken down into a between-groups and a within-groups sum of squares.

The two sums of squares are defined as

$$ss_w = \sum_i \sum_h (y_{ih} - \bar{y}_{\cdot h})^2 \tag{10.4}$$

$$ss_b = \sum_h n_h (\bar{y}_{\cdot h} - \bar{y}_{\cdot\cdot})^2 \tag{10.5}$$

The former sum of squares represents the sum of squared deviations from the group means. In each group, there are n_h such deviations. However, knowing all but one of the deviations in a group would allow one to deduce the other since they must sum to zero. Therefore, only $n_h - 1$ are free to vary. So, we say that there are $n_h - 1$ degrees of freedom (df) in group h. Summing these df across g groups gives a total of $n - g$ degrees of freedom for the within group deviations.

The sum of squares between groups represents deviations of the group means from the grand mean. Only $g - 1$ of these deviations of means are independently free to vary because they must balance out to give a sum of zero when weighted by the group sizes. Thus, ss_b has $g - 1$ degrees of freedom in the same sense that ss_w has $n - g$ degrees of freedom.

The two sums of squares can be divided by their respective degrees of freedom. The resulting ratios are called "mean squares," the mean square within, ms_w, and the mean square between, ms_b:

$$ms_w = ss_w/(n - g) \tag{10.6}$$

$$ms_b = ss_b/(g - 1) \tag{10.7}$$

If the null hypothesis of equal population means is true, the mean squares are both estimates of the population variance σ_y^2, and form the basis of the inferences concerning the means of the groups. This is demonstrated in the next section.

Inferences in Anova

The rationale for the inferential aspect of analysis of variance as a test of hypotheses concerning means rests on the sampling properties of means. The sampling variance of means from the same population depends simply on the population variance and the sample sizes.

$$\sigma_{\bar{y}}^2 = \frac{\sigma_y^2}{n_h} \tag{10.8}$$

In analysis of variance, there are several means. If they all come from the same population (H_0), then their variance is an estimate of $\sigma_{\bar{y}}^2$. This in turn could be used to estimate σ_y^2 itself because of the relationship shown in Equation 10.8. What we call ms_b is actually this estimate; it is based on $g - 1$ degrees of freedom because there are g means, and a degree of freedom is lost for the grand mean.

From each group we can obtain a sample variance, which is also an estimate of the population variance. There are several such estimates, one from each sample. We arrive at one overall estimate of the population variance by pooling these estimates. This pooled estimate is equal to ms_w. Now we have two estimates of the population variance, one derived from the variance of the means and one derived from the within-sample variances. They should be about the same, if the variation in the means is just due to random sampling.

This is the basis for using the means squares to test the null hypothesis H_0: $\mu_1 = \mu_2 = \cdot \cdot \cdot = \mu_g$. Based on the assumptions of random sampling from a normally distributed population, the ratio of the two estimates has the F distribution:

$$ms_b/ms_w = F \tag{10.9}$$

One tests H_0 by comparing this ratio to tables of the F distribution using $g - 1$ and $n - g$ degrees of freedom, rejecting H_0 if the critical value corresponding to the specified alpha level is exceeded.

Strictly speaking, the probabilities in the F table depend on random sampling from a single normal distribution or from several normal distributions with equal means and variances. The effect of departures from normality or equality of variances is often dismissed, but is not trivial. It will be considered again later in this chapter. The calculations of anova are illustrated in Exhibit 10-2.

A Model for the Sums of Squares

A second way to look at analysis of variance is through a *model* for the observations. For the one-way case, this would be

$$y_{ih} = \mu + \alpha_h + e_{ih} \tag{10.10}$$

The parameter μ is interpreted as representing the population grand mean, whereas α_h is the "effect" of treatment h or the mean effect of being in group h. The third term represents a residual or error to account for differences in the scores of individuals who are in the same group. The two parameters μ and α_h can never be observed or estimated separately, so an arbitrary restriction is placed on the α_h. The restriction is that the average effect is zero:

$$\sum_h \alpha_h = 0 \tag{10.11}$$

According to this model, the group means are

$$\bar{y}_{\cdot h} = \mu + \alpha_h + \bar{e}_{\cdot h} \tag{10.12}$$

and the between groups mean square reflects the variance of the α_h plus σ_e^2. This makes the null hypothesis equivalent to saying all $\alpha_h = 0$. This formulation is useful in various ways, including the study of power.

	i	y_{ih}	$y_{ih} - \bar{y}_{..}$	Total $(y_{ih} - \bar{y}_{..})^2$	$y_{ih} - \bar{y}_{.h}$	Within $(y_{ih} - \bar{y}_{.h})^2$	$y_{.h} - \bar{y}_{..}$	Between $(y_{.h} - \bar{y}_{..})^2$
$h = 1$	1	2	−4.0	16.0	−1.0	1.0	−3.0	9.0
	2	4	−2.0	4.0	1.0	1.0	−3.0	9.0
Σy_{11}		6						
$\bar{y}_{.1}$		3.0						
$h = 2$	1	5	−1.0	1.0	−1.0	1.0	0.0	0.0
	2	7	1.0	1.0	1.0	1.0	0.0	0.0
Σy_{12}		12						
$\bar{y}_{.2}$		6.0						
	1	7	1.0	1.0	−1.0	1.0	2.0	4.0
$h = 3$	2	9	3.0	9.0	1.0	1.0	2.0	4.0
	3	8	2.0	4.0	0.0	0.0	2.0	4.0
Σy_{13}		24						
$\bar{y}_{.3}$		8.0						
$\Sigma\Sigma_{yih}$		42	0.0	36.0	0.0	6.0	0.0	30.0
$\bar{y}_{..}$		6.0						

Exhibit 10-2a Calculations for an analysis of variance of the data from Exhibit 10-1. First the deviation of each score from the overall mean is shown, which is then squared. The sum of these squares is ss_t. Then the deviation of each score from the mean of its own group is shown, squared, and summed. This is ss_w. Finally, the deviation of each group's mean is shown, once for each observation, squared, and summed. This is ss_b.

Source	Sum of Squares	*df*	Mean Square	*F*
Between	30.0	2	15.0	10.0
Within	6.0	4	1.50	
Total	36.0	6	6.0	

Exhibit 10-2b Analysis of variance table shows the sums of squares and degrees of freedom, $g - 1$ and $n - g$, respectively. The $F = 10.0$ is significant at the .05 level.

According to this model, the parameter estimates are

$$\hat{\mu} = (1/g)\Sigma\bar{y}_{.g} \tag{10.13}$$

and

$$\hat{\alpha}_h = \bar{y}_{.h} - \hat{\mu} \tag{10.14}$$

It may seem a little odd not to have $\hat{\mu} = \bar{y}_{..}$, but this would make $\hat{\mu}$ depend on the relative sizes of the groups.

This concludes our general outline of the rationale for analysis of variance. Other sources such as Hays (1981), Myers (1979), and Winer (1971) provide greater detail. In other contexts, additional space is devoted to formulas which make the computations more efficient. However, if there are more than a few cases involved, computer programs will do the arithmetic.

PARALLELS WITH REGRESSION

The Constant Vector

We saw that if X_d was composed of binary group membership scores, called dummy variables, the sum of squared deviations from regression corresponded exactly to the sum of squared deviations from the group means, ss_w in anova terminology. In order to see how to get ss_b by a regression interpretation of anova, it is necessary to return to the question of the missing constant vector in the X matrix.

Recall the moonlighting student who tried to use the group-membership variables in his regression analysis, but the computer complained that his covariance matrix was singular. Remember that a multiple regression analysis may be conceived as proceeding in one of two equivalent ways. X may consist of deviation scores on the p predictors, in which case the relevant inverse is that of the matrix of covariances among the predictors. Alternatively, X may consist of the *raw scores,* with an additional column of unities tacked on, that is, X_+. Here, the relevant inverse is of the $p + 1$ by $p + 1$ matrix which consists of the sums of squares and products of the predictors, bordered by an additional row and column which are their sums, with n in the corner (see Chapter 6).

In the case of g dummy variables, the latter matrix would have no inverse because X_+ is linearly dependent. This can be seen in Table 10-1. If we put in the column of ones last, it is clear that *this column will equal the sum of the other three.* This is a very obvious form of linear dependence as can be seen from Table 10-2. Since linear dependence is a mutual relationship among all the vectors, this means that any of the other three vectors could be derived from the constant column and the remaining two.

Formally, the property of linear dependence means that there is a set of constants $c_1, c_2, \ldots, c_g, c_{g+1}$ such that $\Sigma_j c_j x_j = 0$. In the present case, the constants are 1 for x_{g+1} and the other three are -1. Thus the rank of X_+, and therefore of $X_+'X_+$, would be less than its order, and the latter will have no inverse. Due to parallelism, the same will be true of the inverse of the matrix of covariances among the X's.

A consideration of the general problem shows us that linear dependence is not

Table 10-2 Linear Dependence of Dummy Matrix with Constant Vector

x_1	x_2	x_3	*Constants*	*Weights*		
1	0	0	1	-1		0
1	0	0	1	-1		0
1	0	0	1	-1	=	0
0	1	0	1	1		0
0	1	0	1			0
0	0	1	1			0
0	0	1	1			0

peculiar to this example. If X_d is a matrix of scores on g dummy variables, that is, if there is exactly one 1 in each row, and an additional column of ones is appended on this matrix, the latter will always be the sum of the other columns, and consequently, the $g + 1$ columns will be linearly dependent. On the other hand, if the dummy variable for one of the groups is eliminated, as the professor suggested to the student, then the linear dependence will disappear, and the reduced matrix will be of full rank. Correspondingly, the covariance matrix among the $g - 1$ dummy variables will have an inverse.

One may be reluctant to take out one of the dummy variables and replace it with the constant vector because this seems to imply losing the mean of its group as the predicted y. But the prediction is not lost, it just becomes obscured. If $g + 1$ vectors are linearly dependent and any subset of g of the vectors are not, then, in a sense, the information in any $g + 1$ is deducible from the others. Appending the constant vector to the matrix of dummy variables will not improve or in any way change the predictions we would make, because this column does not tell us anything new, so to speak. Conversely, if one of the dummy variables is *deleted*, no information is lost from X_+ because the information in it can be reconstructed from the $g - 1$ dummy variables that remain and the constants column of ones. Due to the linear dependence effect, appending the constant vector cannot make the predictions any better. But after the vector is appended, deleting one of the others does not make the prediction any worse.

This is demonstrated in Exhibit 10-3. Here, when X_+ consists of *two* of the three dummy variables and the constant vector, then $(X_+'X_+)^{-1}$ applied to this X_+ gives $\bar{y}_{.h}$ as predicted y's. The predictions are exactly the same whether X consists of g dummy variables or of $g - 1$ dummy variables and the constant vector. The same conclusion would apply if one used the covariance matrix route for making the predictions.

Weights for the Dummy Variables

One nice aspect of the original, complete dummy variable representation is that both the predictions and the weights turn out to be the group means, $v_h = \bar{y}_{.h}$. This simplicity becomes hidden (but not lost) when the constant vector replaces one of the dummy variables in X_d. Let us examine what the weights will turn out to be in this case, assuming that the constant vector has replaced the dummy variable for the *last* group in X_α, making X_+. Then $X_+'X_+$ will be

$$X_+'X_+ = \begin{bmatrix} n_1 & 0 & 0 & \cdots & n_1 \\ 0 & n_2 & 0 & \cdots & n_2 \\ 0 & 0 & n_3 & \cdots & n_3 \\ \vdots & \vdots & \vdots & \vdots & \vdots \\ n_1 & n_2 & n_3 & & n \end{bmatrix} \qquad (10.15)$$

and $X_+'\mathbf{y}$, presented as a transpose to save space, is

$$(X_+'\mathbf{y})' = \Sigma y_{i1} \quad \Sigma y_{i2} \cdots \Sigma y_{i,g-1} \quad \Sigma\Sigma y_{ih} \qquad (10.16)$$

Group

(a)

$$\begin{array}{c} E1 \\ E1 \\ E2 \\ E2 \\ C \\ C \\ C \end{array}\quad \begin{bmatrix} 1 & 0 & 0 \\ 1 & 0 & 0 \\ 0 & 1 & 0 \\ 0 & 1 & 0 \\ 0 & 0 & 1 \\ 0 & 0 & 1 \\ 0 & 0 & 1 \end{bmatrix} \begin{bmatrix} 1 & 0 & 1 \\ 0 & 1 & 1 \\ 0 & 0 & 1 \end{bmatrix} = \begin{bmatrix} 1 & 0 & 1 \\ 1 & 0 & 1 \\ 0 & 1 & 1 \\ 0 & 1 & 1 \\ 0 & 0 & 1 \\ 0 & 0 & 1 \\ 0 & 0 & 1 \end{bmatrix} \quad \begin{bmatrix} 2 \\ 4 \\ 5 \\ 7 \\ 7 \\ 9 \\ 8 \end{bmatrix}$$

$$\mathbf{X}_d \qquad \mathbf{T} \quad = \quad \mathbf{X}_+ \qquad \mathbf{y}$$

(b)

$$\begin{bmatrix} 2 & 0 & 2 \\ 0 & 2 & 2 \\ 2 & 2 & 7 \end{bmatrix}^{-1} \begin{bmatrix} 6 \\ 12 \\ 42 \end{bmatrix} = \begin{bmatrix} .833 & .333 & -.333 \\ .333 & .833 & -.333 \\ -.333 & -.333 & .333 \end{bmatrix} \begin{bmatrix} 6 \\ 12 \\ 42 \end{bmatrix} = \begin{bmatrix} -5.0 \\ -2.0 \\ 8.0 \end{bmatrix}$$

$$(\mathbf{X}_+'\mathbf{X}_+)^{-1} \qquad \mathbf{X}_+'\mathbf{y} \qquad (\mathbf{X}_+'\mathbf{X}_+)^{-1} \qquad \mathbf{X}_+'\mathbf{y} = \mathbf{b}_+$$

(c)

$$\begin{bmatrix} 1 & 0 & 1 \\ 1 & 0 & 1 \\ 0 & 1 & 1 \\ 0 & 1 & 1 \\ 0 & 0 & 1 \\ 0 & 0 & 1 \\ 0 & 0 & 1 \end{bmatrix} \begin{bmatrix} -5.0 \\ -2.0 \\ 8.0 \end{bmatrix} = \begin{bmatrix} 3.0 \\ 3.0 \\ 6.0 \\ 6.0 \\ 8.0 \\ 8.0 \\ 8.0 \end{bmatrix} \qquad \begin{bmatrix} 1 & 0 & -1 \\ 0 & 1 & -1 \\ 0 & 0 & 1 \end{bmatrix} \begin{bmatrix} 3.0 \\ 6.0 \\ 8.0 \end{bmatrix} = \begin{bmatrix} -5.0 \\ -2.0 \\ 8.0 \end{bmatrix}$$

$$\mathbf{X}_+ \qquad \mathbf{b}_+ = \hat{\mathbf{y}} \qquad \mathbf{T}^{-1} \qquad \bar{\mathbf{y}}_{\cdot h} = \mathbf{b}_+$$

(d) $s_y^2 = \Sigma(y - \bar{y})^2/(n - 1) = 6.0$

$\Sigma(y - \hat{y})^2 = \Sigma(y_{ih} - \bar{y}_{\cdot h})^2 = 6.0 = ss_{dev}$

$\Sigma(\hat{y} - \bar{y})^2 = 30.0 = ss_{reg}$

$s_{reg}^2 = 30.0/6 = 5.0$

$R^2 = s_{reg}^2/s_y^2 = .833$

Exhibit 10-3 Multiple regression using dummy variables. The data from Exhibit 10-1 is used, but the constant vector replaces the dummy variable for Group *C* to make the X_+ matrix shown in part (a) through use of the transformation T. The vector of y scores is also shown. Part (b) shows the products $X_+'X_+$ and $X_+'\mathbf{y}$, followed by the inverse of $X_+'X_+$, and the computation of $\mathbf{b}_+$. In part (c) the weights are used to compute *y*. Note that $y_{ih} = \bar{y}_{\cdot h}$, that is, 3.0, 6.0, or 8.0 for groups *E*1, *E*2, or *C*, respectively. Also shown is T^{-1} and the product $T^{-1}\bar{y}_{\cdot h}$, which is equal to $\mathbf{b}_+$. Part (d) shows the computation of s_y^2, ss_{reg}, and ss_{dev} by comparing $\hat{\mathbf{y}}$ to $\mathbf{y}$. Part (e) computes $\mathbf{b}$ from the variance-covariance matrix of the contrast variables and their covariances with *y*. The coefficients are the same as those in part (b). R^2 is found to be .833, the same as in part (d). Part (f) repeats the process of part (e) using correlation matrices.

(e) $\begin{bmatrix} .238 & -.095 \\ -.095 & .238 \end{bmatrix}^{-1} \begin{bmatrix} -1.0 \\ 0.0 \end{bmatrix} = \begin{bmatrix} -5.0 \\ -2.0 \end{bmatrix}$

$\mathrm{S}_{xx}^{-1} \quad \mathbf{s}_{xy} \quad = \quad \mathbf{b}$

$\mathbf{b}'\mathbf{s}_{xy} = s_{reg}^2 = 5.0 \qquad R^2 = s_{reg}^2/s_y^2 = .833$

(f) $\begin{bmatrix} 1.00 & -.40 \\ -.40 & 1.00 \end{bmatrix}^{-1} \begin{bmatrix} -.836 \\ 0.0 \end{bmatrix} = \begin{bmatrix} -.995 \\ -.398 \end{bmatrix}$

$\mathrm{R}_{xx}^{-1} \quad \mathbf{r}_{xy} \quad = \quad \mathbf{b}^*$

$R^2 = \mathbf{b}^{*\prime}\mathbf{r}_{xy} = .833$

Exhibit 10-3 (continued)

It turns out that when we solve $(\mathrm{X}_+'\mathrm{X}_+)^{-1}\mathrm{X}_+'\mathbf{y} = \mathbf{b}_+$, then

$$\mathbf{b}_+' = (\bar{y}_{\cdot 1} - \bar{y}_{\cdot g} \quad \bar{y}_2 - \bar{y}_{\cdot g} \cdot \cdot \cdot , \bar{y}_{\cdot g-1} - \bar{y}_{\cdot g} \quad \bar{y}_{\cdot g}) \tag{10.17}$$

This shows that the "regression weights" are the differences between the group means and the mean of the deleted group, and "b_0" is the mean of that group. Thus, $\mathrm{X}_+\mathbf{b}_+ = \hat{\mathbf{y}}$ will still give the group means.

Multiple Regression and Anova Sums of Squares

In multiple regression, we partition the total sum of squared deviations from the mean of y into a part due to regression and a residual part due to deviations. The null hypothesis, which states that all the regression coefficients are zero is determined by an F test comparing these two sums of squares. While you no doubt can see some parallels between multiple regression and anova sums of squares. These similarities will be made explicit in the following section.

Since the predictions in the dummy variable regression case are that $\hat{y}_{ih} = \bar{y}_{\cdot h}$, regardless of whether we delete one of the dummy variables, we know that ss_w in analysis of variance is the same as ss_{dev} for a multiple regression where the "predictors" are dummy variables. What about ss_b? It should also be clear that this is the ss_{reg}, of dummy variable regression. In general, ss_{reg}, the sum of squares for the predictions, is $\Sigma_i(\hat{y}_i - \bar{y}_.)^2$ and this is true in dummy variable regression as well. In that case, $\bar{y}_.$ is the grand mean, designated as $\bar{y}_{..}$. The two dots as subscripts replace i and h in this case where the total set of observations is classified into groups. It is just the mean of y, which always is the basis of the sums of squares. The $\hat{y}_i$ in this case are the group means, $\bar{y}_{\cdot h}$. Thus there is an identity between ss_{reg} and ss_b as well as between ss_{dev} and ss_w.

There are other correspondences, too, although they are not always stated. For example, ss_b/ss_t is an R^2. In the early statistical literature, if the subscript h represented a set of rationally ordered categories whose order was defined in terms of some a priori consideration, then this ratio was called a squared "coefficient of nonlinear correlation," designated η^2 (eta-squared). As a sample multiple correlation, this R^2 suffers from the usual problems of bias. It can be corrected by means of Equation 9.3 to furnish an unbiased estimate $\hat{\rho}^2$ of the proportion of variance due to group differences. The coefficient ω^2, suggested by Hays (1982), performs a similar function in the context of the anova sums of squares.

So, we have observed the reasons for the kindly statistics professor's reassurances to the student, including his advice concerning what to do about the singular covariance matrix and his reference to an analysis of variance. We can perform a regression analysis on dichotomous predictors representing group membership, but if a standard regression computer program is used, one of the variables must be deleted, since the full set becomes linearly dependent once the constant vector is included, or once the dummy variables are transformed. If this is not done, the covariance matrix has no inverse, and the computer will balk.

The resulting regression analysis is equivalent to an analysis of variance of the same data. Suppose we analyze the data using an X_d matrix that consists of the g dummy variables, not including the constant vector and bypassing the step of computing the covariance matrix. The results of computing $(\mathrm{X}_d'\mathrm{X}_d)^{-1}\mathrm{X}_d'\mathbf{y}$ would yield a set of "weights" which are the group means. Otherwise, we must delete one of the dummy variables. Then the regression weights represent differences between the means of the groups and the mean of the group whose dummy variable was deleted.

TRANSFORMED VARIATES

The Dummy Variable Case

Analysis of variance goes beyond the simple one-way case described above, moving on to planned contrasts, factorial designs of increasing levels of complexity, including investigations of interactions, hierarchical designs, Latin squares, partial confounding, and the like. To a large extent, the standard variations on analysis of variance can be understood in terms of the use of coded predictors in a multiple regression. Indeed we can sometimes get clearer answers to our research questions more easily using the regression methods rather than anova as it is typically employed.

To understand the meaning of different analysis of variance procedures and their relationship to regression analysis better, we can use the concept of *transforming* the original dummy variables in various ways and then looking at the regression coefficients for the transformed variates. The dummy codes discussed in the previous section already provide an example of this, but now we wish to make a more abstract description of what went on, employing some matrix algebra. To do this, it becomes necessary to introduce some new notation. For example, let us use X_+ to refer to the n by g matrix consisting of scores on $g - 1$ of the dummy variables, with the constant vector replacing the deleted dummy variable, which we will assume for simplicity was the last one.

In terms of X_d we made the predictions about y as follows, using $\hat{\mathbf{y}}_d$ to identify the source of the predictions:

$$X_d\mathbf{v} = \hat{\mathbf{y}}_d \tag{10.18}$$

The expression $\mathbf{v} = (X_d'X_d)^{-1}X_d'\mathbf{y}$ has the $\bar{y}_{.h}$ as its entries. In the same way, if we use X_+ as the basis for prediction, we have

$$X_+b_+ = \hat{y}_b \tag{10.19}$$

We obtain $\mathbf{b}_+$ in the corresponding way, $\mathbf{b}_+ = (X_+'X_+)^{-1}X_+'\mathbf{y}$. From Equation 10.17, its first $g - 1$ entries are $\bar{y}_{.h} - \bar{y}_{.g}$ and the last entry is $\bar{y}_{.g}$. To understand the relationship between $\mathbf{b}_+$ and $\mathbf{v}$, X_+ can be derived from X_d by multiplying X_d by a simple and nonsingular transformation represented by the matrix T:

$$X_+ = X_dT \tag{10.20}$$

where T is the g by g matrix

$$T = \begin{bmatrix} 1 & 0 & 0 & \cdots & 1 \\ 0 & 1 & 0 & \cdots & 1 \\ 0 & 0 & 1 & \cdots & 1 \\ \vdots & \vdots & \vdots & \ddots & \vdots \\ 0 & 0 & 0 & \cdots & 1 \end{bmatrix}$$

That is, $t_{hh} = 1$; $t_{hg} = 1$; and $t_{hh'} = 0$ for $h' \neq h,g$.

The purpose of T is to replace the last column of X_d with the sum of all its columns—that is, the constant—leaving the remaining columns unchanged in forming X_+. The $\hat{y}$ produced by either score matrix are the same, because of the following identities:

$$\mathbf{b}_+ = (X_+'X_+)^{-1}X_+'\mathbf{y} \tag{10.21}$$

Then, using Equation 10.20,

$$\mathbf{b}_+ = (T'X_d'X_dT)^{-1}T'X_d'\mathbf{y} \tag{10.22}$$

Since the inverse of a product is the product of the inverses in the reverse order, then

$$\mathbf{b}_+ = T^{-1}(X_d'X_d)^{-1}T'^{-1}T'X_d\mathbf{y} \tag{10.23}$$

and

$$\mathbf{b}_+ = T^{-1}(X_d'X_d)^{-1}X_d'\mathbf{y} \tag{10.24}$$

But since

$$(X_d'X_d)^{-1}X_d'\mathbf{y} = \mathbf{v} \tag{10.25}$$

where v is the group means, we will have

$$\mathbf{b}_+ = T^{-1}\mathbf{v} = T^{-1}\bar{\mathbf{y}}_{\cdot h} \tag{10.26}$$

Then the predictions are also $\hat{y}_{ih} = \bar{y}_{\cdot h}$ because

$$\hat{\mathbf{y}}_b = X_+\mathbf{b}_+ \tag{10.27}$$

By substitution,

$$\hat{\mathbf{y}}_b = (X_dT)\,(T^{-1}\mathbf{v}) = X_d\mathbf{v} = \hat{\mathbf{y}}_{d\cdot} = \bar{\mathbf{y}}_{\cdot h} \tag{10.28}$$

Thus the predictions are identical whether we use $\mathbf{b}_+$ or $\mathbf{v}$.

Generality of the Principle

Although we started with a particular kind of transformation of the dummy code matrix, the algebra shows that we are not limited to this one. T *could be any nonsingular transformation* of the dummy code matrix. This illustrates an important principle, one that simplifies the understanding of the operation of a large number of different analyses of variance. If weights for predicting **y** are found for one form of the predictor matrix, and another form of predictor matrix is formed by means of a nonsingular transformation of the first, then the weights for the second set may be obtained by premultiplying those for the first by the inverse of the transformation. Moreover, if X_+ is *any* score matrix formed by transforming the dummy variables, the resulting regression coefficients are the inverse transformation of the groups means, provided such an inverse exists. The predicted y scores are then always the group means; ss_{reg} is identical to ss_b and ss_{dev} is identical to ss_w.

This is important because it shows that we can transform the dummy variables in a variety of ways, depending on how we wish to define the variables of interest and get equivalent results. Additionally, we can immediately figure out what the weights would be without having to solve the least-squares equations. In many common cases it is easy to come up with these transformations because they correspond directly to the contrasts. These relationships also give insight into what is going on in various cases of analysis of variance.

The regression weights, the predictions, and the sums of squares are not the only quantities of interest, however. In multiple regression, we are also interested in assessing the effects or importance of individual variables or groups of variables. Similarly, in anova, we want to know whether certain groups, or combinations of groups, have significantly different means. Such questions can be assessed in the regression approach to anova through the transformed dummy variables. These transformed variables represent the comparison of means or *contrasts* that are of interest.

In regression, there are two main ways to evaluate the usefulness of a particular variable. One is the increase in squared multiple correlation when that variable is added last to the regression equation. The other measure is the squared simple correlation of that variable with the dependent variable. The former measures the proportion of variance that the variable is *necessary* to predict, and the latter measures the proportion that the variable is *sufficient* to predict. The tests of hypotheses concerning these quantities are the significance of the regression coefficient and the significance of the simple correlation, respectively. These quantities and test are generalized to groups of several variables each. The same ideas apply in analysis of variance.

Correlations of Dummy Variables

We have seen that the quantities that function as "regression coefficients" for dummy variables represent differences between the mean of the group represented by the dummy variable and the mean of the deleted group. A test of significance of this regression coefficient represents a test of this mean difference. We can inquire about what the correlations between dummy variates and the dependent variable might represent.

It turns out that these correlations reflect the differences between the mean of group h and the mean of all other groups. Another way of saying this is that r_{hy} is the correlation of a dichotomous variable representing membership-nonmembership in group h and the dependent variable. Such correlations between a dichotomous variable and a continuous one are called *point-biserial* correlations. In such cases, the ordinary Pearson correlation (see Chapter 6) can be expressed in the following way:

$$r_{hy} = \frac{\overline{x}_{\cdot h} - \overline{x}_{\sim h}}{[(p_h(1 - p_h)s_y]^{1/2}} \tag{10.29}$$

In this equation, $\overline{x}_{\cdot h}$ is the mean of group h and $\overline{x}_{\sim h}$ is the mean of all other observations not in h, and $p_h = n_h/n$. The test of significance of this correlation is the same as the t test (or F test) of the corresponding means.

The correlations between different dummy variables h and h' are simple functions of the group frequencies:

$$r_{hh'} = \frac{-(n_h n_{h'})^{1/2}}{[(n - n_h)(n - n_{h'})]^{1/2}} \tag{10.30}$$

Note that the correlations are all negative; this is understandable since membership in one group must be negatively correlated with membership in another. Also, when the groups are of equal size, the equation reduces to $-n_h/(n - n_h)$.

The availability of these equations means that the analysis of variance could be performed in terms of a correlation matrix R_{xx} and a vector of correlations with the dependent variable $\mathbf{r}_{xy}$, where the x variables are the $g - 1$ dummy variables. This is not ordinarily done, but it serves to relate regression and anova. Here, the multiple correlation R^2 would be the ratio ss_b/ss_t, and the usual F tests of multiple correlations could be carried out. The latter would be numerically the same as the overall F ratios in anova.

RATIONALIZED VARIABLES IN ANOVA

Contrasts and Comparisons

The preceding discussion demonstrates a number of direct correspondences between anova and multiple regression. Anova could be performed as a multiple regression involving dummy variables. However, the fact that the computations depend on only the group means and the within-group sum of squares makes such an approach inefficient. Such an analysis tells us relatively little. It is true that we can test the hypothesis that all the means are equal. Similarly, testing the regression coefficients for significance corresponds to comparing the mean of a group to the mean of the deleted group, and testing the dummy variable correlations reveals the difference between the mean of a group and the mean of all the other groups. Such tests may not be particularly interesting. In fact, it is typically the case that there are other group comparisons of more direct concern to the investigator.

Such comparisons can almost always be expressed in terms of *contrasts* between means. A contrast is a vector of coefficients **c** to be applied to the groups means such that (a) they represent some set of comparisons of interest and (b) to provide simplification, $\Sigma_h n_h c_h$ is set equal to zero, where c_h is the coefficient applied in group h.

For the present, we will assume that the investigator can define $g - 1$ contrasts that are of interest, but we will see that it is possible to work with fewer, and even possible to work with more than $g - 1$ under certain circumstances. The reason for having $g - 1$ contrasts is that then we can think of each contrast defining an independent variable on which each individual obtains a score that is equal to the contrast coefficient for his group. This process is similar to constructing an $\mathbf{X}_+$ matrix as a transformation of the dummy code matrix. The first $g - 1$ columns of $\mathbf{X}_+$ each consist of a contrast variable x_c. An individual in group h will have a score on x_c equal to t_{hc}. Then everything we found out in the previous section will apply directly. The various sums of squares remain the same; all that changes are the regression coefficients and the correlations with y of the contrasts.

Often the contrasts are defined in such a way as to be *orthogonal.* This is because, under some circumstances, this makes the corresponding variables in the $\mathbf{X}_+$ matrix uncorrelated. Recall that in the uncorrelated case such quantities as the regression coefficients, the simple correlations, amounts of variance accounted for, and the significance of both the simple correlations and the regression coefficient were very simply and directly related. Unfortunately, the circumstances under which orthogonal contrasts result in uncorrelated variables are almost entirely limited to cases in which all the group sizes are equal. This has led to a good deal of confusion and controversy concerning the legitimacy of certain kinds of analysis.

In the following sections we will take up the case of orthogonal contrasts that do lead to uncorrelated predictors, and then move on to see how they are affected when the groups are unequal, before addressing the general case that includes nonorthogonal contrasts.

Orthogonal Contrasts

While there is an infinite variety of possible orthogonal contrasts, three types are the most common: nested comparisons, orthogonal polynomials, and factorial designs. The latter may be combined with some form of the other two in multilevel situations. In orthogonal contrasts, the transformation of the dummy coded matrix into its new form is orthogonal. That is, $T'T = D$, a diagonal matrix. The last column of X_+ remains the constant vector, so the last column of T must consist of all unities. To preserve the orthogonality, the sums of the coefficients in all the other columns must be zero.

Consider a situation in which nested contrasts are appropriate. Suppose a clinical psychologist wanted to compare two forms of therapy to a control condition and to each other. First she compares the average of the two therapy conditions to the control condition, testing $\frac{1}{2}(\mu_1 + \mu_2) - \mu_3 = 0$. Then she compares one therapy to the other, testing $\mu_1 - \mu_2 = 0$. Then the transformation matrix would be

$$T = \begin{bmatrix} .5 & 1 & 1 \\ .5 & -1 & 1 \\ -1 & 0 & 1 \end{bmatrix}$$

If there are more groups, then $g - 1$ groups are compared to the gth group. Of the $g - 1$ groups, $g - 2$ groups are compared to the $g - 1$st group, then the $g - 3$ groups are compared to the $g - 2$nd group, and so on, until the first group is compared to the second. These are sometimes called Helmert or reverse Helmert contrasts. In each successive contrast, the odd group in the previous contrast is left out. The groups not involved in a contrast receive zero coefficients in that column of the transformation.

Orthogonal polynomials occur when the levels of the classification correspond to equally-spaced levels of some quantitative variable. For example, there might be four levels of intensity of shock, three levels of dosage of a drug, or five lengths of lists of words to be studied, and so on. Orthogonal polynomials are used in order to describe the relationship between the group means on y and the levels of x in a simple, standard way. Call this multilevel variable a; then each contrast corresponds to a power of a, but with all lower powers partialed out. The coefficients that result are called orthogonal polynomials, and many books on statistics or experimental design contain tables of these coefficients. With g levels of a, the terms go up to a^{g-1}. Figure 10-1 illustrates their relationship to x up to a^5. Note that for each increase in the exponent there is one more peak or valley in the curves. Figure 10-2 illustrates some curves that are composites of these.

Factorial designs are the most common type of orthogonal contrasts in behavioral science research. In factorial design, there is more than one independent variable. The category or level of each independent variable is observed under all possible combinations of levels of the others. The simplest case is a two-by-two. An example of two-by-two is gender by treatment in which there are

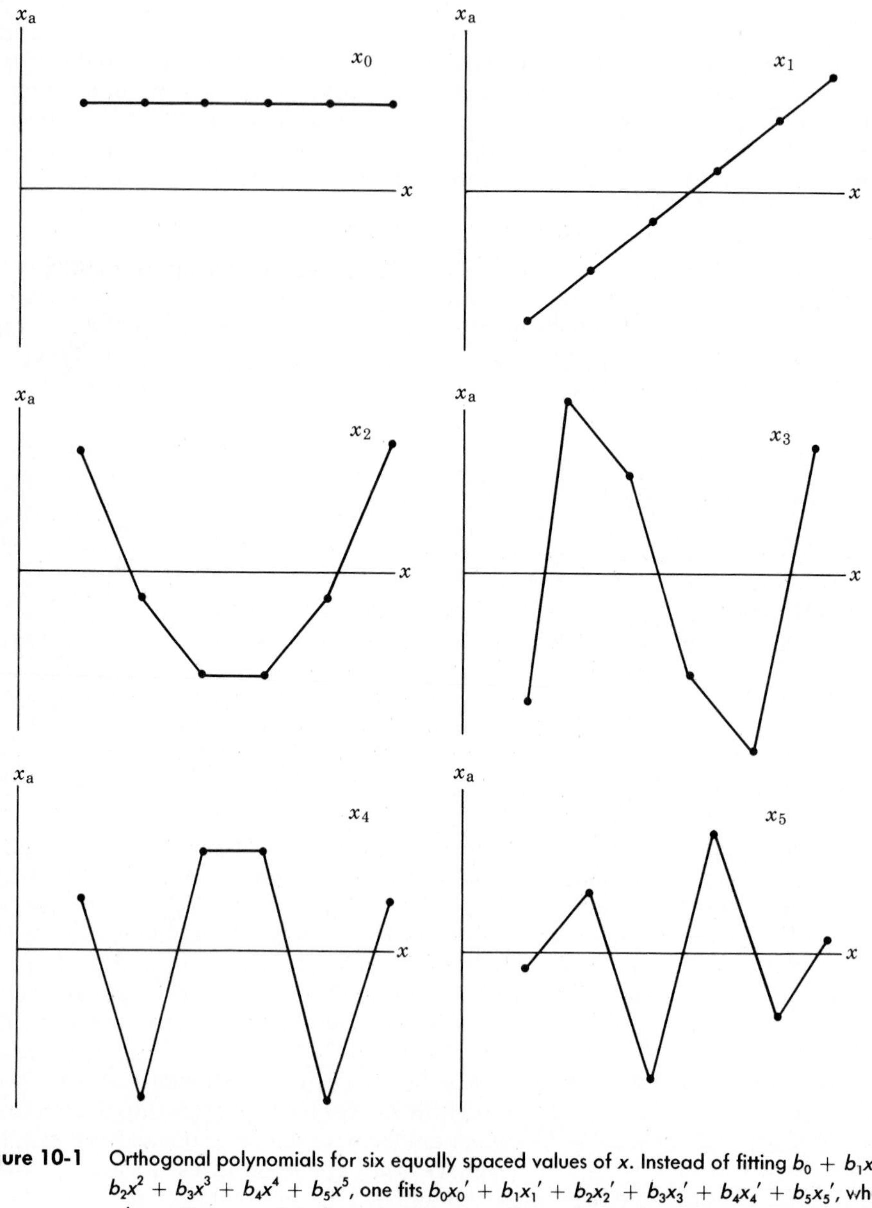

Figure 10-1 Orthogonal polynomials for six equally spaced values of x. Instead of fitting $b_0 + b_1x + b_2x^2 + b_3x^3 + b_4x^4 + b_5x^5$, one fits $b_0x_0' + b_1x_1' + b_2x_2' + b_3x_3' + b_4x_4' + b_5x_5'$, where x_a', $a = 0, 1, \ldots, 5$, is determined by the ordinate of each function.

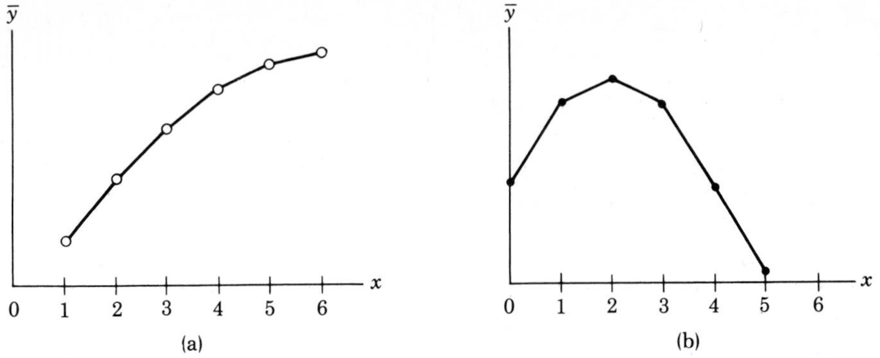

Figure 10-2 Part (a) illustrates a curve that will have a large positive coefficient for the linear term and a slight negative one for the quadratic (x^2) term. Part (b) illustrates a curve that will have a slight negative b_1 coefficient and a substantial negative b_2 coefficient.

four groups: males in the treatment condition, males in the control, females in the treatment, and females in the control. The factorial design allows the investigator to assess the importance and significance of "main effect," in this case representing treatment versus control averaged across genders, and gender differences averaged across treatment versus control. It is also possible to assess the "interaction" of the gender and treatment variables, that is, the differential effect of treatment, depending on gender—or, viewed the other way, the degree to which gender differences differ, depending on treatment versus control.

Factorial designs can generalize to more than two levels of the factors and to more than two factors, and we will have considerably more to say about their analysis later. For the moment, we will consider factorial designs as a form of orthogonal comparisons. Using an orthogonal transformation, they can be understood in terms of transforming the dummy variables into a set of scores that correspond to the effects in the design.

When groups are of equal size, the orthogonal comparisons make life quite simple. In discussing multiple regression, the special case of uncorrelated predictors was mentioned. Under these circumstances the standard score regression weights were the same as the correlations with the criterion because $\mathrm{R}_{xx} = \mathrm{I}$, so $\mathrm{R}_{xx}^{-1}\,\mathbf{r}_{xy} = \mathbf{r}_{xy}$. This means that $R^2 = \Sigma_j r_{jy}^2$ and therefore

$$\hat{\sigma}_e^2 = \frac{n-1}{n-p}\, s_y^2\Big(1 - \sum_j r_{jy}^2\Big) \tag{10.31}$$

The difference in Equation 9.14 used to test the significance of a regression coefficient becomes r_{jy}^2:

$$R_p^2 - R_{p-1}^2 = r_{jy}^2$$

Here, $p = g - 1$ and j refers to one of the contrasts. Similarly, the contribution of a set of contrasts (see Equations 9.35 and 9.36) can be tested very simply by summing the squared correlations of the corresponding orthogonal contrasts, which are used in the numerator of Equation 9.36. The analysis can also be translated rather simply into one involving sums of squares using Equation 9.35. We will not go into the details here, however.

Thus, the use of orthogonal contrasts on equal-sized groups results in columns of X that have means of zero and zero intercorrelations. If any of these conditions fails, then the correlations are not zero except in very special cases. Speaking practically, one cannot test the contribution of a contrast—the amount of variance it uniquely contributes—in a simple way unless one is doing orthogonal contrasts on equal-sized groups. This accounts for a large part of the controversy that has surrounded methods for testing the significance of effects in factorial designs involving unequal groups.

Orthogonal Contrasts Expressed in Terms of Means

The regression coefficients for orthogonal contrasts are very readily available. Equation 10.26 demonstrated that the regression coefficients for any transformed dummy codes are the inverse transform of the group means. The inverse of an orthogonal transformation is directly related to the transformation itself. Since $T'T = D_T$, where D_T is a diagonal, $D_T^{-1}T'T = I$. According to the definition of an inverse, this means that $D_T^{-1}T' = T^{-1}$. Therefore, $T^{-1}\bar{\mathbf{y}}_{\cdot h}$ has $g - 1$ elements:

$$b_c = \sum_h t_{hc}\bar{y}_{\cdot h} \Big/ \sum_h t_{hc}^2, \qquad (10.32)$$

and,

$$b_0 = \sum_h \bar{y}_{\cdot h} \Big/ g. \qquad (10.33)$$

since the last column of T is always a column of unities. Since Equation 10.26 holds under all conditions, these equations also hold very generally.

The covariance between the contrast variables and y can be obtained quite simply in the orthogonal contrast, equal-group case. They are

$$\mathbf{s}_{cy} = \frac{n_h}{n-1} T'\bar{\mathbf{y}}_{\cdot h} \qquad (10.34)$$

In scalar notation, this becomes

$$s_{cy} = \frac{n_h}{n-1} \sum_h t_{hc}\bar{y}_{\cdot h} \qquad (10.35)$$

Since the variance of a contrast variable x_c is $\Sigma_h n_h t_{hc}^2/(n-1)$, the correlations of the contrast variables with y are readily obtained from Equations 10.34 or 10.35. We know that the formula for any regression coefficient is $\mathbf{b} = S_{xx}^{-1}\mathbf{s}_{xy}$. In scalar terms, the elements of $\mathbf{b}$ can be extremely complex to express. Here, however, b_c and s_{cy} are identical except for the denominator in the former. The difference is that in the present case, S_{xx} is diagonal. The last column of T is unities, and all the other columns sum to zero. As a result, the first $g - 1$ columns of X_+ also sum to zero; thus they are deviation scores. Now, $X_+'X_+$ is diagonal, because while

$$X_+'X_+ = T'X_d'X_dT = T'D_{n_h}T$$

here D_{n_h} is a scalar that can be factored out as $n_hT'T = n_hD_t$. Therefore, $X_+'X_+$ is diagonal, meaning that the different x_c are uncorrelated.

If desired, we can drop the last column of X_+ and simply call the remainder a deviation score matrix X from which we can calculate a covariance matrix. This is simple because all the off-diagonal terms are zero, and because the variances s_c^2 will be

$$s_c^2 = \frac{n_h}{n-1}\Sigma t_{hc}^2 \tag{10.36}$$

Here, S_{xx}^{-1} is a diagonal matrix with the reciprocals of these variances, and the expression $S_{xx}^{-1}\mathbf{s}_{xy}$ reduces to dividing each covariance in Equation 10.35 by the corresponding variance in Equation 10.36. Canceling out like terms in numerator and denominator and remembering that $gn_h = n$, gives Equation 10.32 as b_c. So it all fits together.

We already know how to find the sum of squares for "regression," ss_b, as the sum of squares between groups. We can also find it as $\mathbf{b}'\mathbf{s}_{xy}$, which is also

$$ss_b = \sum_c\left(\sum_h t_{hc}\bar{y}_{\cdot h}\right)^2 \Big/ \sum_h t_{hc}^2 \tag{10.37}$$

These relationships are illustrated in Exhibit 10-4.

An important consequence of this simplicity lies in the testing of the significance of regression coefficients and groups of regression coefficients. Remember that in multiple regression these tests are conducted by "leaving out" those variables from the regression equation and observing the amount of variance lost. With uncorrelated predictors, such as orthogonal comparisons based on equal groups, the variance due to a predictor c is just the product of b_c and s_{cy}. These are the same except for the denominator. Then the test of a contrast, which is the same as the test of a regression coefficient, becomes

$$F = \frac{n_h(\Sigma t_{hc}\bar{y}_{\cdot h})^2}{ms_w\Sigma t_{hc}^2} \tag{10.38}$$

This equation tests the significance of contrast c; it is numerically identical to the F that would result if contrast variables were literally constructed and the regression coefficients were tested for significance by Equation 9.33. It has $n - g$ *df*, and ordinarily the alpha level is treated in a parameterwise (or contrastwise) fashion.

In a similar way, several contrasts could be tested simultaneously, summing the $n_h(\Sigma t_{hn}\bar{y}_{\cdot h})^2/\Sigma t_{hc}^2$ terms corresponding to each, and dividing by m_{sw}. The resulting F is identical to Equation 9.33, where q represents the number of contrasts whose effect is being pooled. One may well wonder why this is necessary, since the clarity of interpretation of separate contrasts would be lost. However, this is just

Group

(a)

$$
\begin{matrix} 1 \\ 1 \\ 2 \\ 2 \\ 3 \\ 3 \\ 4 \\ 4 \end{matrix}
\quad
\underset{\mathbf{X}_d}{\begin{bmatrix} 1 & 0 & 0 & 0 \\ 1 & 0 & 0 & 0 \\ 0 & 1 & 0 & 0 \\ 0 & 1 & 0 & 0 \\ 0 & 0 & 1 & 0 \\ 0 & 0 & 1 & 0 \\ 0 & 0 & 0 & 1 \\ 0 & 0 & 0 & 1 \end{bmatrix}}
\underset{\mathbf{T}}{\begin{bmatrix} 1 & 1 & 1 & 1 \\ 1 & 1 & -1 & 1 \\ 1 & -2 & 0 & 1 \\ -3 & 0 & 0 & 1 \end{bmatrix}}
=
\underset{\mathbf{X}_+}{\begin{bmatrix} 1 & 1 & 1 & 1 \\ 1 & 1 & 1 & 1 \\ 1 & 1 & -1 & 1 \\ 1 & 1 & -1 & 1 \\ 1 & -2 & 0 & 1 \\ 1 & -2 & 0 & 1 \\ -3 & 0 & 0 & 1 \\ -3 & 0 & 0 & 1 \end{bmatrix}}
\quad
\underset{\mathbf{y}}{\begin{bmatrix} 1 \\ 3 \\ 4 \\ 8 \\ 5 \\ 7 \\ 2 \\ 6 \end{bmatrix}}
$$

(b)

$$
\underset{(\mathbf{X}_+'\mathbf{X}_+)^{-1}}{\begin{bmatrix} 24 & 0 & 0 & 0 \\ 0 & 12 & 0 & 0 \\ 0 & 0 & 4 & 0 \\ 0 & 0 & 0 & 8 \end{bmatrix}^{-1}}
\underset{\mathbf{X}_+'\mathbf{y}}{\begin{bmatrix} 4 \\ -8 \\ -8 \\ 36 \end{bmatrix}}
=
\underset{\mathbf{b}_+}{\begin{bmatrix} .167 \\ -.667 \\ -2.000 \\ 4.500 \end{bmatrix}}
$$

(c)

$$
\underset{\hat{\mathbf{y}} = \mathbf{X}_+\mathbf{b}_+}{\begin{bmatrix} 2.0 \\ 2.0 \\ 6.0 \\ 6.0 \\ 6.0 \\ 6.0 \\ 4.0 \\ 4.0 \end{bmatrix}}
\quad
\underset{\overline{\mathbf{y}}_{\cdot h}}{\begin{bmatrix} 2.0 \\ 6.0 \\ 6.0 \\ 4.0 \end{bmatrix}}
\quad
\underset{\mathbf{T}'\mathbf{T} = \mathbf{D}_\mathrm{T}}{\begin{bmatrix} 12 & 0 & 0 & 0 \\ 0 & 6 & 0 & 0 \\ 0 & 0 & 2 & 0 \\ 0 & 0 & 0 & 4 \end{bmatrix}}
\quad
\underset{\mathbf{T}^{-1} = \mathbf{D}_\mathrm{T}^{-1}\mathbf{T}'}{\begin{bmatrix} .0833 & .0833 & .0833 & .2500 \\ .1677 & .1677 & -.3333 & 0 \\ .5000 & -.5000 & 0 & 0 \\ .2500 & .2500 & .2500 & .2500 \end{bmatrix}}
\quad
\underset{\mathbf{b}_+ = \mathbf{T}^{-1}\overline{\mathbf{y}}_{\cdot h}}{\begin{bmatrix} .167 \\ -.667 \\ -2.000 \\ 4.500 \end{bmatrix}}
$$

(d) $s_y^2 = \Sigma(y - \overline{y})^2/(n - 1) = 6.0$

$\Sigma(y - \hat{y})^2 = \Sigma(y_{ih} - \overline{y}_{\cdot h})^2 = 20.0 = ss_{dev}$

$\Sigma(\hat{y} - \overline{y})^2 = 22.0 = ss_{reg}$

$s_{reg}^2 = 22.0/7 = 3.143$

$R^2 = s_{reg}^2/s_y^2 = .524$

Exhibit 10-4 Multiple regression on orthogonal contrast variables. Section (a) shows the dummy matrix X_d for four groups, each with two observations and an orthogonal contrast transformation T; this example shows nested contrasts. $\mathrm{X}_d\mathrm{T}$ gives X_+; also shown is y. Part (b) shows the standard computation of $\mathbf{b}_+ = (\mathrm{X}_+'\mathrm{X}_+)^{-1}\mathrm{X}_+'\mathbf{y}$. Part (c) gives first $\hat{\mathbf{y}}$ and then the $\overline{\mathbf{y}}_{\cdot h}$ for comparison; then it shows the computation of first T^{-1} and then b_+ via $b_+ = \mathrm{T}^{-1}\overline{\mathbf{y}}_{\cdot h}$. Part (d) shows the computation of s_y^2, ss_{reg}, and ss_{dev} by comparing $\hat{\mathbf{y}}$ to $\mathbf{y}$. Part (e) computes $\mathbf{b}$ from the variance-covariance matrix of the contrast variables and their covariances with $\mathbf{y}$. The coefficients are the same as those in part (b). The R^2 is found to be .524, the same as from part (d). Part (f) repeats the process of part (e) using correlation matrices.

(e) $$\begin{bmatrix} 3.43 & 0 & 0 \\ 0 & 1.71 & 0 \\ 0 & 0 & .571 \end{bmatrix}^{-1} \begin{bmatrix} .571 \\ -1.143 \\ -1.143 \end{bmatrix} = \begin{bmatrix} .167 \\ -.667 \\ -2.000 \end{bmatrix}$$

$$\mathrm{S}_{xx}^{-1} \qquad \mathbf{s}_{xy} = \mathbf{b}$$

$$\mathbf{b}'\mathbf{s}_{xy} = s_{reg}^2 = 3.143 \qquad R^2 = s_{reg}^2/s_y^2 = .524$$

(f) $$\begin{bmatrix} 1.00 & 0 & 0 \\ 0 & 1.00 & 0 \\ 0 & 0 & 1.00 \end{bmatrix} \begin{bmatrix} .126 \\ -.356 \\ -.617 \end{bmatrix} = \begin{bmatrix} .126 \\ -.356 \\ -.617 \end{bmatrix}$$

$$\mathrm{R}_{x\dot{x}} \qquad \mathbf{r}_{xy} = \mathbf{b}^*$$

$$R^2 = \mathbf{b}^{*\prime}\mathbf{r}_{xy} = 5.24$$

Exhibit 10-4 (continued)

what one is doing when a multilevel effect is tested in a factorial design. We will see how this is done later when factorial designs are discussed in more detail.

It may be worthwhile to review the different aspects discussed so far in the case of planned contrasts. If there are $g - 1$ contrasts, they may be expressed as a transformation matrix T, $\mathrm{X}_d\mathrm{T} = \mathrm{X}_+$. Then the regression coefficients are obtainable as $\mathrm{T}^{-1}\bar{\mathbf{y}}_{.h}$ provided T^{-1} exists, that is, the contrasts are not linearly dependent. The significance of a contrast can be tested as the significance of the corresponding regression coefficient. In the special case of orthogonal contrasts on equal-sized groups, this test is particularly simple since the regression weights are obtainable directly from the contrasts, and computing the variance due to a contrast need not involve possible correlations with other variables. Matters become considerably more complicated when the group sizes are unequal or the contrasts nonorthogonal.

Reduced Rank Cases

Sometimes an investigator may not have $g - 1$ contrasts that are interesting. What then? Considered as a regression problem, this should pose no particular difficulties. There will still be an X_+ equal to $\mathrm{X}_d\mathrm{T}$, and we could still find $(\mathrm{X}_+'\mathrm{X}_+)^{-1}\mathrm{X}_+'\mathbf{y} = b_+$. The difference is that we no longer have the shortcut $\mathbf{b}_+ = \mathrm{T}^{-1}\bar{\mathbf{y}}_{.h}$, because there is no T^{-1}. Instead, we have to solve

$$(\mathrm{T}_-'\mathrm{D}_{n_h}\mathrm{T}_-)^{-1}\mathrm{T}_-'\bar{y}_{.h} = b_+ \tag{10.39}$$

where T_- is a transformation with fewer columns than rows without any shortcuts except in the orthogonal equal group case. In addition, only part of the ss_b will be accounted for, because $\hat{y}_{ih}$ will not be $\bar{y}_{.h}$. Instead, it will be the mean *predicted* on the basis of the reduced set of contrasts, $\hat{\bar{y}}_{.h} = \Sigma_c b_c t_{hc}$. There will be a sum of squares for predictions, ss_{reg}, $\Sigma_h n_h(\hat{\bar{y}}_{.h} - \bar{y}_{..})^2$, and a sum of squares for deviations, which will consist of ss_w plus whatever variation in means is not accounted for by the contrasts, $\Sigma_n n_h(\hat{\bar{y}}_h - \bar{y}_{.h})^2 = ss_b - ss_{reg}$. The total sum of squares thus breaks down as

$$ss_t = ss_{reg} + (ss_b - ss_{reg}) + ss_w \qquad (10.40)$$

If we really feel beforehand that the contrasts used should account for the variation in means except for random deviations, the appropriate omnibus test would be

$$F = \frac{ss_{reg}/m}{(ss_t - ss_{reg})/(n - m - 1)} \qquad (10.41)$$

where there are m main contrasts. Note that the assumption that the deviations $\hat{\bar{y}}_{\cdot h} - \bar{y}_{\cdot h}$ are random puts that variance in the denominator. If this F leads to rejection, it means that significant variance is attributable to the m contrasts taken as a whole. They could subsequently be tested individually either as contrasts on the means (that is, as correlations), or as regression coefficients, depending on whether one wishes to partial out all other contrasts.

If H_0 is rejected, one could then test the assumption that all other variation in means are random:

$$F = \frac{(ss_b - ss_{reg})/(g - 1 - m)}{ss_w/(n - g)} \qquad (10.42)$$

Modern computer programs make it relatively easy to carry out such tests.

Finding the sum of squares that corresponds to a reduced tank regression is simplified in the orthogonal equal groups case because the contrast variables remain uncorrelated. By summing the elements of Equation 10.37 that correspond to the contrasts used, we obtain the ss_{reg} of Equation 10.41 and easily make the omnibus test. The residual between-groups can be tested using Equation 10.42.

Summary of Contrast Testing

There are perhaps half a dozen major things that an investigator routinely wants to know in a multiple regression analysis. For example, there is the simple correlation of each variable with the criterion from which the amount or proportion of variance that a variable is sufficient to predict can be determined. There are also the multiple correlation, its significance, the raw and standard score regression weights, and their individual significances. Finally, there is the significance of individual subsets of variables. This may be tested in one of two ways depending on the particular investigative strategy that seems appropriate. We could use either a hierarchical strategy, or find the amount of variance lost when the set is removed from the regression equation. On occasion, there may be other issues, but these are the most typically salient.

These quantities are all easy to find in the case of orthogonal comparisons on equal-sized groups, as we have seen. The simple correlations are fairly easy to obtain from the group means, as are the contrasts that define the weights. Due to orthogonality, the multiple correlation can be obtained by simply squaring and summing the simple correlations. The overall significance can be tested quite directly by comparing the between mean square to the within mean square. The significance of individual regression weights is tested by means of the F test for the contrast, comparing that weighted combination of means to the within mean square. The significance of groups of variables can be tested by squaring and summing the amounts of variance attributable to them, or their correlations. Then the amount of variance is tested against ms_w if the contributions of all other variables are being removed (a "these variables last" test) or, if a hierarchical approach is being taken, by pooling the variance due to the contrasts lower in the hierarchy with ss_w. All of these procedures are straightforward in the orthogonal equal groups case. This is why they have been cherished by investigators over the years, particularly in the somewhat-disguised form of factorial and related designs.

Some things remain simple when the groups are unequal. The weights are always obtainable from Equation 10.32, and the overall F test remains the same. The correlations can be obtained fairly simply provided one corrects for the possible nonzero mean of the contrast variables. Assessing the significance of individual variables or sets of variables, however, becomes considerably more complicated.

Testing the Contrasts Directly

In the preceding discussion of the nonorthogonal case, two different ways of testing a contrast were mentioned. That is, the importance of a predictor variable can be assessed by testing either its simple correlation or its regression coefficient. In traditional analysis of variance, there is a philosophy that may yield a different answer in the nonorthogonal case. This philosophy focuses on contrasts involving the *means,* avoiding any consideration of contrast *variables.* In essence, it considers any hypothesis about means separately from every other hypothesis, whereas testing the regression coefficient for a contrast variable takes into consideration the fact that the corresponding contrast variable may be correlated with other contrast variables.

Based on this philosophy, we test whatever hypotheses H_0: $\mathbf{c}'\boldsymbol{\mu} = 0$ come to mind (where $\mathbf{c}$ is a vector of coefficients defining a contrast and $\boldsymbol{\mu}$ is a vector of group means), using the well-understood sampling behavior of means. This is based on the knowledge that the expected value of the sample mean is the population mean: $\mathrm{E}(\bar{y}_{.h}) = \mu_h$, and that the sampling variance of a mean is the variance of the variable divided by the sample size: $\sigma^2_{\bar{y}.h} = \sigma^2_e/n_h$. Then the expected value of a combination of sample means weighted by the coefficients of a contrast is $\mathrm{E}(\mathbf{c}'\bar{\mathbf{y}}_{.h}) = \mathbf{c}'\boldsymbol{\mu}$, and the variance of that contrast is

$$\sum_h c_h^2\sigma_e^2/n_h = \sigma_c^2 \tag{10.43}$$

The ms_w can be used to estimate σ_e^2, and any hypothesis about a contrast can be tested using the two sample values.

If the hypothesis has a priori status and we wish to take a parameterwise stance (here, contrastwise corresponds to parameterwise) then the test of any linear combination of means is

$$\frac{\mathbf{c}'\bar{\mathbf{y}}_{.h}}{ms_w\Sigma c_h^2/n_h} = F \tag{10.44}$$

where the F has one and $n - g$ df. If the test is of a post hoc nature, then remember that the criterion must be adjusted. Here, there are a number of approaches. For example, we could use a Bonferroni correction on the nominal alpha level, correcting by a factor equal to the number of *possible* tests. If we are only interested in comparing groups in pairs, then the criterion of Tukey's "honest significant difference" (hsd) can be used, as explained in most anova texts. Alternatively, if any combination of groups could be considered, and the tests are post hoc (that is, based on differences observed in the data), then the Scheffé procedure is appropriate, as it was in the case of regression coefficients. This simply means here that the critical criterion F to be used is $(g - 1)F_{g-1,n-g}$, not $F_{1,n-g}$. This allows us to test any conceivable combination of groups and still have a probability of α or less of making any Type I errors when all H_0 are true.

In the orthogonal equal groups case, Equation 10.44 yields the same answers as Equation 10.41. In other cases, the similarity of the answers will depend on how unequal the group sizes are and/or on how nonorthogonal the contrasts are.

The above procedure can be used to construct confidence intervals for the contrasts. One simply forms $\mathbf{c}'\bar{\mathbf{y}}_{.h} \pm (ms_w)(\Sigma_h c_h^2/n_h)F$ using whichever F is appropriate for the stance being taken toward Type I error probabilities.

The difference between the F test of Equation 10.44 and testing the regression coefficient for a contrast variable is partly one of semantics. However, there can also be a real distinction, depending on the context. For example we could define Equation 10.44 as a contrast on means. We could use exactly the same coefficients to define a contrast variable. The variable's regression coefficient would also be a function of the group means, but not necessarily the same qualitative function unless the *other* contrasts are orthogonal to it. That is, the regression coefficient depends on the other contrasts being studied. When we evaluate the significance of that coefficient, we are asking, as in regression, Does this contrast represent a real difference that is separate from the *other* differences being evaluated? On the other hand, Equation 10.44 does not depend on the other contrasts being tested.

Heterogeneity of Variance

The equation for sampling variance of a contrast on means, (Equation 10.44) is correct when the within-group variances are the same in all groups. There are a variety of situations in which this will not be the case, however. For example, for many response scales, larger means tend to go with larger variances. Also, some treatments may well have an effect on variability as well as, or instead of, an effect on means. Where possible, the dependent variable can be transformed by some function that removes the association between means and variances, such as log y, $y^{1/2}$, $1/y$, or arcsine y. If any y is zero, then the transformation is indeterminate in the case of log y or $1/y$. Adding a y_0 to all the y_i circumvents the problem, where y_0 is between the lowest observed y and zero.

If transformation is ineffective or undesirable for interpretive reasons, then it becomes necessary to correct for the heterogeneity of variance if the probabilities are to be preserved. The true sampling variance of a combination of means is

$$\sigma_c^2 = \sum_h c_h^2 \sigma_{e_h}^2 / n_h \tag{10.45}$$

An unbiased estimate of σ_c^2 can be found by substituting the individual cell variances for the $\sigma_{e_h}^2$. This is particularly important in the case of unequal groups.

The F tests of regression coefficients for contrasts should use this estimate rather than the pooled-within-groups mean square wherever heterogeneity is suspected.

This procedure is particularly important when some groups in a contrast are receiving zero weights. If their variances are different from the groups involved in the contrast, then they are contributing misleading information about the sampling instability of the contrast.

The within-group variances in the sample never will be the same. How do we know whether these differences reflect true underlying differences? We could perform a statistical test of homogeneity of variance, but this turns out to be counterproductive since it is sensitive to the sample size in the usual way. The hypothesis of homogeneity tends to be accepted when the n_h are small, leading one to pool the within-cells variances into one ms_w, even when the variances are unequal in the population and perhaps quite different. The small-sample case is exactly the one where heterogeneity is most disruptive, so one tends to pool in exactly those cases where one should not.

Two strategies can be suggested. The first strategy is to never pool. This sacrifices a little power in those cases where variances are in fact equal, but that is a small price to pay. The other strategy is to perform both the pooled and unpooled analyses. If they are very similar in their results—that is, in terms of the F ratios for the contrasts—then pooling is probably justified.

With heterogeneity of variance, there is not only a problem with what estimate to use for the error variance, but with what denominator degrees of freedom to use for the critical value of F. This degrees of freedom figure measures the accuracy with which the variance has been estimated. With homogeneity of variance, we can count all the degrees of freedom in all the cells, but with heterogeneity, the degrees of freedom from each cell should be weighted differently. Exactly how to do this is impossible to know since it depends on the true cell variances. However, we do know that the appropriate number of degrees of freedom is somewhere between the df for the smallest group and $n - g$. We could test our obtained F against the value tabled for these numbers of denominator degrees of freedom. Using the size of the smallest group can be highly conservative, of course. But existing formulas for estimating the appropriate number of degrees of freedom are complex, and their validity is somewhat controversial. Unless the smallest group is quite small—say, fewer than 10 observations, this choice of degrees of freedom is rather unimportant, as can be seen by inspecting the F table. Critical values of F change rather slowly as a function of degrees of freedom beyond that point. Wilcox (1986) provides a thorough discussion of these and other issues.

SUMMARY

This chapter first considered predicting y from a matrix of binary dummy codes X_d representing group membership in one of g groups. We found that $\mathbf{v} = \bar{\mathbf{y}}_{.h}$, the vector of group means, gives the best prediction, and that $\hat{y} = \mathrm{X}_d\mathbf{v}$ indicates that the "predicted" scores are the group means. Therefore, the sum of squared deviations from the predictions is the same as the anova within-groups sum of squares, and the sum of squared "predictions" is the sum of squares between groups.

We cannot input the dummy code matrix and perform multiple regression on the g dummy variables because the deviation scores will be linearly dependent.

However, if one dummy variable is deleted, the linear dependence is removed. We can therefore think of an X_+ matrix where the constant column replaces one of the dummy codes. Then the regression weights $\mathbf{b}_+ = (X_+'X_+)^{-1}X_+'\mathbf{y}$ will consist of $g - 1$ "regression weights," each of which is the difference between the mean of the corresponding group and the mean of the deleted group. The "intercept" is the mean of the deleted group. The predicted values of y are still the group means, and so the sums of squares remain the between-group *ss* and within-groups *ss*.

Deleting one dummy code and replacing it with the constant column was shown to be a transformation of the X_d matrix. This leads to an important general principle. Any set of group contrasts that can be regarded as a nonsingular transformation of the dummy codes corresponds to a vector of regression weights that is the inverse transformation of the group means:

$$\mathbf{b}_+ = T^{-1}\bar{\mathbf{y}}_{\cdot h}$$

The T matrix can be composed as a set of $g - 1$ columns of comparisons of means, plus the constant column, where each comparison represents a null hypothesis about the means. These comparisons can take nearly any form, but some are quite commonly used. Sets of *orthogonal contrasts* are particularly useful, including orthogonal polynomials, nested or Helmert contrasts, and the factorial design. In orthogonal contrasts, T^{-1} is easy to find. These procedures were employed before computerized analyses were common. If the groups are equal in size, it is easy to test the significance of the regression coefficients that correspond to the comparisons. However, this simplicity breaks down when groups are unequal.

It is possible with any set of contrasts to employ the same kinds of analyses and tests used in multiple regression. One can test the significance of individual contrasts, or groups of contrasts, and the like. A careful distinction needs to be made, however, between testing the *correlation* of the variable representing a contrast and testing its regression coefficient, even though these may look very similar.

One can formulate contrasts as comparisons of means without applying any regression concepts, simply by using direct formulas for the sampling variances of means. The difference between this and regression analysis is that in regression analysis, we are taking into account the other comparisons being made in the analysis.

In all comparisons of means, the sampling variance of the contrast depends on the within-group variances, the group sizes, and the squared weights as demonstrated in Equation 10.45. Heterogeneity of variance needs consideration, particularly where the group sizes are unequal.

COMPUTER APPLICATIONS

Analyses of variance can be done using a multiple-regression program. However, these analyses are cumbersome and time-consuming to perform due to the necessity of generating dummy codes and/or contrast variables. More sophisticated computer procedures for performing anova are also available, and can be used to perform the analyses described in this chapter. To avoid repetition, however, such implementation will be discussed in the next chapter.

PROBLEMS

1. Perform the one-way anova using the following data:

Group C1	Group C2	Group E1
1	2	8
4	0	6
1		10

2. Make the complete dummy code matrix X_d for the data in Problem 1. Find $\mathbf{v} = (X_d'X_d)^{-1}X_d'\mathbf{y}$ and $\hat{\mathbf{y}} = X_d\mathbf{v}$. Then find the sums of squares for "regression" and for deviations.

3. Replace the third column of X_d by the constant vector to make X_+ and find $\mathbf{b}_+ = (X_+'X_+)^{-1}X_+'\mathbf{y}$. Note that this may be done without actually finding the inverse of $X_+'X_+$. Find $\hat{\mathbf{y}} = X_+\hat{\mathbf{b}}_+$ and verify that it is the same as in Problem 2.

4. What do the regression coefficients for Problem 3 reflect? What null hypothesis would their significance reflect?

5. Write out the score matrix corresponding to the orthogonal transformation of X_d that gives a comparison of the two controls to the experimental as x_1 and a comparison of the controls to each other as x_2. Make the corresponding transformation of v yield the regression weights.

ANSWERS

1. $ss_w = 16; ss_b = 78$ $\quad \bar{x}_{.1} = 2.0; \bar{x}_{.2} = 1.0; \bar{x}_{.3} = 8.0; \bar{x}_{..} = 4.0$ $\quad F = \dfrac{78/2}{16/5} = 12.19.$

2. $$X_d' = \begin{bmatrix} 1 & 1 & 1 & 0 & 0 & 0 & 0 & 0 \\ 0 & 0 & 0 & 1 & 1 & 0 & 0 & 0 \\ 0 & 0 & 0 & 0 & 0 & 1 & 1 & 1 \end{bmatrix} \qquad (X_d'X_d)^{-1} = \begin{bmatrix} 1/3 & 0 & 0 \\ 0 & 1/2 & 0 \\ 0 & 0 & 1/3 \end{bmatrix}$$

$$\mathbf{v}' = [2.0 \quad 1.0 \quad 8.0] \qquad \hat{\mathbf{y}}' = [2.0 \quad 2.0 \quad 2.0 \quad 1.0 \quad 1.0 \quad 8.0 \quad 8.0 \quad 8.0]$$

3. $$T_d = \begin{bmatrix} 1 & 0 & 1 \\ 0 & 1 & 1 \\ 0 & 0 & 1 \end{bmatrix} \qquad T_d^{-1} = \begin{bmatrix} 1 & 0 & -1 \\ 0 & 1 & -1 \\ 0 & 0 & 1 \end{bmatrix} \qquad \mathbf{b} = \begin{bmatrix} -6.0 \\ -7.0 \\ 8.0 \end{bmatrix}$$

$$\hat{\mathbf{y}}' = [-6.0 + 8.0 \quad -6.0 + 8.0 \quad -6.0 + 8.0 \quad -7.0 + 8.0 \quad -7.0 + 8.0 \quad 8.0 \quad 8.0 \quad 8.0]$$

4. They reflect $\bar{y}_{C1} - \bar{y}_{E1}$ and $\bar{y}_{C2} - \bar{y}_{E2}$, respectively. H_0: $\mu_{\bar{E}1} = \mu_{C1}$, and H_0: $\mu_{E1} = \mu_{C2}$.

5. $$X_c' = \begin{bmatrix} 1 & 1 & 1 & 1 & 1 & -2 & -2 & -2 \\ 1 & 1 & -1 & -1 & -1 & 0 & 0 & 0 \\ 1 & 1 & 1 & 1 & 1 & 1 & 1 & 1 \end{bmatrix} \qquad T_c = \begin{bmatrix} 1 & 1 & 1 \\ 1 & -1 & 1 \\ -2 & 0 & 1 \end{bmatrix}$$

(Signs may be reflected in X_c and T_c.)

$$\mathbf{T}_c^{-1} = \begin{bmatrix} 1/6 & 1/6 & -1/3 \\ 1/2 & -1/2 & 0 \\ 1/3 & 1/3 & 1/3 \end{bmatrix} \qquad \mathbf{b}_c = \begin{bmatrix} -13/6 \\ 1/2 \\ 11/3 \end{bmatrix}$$

READINGS

Kempthorne, O. (1950). *Design and analysis of experiments.* New York: Wiley, pp. 38–67.

Kerlinger, F. N., and E. J. Pedhazur. (1973). *Multiple regression in behavioral research.* New York: Holt, Rinehart, and Winston, pp. 107–153.

Myers, J. L. (1979). *Fundamentals of experimental design.* Boston: Allyn & Bacon, pp. 378–389.

Tatsuoka, M. M. (1975). *The general linear model; a "new trend" analysis of variance.* Champaign, IL: Institute for Ability and Personality Testing, pp. 1–64.

Wilcox, R. (in press). New designs in the analysis of variance. *Annual Review of Psychology.*

Winer, B. J. (1971). *Statistical principles in experimental design* (2nd ed.). New York: McGraw-Hill.

11

Factorial Analysis of Variance and Regression

FACTORIAL DESIGNS

Fully Crossed Variables

Multiple regression is concerned with investigating the effect of a number of independent variables on a dependent variable. More often than not, this is also the case in analysis of variance applications. The most common use of multiple independent variables takes the form of a factorial design. Here, several different variables or *factors* have several different values or *levels*, and the observations of a dependent variable take place under all possible combinations of levels of the different factors. In factorial design, it is common to label the factors with capital letters such as A, B, and soon, whereas the levels within a factor are defined by subscripted lowercase letters such as a_1, a_2, and so on. Particular combinations of conditions are labeled as combinations of these lowercase letters:

$$a_1b_1, a_1b_2, \ldots, a_2b_1, a_2b_2, \ldots,$$

For example, in a study of the effect of list length on memory, we might also be interested in the effect of different amounts of previous exposure to the lists, and might allow subjects one, two, or three minutes of study of the lists before the probe words are presented. In a factorial design, for each of the three levels of list length (A), three levels of study time (B) would be used, making nine conditions in all. Alternatively, one might use a characteristic of the *subjects* as a factor in the design, for example, the handedness of the subject as B. Then we would have left-handed subjects using two-, four-, or six-word lists (a_1b_1, a_2b_1, a_3b_1), and the same for right-handed subjects (a_1b_2, a_2b_2, a_3b_2), a 2×3 design. We could also combine all three factors, as shown in Table 11-1. Here, the upper part of the table has three levels of list length and two levels of handedness. The lower part illustrates the addition of a third factor, study time, with three levels. Each box represents a condition of the experiment.

Many other types of experiments follow this form: this is by far the most common experimental design in psychology. Suppose we are studying types of psychotherapy and their effects on different types of anxiety. There might be four different types of patients and three different therapies. Some of each type of patient are treated by each kind of therapy.

In a social psychology experiment, different types of persuasive messages (one factor) might be presented under different types of conditions (the other). A very common type of study uses age as one factor of the experiment. For example, there are different conditions, and people of each of several age groups are observed under each of the possible conditions. The crucial characteristic of a factorial design study is that all levels of one variable are observed under all levels of the other.

This concept can be extended to include more than two factors. For example, both handedness and study time could be included in the list length study. The design is still factorial as long as all possible levels of each factor exist under all combinations of levels of the others. It is convenient to illustrate a factorial design in terms of boxes, as in Table 11-1.

If there are g_A levels of one factor and g_B levels of a second, then the total

Table 11-1 Schematic Diagram of Factorial Designs

		Handedness	
		Left	Right
List Length	2 Words		
	4 Words		
	6 Words		

		Study Time					
		5 seconds		*10 seconds*		*15 seconds*	
		Left	*Right*	*Left*	*Right*	*Left*	*Right*
List Length	2 Words						
	4 Words						
	6 Words						

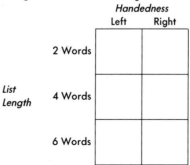

number of groups is $g_A \times g_B = g$, and the between-groups sum of squares ss_b is broken down into some specific components in a factorial design. In a two factor design, these components are one *main effect* for each factor and one *interaction* effect. The main effect for one factor is the overall effect it has across the levels on the other factor; the interaction effect accounts for the way the effect of a given factor differs as a function of the specific level of the other factor.

The Factorial Model

In the usual analysis of a factorial design, we assume that a given observation, y_{ijk}, the ith observation at level j of factor A and level k of factor B, can be described as follows:

$$y_{ijk} = \mu + \alpha_j + \beta_k + \alpha\beta_{jk} + e_{ijk} \tag{11.1}$$

where

μ is the overall population mean

α_j is the main effect for level j of factor A

β_k is the main effect for level k of factor B

$\alpha\beta_{jk}$ is the interaction of level j of A with level k of B

e_{ijk} is the residual or error term assumed to have a population mean of zero and a variance σ_e^2, which is the same at all levels j and k

If there is a constant number of observations, r, in each cell of the design, we can define the means of the cells and the means for the different factors as follows:

$$\bar{y}_{.jk} = \sum_i y_{ijk}/r \tag{11.2}$$

$$\bar{y}_{..k} = \sum_i \sum_j y_{ijk}/rg_A \tag{11.3}$$

$$\bar{y}_{.j.} = \sum_i \sum_k y_{ijk}/rg_B \tag{11.4}$$

$$\bar{y}_{...} = \sum_i \sum_j \sum_k y_{ijk}/rg_Ag_B \tag{11.5}$$

A tabular presentation of an example is shown in Table 11-2.

We can specify the relationship of the means to the model, provided we make some kind of restrictions on the terms in the model. These are similar to the ones placed on the α_j when the one-way models were introduced. Specifically,

$$\Sigma\alpha_j = 0 \tag{11.6}$$

Table 11-2 Observations y_{ijk}, Cell Means $\bar{y}_{ijk}$, Row Means $\bar{y}_{.j.}$, Column Means $\bar{y}_{..k}$, and Grand Mean $\bar{y}_{...}$ for a 2 × 3 Design with Four Replications per Cell

y_{111}	y_{121}	y_{131}	
y_{112}	y_{122}	y_{132}	
y_{113}	y_{123}	y_{133}	
y_{114}	y_{124}	y_{134}	
$\bar{y}_{.11}$	$\bar{y}_{.12}$	$\bar{y}_{.13}$	$\bar{y}_{.1.}$
y_{211}	y_{221}	y_{231}	
y_{212}	y_{222}	y_{232}	
y_{213}	y_{223}	y_{233}	
y_{214}	y_{224}	y_{234}	
$\bar{y}_{.21}$	$\bar{y}_{.22}$	$\bar{y}_{.23}$	$\bar{y}_{.2.}$
$\bar{y}_{..1}$	$\bar{y}_{..2}$	$\bar{y}_{..3}$	$\bar{y}_{...}$

$$\Sigma\beta_k = 0 \tag{11.7}$$

$$\sum_j \alpha\beta_{jk} = \sum_k \alpha\beta_{jk} = 0 \tag{11.8}$$

That is, the sum of the row effects is zero, the sum of the column effects is zero, and within any row or column, the sum of the interaction effects is zero. We may ask whether these restrictions are reasonable; the answer is that they are indeed reasonable, but not necessary. Some restrictions must be made, otherwise the parameters are not identifiable, but other restrictions could have been made. Indeed, it is the nature of these restrictions that leads to a certain amount of controversy in anova. We will return to this issue later. These particular restrictions make the estimation of parameters in analysis of variance as simple as possible.

If we substituted the model in the equations for the various means, these specifications would create the following relationships between the model terms and the means:

$$\overline{y}_{.jk} = \mu + \alpha_j + \beta_k + \alpha\beta_{jk} + e_{.jk} \tag{11.9}$$

$$\overline{y}_{..k} = \mu + \beta_k + e_{..k} \tag{11.10}$$

$$\overline{y}_{.j.} = \mu + \alpha_j + e_{.j.} \tag{11.11}$$

$$\overline{y}_{...} = \mu + e_{...} \tag{11.12}$$

These equations can also be used to estimate the effects, employing principles similar to those used in one-way design in Chapter 10. In effect, we replace all the terms involving mean errors by their expected values—zero—and then solve for the parameters:

$$\hat{\alpha}_j = \overline{y}_{.j.} - \overline{y}_{...} \tag{11.13}$$

$$\hat{\beta}_k = \bar{y}_{\cdot\cdot k} - \bar{y}_{\cdots} \tag{11.14}$$

$$\hat{\mu} = \bar{y}_{\cdots} \tag{11.15}$$

$$\widehat{\alpha\beta}_{jk} = \bar{y}_{\cdot jk} - \bar{y}_{\cdot j\cdot} - \bar{y}_{\cdot\cdot k} + \bar{y}_{\cdots} \tag{11.16}$$

These are also least-squares estimates of these parameters.

Sums of Squares and Tests of Hypotheses

So far, we have identified the basic definitions on which analysis of variance in the factorial case is based. Various substitutions lead to the specification of several different sums of squares and their interpretation in terms of the model. First, we define ss_w, the within-cells variation:

$$ss_w = \sum_i \sum_j \sum_k (y_{ijk} - \bar{y}_{\cdot jk})^2 \tag{11.17}$$

This ss_w is the same as that described in the one-way case in Chapter 10; it is the sum of squared deviations from the group means. The difference is that now we have an organizational structure for the group means, which leads us to designate them by means of two subscripts instead of one. Similarly, we can define ss_b:

$$ss_b = \sum_i \sum_j \sum_k (\bar{y}_{\cdot jk} - \bar{y}_{\cdots})^2 = \sum_j \sum_k r(y_{\cdot jk} - y_{\cdots})^2 \tag{11.18}$$

These two sums of squares, Equations 11.17 and 11.18, are converted into mean squares by dividing them by $n - g$ and $g - 1$, respectively. Under the null hypothesis that *all* the α's, β's, and $\alpha\beta$'s are equal to zero, the ratio of the two mean squares has the F distribution with the given degrees of freedom, because then both are estimates of the error variance σ_e^2. Thus, the value of the ratio leads to acceptance or rejection of the null hypothesis of equal means in the usual way.

This F test is not usually carried out in the case of a factorial design, although perhaps it should be in order to provide an experimentwise control of Type I error. Instead ss_b is broken down further into components which separately test the hypotheses concerning the A effect, the B effect, and the interaction effects. Algebraic manipulation of ss_b shows that it can be broken down into three additive components:

$$\begin{aligned} ss_b = {} & \underbrace{g_A r \sum_k (y_{\cdot\cdot k} - y_{\cdots})^2}_{B \text{ main effect}} + \underbrace{g_B r \sum_j (y_{\cdot j\cdot} - y_{\cdots})^2}_{A \text{ main effect}} \\ & + \underbrace{r \sum_j \sum_k (y_{\cdot jk} - y_{\cdot\cdot k} - y_{\cdot j\cdot} + y_{\cdots})^2}_{AB \text{ interaction}} \end{aligned} \tag{11.19}$$

The three components are referred to as the sum of squares for columns (ss_B, the B main effect), the sum of squares for rows (ss_A, the A main effect), and the sum of squares for interaction, respectively. Note that the multiplying factor for the *column effect* is g_A, the number of rows, and the factor for the *row effect*, g_B, is the number of columns. The constants r, $g_A r$, and $g_B r$ represent the numbers of like terms that are summed in deriving these two sums of squares. Thus we have

$$ss_b = ss_B + ss_A + ss_{AB} \tag{11.20}$$

When the expressions for the means are substituted in terms of the model elements, we can see that

$$ss_A = g_B r \Sigma(\alpha_j + e_{.j.} - e_{...})^2 = g_B r \Sigma \alpha_j^2 + g_B r \Sigma(e_{.j.} - e_{...})^2 \tag{11.21}$$

Observations

B	A = 1	A = 2	A = 3
1	7, 9	6, 10	12, 10
2	10, 8	8, 10	10, 14
3	6, 8	5, 7	15, 13
4	9, 7	14, 12	16, 14

Cell Means

B	A = 1	A = 2	A = 3	B Means
1	8.0	8.0	11.0	9.0
2	9.0	9.0	12.0	10.0
3	7.0	6.0	14.0	9.0
4	8.0	13.0	15.0	12.0
A Means	8.0	9.0	13.0	10.0

Deviations from Cell Means

B	A = 1	A = 2	A = 3
1	−1, 1	−2, 2	1, −1
2	1, −1	−1, 1	−2, 2
3	−1, 1	−1, 1	1, −1
4	1, −1	1, −1	1, −1

Deviations of Means from Grand Mean

B	A = 1	A = 2	A = 3	B Means
1	−2.0	−2.0	1.0	−1.0
2	−1.0	−1.0	2.0	0.0
3	−3.0	−4.0	4.0	−1.0
4	−2.0	3.0	5.0	2.0
A Means	−2.0	−1.0	3.0	0.0

Exhibit 11-1(a) Analysis of variance of 4 × 3 design with two replications per cell. The first four panels show the observations, the cell means, the deviations from the cell means, and the deviations of the cell and marginal means from the grand mean.

Total Sum of Squares $(7 - 10)^2 + (9 - 10)^2 + (6 - 10)^2 + \cdots + (14 - 10)^2 = 224$

Between-Groups Sum of Squares $2\,(8 - 10)^2 + 2\,(8 - 10)^2 + 2\,(11 - 10)^2 + \cdots + 2\,(15 - 10)^2 = 188$

Within-Groups Sum of Squares $(7 - 8)^2 + (9 - 8)^2 + (6 - 8)^2 + \cdots + (14 - 15)^2 = 36$

$ss_b + ss_w = ss_t \qquad 188 + 36 = 224$

A Main Effect $4 \times 2\,(8 - 10)^2 + 4 \times 2\,(9 - 10)^2 + 4 \times 2\,(13 - 10)^2 = 112$

B Main Effect $3 \times 2\,(9 - 10)^2 + 3 \times 2\,(10 - 10)^2 + 3 \times 2\,(9 - 10)^2 + 3 \times 2\,(12 - 10)^2 = 36$

AB Interaction $2\,[(8 - 10) - (9 - 10) - (8 - 10)]^2 + 2\,[(8 - 10) - (9 - 10) - (9 - 10)]^2 + \cdots + 2\,[(15 - 10) - (12 - 10) - (13 - 10)]^2 = 2\,(1)^2 + 2\,(0)^2 + \cdots + 2\,(0)^2 = 40$

$ss_A + ss_B + ss_{AB} = ss_b$: $112 + 36 + 40 = 188$

Exhibit 11-1(b) Computation of sums of squares by Equations 11.21, 11.22, and 11.23.

Source	*Sum of Squares*	*df*	*Mean Square*	*F*
A Main Effect	112	$g_A - 1 = 2$	56.0	18.67**
B Main Effect	36	$g_B - 1 = 3$	12.0	4.00*
AB Interaction	40	$(g_A - 1)(g_B - 1) = 6$	6.67	2.22
Total Between Groups	(188)	$(g - 1 = 11)$	(17.09)	(5.70)**
Within-Groups	36	$n - g = 12$	3.00	

*Reject at α less than .05.
**Reject at α less than .01.

Exhibit 11-1(c) Analysis of variance table including *F* ratios. The *A* main effect is significant at $\alpha = .01$, the *B* main effect at $\alpha = .05$, and the *AB* interaction is not significant. The total between groups is also tested to provide control of analysiswise alpha, and this too is significant at $\alpha = .01$. All the effect mean squares are tested against the within-groups mean square.

$$ss_B = g_A r \Sigma (\beta_k + e_{..k} - e_{...})^2 = g_A r \Sigma \beta_k^2 + g_A r \Sigma (e_{..k} - e_{...})^2 \tag{11.22}$$

$$ss_{AB} = r \Sigma\Sigma (\alpha\beta_{jk} + e_{.jk} - e_{.j.} - e_{..k} + e_{...})^2 = r\Sigma\Sigma\alpha\beta_{jk}^2 + r\Sigma\Sigma(e_{.jk} - e_{.j.} - e_{..k} + e_{...})^2 \tag{11.23}$$

Under the various null hypotheses concerning *A*, *B*, and interaction effects, the terms involving α_j, β_k, and $\alpha\beta_{jk}$ become zero. These sums of squares are reduced to the terms involving the mean errors such that

$$E(ss_A) = (g_A - 1)\sigma_e^2 \tag{11.24}$$

$$E(ss_B) = (g_B - 1)\sigma_e^2 \tag{11.25}$$

$$E(ss_{AB}) = (g_A - 1)(g_b - 1)\sigma_e^2 \tag{11.26}$$

When the error distributions are normal with constant variance, these quantities have the $\chi^2\sigma_e^2$ distribution, so that when they are converted to mean squares:

$$ms_A = ss_A/(g_A - 1) \tag{11.27}$$

$$ms_B = ss_B/(g_B - 1) \tag{11.28}$$

$$ms_{AB} = ss_{AB}/(g_A - 1)(g_B - 1) \tag{11.29}$$

the expected values of the mean squares is σ_e^2. Then the ratio of these mean squares to ms_w has the F distribution. Thus, the null hypotheses concerning each effect may be tested.

These tests are usually made by means of a hypothesiswise alpha level for each effect. Note that by comparing each effect to the residual, these tests are analogous to testing a regression coefficient, or group of coefficients.

These various quantities and statistics for the various mean squares are summarized in Exhibit 11-1 along with an example showing the various computations in a factorial anova.

Factorial designs readily generalize to more than two factors. Then there is one main effect for each factor, and a larger number of interaction terms. There is one *first-order* interaction for each pair of factors, *AB*, *AC*, *BC*, and so on; one second-order interaction for each triple of factors, *ABC*, *ABD*, *ACD*, and so on; and one third-order interaction for every quadruple, and so on. So, in general, the number of different types of interactions is one less than the number of factors. Although the details of the analysis for the higher levels of designs will not be discussed here, there are several good texts on analysis of variance, including Kirk (1982), Myers (1979), Winer (1971).

Design of Experiments and Interpretation of Effects

Analysis of variance was invented by R. A. Fisher in the early 1920s. His text on analysis of variance was called *Design of Experiments*, and one of anova's main contributions to research is that it has served as a useful *conceptual* framework, a way to think about research. Factorial experiments are a major feature of analysis of variance. They have been used in psychology and other experimental sciences more than any other type.

The beauty of factorial designs is the way in which they allow the experimenter to control variables and simultaneously explore the generality of their effects. Suppose in the list-length study that all we varied was list length, having different groups study lists of different lengths. We might find relatively small, and possibly confusing, differences. Someone may come along and point out that perhaps those who had the longer lists spent more time studying them and were more familiar with them, thus facilitating their recognition of the words when they were presented. If we had anticipated this possibility, we could have required the subjects to study the lists for a fixed time, thus holding the variable constant. In this case, however, someone might dismiss the results as lacking in generality; the results might not hold if another study time had been used. We might in turn dismiss this objection as nit-picking, but still feel that it would be nice to be able to answer more directly.

Suppose that on another occasion we decide to study the effect of study time itself. Perhaps we would pick a list of a certain length and have different groups study it for different times. In this case, someone may claim that *these* results are specific to the particular *length of list* that we used.

The way to satisfy both types of critics is to employ a complete factorial design. Then we would find out what the differential effects of list length are at several different study times, and vice versa. If these differences on one variable are more or less the same, regardless of the other variable, then a firm conclusion concerning the operation of the variable can be reached. Furthermore, just by looking at the cell means, we could get an overall impression of the relative effects of list length and study time on the time it takes to locate a word. On the other hand, the effects of one variable may be quite different, depending on the level of the other. This is what is meant by interaction—the characteristics associated with one variable interact with those associated with the other variable to produce an effect that is unexpected, given the overall effects that they typically have.

The various different effects are symbolized by the α's, β's, and $\alpha\beta$'s in the model. If there is no interaction—that is, if all the $\alpha\beta$'s are zero, or nearly so—then we can get an accurate estimate of the mean in any cell by adding the estimate of the row effect (Equation 11.13), the column effect (Equation 11.14), and the grand mean (Equation 11.15). Negligible interaction means there is "additivity" for the effects of the independent variables. Where interaction is present, a distinction is made between "ordinal" and "disordinal" interactions. If the interaction is ordinal, the cell means are in the same order looking across one variable, regardless of the level of the other variable, although the differences are not the same. Thus, the differences are the same *qualitatively* regardless of the other variable, but the presences of interaction would mean that their magnitude is different, depending on the other variable.

Table 11-3 illustrates several different examples. In the first section, there are no effects at all, so all the cell and marginal means are the same as the grand mean. In the second section, there is only one effect for one variable. The third section illustrates a case in which there are both main effects but no interaction, with the A effect being larger than the B. The fourth section shows interaction. The fifth section of the table gives the values for the effects that were used in constructing the fourth section; internal entries are interactions, whereas the last row and column are main effects. Note that all the effect values sum to zero.

For a slightly different slant on factorial designs, consider the situation in which one of the variables is a characteristic of the subjects involved rather than some variable that the experimenter actively manipulates. Again, the considerations of control and generality apply. Suppose there were some reason to expect that the educational level of the subjects was associated with performance on the task—for example, the response time in the list-length study.

The idea of controlling educational level has several aspects. Suppose a critic hypothesizes that the list-length effects were an accidental result of differences in the compositions of the groups, since there are more people of a higher educational level in those groups with higher means. (Perhaps this subtly suggests that we intended it to support our hypothesis spuriously.) We have two possible defenses against the argument: randomization and balancing. The randomization defense consists of using a random device to decide which group each individual person will belong to. This ensures that any differences in the groups, aside from the treatments, are differences that literally occurred by chance. The usual tests of significance, which are based on the notion of random

Table 11-3 Examples of Anova Effects

1. No Effects

		A 1	2	3	*B Effect*
B	1	10	10	10	10
	2	10	10	10	10
	3	10	10	10	10
	4	10	10	10	10
A Effect		10	10	10	10

2. A Effect Only

		A 1	2	3	*B Effect*
B	1	8	9	13	10
	2	8	9	13	10
	3	8	9	13	10
	4	8	9	13	10
A Effect		8	9	13	10

3. A and B

		A 1	2	3	*B Effect*
B	1	7	8	12	9
	2	8	9	13	10
	3	7	8	12	9
	4	10	11	15	12
A Effect		8	9	13	10

4. A, B, and AB

		A 1	2	3	*B Effect*
B	1	8	8	11	9
	2	9	9	12	10
	3	7	6	14	9
	4	8	13	15	12
A Effect		8	9	13	10

5. A, B, and AB Effect Values Underlying the Tables Above

AB Values

		A 1	2	3	*B Effect*
B	1	1	0	−1	−1
	2	1	0	−1	0
	3	0	−2	2	−1
	4	−2	2	0	2
A Effect		−2	−1	3	$10 = \mu$

sampling from a population, are virtually equivalent to corresponding tests based on randomization of assignment. Thus randomization provides protection against all possible subject variables—those that we have not considered as well as those that we have considered.

Still, randomization leaves the subject variables free to exert random effects. For this reason, they may obscure the effects that we are mainly concerned with. A way around this is to balance (equalize) the effect of the subject variable. In this example, we might divide the group into college graduates, people with some college education, and people with no college education. Within these groups we assign people to the experimental groups (often called "treatments," whether or not the experimental variables are treatments in the medical sense) at random. The subject or control variable is called a *blocking* variable, and the design is called randomized *blocks*. This method ensures that each group contains equal numbers of subjects of each kind. Thus any differences among groups on this blocking variable do not contribute in any way to the group differences. It is possible to do this with more than one variable, although it tends to become increasingly difficult as the number of variables increases, due to the likely scarcity of subjects with some specified combinations of characteristics.

If we have gone to the trouble of balancing, then it is wise to make use of it in our analysis of variance, turning a one-way analysis in which the experimental variable is the only one used, into a two-way analysis in which both the blocking variable and the experimental variable are used. The use of a blocking variable as a dimension of the design has several benefits. First of all, it reduces the within-cells variance by separating it from the main effect for the subjects variable and any interactions it may have with the experimental variable, thus increasing the power of the test of hypothesis. Even more important is the fact that two-way analysis allows us to see whether the blocking variable is in fact associated with the dependent variable. This will be revealed by its sums of squares for main effect and for interaction with the other variable. The latter may be of particular interest because the presence of interaction will indicate that the experimental variable is having a differential effect, depending on which level of the blocking variable we are dealing with—for example, that the effect of list length was different depending on the educational level of the subjects, or that the relative efficacies of different types of therapeutic treatments were different depending on the type of patient, or any one of an unlimited list of possible studies.

The number of factors can, of course, expand greatly. It is not at all uncommon to have three or four experimental factors in combination with two or more subject variables, giving a total of five or six factors. Contrary to popular belief, three- and four-factor interactions often have ready interpretations. When some of the interactions are felt to be of little interest, it is possible to select only certain combinations of the factors, thus avoiding the proliferation of cells in the design: a $2 \times 3 \times 3 \times 4 \times 4$ design has 288 cells. Advanced books on experimental design such as the texts cited earlier (see also Cochran and Cox, 1957) can be consulted on how to do this.

For our purposes, the main reason we use factorial designs is the control and generality they allow. Once we become familiar with the terminology and concepts involved, factorial anova can furnish a very effective framework for thinking about research design and the analysis of experiments. Thinking about which variables in an area are likely to have effects, which are likely to interact with each other, and how these things will manifest themselves in data, have all proven extremely useful to experimenters in a number of areas.

CONTRAST VARIABLES AND FACTORIAL DESIGNS

Preliminary Cautions

Because of its simplicity, factorial analysis of variance helps in organizing both our thoughts and our data. It may also serve to distort or obscure the truth—in the sense of the most plausible, meaningful, and parsimonious interpretation of the observations. In many respects, these distortions and obscurities may be reduced if one looks at factorial analysis of variance from the regression point of view. The analysis may then be more precisely directed toward the questions that the experimenter really has in mind. Moreover, such consideration offers the only general remedy when the number of observations in each cell are unequal.

The Contrasts in Factorial Designs

The analysis of a factorial design is another example in which the group membership variables are transformed to correspond to a particular set of contrasts that conform to the interests of the investigator. The simplest possible factorial design is the 2×2.

The main effect for variable A reflects the difference in the effect of being at level 1 compared to being at level 2. This can be shown by defining a variable such that those observations are at level 1 of A—that is, the first four scores—are 1, and the last four are -1. (Or -1 and 1; the order we choose depends on which level of A is most reasonably considered the "positive" or "high" level of the variable.) Another variable which corresponds to the B main effect may be defined by having the second variable be equal to 1 for groups a_1b_1 and a_2b_1 and -1 for a_1b_2 and a_2b_2. Finally, the interaction contrast is defined by giving a_1b_1 and a_2b_2 scores of 1 and a_1b_2 and a_2b_1 scores of -1. The fact that this does correspond to the interaction may be more obvious if we think of it as giving the "difference of the differences"—that is, $(\bar{y}_{11} - \bar{y}_{12}) - (\bar{y}_{21} - \bar{y}_{22})$. Thus we have three variables, each of which corresponds to one of the effects in the factorial design. Each is a contrast among the group means, as will be demonstrated.

This set of variables can be obtained from X_d simply by using the coefficients we used to define the corresponding contrast as the columns of the transformation. With two observations per group, and labeling the groups as a_1b_1 and so on, we have the following transformed version of X_d:

$$
\begin{matrix} a_1b_1 \\ \\ a_1b_2 \\ \\ a_2b_1 \\ \\ a_2b_2 \\ \\ \end{matrix}
\begin{bmatrix} 1 & 0 & 0 & 0 \\ 1 & 0 & 0 & 0 \\ 0 & 1 & 0 & 0 \\ 0 & 1 & 0 & 0 \\ 0 & 0 & 1 & 0 \\ 0 & 0 & 1 & 0 \\ 0 & 0 & 0 & 1 \\ 0 & 0 & 0 & 1 \end{bmatrix}
\begin{bmatrix} 1 & 1 & 1 & 1 \\ 1 & -1 & -1 & 1 \\ -1 & 1 & -1 & 1 \\ -1 & -1 & 1 & 1 \end{bmatrix}
=
\begin{matrix} A & B & AB & C \end{matrix}
\begin{bmatrix} 1 & 1 & 1 & 1 \\ 1 & 1 & 1 & 1 \\ 1 & -1 & -1 & 1 \\ 1 & -1 & -1 & 1 \\ -1 & 1 & -1 & 1 \\ -1 & 1 & -1 & 1 \\ -1 & -1 & 1 & 1 \\ -1 & -1 & 1 & 1 \end{bmatrix}
$$

$$X_d \quad T_f \quad = \quad X_f$$

If the transformation is applied to the group means, then each of the variables in X_f yields the appropriate comparisons among the means. In Chapter 10 we saw that the regression coefficients corresponding to the contrast scores defined by orthogonal transformations are the corresponding contrasts among the group means (see Equation 10.32).

Labeling these regression coefficients by the effects they represent and using double subscripts for the means yields the following:

$$b_A = (\bar{y}_{\cdot 11} + \bar{y}_{\cdot 12} - \bar{y}_{\cdot 21} - \bar{y}_{\cdot 22})/4 \tag{11.30}$$

$$b_B = (\bar{y}_{\cdot 11} - \bar{y}_{\cdot 12} + \bar{y}_{\cdot 21} - \bar{y}_{\cdot 22})/4 \tag{11.31}$$

$$b_{AB} = (\bar{y}_{\cdot 11} + \bar{y}_{\cdot 22}) - (\bar{y}_{\cdot 12} + \bar{y}_{\cdot 21})/4 \tag{11.32}$$

$$b_o = \hat{\mu}. = (\bar{y}_{\cdot 11} + \bar{y}_{\cdot 12} + \bar{y}_{\cdot 21} + \bar{y}_{\cdot 22})/4 \tag{11.33}$$

When these coefficients are applied to the score matrix X_f, the prediction is the mean for the group to which an observation belongs, as it is in all the other cases of transformation of the dummy matrix. This must be true on general principles because the factorial contrast variables X_f are just a transformation of X_d.

Constructing the contrast variables is quite straightforward in the case of the 2×2 design. The same principles apply in terms of the correspondence between the anova sums of squares and the amounts of variance attributable to certain contrast variables when there are more levels. There is a degree of complication in how we define the contrast variables, however. The complication arises out of the increased freedom that we have when there is more than one level to a factor. It is not that the contrast variables cannot be defined; it is just that we must choose from among the numerous possible ways to do it.

The next simplest case is the 3×2 design, which illustrates some additional principles. The variable with two levels, B say, presents no difficulties. We simply define the contrast variable as before. However, the three-level variable forces us to decide how we wish to represent it. Any of the various contrasts among three variables may be chosen. For example, if the levels are quantitative, we could use the orthogonal polynomial contrasts—that is, 1, 0, -1 for one contrast variable, and 1, -2, 1 for the other. We may think in terms of the transformation matrix without considering the whole X_+; the transformed score matrix will have the same character as the transformation since each of its rows is a replication of the appropriate row of the transformation matrix. The first three columns of the transformation matrix are listed below. The rows are labeled according to the group they represent and the columns label the contrast:

	Linear A	*Quadratic A*	*B*
a_1b_1	1	1	1
a_1b_2	1	1	-1
a_2b_1	0	-2	1
a_2b_2	0	-2	-1
a_3b_1	-1	1	1
a_3b_2	-1	1	-1

Now it is necesary to pick the contrast variables for the interaction. There are an unlimited number of ways to do this, but the simplest and almost always the most meaningful, is to define them so that they are conceptually parallel to the main effects while remaining numerically orthogonal. This can be done by defining the coefficients in the interaction contrasts by multiplying the coefficients in the A contrasts with the coefficients in the B contrasts.

In this example, we first define a "linear A by B effect." This is done by multiplying in each row the coefficient in the "linear A" column by the one in the B column. Similarly, a "quadratic A by B" contrast is defined by multiplying the elements in the A and B columns. In this case, the result would be:

	Linear A × B	*Quadratic A × B*
a_1b_1	$1 \times 1 = 1$	$1 \times 1 = 1$
a_1b_2	$1 \times -1 = -1$	$1 \times -1 = -1$
a_2b_1	$0 \times 1 = 0$	$-2 \times 1 = -2$
a_2b_2	$0 \times -1 = 0$	$-2 \times -1 = 2$
a_3b_1	$-1 \times 1 = -1$	$1 \times 1 = 1$
a_3b_2	$-1 \times -1 = 1$	$1 \times -1 = -1$

The first of these two contrasts compares the linear effect of A at the two levels of B, and the second contrasts the quadratic effect of A at the two levels of B.

The transformation matrix for carrying the dummy variables into the contrasts that correspond to the A effect—divided into two orthogonal contrasts—the B effect, and the AB interaction—also divided into two orthogonal contrasts—can now be presented by combining the two sections and the constant as follows:

	Linear A	*Quadratic A*	*B*	*Linear A × B*	*Quadratic A × B*	*Constant*
a_1b_1	1	1	1	1	1	1
a_1b_2	1	1	−1	−1	−1	1
a_2b_1	0	−2	1	0	−2	1
a_2b_2	0	−2	−1	0	2	1
a_3b_1	−1	1	1	−1	1	1
a_3b_2	−1	1	−1	1	−1	1

T_f

Each row of X_f would be the same as the row of T_f corresponding to its group's designation, so, there would be r repetitions of that row. In X_f, x_1 will represent the linear effect of the A variable and x_2 will represent departures from linearity. Taken together, they will give us the total amount of variance for the A main effect. The significance of these individual components of the A main effect can be evaluated by testing the significance of the corresponding regression coefficients. The coefficients are given by Equation 10.32 and can be tested using Equation 10.38 if all n_h are equal. If all n_h are not equal, then it is necessary to adapt Equation 10.42. In this equation, ss_{reg} would be the sum of squares for regression using all contrast variables *except* the one being tested.

Since there are only two levels of B the variables operate just the same as in the 2×2 cases. The interaction terms bear some examination, however. The way they have been constructed, the linear $A \times B$ contrast corresponds to the *difference* in the linear effect of A at level 1 of B and this linear effect at level 2 of B. We are literally subtracting the linear A effect at level 2 of B from the linear A effect at level 1 of B. Thus, this will give us the difference in slopes of the function relating y to A, depending on B. For example, y might be a measure of performance, A might represent different amounts of practice, and B might be the distinction between high-aptitude and low-aptitude students. Then the linear $A \times B$ term tells us whether the linear trends resulting from practice are the same in the two ability groups or not. Similarly, the other interaction term indicates whether the quadratic trends are the same; here it represents whatever curvilinearity is present. Taken together, the variance accounted for these two variables is the same as the interaction in the corresponding analysis of variance. By sorting it out this way we can find out more about the nature of the interaction if it is significant or appreciable in magnitude. Are they just slope differences or do they represent differences in the shape of the curves as well?

Thus, structuring the problem in this way allows a complete partition of the effects into components that correspond to each of the degrees of freedom in the group differences. The overall ss_{reg} for all five variables is, of course, the same as the overall ss_b in the analysis of variance. The combined variance predicted by the two A contrast variables gives us the proportion of variance for the A main effect, and the B contrast is the same as the B main effect. Finally, the combined sums of squares due to regression on the two interaction contrasts is the same as the interaction effect in the analysis of variance. When group sizes are equal, the total variance attributable to all the variables in an effect can be gotten simply by summing the amounts of variance attributable to the two components of the effect. This is because $X_f'X_f$ is diagonal. However, when group sizes are not equal, it is almost certain that there are correlations between the variables that make up the effect together. In this case, it is usually best to fall back on using the regression approach, finding $X_f'X_f$ by $T_f'D_nT_f$ and using its inverse to find $\mathbf{b}$.

If the A variable were of a different type, we might want to make different types of contrasts for it. For example, if we had two experimental groups and a control, we might construct the contrasts to be of the same type used for this situation in the one-way case. There, we used 1, 1, and -2 to compare the experimentals to the control and 1, -1, and 0 to compare the two experimentals. We do just the same here to represent the A effect, except that each of the coefficients must be repeated at each level of B. The B contrast is the same as before, and it would be most reasonable to define the interaction contrasts by the multiplying-coefficients approach used earlier. Now, however, the first contrast represents the difference in the experimentals-versus-control comparison at the two levels of B and the second contrast is the difference in the comparisons of the two experimentals. The complete matrix of coefficients needed to go from the dummy variables to these contrasts is $T_{b'}$:

	Exp./Cont.	*Exp./Exp.*	*B*	*Exp./Cont.* × *B*	*Exp./Exp.* × *B*	*Const.*
a_1b_1	1	1	1	1	1	1
a_1b_2	1	1	-1	-1	-1	1
a_2b_1	1	-1	1	1	-1	1
a_2b_2	1	-1	-1	-1	1	1
a_3b_1	-2	0	1	-2	0	1
a_3b_2	-2	0	-1	2	0	1

$T_{f'}$

From a dimensional or geometric point of view, what happens here may be explained as follows. The six columns of the dummy matrix define an orthonormal basis for a six-dimensional space. The X_f defines a different orthogonal basis for the same space, and T_f defines how to go from one set of coordinates to the other. The ss_b defines the extent to which a seventh vector $\mathbf{y}$ projects into the five dimensions of the space which is defined by the contrast vectors. The second set of contrasts, $X_{f'}$, has the same constant vector and its five dimensions define the same space as the other five, but the axes are oriented differently. Thus ss_b is the same no matter which set of variables are used. This is true whether we use a factorial analysis or not, but in this case, the definitions of the two spaces are quite similar. The two different ways of dividing up the A main effect are constrained to be within the same two-dimensional subspace of the five dimensions, and the same is true of the interaction effects. So, neither change affects how we assess the overall amounts of these effects, only how they are divided up.

Greater Complexity of Design

Analysis of variance clearly includes more complex designs than have been considered so far. Perhaps the one we should consider next is the $2 \times 2 \times 2$ design. Here, there are three main effects, A, B, and C—for example, list length, study time, and handedness of subject. Each effect has two levels, so there is only one main effect contrast variable for each factor. There are three types of first order interaction, AB, AC, and BC, each with one degree of freedom, and a single third-order interaction, ABC. This makes a total of seven possible orthogonal contrasts, the same as the number of degrees of freedom for the between groups *ms* when considered all at once.

The contrast variables are constructed in the same way as before, extended to meet the new complexity. This is illustrated in the following table:

	A	*B*	*B*	*AB*	*AC*	*BC*	*ABC*	*Constant*
$a_1b_1c_1$	1	1	1	1	1	1	1	1
$a_1b_1c_2$	1	1	−1	1	−1	−1	−1	1
$a_1b_2c_1$	1	−1	1	−1	1	−1	−1	1
$a_1b_2c_2$	1	−1	−1	−1	−1	1	1	1
$a_2b_1c_1$	−1	1	1	−1	−1	1	−1	1
$a_2b_1c_2$	−1	1	−1	−1	1	−1	1	1
$a_2b_2c_1$	−1	−1	1	1	−1	−1	1	1
$a_2b_2c_2$	−1	−1	−1	1	1	1	−1	1

$T_{f''}$

To define the A main effect, members of all four groups at one level of A would receive scores of 1 and the other four would receive scores of -1. The B and C main effects would be defined correspondingly. The first-order interactions are set up the same way as the two factor case, multiplying together the coefficients of the appropriate levels of the factors involved. When it comes to the second-order interaction, the contrast may be constructed by taking one of the first-order interaction contrasts and multiplying its coefficients by those of the missing main effect. For example, there is a coefficient of $+1$ in the last column of the transformation in the next-to-last line that corresponds to groups $a_2b_2c_1$. This can be obtained by multiplying together the coefficient in the same line for AB $(+1)$ and that for C (also $+1$), or by multiplying together the one for AC (-1) and that for B (also -1), or by multiplying the -1 of the BC interaction by the -1 for A.

When there are more than two levels for a factor, the set of contrasts on the main effect that is most meaningful should be made. The same is true of the interactions, although these are most easily constructed and understood when they relate to the main effects in the manner used here. In the score matrix, the rows for the members of a given group would be repetitious of the corresponding row of the transformation.

There is no need to literally construct the score matrix. It is primarily a heuristic device that allows one to relate anova to regression. Indeed, there is no need to construct the transformation matrix T_f, or even to decide on what contrasts are appropriate. In the equal-groups case, traditional methods of analysis of variance separate the main effects and interactions as orthogonal components that represent any possible system of contrasts *within* these effects. Where the group sizes are unequal, there is more than one possible approach. Modern machine computation methods provide the sums of squares that are appropriate to whichever interpretational strategy we think is appropriate.

Conclusion

In this section we have tried to show how the standard orthogonal designs correspond to score matrices that embody particular contrasts. When there is only a single degree of freedom for a main effect or interaction, the nature of the contrast variable is completely determinate, except that the sign of all the elements may be reversed. When there is more than one degree of freedom for an effect, there is exactly the same latitude in the contrast variables as in the one-way case. In fact, a good way to simplify the process of constructing a contrast variable is to come up with coefficients that we would use if we were making a one-way analysis of the variable defining one factor. Then these coefficients would be repeated at all levels of the other factor. The interaction contrast variables are best defined by multiplying together the coefficients for contrast variables that belong to the factors involved in the interaction. The main effect and interaction contrast variables are orthogonal, with mean zero. In their simplest forms, they may not have equal variances, but if we wished, it would be simple to standardize them.

The factorial analysis represents one of the many full-rank orthogonal contrast score matrices. That is, it is a transformation of the dummy variable matrix and also of any of the other possible complete sets of contrasts for a given set of data. The variance accounted for is the same no matter which of the sets of contrast variables is used. The fact that they are orthogonal means that the zero-order relations may be tested directly without worrying about whether they can be accounted for by other predictors.

When the factorial design is chosen the orthogonal decomposition of the between-groups mean square is presumably of the most interest. This presumption is so often true that the factorial design breakdown tends to be chosen automatically. This choice may deserve more attention than it often gets. This is among the questions to be discussed in the next section.

DESIGN, INTERPRETATION, AND INFERENCE IN FACTORIAL DESIGNS

What Is the Question?

We gather data with certain ideas in mind. These may be only vague notions that a certain roughly defined variable ought to be related to another equally roughly defined one, or that manipulation of one variable might have an unknown effect on another. Alternatively, our ideas may be very precise in the clarity of definition of variables and in the expected nature of their relationships. There may be only a single question that is of any interest, or there may be a dozen or more questions of roughly comparable salience. Most important is to have the right sort of data to shed light on our questions, be they vague notions or exquisitely precise hypotheses. Given that we have some data, we want to make the analysis correspond as closely as possible to the empirical questions. The data should call attention to what is happening when our preconceptions are amorphous and should sharply distinguish among alternatives when our ideas are well defined. Perhaps most important, our interpretation of the data should correspond closely to the facts represented in the data. For all of these reasons, it is important to have as accurate an idea as possible of what is and what is not in a factorial design analysis of differences between means.

Perhaps the major source of the difficulty is that, as behavioral scientists, we tend to think in verbal concepts, whereas data analysis is a quantitative mathematical process. It is not easy to translate a verbal concept into the proper design and analysis, and it is correspondingly difficult to make the appropriate verbal interpretation of a particular analytic result. Interpretations and analyses that are only approximately correct are adequate for many purposes, but for some data, only one interpretation is appropriate. There is no guaranteed way of ensuring that the analysis made is the best possible course of action, or even that it is good enough for the purpose in mind. In this section we will attempt to translate just what the data are saying, point out some of the more common ambiguities of design or analysis, and give advice on how to handle a few common situations. Remember that in some cases what is provided is one person's opinion, and that there is often disagreement among experts, whether they be self-styled or world-renowned.

Probably the most important thing to bear in mind in interpreting the analysis of variance is the model, expressed in Equation 11.1 *Every two-factor analysis of variance fits this model to the data.* The contrast variable approach emphasized here used the linear dependence of the parameters to specify one or the other particular form for them, but the model is still there, and we can go back and estimate the α's, β's, and $\alpha\beta$'s from the contrast variables and their regression coefficients.

The analysis of variance applies most directly to a situation in which there are only a few *possible* values of the A variable and only a few *possible* values of the B variable, and where these are variables under the direct control of the experimenter, who has used all the possible combinations of the two variables an equal number of times. In this case there is no ambiguity about the results of the analysis except those that result from sampling of the error terms. The experimenter does not have to worry about someone—including his or her own alter ego—coming around after the experiment asking "What if . . . "? As long as any given study approximately conforms to this ideal, there should be no ambiguity about the study's interpretation. Sometimes, ambiguity can be eliminated by adjusting the analysis process; other times it remains. Sometimes it is traditional to ignore the ambiguity—and who are we to go against tradition? Nevertheless, it may be interesting to try to understand what the ambiguities are so that they may be avoided when new ground is broken.

Besides the central issue of how conclusions must be modulated when they depart from the ideal model, there are two other recurring concerns in analysis of variance. One is the problem of nonorthogonality—the fact that unequal cell sizes lead to correlations among the effects. In some respects, this is related to the departure from the ideal model, although it can arise in the ideal case as well. The other has to do with the control of alpha levels and how ambiguities about what to do in this respect arise in anova. These issues concerning alphas will be considered first.

Alpha Levels in Analysis of Variance

In multiple regression, several different approaches may be taken in controlling Type I error probabilities at the desired value (see Chapter 9). First, there is the distinction between parameter or hypothesiswise alpha levels and experiment or analysiswise levels. In the former we are willing to wrongly reject a hypothesis about a given parameter α percent of the time: If we are testing 20 regression coefficients and the null hypothesis is true about all of them, we are willing to reject (falsely) one of the null hypotheses, on the average. In the latter case we want there to be only one chance in 20 of rejecting *any* of the null hypotheses. There are various ways of testing this, including Bonferroni, Scheffé, hierarchical tests, and protection levels. Each of these tests has a cost in terms of decreased ability to reject the null hypothesis.

Alpha Levels in Factorial Designs

In analysis of variance as well as regression, there is a fundamental distinction between hypothesiswise or parameterwise alpha levels on the one hand and experimentwise or analysiswise alpha levels on the other. In factorial designs, the investigator traditionally tests each effect using the nominal, hypothesiswise alpha level. This is reasonable since there is usually a separate interest in the presence or absence of each effect. However, because multiple tests are being made, we enhance the probability—perhaps considerably—of rejecting a null hypothesis about one of the main effects or interactions when all hypotheses are true. With f factors, there are $2f - 1$ effects, or 15 effects in the case of four factors. When there is serious concern that all the effects are zero, it is reasonable to begin the analysis with an omnibus test of all the effects at once, $F = ms_b/ms_w$, going on to test separate effects only if this test is significant. For the most part, however, testing each effect separately is more appropriate, because we believe it is unlikely that all the effects are zero.

If a given effect is significant, and where there is more than one degree of freedom, it is natural to wonder just which means are producing the effect. Here, we have emphasized the contrast approach, where the most meaningful comparisons are decided upon a priori. In this case, each contrast in an effect can be tested separately, just as in the one-way case. In the absence of such a priori contrasts, we would fall back on multiple comparison procedures, bearing in mind that the Tukey and Scheffé methods are overconservative when the overall test leads to rejection.

Nonorthogonality

Probably no issue in analysis of variance causes more head-scratching, nail-biting, dog-kicking, wrist-slashing, name-calling, and finger-pointing than nonorthogonality. For decades, almost no one knew what to do in a factorial anova when the cell frequencies were unequal. When textbook formulas for sums of squares—valid in the equal-frequency case—led to negative sums of squares for interaction, investigators sat blinking at them in frustration, feeling certain that there was an error in computation since squared numbers cannot be negative. Today, many investigators *know* what to do in such cases, although they may end up doing different things in the same situation. The alternative solutions, and when to employ them, become more understandable when anova is thought of as a regression problem. Indeed, this is a major reason for introducing the regression mode of conceptualizing anova.

The nature of the problem can be understood by considering a dummy code that denotes membership in a particular level of one factor—a_1, say—and then another that denotes membership in one level of the other factor—b_1. If there are $g = g_A g_B$ equal-sized groups, then the two dummy variables are uncorrelated in the sense that if we compute the correlation between the a_1 dummy variable and the b_1 dummy variable, it will be zero. Generally, inequality of groups means that a dummy or contrast code representing one factor will be correlated with that representing another factor. (The correlation will also be zero in some cases of unequal groups that are described below.) The occurrence of these correlations means that an apparent difference in the means on one factor might be accounted for in terms of differences on another factor. Suppose, for example, that in a sex-by-treatment study, there is a known sex effect, and an imbalance of males in one condition; then differences in means for conditions could be attributable to this imbalance. It is not always clear how to handle such cases.

In Chapter 10, it was shown that the multiple regression process is quite simple for uncorrelated predictors of which orthogonal comparisons with equal groups is an example. The regression weights are directly related to the correlations and it is quite simple to assess the amount of variance that is attributable to a given variable or group of variables. In a factorial design, the main effects and interactions are all mutually uncorrelated in the equal-group case—even if within an effect nonorthogonal contrasts are used. When the groups are unequal, this orthogonality between effects is lost except in special cases. Then different strategies in multiple regression are possible, depending on the models that are thought to be relevant.

The issues usually revolve around testing effects for significance. In the regression case, this means testing the correlations involving a subset of variables or their regression coefficients. If the regression coefficients are tested, different strategies can be adopted, particularly the hierarchical or the "subset-last" approaches.

As described earlier, the traditional procedure for analyzing balanced factorial experiments in psychology has been to find the sums of squares attributable to each effect which are then tested against the within-groups sum of squares. This corresponds in regression to the "subset-last" approach where the hypothesis that a subset of regression coefficients is zero is tested by comparing the error that results from including the subset in the equation to the error that results from including all regression coefficients minus the subset. This procedure can likewise be followed in unbalanced factorial experiments, testing each main effect and interaction against the within-groups. The difference in the unbalanced case is that the simple methods for finding the sums of squares for the various effects do not apply. Instead, we must fit all the effects, except the one being tested, in order to find the ss_{dev} for all but the effect being tested. Modern computer programs do this routinely, if correctly instructed by the user, so there is no need to go into the details of this procedure.

The alternative is to take a hierarchical approach, but there are differences of opinion as to the most appropriate order to fit the effects. If the null hypothesis is true for some effects and not others, then power for detecting the true effects is enhanced if the sums of squares and degrees of freedom for the null effects are combined with those for within-cells. Furthermore, if there is a concern with alpha levels when multiple effects are tested, then a hierarchical strategy that tests the most likely effects first, or those which are most crucial to the study, will control the analysiswise alpha level while answering the most important questions. Usually this hierarchical approach focuses on fitting the main effects first, and chooses some order for them.

Another point of view focuses on the interaction between variables. For example, Cramer and Appelbaum (1980) argue that if there is significant interaction, then all other tests are irrelevant. They contend that if interaction is present, it is necessary to know all the cell means anyway, and that main effects lose their simple interpretations in the presence of interaction. While this argument has some validity, it represents too narrow a view. Most investigators are interested in the main effects even when there is interaction, because main effects represent the effect of a factor averaged across the levels of the other(s). Furthermore, an interaction, while significant, may be small in comparison to the main effects and may even be ignored for many purposes. The presence of interaction does imply that the main effects must be examined in detail before the conclusions about a study are made, but it seems unnecessarily restrictive to let it preclude testing main effects at all.

We can summarize this discussion of the various tests and analyses in factorial designs using multiple correlation terminology, thinking of various multiple correlations that could be found using the various anova effects. R_A^2 represents the proportion of variance accounted for just using the A main effect, ignoring all other information. Similarly, R_B^2 is the proportion accounted for by the B main effect, and R_{AB}^2 is the proportion accounted for by just the interaction. $R_{A,B}^2$ is the proportion that is predicted using both main effects but not the interaction, whereas $R_{A,AB}^2$ and $R_{B,AB}^2$ use one main effect and the interaction. The proportion $1 - R_{all}^2$ would correspond to the ratio ss_w/ss_t and $R_{all}^2 = ss_b/ss_t$.

In the equal-groups case, $R_{all}^2 = R_A^2 + R_B^2 + R_{AB}^2$ and $R_{A,AB}^2 = R_A^2 + R_{AB}^2$; $R_{A,B}^2 = R_A^2 + R_B^2$; $R_{B,AB}^2 = R_B^2 + R_{AB}^2$. More generally, though, the variance attributable to a group of predictors is the difference between the variance using all the predictors minus the variance using all but that group. This is how we test for zero regression weights for that group:

$$F = \frac{(R_{all}^2 - R_{\sim q}^2)/q}{(1 - R_{all}^2)/(n - p - 1)} \tag{11.34}$$

Because of the additivity in the equal-groups case, $R_{all}^2 - R_{A,AB}^2 = R_B^2$, and similarly for the other effects. Therefore, direct testing of each effect against interaction is the same as testing the increase in squared multiple correlation that results from using that effect. When the groups are unequal, the additivity breaks down and $R_b^2 \neq R_{all}^2 - R_{A,AB}^2$. The latter R^2 have to be computed and used directly, and we must do similarly for the other effects.

Using the hierarchical approach, on the other hand, we would choose one effect or set of effects and test this first, pooling the others with the ss_w. For example,

$$F = \frac{R_A^2/(g_A - 1)}{(1 - R_A^2)/(n - g_A)}$$

followed by

$$F = \frac{(R_{A,B}^2 - R_A^2)/(g_B - 1)}{(1 - R_{A,B}^2)/(n - g_A - g_B + 1)}$$

and then by

$$F = \frac{(R^2_{all} - R^2_{A,B})/(g_A - 1)(g_B - 1)}{(a - R^2_{all})/(n - g_A g_B)}$$

This order assumes that the A main effect is most important conceptually, followed by the B main effect, and then their interaction. Other conceptual orders would result in the corresponding sequences. In each case a nonsignificant test would result in subsequent tests not being carried out, thus conserving the alpha level.

The Cramer and Appelbaum (1980) approach is somewhat different. It tests R^2_{AB} *first;* if it is significant, then testing stops. If interaction is not significant, then the main effects are tested hierarchically. Their reasoning is that if the interaction is present, then the main effects are irrelevant. With interaction, one might as well use the cell means as directly, rather than fitting a model using separate main effect and interaction terms. Furthermore, according to Cramer and Appelbaum, the main effects lose their interpretability in the presence of interaction.

Although these points are valid, many feel that they take too narrow a view of the concerns of investigators. Many investigators have a strong concern with the main effects whether or not there is interaction. These investigators view a main effect as the value a variable has averaged across the levels of the other factor. Furthermore, an interaction, while significant, may not be disordinal. That is, the means for different levels of one factor are always in the same order, regardless of the levels of the others. It may even be quite small, and in many instances could be ignored. Thus, although the presence of interaction implies that the effects must be examined in detail before conclusions can be made, it seems unnecessarily restrictive that its presence should prohibit testing main effects at all.

The Cramer and Appelbaum view, however, raises some useful interpretative issues. It points to the uncertainty that surrounds the estimation of main effects in the presence of interaction. Essentially, the presence of interaction means that the "effect" of a certain level of A depends on the level of B with which it is combined. Indeed, if only certain levels of B have been selected out of a large number of possibilities, the overall A main effect could be quite different from the one observed. The α_j will in fact take on different values, depending on which levels of B are present. This will not be the case when the levels of B that are used are exhaustive, which occurs in a minority of applications.

Appelbaum and Cramer introduce the distinction between main effects and *marginal* effects—the term marginal referring to the fact that the means tested by the usual main effects are the marginal ones, as opposed to the cell means, whereas in their terminology "main" effects refer to the parameters. The authors would admit, perhaps, investigators' concern with the equality of the marginal means in a factorial analysis of variance, but apparently they feel that this concern is generally misplaced, and that it should be distinguished from a concern with the main effect parameters. Several of the readings listed at the end of this chapter, notably Keren (1982) and Overall, Lee, and Hornick (1982) present different opinions on this issue, which has a long history and a large and rather acerbic literature.

Before leaving this issue, it should be noted that there are fundamental principles in the abstract theory of measurement that make it desirable that the presence of interactions be zero. Otherwise, our understanding of the variables involved in a study is uncertain. Krantz (1972), and Krantz et al. (1971), present very formal treatments of this issue. A somewhat less abstract discussion is given by Cliff (1982).

The reader may finish this section with a sense of bewilderment at the necessity of choosing among the various options. The ideal approach is to consider every alternative carefully so that the right one can be chosen. Sometimes this may be more effort than it is worth, in which case it is better to adopt one reasonable strategy and follow it all the time than to be paralyzed with indecision, or wait hat in hand for the specialist consultant to arrive. Testing each effect against ms_w, whether through an ordinary anova or the corresponding regression analysis, is generally satisfactory and corresponds to the typical practice. Bear in mind, however, that this is only appropriate when the independent variables are fixed, rather than random effects, and when the analysis is "between subjects." In other cases, more complicated strategies are necessary.

OTHER ISSUES

Discrete or Continuous Factors?

Analysis of variance is applied in a wide variety of situations and is used on factors of quite different natures. When applying analysis of variance, one should remember the idealized case presented earlier and note any departure from it.

Ideally, anova is applied to truly discrete or qualitative factors. The model says that the effect α_j is a *constant* at level j. Very often, however, one or more factors in an anova are continuous variables such as exposure time for a stimulus, years of education for an individual, and so on. These variables may be treated as discrete by selecting only a few—often only two—of their values. This can happen in one of two ways. In one, the investigator selects a few specific values—100 and 500 milliseconds, for example—and uses only those values. Or, if the variable is a natural characteristic of subjects, such as education or I.Q., it is much more likely that values of the variable are lumped into coarse groups—for example, all those with no college versus all those with some. Each of these possibilities carries its own interpretational ambiguities.

Strictly speaking, the use of 100 and 500 milliseconds of exposure only allows the estimation of the effects of exactly those exposures, even though the investigator is interested in the overall variable of exposure time. Indeed, when we read his conclusions, it is likely to appear that he used all possible values. These results may have been quite different, however, had he used 50, 300, and 600 milliseconds, or even 101 and 499 milliseconds. So, the investigator is letting two discrete values represent the whole variable, and asking others to accept on faith his conclusions in this regard. The validity of this interpretation may be questionable, or it may be a reasonable approximation. The issue becomes particularly important when interactions are present. Here, the use of a different set of values for the variable might lead to *different* interaction effects and even possibly to the elimination or reversal of obtained main effects, since they are averaged over levels. These possibilities may seem remote and in some cases they are, but in others they may not be, and the investigator should be aware of them. As a general rule, one is better off running more than two or three levels of a variable to get a clearer understanding of the exact nature of an effect. The extra time and effort may be worth it, and the use of more levels may be partially compensated by using fewer replications per cell.

When a more or less continuous variable is broken up into a few coarse categories, the situation is similar, except that now the model specifies α_j as a constant within level j, when in fact it must be a variable, depending on the exact level. For example, level of education, or whatever complex of psychosocial influences underlies it, presumably operates incrementally rather than in a single discrete jump. Treating it as a discrete variable adds to the within-cells variance in an artificial way, thus weakening the analysis. Additional uncertainties are introduced when interaction is present because it is not clear just what effect the particular choice of break-points has had.

If one is going to use a regression program anyway, it may make more sense to leave a variable in its continuous state. If desirable, several different powers of the variable may be used and interaction variables may be constructed by forming the products of the polynomials with other factors. One or the other approach of this type may be necessary if one is interested in gaining a fairly accurate idea of *how* a variable has its effect, rather than just an approximate notion that it has an effect of some sort.

Manipulated Versus Natural Variables

Another distinction that may result in confusion or controversy is that between manipulated and natural variables. With manipulated variables it makes perfectly good sense to set things up with equal numbers of observations per cell and then force orthogonality on the factors. But anova is often used on natural variables such as gender, marital status, ethnic group, place of residence, handedness, eye color, use of false teeth, and the like. These variables are used either individually, or in combination, or with manipulated variables. We speak here particularly of variables that have only a few distinct categories, at least according to some widely accepted convention. The question becomes whether these variables, treated as factors in an anova, should be forced into an equal cell frequency mold. The answer, as usual, is "maybe."

Suppose one variable in a factorial design is a subject characteristic such as handedness and the other is a manipulated variable such as which visual field a stimulus is presented to. What proportion of left-handed subjects should there be? Two possible arguments could be made. According to one argument, the best thing to do is use equal numbers of left- and right-handed subjects because it is the most efficient method of assessing the importance of this variable. According to the other argument, we should approximate the population proportions by having several times as many right-handers as left-handers. (Naturally we would use equal numbers of a given handedness under each eye-presentation condition in either case.) Now suppose we took the equal-numbers approach and coded contrast variables for the two main effects and interaction in the usual way. Let us assume for the moment that all three effects are significant and appreciable. We would find a certain regression equation along with a certain error variance. Using the same data, suppose we discarded at random some of the lefthanders, reducing their number down to the population fraction. For the data that remain, nothing fundamental has changed. The phenomenon is the same, so the model that describes it, and the population accuracy of prediction (the error variance) are the same. Assuming that the discarding was random and the data were fit by a regression model, the only difference will be of the statistical sampling type. This means that *in principle, it doesn't make any difference which approach was taken*. The resulting model should be the same, except for statistical fluctuation. The model accounts for the *cell* means, and their expected values are not changed.

In practice, it may seem to make a difference because we are changing the variance due to handedness. When the fraction of left-handers is reduced, the handedness variance is correspondingly reduced; the variance of the handedness variable is $p_h(1 - p_h)$, where p_h is the proportion with a given handedness, which has a maximum when $p_h = .5$. Consequently, the variance *due to* handedness will also decrease when there are fewer left-handers. It will decrease to zero if all the left-handers are eliminated. The handedness effect may no longer even be significant due to this reduction, perhaps leading to a qualitatively different conclusion (which happens to be a Type II error). In any event, the proportion of the total variance in y that is apparently due to handedness will be reduced. This estimate of variance is the "correct" estimate since it reflects the population distribution of handedness.

For most purposes the equal-numbers design is the most efficient way of testing or estimating effects. It gives the best estimates per unit of observation, and should give the same model of the phenomenon as proportional representation. The only thing we need to be cautious about is the interpretation of the importance of the variables. Even in this respect it would be possible to use the data from the equal-numbers design to estimate the appropriate proportions of variance in the population, but the details of how to do that are beyond the scope of this discussion.

The same sort of principles apply when both factors are organismic variables. Suppose we take marital status—married, single, widowed, and divorced—and type of residence—single versus multiple family—as the factors and study their relationship to attitude on a taxation question. Suppose there are a large pool of prospective respondents and we know in advance their status on these two factors and wish to select a sample to survey on the taxation question. We could either select at random, which would undoubtedly lead to disproportionate frequencies in different cells, or we could randomly select within each cell the same number. The two main effect variables are bound to be correlated in the former case, which is bound to lead to both main effects being nonorthogonal with respect to the interaction. Again, however, the phenomenon does not change; the expected value of the *cell* means is surely the same under each scheme, and this is all that we can account for in any case. Therefore, we should end up with the same regression model and estimate of error variance no matter which method is used. Again, all that can change is the amount of variance associated with an effect because the variance of the effect changes when the proportions in different cells of the design change. Therefore, an effect that was significant in the equal-cells case may not be under proportionate sampling, assuming that the total n is the same.

One qualifying statement should be made about the preceding discussion. The expected cell means and the within-cell variances are not affected by the number of observations per cell. However, if there is heterogeneity of variance among the cells, the overall estimate of the ms_w may change with different cell proportions due to different weighting of the within-cells variances to arrive at the sample ms_w.

As a general rule it may be simplest to use equal cell frequencies when the composition of the sample is under simple and direct control of the investigator, but to take the observed, natural frequencies when data are of the survey kind. In the latter case, a regression analysis would be necessary; using observed frequencies avoids any complication about estimating the size of effects and is easier to justify to the statistically unsophisticated—such as a boss. The greater efficiency lies with the equal cell sizes, however. As noted, complications arising out of heterogeneity of variance are less likely to confound significance tests in the equal-cell case (although we would end up with a biased estimate of the actual average within-cells variance).

One other point must be addressed with respect to nonorthogonality: the reason for its existence. Perhaps the most common reason is that while the investigator intends to have equal-cell frequencies, "accidental" reasons lead to some relatively minor differences. Here the most important question is whether these accidents are unrelated to the experiment. For example, if some animals die or some subjects did not show up for part of the experiment, was this because of the treatment in that cell? If so, then *any* analysis of the data may be questionable. If not, then it is legitimate to go ahead and analyze them. Missing cell frequencies call for honest investigation into the causes. It would be embarrassing to conclude that some form of therapy led to shorter stays in your hypothetical hospital when further investigation revealed that it was really because patients who received that therapy decided your hospital was bad for their health and, if they survived, escaped as soon as possible!

One final bit of advice may be given regarding the differences in cell frequencies in a factorial study. When these differences are small and seemingly unrelated to the factors, little information is likely to be lost if data are discarded (randomly!) to make the cells equal. This is assuming that only a small proportion of the data needs to be discarded in order to have this effect. Only a little power is lost because it is the size of the *smallest* cells which is most important in determining power. A lot of headaches can be avoided this way. If you do not trust yourself to do the discarding randomly, get your colleague across the hall to do it for you.

Fixed and Random Effects

One issue that the regression approach to anova does not address at all directly is the distinction between fixed and "random" effects. In many cases one of the factors in a factorial design must be considered to be a relatively small sample from a larger universe. For example, a factor of the design might be different word lists or different stories in a verbal learning experiment; in an educational experiment it might be schools, each of several treatments being tried in each of several schools; in a psychotherapy study each of several therapists might perform each of several treatments; or in a social psychological study there might be several different assistants who ran the study, using the same number of groups under each of the experimental conditions. In all of these cases, the elements of the factor—lists, stories, schools, therapists, and experimenters—in the experiment are only a small fraction of a larger universe to which the investigator would like to generalize. Thus the effects of the corresponding variable, and its interactions with others, is a *sample* from a universe of other possible variables.

Suppose that while testing against ms_w a significant treatment effect *and* a significant interaction were found with word list, school, therapist, or assistant. The interaction means that the value of a treatment effect is sometimes smaller and sometimes larger, depending on the other variable. Perhaps it even reverses direction. This tells us that the particular main effect for treatment is at least to some extent a function of the particular word lists, and so on, that we happened to use in the experiment. Considering the overall population of possible word lists, it is possible that the treatment effect is in fact quite different than the effect observed in our little study. This means that in order to assess the significance of the treatment effect it must be tested against the *interaction,* not the within-cells mean square. If this test is significant, we can be reasonably sure that the treatment effect we observed would hold up across the use of a different set of word lists, schools, and so on. If the test is not significant, then logically we cannot be sure of the treatment's significance. When the number of degrees of freedom for the interaction is small, it is clearly difficult to find significance. Look in the F table in the Appendix for 1 and 1 *df.*

Beyond the simplest cases such as those we have discussed, the complications on which F tests are legitimate can ramify alarmingly; we will not go into them here. A number of anova progams will now make decisions concerning error terms for us if we specify which variables are random. Further details can be found in texts on anova.

Interpreting Interactions and Alternatives to Factorial Analyses

Sometimes the questions answered in factorial anova are not exactly those the investigator had in mind. There may be a more appropriate set of contrasts than the ones underlying the factorial design. This is frequently true when one factor of the design is included for purposes of greater generalizability. Suppose, for example, we have two different problem-solving conditions and use subjects of both genders. The major interest centers around which condition leads to better problem solving, and whether the effect holds for both men and women. Possible main effect differences for sex are of incidental interest.

The usual way to analyze these data is with a factorial design. If there is a main effect for methods, this implies on overall superiority of one method over the other. An absence of interaction implies that there is no difference in the relative effectiveness for men and women, whereas a significant interaction supports a differential effect. A significant interaction, however, may have several different sources. It could be that the same condition was most effective for both genders, but that the amount of difference was not the same. Or, it could be that there was no difference for one sex and a difference for the other, or even that the direction reversed in the different genders. When there is interaction, one has to see what form it takes. This puts us in the position of forming post hoc inferences, a process that is considerably more uncertain than direct tests of hypotheses which correspond to the investigator's ideas.

In the typical case, the investigator has exactly as many questions to ask of the data as there are degrees of freedom among groups. For example, the three questions in the case of the 2×2 design are: Is there an A main effect? Is there a B main effect? Is there an interaction? Sometimes, however, the investigator may have *more* questions than degrees of freedom. The present example involving relative effectiveness for men and women is a case in point. Here, the investigator wants to know, "Is there a treatment effect for males?" "Is there a treatment effect for females?" "Is the effect the same?" "Is there an overall gender difference?" The factors here will be labeled mnemonically, T for treatment and G for gender. The standard analysis answers the last two questions through the

TG interaction and the *G* main effect, but the first two are not answered directly. Instead, the *T* main effect answers the question, "Is there a treatment effect averaged across sexes?"

The first two questions could be answered directly if we were willing to substitute them for the tests of the overall treatment effect and the interaction. All we need to do is test the treatment effect in each gender. This is called a "simple effects" test, which is occasionally employed. For this test, the dummy code transformation would be:

Group	Treatment Within g_1	Treatment Within g_2	Gender	Constant
t_1g_1	1	0	1	1
t_2g_1	−1	0	1	1
t_1g_2	0	1	−1	1
t_2g_2	0	−1	−1	1

The transformation is clearly orthogonal, and so it accounts for the means in exactly the same way as the usual main-effects-and-interaction does. The mean differences are just being expressed in a different way.

This principle generalizes readily to more than two levels on either or both factors. Essentially, we would perform a one-way analysis of the variable which is of principal interest within each level of the other. Each of these tests uses $g_T - 1$ degrees of freedom and there are g_G of them, so

$$g_G(g_T - 1) = g_G g_T - g_G$$

degrees of freedom are used in all. This leaves $g_G - 1$ to test the overall effect of the other variable.

Because both the simple effects and factorial analyses are the outcome of nonsingular transformations of the dummy code variables, it would appear that they must be nonsingular transformations of each other, which is indeed the case. The between-group variance is just being differently distributed. In fact, since the gender effect is still represented, what is happening is that, in the simple effects analysis, the variance due to the other main effect and to the interaction is being redistributed into the simple effects. We can look at this the opposite way: All the simple effects variance is being redistributed as a main effect and an interaction.

Because the variance is being distributed differently, the outcome of the two analyses is fairly consistent conceptually: Where there is a strong main effect, at least one of the simple effects should be significant, and vice versa. However, a borderline effect may be significant in one analysis and not in the other.

Note also that heterogeneity of variance should be a concern here. As with all analyses where the cells are differently weighted, there may be different conclusions, depending on whether the within-cells variances are equally weighted or weighted according to the squared weights that are actually used in the analysis. In our example, this weighting is easy to do. We simply use the variances from only those cells that enter into the test of a given simple effect. For example, in testing a treatment effect within males, we would use only the within-cell variances for males.

Simple effects analysis is only one example of the general idea that the investigator should try to formulate the questions that are most salient and tailor the analysis accordingly. If there is plenty of variance to go around, then this is not as important. We can then test different questions separately, even though they are not mutually independent. But if the effects are borderline, then it is important to ensure that the most important effects have first chance at whatever variance is in the analysis.

SUMMARY

Factorial design is the most common form of analysis of variance, and has proved to be of considerably heuristic value to researchers in formulating their studies and in the operationalization of theories. From the viewpoint of data analysis, a factorial design is a particular set of orthogonal contrasts. That is, the main effects and interactions of the factorial analysis of variance correspond to orthogonal subsets of contrasts among the group means. This is easiest to see where there are only two levels to each factor and only one degree of freedom for each main effect and interaction. It generalizes to the multilevel case where there is the same sort of latitude in constructing the orthogonal contrasts for the main effect as in the one-way case. The significance tests in factorial designs then correspond to tests of the proportions of variance due to those contrast variables.

In the case of multiple regression, there are a number of ways of looking at the data, and corresponding ways of doing so in factorial anova. In the case of equal groups, several of these ways are equivalent because the effects are then orthogonal. The regression analysis that corresponds to the usual factorial anova in psychology is the one that tests each subset of regression coefficients by entering it last into the regression equation. This type of analysis seems to be the most likely when the groups are unequal, but there are other possibilities, depending on the investigator's goals and preconceptions.

The factorial anova model and the corresponding analysis apply most directly to the case where the levels of the factors take on only a few fixed values and all these values are represented in the experiment. However, anova is also applied in situations that depart in various ways from this paradigm. Such departures make the interpretation of results more ambiguous. In addition, the factorial analysis may not correspond to the questions that the investigator has in mind. In such cases, alternatives such as the "simple effects" model may be more appropriate.

COMPUTER APPLICATIONS

SAS

If the factorial design is balanced, then either PROC ANOVA or PROC GLM can be used, although PROC GLM is necessary in the unbalanced case. The conventions used in these two programs are much the same, so our discussion will be confined to GLM. The reader should bear in mind that ANOVA is more efficient, and somewhat easier to use, in cases where the cell sizes are equal.

The GLM procedure is used in the regression type of analysis when some of the variables are categorical. There are numerous control words in PROC GLM; we will consider only those that are most relevant to the discussion in this chapter.

The first thing to specify in PROC GLM is the variables. This is done in a CLASSES statement: CLASSES A B C specifies that A, B, and C—which the program must already know about—are categorical variables. Otherwise they are treated as numerical. The next thing to specify is how the variables are to be included in the analysis. MODEL Y = A B A*B indicates that both main effects and the interaction are to be included, and that they are to be fitted in that order. MODEL Y = A A*B B specifies that the interaction is to be included before the B main effect. This makes no difference in the balanced case, only in the unbalanced case.

These commands are sufficient to do the analysis, although a variety of additional statements are useful. For example, the investigator would presumably like to know the means corresponding to various conditions. In the case of three factors, the statement MEANS A B C A*B A*C would print the marginal means for all three factors and for all combinations of A with B and A with C, but not B with C nor the triple classification. A number of other statements will print or look at various aspects of the data.

Two useful statements are CONTRAST and ESTIMATE. The former computes contrasts on any factor. For example, if factor A has four levels, the statement

CONTRAST

A 3 1 −1 −3

will compute the linear contrast on it. The ESTIMATE statement estimates the parameters of the model. In a g_A by g_B set of data, there are $1 + g_A + g_B + g_A g_B$ parameters. The ESTIMATE statement makes use of a vector that consists of all of the parameters of the anova model $y_{ijk} = \mu + \alpha_j + \beta_k + \alpha\beta_{jk}$ listed in a certain order. This order is μ, followed by the g_A α's, the g_B β's, and the $g_A g_B$ $\alpha\beta$'s. That is, for a 3×2 design, the vector is

$$(\mu, \alpha_1, \alpha_2, \alpha_3, \beta_1, \beta_2, \alpha\beta_{11}, \alpha\beta_{12}, \alpha\beta_{21}, \alpha\beta_{22}, \alpha\beta_{31}, \alpha\beta_{32},)$$

Terms that are not used in the model—for example, the $\alpha\beta$'s if MODEL A B is used—are not included in the vector. The ESTIMATE statement allows one to estimate linear combinations of these parameters. At its simplest, it can be used to find the estimates of the parameters themselves. ESTIMATE INTERCEPT 1 will estimate. μ; ESTIMATE A 1 gets α_1; ESTIMATE A 0 1 gets α_2; ESTIMATE A*B 0 0 0 1 would provide $\alpha\beta_{31}$. The last two estimates may seem peculiar, but recall that we are getting linear combination of parameter estimates. The program automatically adds zeros at the end, so that ESTIMATE A 1 is really ESTIMATE A 1 0 0. If other linear combinations of the parameters are of interest, these too may be found. Note that this feature is also of use in multiple regression in cases where the investigator might want to look at combinations of the weights.

A potentially confusing aspect of PROC GLM is the reference to Type I, II, III, or IV sums of squares, a terminology that is not widespread. The only relevant sums of squares to use here are Types I and III: A Type I *ss* for an effect is the *ss* contributed by a given effect when it is added to the model in the specified order. For example, if MODEL Y = A B A*B is used, the Type I *ss* for *B* is the variance added by using *B* when *A* is already in the equation but *A***B* is not. A Type III *ss*, on the other hand, is the *ss* contributed when that effect is added last. The Type III *ss* for *B* is the amount contributed by *B* when both *A* and *A***B* are in the equation. Testing the Type I *ss* for *B* is analogous to testing *B* in a hierarchical analysis in which *B* comes second, whereas the Type III is the variance for which *B* is *necessary* in our terminology. In the balanced design, there is no difference between these two.

BMDP

BMDP offers a variety of anova programs, but P1V and P2V are sufficient for the analyses described in this chapter. A third program, P4V, is a very general program that will be discussed in Chapter 16 on multivariate analysis of variance.

The simpler program, P1V, does one-way analysis of variance only. After the statements that specify the input and the variable names and several other data-specification sentences, the independent variable is specified in a statement of the form GROUPING IS X1, where X1 is an already named variable that has the categories of the independent variable as its values. If only some of the values are to be used, then these are specified with a CODES ARE statement. A continuous independent variable may be made categorical by means of a CUTPOINTS statement. (These procedures are described in some detail in the introductory section of the manual.) Then the dependent variables are specified in the DESIGN paragraph, for example, DESIGN DEPENDENT ARE Y1, Y2. The one-way analysis of each dependent variable is performed in a straight forward manner, and the output is quite self-explanatory.

The P1V program allows the specification of contrasts. Assuming that the groups have been specified in the GROUPING IS statement, part of the DESIGN paragraph can be the specification of one or more CONTRAST statements. For example,

CONTRAST IS 3, 0, −1, −1, −1

for five groups, means that Group 1's mean is to be contrasted with the means of the last three groups. One may use as many contrasts as desired, and the *t*-values and associated two-tail probabilities are printed in the output.

A somewhat more elaborate program is P2V, which permits a number of anova designs, including repeated measures and a number of the variations on factorials that allow the investigator to omit many of the conditions from the experiment. The discussion here will be confined to the factorial design of fixed variables.

The problem specification proceeds much as in P1V with input and variable specifications. The problem is specified in the DESIGN paragraph with DEPENDENT IS (or ARE) to name the dependent variable or variables. Then the factors are specified using GROUPING ARE followed by the names of the variables that define the factors. The output is quite clear; it includes the mean, standard deviation, and frequency for each cell, as well as the anova table.

The P2V program allows only one approach to the nonorthogonal case, the one that corresponds to the same analyses in both orthogonal and nonorthogonal cases. That is, the sum of squares for each effect is computed by partialing out all other effects, as in TYPE III sums of squares in SAS GLM, and all are tested against the within-cells. This corresponds to the approach recommended for most analyses.

SPSS^x

SPSS^x offers three programs that can be used to do anova: ANOVA, ONEWAY and MANOVA. Two of these, ANOVA and MANOVA, will be discussed here. ANOVA permits the analysis of the types of simple designs where all variables are "between subjects," and all can be treated as fixed effects. It allows the possibility of covariates, discussed in Chapter 12. MANOVA permits the analysis of multiple dependent variables simultaneously, as in multivariate analysis of variance, discussed in Chapter 16.

In SPSS^x analyses, it is assumed that the data exist in an internal file, and that the locations and names of the variables have been specified prior to the statements that control the analyses. The details of this procedure are discussed in the manual.

The use of ANOVA requires the specification of the dependent variable; the factors or independent variables, including their levels; the covariates, if any; and possibly some optional details of the analysis. The statement

```
ANOVA SCORE BY FACA (1, 3) FACB, FACC (1, 2) WITH AGE
```

beginning in column 1, tells SPSS^x to perform an anova of the dependent variable called SCORE using the factors called FACA, FACB, and FACC using AGE as a covariate. FACA has three levels, entered 1, 2, or 3, and the others have two. Additional anovas can be specified simply by separating the specifications with slashes:

```
ANOVA SCORE BY FACA (1, 3)/DEP BY SEX (1, 2), SCHOOL (1, 4).
```

Multiple dependent variables can be specified, but are analyzed separately. The statement

```
Y1, Y2 BY FACA, FACB (1, 2)
```

will analyze both Y1 and Y2 using factors A and B.

In SPSS^x, choices among alternative analyses are governed by default values unless they are overridden by commands or options. If so, then the overrides stay in force until they in turn are overridden, or a new procedure is called. The main alternatives in anova have to do with error terms and the order of fitting effects. One can also govern the specifics of output.

In ANOVA, all effects are tested against within-cells unless the user requests pooling of interactions through the use of one of the OPTIONS statements. The latter can specify the level of interactions to be pooled with the within-cells, called RESIDUAL. One can, of course, do the pooling and corresponding F tests by hand after a default analysis. Otherwise, such strategies as the hierarchical analysis must be carried out by several analyses of the same data.

Where the cell frequencies are unbalanced, the sum of squares for an effect depends on what other effects have been "adjusted for." The default option, called the "classical experiment," adjusts main effects for other main effects, interactions for main effects, and other interactions at the same or lower levels. That is, for an A by B by C design, the A effect corresponds to

$$R^2_{A,B} - R^2_B$$

and the B effect corresponds to

$$R^2_{A,B} - R^2_A$$

The AB interaction is

$$R^2_{A,B,C,AB,AC,BC} - R^2_{A,B,C,AC,BC}$$

which is similar for the AC and BC, and the ABC interaction is

$$R^2_{\text{all}} - R^2_{A,B,C,AB,AC,BC}$$

The "regression approach" (OPTION 9) is used when any effect is adjusted for all other effects. OPTION 10 takes into consideration the order in which the factors are listed, adjusting the main effects for main effects listed earlier, for example, B is adjusted for A but not vice versa. All effects use the same error term, contrary to our discussion of various alternative strategies.

ANOVA does not implement any procedures for making particular comparisons or contrasts, nor does it allow an "unpooled" estimate of within-cells variance.

THE SPSS[x] MANOVA offers additional flexibility at the cost of considerable complication in controlling the program. The same structure is used as in ANOVA: "MANOVA *dependent(s)* BY *independents* WITH *covariates*" is the basic command. (Be sure to end the MANOVA statement with a slash!) One can specify subsets of variables to be used in different analyses with one or more (indented) ANALYSIS= subcommands. Otherwise the analysis will use all the variables specified. The DESIGN subcommand specifies effects that are to be included in the analysis. For example,

```
MANOVA Y1 Y2 BY A,B,C,/
    ANALYSIS=Y1 BY A,C/
    DESIGN=A,C/
    ANALYSIS=Y2 BY A,B/
    DESIGN=A,B,AB/
```

would first analyze the dependent variable Y1 using factors A and C but not the AC interaction. Then it would analyze Y2 using factors A and B considering both main effects and interaction. The default (with no DESIGN subcommand) is to consider all effects, but once a DESIGN subcommand has been used, the defaults are changed.

MANOVA offers the CONTRAST subcommand to specify a priori comparisons the user can specify using such standard names as POLYNOMIAL or a name specified by the user using SPECIAL, followed by a particular contrast vector.

Some flexibility in the denominators of F tests is allowed by the ERROR subcommand, but the details must be carefully examined beforehand. Overall, it may be simplest to use printed results from a default analysis to combine various error terms by hand calculation if a nondefault treatment of error is desired.

MANOVA includes many additional features such as a number of plotting and printing options, several subcommands relevant to the multivariate aspects, incomplete factorial designs of various kinds, mixed between and within subjects designs, and so on. Unfortunately, the manual is not always as complete as it might be, so a trial and error process may be required for analyses that go beyond the most straightforward.

PROBLEMS

1. Below are the data for a two-way design, with two observations per cell. Perform an anova by traditional methods.

	b_1	b_2
a_1	2, 2	3, 5
a_2	8, 4	11, 13

2. Convert the data to an X_+ matrix. Find $\mathbf{b}_+$, ss_{reg}, and ss_{dev}.

3. Test each of the regression coefficients from Problem 2 for significance and compare your results to those of Problem 1.

4. Show that the regression coefficients are the same as the product of the appropriate T^{-1} matrix and the group means.

5. A study has two different treatment conditions and a control. Each condition has 16 subjects; half of these are men and half are women. Assume the relevant contrasts are the two experimental groups versus the control group and the two experimentals versus each other, and the interactions of these contrasts with gender. Show the appropriate T matrix. The groups are listed in the following order: e_1 males, e_1 females, e_2 males, e_2 females, c_1 males, c_1 females. Name each of the resulting contrast variables.

6. The group means from the preceding problem are $\overline{\mathbf{y}}' = (4.0, 6.0, 8.0, 4.0, 12.0, 8.0)$, and $ss_w = 772.8$. Find the regression coefficients for each of the contrast variables and test each for significance.

7. What is the T matrix for a 3 × 3 design where the levels represent equally spaced quantitative variables and the contrasts represent the orthogonal polynomials? Indicate the nature of each of the contrasts, for example, linear A, quadratic A- linear B, and so on.

8. The values of the dependent variable for 21 subjects in a 2×2 design are as follows:

a_1b_1: 7, 14, 6, 11, 11, 10;

a_1b_2: 4, 7, 5, 7, 10;

a_2b_1: 9, 11, 6, 7, 6;

a_2b_2: 1, 7, 8, 7, 0

Use a factorial anova program that allows for unequal cell sizes to analyze the data. Then analyze the data as a regression problem where the three predictors are the contrast variables corresponding to the two main effects and interactions.

9. Find a study in the research literature, or in your own research, in which anova was used. Identify the dependent variable and the research questions that were studied. Show the T matrix that would be used to provide the contrast variables that could have been used to test the hypotheses of the study.

ANSWERS

1.

Source	*ss*	*df*	*ms*	*F*
A	72	1	72.00	24.00
B	32	1	32.00	10.67
AB	8	1	8.00	2.67
Total Between	112	3	37.33	12.44
Within	12	4	3.00	

Means

	b_1	b_2	
a_1	2.0	4.0	3.0
a_2	6.0	12.0	9.0
	4.0	8.0	6.0

$\bar{y}_{1\cdot} - \bar{y}_{2\cdot} = -6.0$

$\bar{y}_{\cdot 1} - \bar{y}_{\cdot 2} = -4.0$

$\bar{y}_{11} - \bar{y}_{21} - \bar{y}_{12} + \bar{y}_{22} = 4.0$

2.

$$\mathbf{X}_+' = \begin{bmatrix} 1 & 1 & 1 & 1 & -1 & -1 & -1 & -1 \\ 1 & 1 & -1 & -1 & 1 & 1 & -1 & -1 \\ 1 & 1 & -1 & -1 & -1 & -1 & 1 & 1 \\ 1 & 1 & 1 & 1 & 1 & 1 & 1 & 1 \end{bmatrix}$$

$$\mathbf{y}' = [2 \quad 2 \quad 3 \quad 5 \quad 8 \quad 4 \quad 11 \quad 13]$$

$$(\mathbf{X}_+'\mathbf{X}_+)^{-1} = \begin{bmatrix} .125 & 0 & 0 & 0 \\ 0 & .125 & 0 & 0 \\ 0 & 0 & .125 & 0 \\ 0 & 0 & 0 & .125 \end{bmatrix}$$

$$(\mathbf{X}_{+}'{}_{\mathbf{y}})' = [-24 \quad -16 \quad 8 \quad 48]$$

$$\mathbf{b}_+' = [-3.0 \quad -2.0 \quad 1.0 \quad 6.0]$$

$ss_t = 124.0$; $\quad ss_{dev} = 412 - 400 = 12.0$; $\quad ss_{reg} = 112.0$

3. $\sigma^2_{b_A} = \sigma^2_{b_B} = \sigma^2_{b_{AB}} = (.125)\,(3.0) = .375 \qquad F_A = 9/.375 = 24.0;\ F_B = 10.67;\ F_{AB} = 2.67$

4.
$$\mathrm{T} = \begin{bmatrix} 1 & 1 & 1 & 1 \\ 1 & -1 & -1 & 1 \\ -1 & 1 & -1 & 1 \\ -1 & -1 & 1 & 1 \end{bmatrix} \qquad \bar{\mathbf{y}}_{\cdot h} = \begin{bmatrix} 2.0 \\ 4.0 \\ 6.0 \\ 12.0 \end{bmatrix}$$

$$\mathrm{T}^{-1} = \begin{bmatrix} 1/4 & 1/4 & -1/4 & -1/4 \\ 1/4 & -1/4 & 1/4 & -1/4 \\ 1/4 & -1/4 & -1/4 & 1/4 \\ 1/4 & 1/4 & 1/4 & 1/4 \end{bmatrix} \bar{\mathbf{y}}_{\cdot h} = \begin{bmatrix} -3.0 \\ -2.0 \\ 2.0 \\ 6.0 \end{bmatrix}$$

5.
$$\mathrm{T} = \begin{bmatrix} 1 & 1 & 1 & 1 & 1 & 1 \\ -1 & 1 & 1 & -1 & -1 & 1 \\ 1 & 1 & -1 & 1 & -1 & 1 \\ -1 & 1 & -1 & -1 & 1 & 1 \\ 1 & -2 & 0 & -2 & 0 & 1 \\ -1 & -2 & 0 & 2 & 0 & 1 \end{bmatrix}$$

Contrast variables are, respectively: gender, experimentals versus control, difference in experimentals, gender by experimental-control, gender by experimental 1-experiment 2, and constant. The columns of T may be listed in a different order, but the names must correspond. Also, any column can be multiplied by a constant, including negative constants. For example, column 2 could be $(-1/2, -1/2, -1/2, -1/2, 1, 1)$.

6. For the order used here, $b_1 = 1.0$, $b_2 = -1.5$, $b_3 = -.5$, $b_4 = -.5$, $b_5 = -1.5$; $b_0 = 7.0$; $F_1 = 2.61$; $F_2 = 11.74$; $F_3 = .43$; $F_4 = 1.30$; $F_5 = 3.91$

7.
$$\mathrm{T} = \begin{array}{c} a_1b_1 \\ a_1b_2 \\ a_1b_3 \\ a_2b_1 \\ a_2b_2 \\ a_2b_3 \\ a_3b_1 \\ a_3b_2 \\ a_3b_3 \end{array} \begin{bmatrix} 1 & 1 & 1 & 1 & 1 & 1 & 1 & 1 & 1 \\ 1 & 1 & 0 & -2 & 0 & -2 & 0 & -2 & 1 \\ 1 & 1 & -1 & 1 & -1 & 1 & -1 & 1 & 1 \\ 0 & -2 & 1 & 1 & 0 & 0 & -2 & -2 & 1 \\ 0 & -2 & 0 & -2 & 0 & 0 & 0 & 4 & 1 \\ 0 & -2 & -1 & 1 & 0 & 0 & 2 & -2 & 1 \\ -1 & 1 & 1 & 1 & -1 & -1 & 1 & 1 & 1 \\ -1 & 1 & 0 & -2 & 0 & 2 & 0 & -2 & 1 \\ -1 & 1 & -1 & 1 & 1 & -1 & -1 & 1 & 1 \end{bmatrix}$$

As listed, the variables represent linear A, quadratic A, linear B, quadratic B, linear A-linear B, linear A-quadratic B, quadratic A-linear B, quadratic A-quadratic B, and constant.

8. Using SPSS^x ANOVA and choosing Option 7 yields the following anova table:

Source of Variation	*Sum of Squares*	*df*	*Mean Square*	*F*	*Alpha Level*
Main Effects					
A	21.219	1	21.219	2.576	.127
B	53.984	1	53.984	6.554	.020
2-Way Interaction	.001	1	.001	.000	.990
Explained	78.633	3	26.211	3.182	.051
Residual	140.033	17	8.237		

Using SAS GLM as the *regression* program yields:

Variable	*TYPE IV SS*	*F Value*	*PR F*
A Main	21.219	2.58	.127
B Main	53.984	6.55	.020
Interaction	.001	.00	.990

Note that the sums of squares are exactly the same as for the effects in the SPSS ANOVA. The explained and residual sums of squares are also identical.

READINGS

Bock, R. D. (1975). *Multivariate statistical methods for behavioral sciences.* New York: McGraw-Hill. pp. 269–355.

Cliff, N. (1982). What is and isn't measurement. In *Statistical and methodological issues in psychology and social science research.* Edited by G. Keren. Hillsdale, NJ: Erlbaum. pp. 3–38.

Cramer, E. M., and M. I. Appelbaum (1980). Non-orthogonal analysis of variance—once again. *Psychological Bulletin* 87:51–59.

Horst, P., and A. L. Edwards (1982). Analysis of non-orthogonal designs: The 2^k factorial experiment. *Psychological Bulletin* 91:190–92.

Keren, G. (1982). A balanced approach to unbalanced designs. In *Statistical and methodogical issues in psychology and social sciences research.* Edited by G. Keren. Hillsdale, NJ: Erlbaum. pp. 155–186.

Kerlinger, F. N., and E. J. Pedhazur (1973). *Multiple regression in behavioral research.* New York: Holt, Rinehart, and Winston, pp. 154–98.

Kirk, R. E. (1982). *Experimental design: procedures for the behavioral sciences* (2nd ed.). Monterey, California: Brooks/Cole.

Krantz, D. H. (1972). Measurement structures and psychological laws. *Science,* 175, 1427–35.

Myers, J. L. (1979). *Fundamentals of experimental design* (3rd ed.). Boston: Allyn and Bacon, pp. 378–405.

Overall, J. E., D. M. Lee, and C. W. Hornick (1981). Comparison of two strategies for analysis of variance in non-orthogonal designs. *Psychological Bulletin* 90:367–75.

Winer, B. J. (1971). *Statistical principles in experimental design* (2nd ed.). New York: McGraw-Hill.

12

Analysis of Covariance

PURPOSES OF ANALYSIS OF COVARIANCE

Statistical "Control" and Correction

The statistical procedure called analysis of covariance—or "ancova"—is actually a kind of partial correlation analysis. It is used in the context of an analysis of variance, that is, with qualitative independent variables. Ancova has one of two goals. The first is to increase the power of an anova through the reduction of the error mean square. Thus ancova is trying to *enhance* the relationship between dependent and independent variables. This applies particularly in the context of an experimental study where subjects are assigned to conditions at random. For example, we may use a verbal ability score as a covariate or control variable in an information processing experiment in order to reduce the within-conditions mean square.

The second goal of ancova applies when the independent variable in an analysis of variance is a natural or observed characteristic of the subjects. For example, if different educational procedures are used in different schools and we wish to study their effect on the students' achievement, there are likely to be differences among the schools on many extraneous characteristics. The differences in educational treatments are confounded by these differences such that observed differences in achievement *may* be due to differences on the other variables. Here, analysis of covariance attempts to correct for these other extraneous differences, and to rule out these alternative explanations of the findings.

These two goals just described parallel those that motivate partial correlation analysis. Analysis of covariance is a partial correlation analysis involving a treatment or other qualitative variable as the predictor and a continuous control variable. We remove the variance associated with the control variable and examine the relationship that remains between the predictor and dependent variable. The essence of this description does not change, even when there are multiple covariates and when the predictor variables are the factors in a complex design, although there are obviously some additional complications in the procedures. These will be dealt with later in this chapter.

In another sense, the above description is somewhat simplistic. It could be that the *relationship* between the dependent variable and the covariate is different under different treatments. Such occurrences tend to invalidate the interpretation of the simple partial correlations described above. A complete ancova is designed to reveal such differences when they occur and take appropriate action.

ANCOVA AND DEVIATIONS FROM REGRESSION

Simple Cases Illustrated

Probably the clearest and most accurate appreciation of the major aspects of ancova can be derived from consideration of the regression of the *dependent* variable on the *control* variable in the different groups, defined by the *independent* variable. Suppose we find a moderate difference in favor of an experimental group on y, but considerable overlap in the distributions for E and C in the randomized groups case. Two possibilities are shown in Figure 12-1. In the first case, where there is a fairly strong relationship between y and x_2, there is very little overlap in the *bivariate* distributions for the two groups. For any given value of x_1, if we take at random a member of E and a member of C, the score of E on y is almost always the higher. In the second case, there is no relationship between y and x_1, so for any particular value of x_1, the likelihood of the y value of a member of the E group being larger than the y value for a member of the C group is the same as the degree of overlap of the complete distributions. In this case there is no advantage to the ancova, unlike in the first case.

Figure 12-2 illustrates three possible situations in which there are differences between the two groups on x_1, the control variable, as would be the case if E and C were intact groups. In all three cases there is an apparent difference in favor of E.

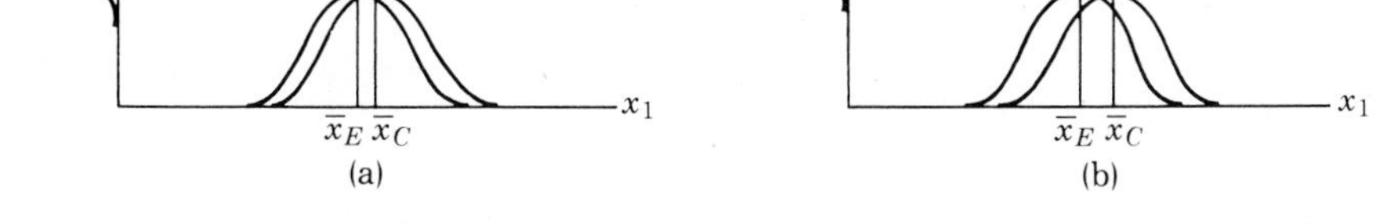

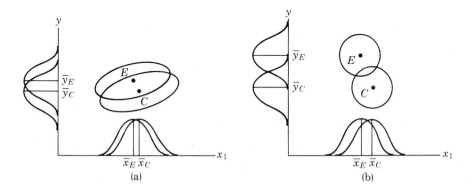

Figure 12-1 Analysis of covariance in the randomized groups (no population differences on the control variable) case; x_1 is the control variable and y is the *d.v.* In (a), there is a strong relationship between x_1 and y, so for any fixed value of x_1, there is little overlap between the E and C distributions. In (b), there is no relationship between x and y, so consideration of the value of x does not make the overlap smaller.

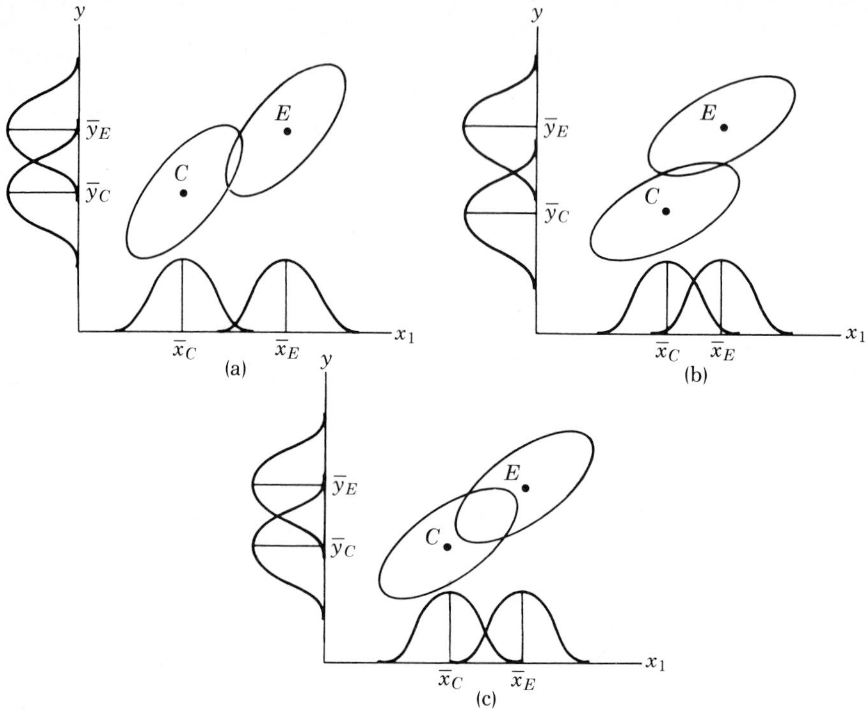

Figure 12-2 Analysis of covariance in intact groups where there are differences on the control variable. In (a), an apparent difference in *E* and *C* on *y* disappears when x_1 is considered. In (b), a moderate difference is enhanced when x_1 is considered, whereas in (c), the moderate difference in favor of *E* on *y* is reversed when the even larger difference on x_1 is considered.

However, in (a) this difference is completely accounted for by the large difference in x_1 for the two groups. In (b) the same sort of difference in y is *enhanced* when x_1 is considered because the difference on x_1 is small, much as in the randomized groups case. In (c) the difference on y is *reversed* when x_1 is considered, because the difference in x_1 is much larger than the one in y, taking into account the correlation. In all three cases, the true nature of the relationship is revealed when one examines those portions of the bivarate distributions for the range of values of x_1 where the two distributions overlap.

Deviations from Regression

One way of understanding ancova is in terms of deviations from a common regression line of y on x_1: ancova is an anova of these deviations. The question answered by the ancova is whether the deviations from this regression line tend to be in one direction in one group and the other direction in the other group. Usually, however, it is expressed in terms of whether one or more than one regression line is needed. Suppose we pool the data from all groups and fit a single regression line to the relationship between y and x, deriving the usual regression equation for observation i in group h as follows:

$$\hat{y}_{ih} = b_1 x_{ih} + b_o \qquad i = 1, 2, \ldots, n_h, h = 1, 2, \ldots, g \tag{12.1}$$

We could find the deviations

$$e_{ih} = y_{ih} - \hat{y}_{ih} \tag{12.2}$$

and do an anova, assessing group differences in mean deviations, but in ancova the analysis is usually formulated in terms of comparing the sum of squares of these deviations to the sum of squares resulting from fitting *different* regression lines in different groups. In order for the interpretation to be simple, the different lines must have the same slope but different intercepts. That is, instead of one regression equation for all the observations, as in 12.1, we have a separate system for each group:

$$\hat{y}_{ih}' = b_1 x_{ih} + b_{ho} \qquad i = 1, 2, \ldots, n_h; h = 1, 2, \ldots, g \tag{12.3}$$

Correspondingly, there are deviations from the separate lines:

$$e_{ih}' = y_{ih} - \hat{y}_{ih}' \tag{12.4}$$

So, the idea that there may be different mean deviations from the single regression line is equivalent to the idea that in different groups there are different regression equations having a common slope but different intercepts. The reasoning involved in this simple sort of ancova is that there are two equivalent ways of expressing what is going on. We could say that we are fitting a single regression equation to the data for all the groups and then doing an anova of the deviation from the regression line. Or, we could say that we are fitting a separate regression line for each group, with a common slope, and examining the different intercepts that result.

Partitioning the Variance

As usual, the main goal is to account for the total sum of squared deviations from the mean of y. Because of the logic of ancova, we would first look at the overall regression of y on x_1, the control variable, and then separate out the sum of squares into a portion due to regression ss_{reg} and one due to deviations, ss_{dev}. A test of significance may be performed in the usual way if we are interested in seeing whether any relationship between y and the covariate exists, but this is not our major interest.

The main purpose is to analyse the residual further. That is, to break it down into a between-groups and a within-groups portion. The between-groups portion is the sum of squares for intercept differences, whereas the remainder is the sum of squares for deviations from the individual regression lines. The former may be tested against the deviations as a test of the null hypothesis that the regression lines for the individual groups all have the same intercept. At this point, however, we must consider that the full story has not yet been recounted, and another issue must be considered before the analysis is carried out.

Slope Differences and Interaction

The complication that needs to be considered is the possibility that the *slopes* of the regression lines as well as the intercepts are different for different groups. If the slopes are different, then the differences in intercepts are much more difficult to interpret since the regression lines are then bound to cross. If they cross, then the answer to the question of which group is "higher" will depend on the value of the control variable; for some values one group will be higher; for others, another group will be higher. The mere presence of slope differences calls into serious question the relevance of any test of means or intercepts. There is still something that can be done, but at best the situation is complicated.

The connection between slope differences in ancova and interaction in anova may be noted. If the regression lines of y versus the control variable are not parallel for different treatments or groups, this is equivalent to saying that the amount of the effect of the treatment is different, depending on the value of the control variable. This is very similar to one of the ways of verbalizing the effect of the interaction in anova. The anova interaction may also be described as a lack of parallelism in the curves representing the effects of a variable, controlling for levels of the other variable. Thus different regression slopes in ancova is similar to interaction in anova.

Separate Regressions and the Sums of Squares

If the slopes of the regression lines are different, then a separate regression equation is needed for each group, different slopes as well as intercepts. The prediction should be

$$\hat{y}_{ih}'' = b_{1h}x_{ih} + b_{ho} \tag{12.5}$$

with a corresponding error

$$e_{ih}'' = y_{ih} - \hat{y}_{ih}'' \tag{12.6}$$

The sum of squared deviations from the group regression lines is the final ss_{dev}, $\Sigma_{i,h}(e_{ih}'')^2$. The difference between this and the previous equation $\Sigma_{i,h}(e_{ih}')^2$, that assumed different intercepts but a common slope, would then represent the sum of squares for slope differences. Symbolically, the different sums of squares may be represented as follows:

$$ss_{dev_1} = \Sigma\Sigma(y_{ih} - \hat{y}_{ih})^2 = \Sigma\Sigma e_{ih}^2 = \text{Deviations from common regression} \tag{12.7}$$

$$ss_{dev_2} = \Sigma\Sigma(y_{ih} - \hat{y}_{ih}')^2 = \Sigma\Sigma e_{ih}'^2 = \text{Deviations from regression lines with common slope} \tag{12.8}$$

$$ss_{dev_3} = \Sigma\Sigma(y_{ih} - \hat{y}_{ih}'')^2 = \Sigma\Sigma e_{ih}''^2 = \text{Deviations from separate regressions} \tag{12.9}$$

The differences between these sums of squares represent the effects of fitting the additional constants. First,

$$ss_{dev_1} - ss_{dev_2}$$

is the effect of allowing different intercepts, Equation 12.3 instead of 12.1. Then

$$ss_{dev_3} - ss_{dev_2}$$

is the effect of allowing different slopes, Equation 12.5 instead of Equation 12.3. The difference between ss_{dev_1} and ss_{dev_2} is the same as the $ss_{between}$ of the deviations, Equation 12.2.

These differences can be formulated in terms of multiple regressions on predictor matrices that contain both the qualitative and quantitative variables. In the simplest case of no group membership variables we have the standard univariate X_+ matrix, as shown in Exhibit 12-1, labeled X_{+r} to distinguish it. The only difference between this and the X_+ matrix described at the beginning of Chapter 6 is that x has been doubly subscripted to indicate group membership. We can obtain the regression of y on this x in the usual way, by finding b_1 and b_0. In the second X_+ matrix of the table, labeled X_{+i}, to show it allows for different intercepts, we have added $g - 1$ group membership (dummy) variables, as in analysis of variance. For purposes of the illustration, $g = 3$. The final X_+ matrix, called X_{+s} to show it allows for different slopes, spreads out the x variable so that there is a different variable within each group. This matrix is simple to construct:

$$X_{+r} = \begin{bmatrix} x_{11} & 1 \\ x_{21} & 1 \\ \vdots & \vdots \\ x_{n_1 1} & 1 \\ x_{12} & 1 \\ x_{22} & 1 \\ \vdots & \vdots \\ x_{n_2 2} & 1 \\ x_{13} & 1 \\ x_{23} & 1 \\ \vdots & \vdots \\ x_{n_3 3} & 1 \end{bmatrix} \quad X_{+i} = \begin{bmatrix} x_{11} & 1 & 0 & 1 \\ x_{21} & 1 & 0 & 1 \\ \vdots & \vdots & \vdots & \vdots \\ x_{n_1 1} & 1 & 0 & 1 \\ x_{12} & 0 & 1 & 1 \\ x_{22} & 0 & 1 & 1 \\ \vdots & \vdots & \vdots & \vdots \\ x_{n_2 2} & 0 & 1 & 1 \\ x_{13} & 0 & 0 & 1 \\ x_{23} & 0 & 0 & 1 \\ \vdots & \vdots & \vdots & \vdots \\ x_{n_3 3} & 0 & 0 & 1 \end{bmatrix} \quad X_{+s} = \begin{bmatrix} x_{11} & 0 & 0 & 1 & 0 & 1 \\ x_{21} & 0 & 0 & 1 & 0 & 1 \\ \vdots & \vdots & \vdots & \vdots & \vdots & \vdots \\ x_{n_1 1} & 0 & 0 & 1 & 0 & 1 \\ 0 & x_{12} & 0 & 0 & 1 & 1 \\ 0 & x_{22} & 0 & 0 & 1 & 1 \\ \vdots & \vdots & \vdots & \vdots & \vdots & \vdots \\ 0 & x_{n_2 2} & 0 & 0 & 1 & 1 \\ 0 & 0 & x_{13} & 0 & 0 & 1 \\ 0 & 0 & x_{23} & 0 & 0 & 1 \\ \vdots & \vdots & \vdots & \vdots & \vdots & \vdots \\ 0 & 0 & x_{n_3 3} & 0 & 0 & 1 \end{bmatrix}$$

Regression Only — Common Slope, Different Intercepts — Different Slopes and Intercepts

Exhibit 12-1 Difference X_+ matrices for different covariance models.

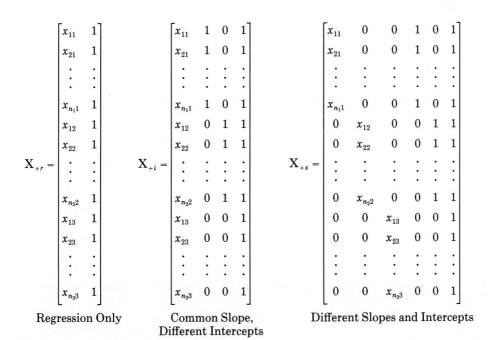

the within-group control variables are simply the product of the control variable with the corresponding dummy variable.

Now we need to think of the different $\mathbf{b}_+$ vectors that go with each of these X_+ matrices in Exhibit 12-1. With the case of the first X_{+r}, $\mathbf{b}_{+r} = (b_1, b_0)'$. With the second, X_{+i}, $\mathbf{b}_{+i} = (b_1, b_{01} - b_{03}, b_{02} - b_{03}, b_{03})'$. In the third, $\mathbf{b}_{+s} = (b_{11}, b_{12}, b_{13}, b_{01} - b_{03}, b_{02} - b_{03}, b_{03})'$. Recall that the numerical values of regression coefficients will be different when different "variables" are added to the equation. Applying the appropriate $\mathbf{b}_+$ vector to its corresponding X_+ matrix will clearly give us y, $\hat{y}'$, or $\hat{y}''$. For example, in Exhibit 12-1 consider the row for x_{22}. Multiplying this row of X_{+r} by $\mathbf{b}_{+r}$ gives $b_1x_{22} + b_0$. Multiplying the same row of X_{+i} by $\mathbf{b}_{+1}$ gives $b_1x_{22} + b_{02}$. Similarly, that row of X_{+s} by $\mathbf{b}_{+s}$ gives $b_{12}x_{22} + (b_{02} - b_{03}) + b_{03}$.

Note also that the first column of the middle matrix is the sum of the first three columns of the last one, and that the second column of the first matrix is the sum of the last three columns of the middle matrix. Thus, the information in each of the matrices is nested within the matrix on its right; the models corresponding to them are also nested, one within the other.

An example of this type of analysis might be a word-recognition experiment in which y is the time taken to respond to a word flashed on a screen, and subjects are divided into three groups representing different experimental conditions. The control variable is the score on a vocabulary test. Then, the complete X_{+s} matrix has five predictors. The first predictor contains the vocabulary scores of the first group with zero for everyone else; the second is zero for everyone in the first group, the vocabulary scores of those in the second group, and zero for everyone in the third group; the third is zero for everyone in the first two groups and the vocabulary scores of the third group. The fourth variable is a dummy variable indicating membership in the first group, and the fifth indicates membership in the second group. There is also the constant column.

A regression analysis on all five predictors is carried out. A second analysis is done, using the two indicator variables and only a single vocabulary variable that contains the scores of all individuals. Finally, a simple regression is carried out using only the vocabulary scores. The error mean squares and multiple correlations from each of the three analyses indicate the degree of fit of the various models.

The process described above is illustrated in detail in Exhibit 12-2 for a small sample. The various regression lines and deviations for those data are shown in Figure 12-3.

Multiple Covariates

In many problems it is reasonable to use numerous covariates. However, some discretion is advised. Filling the old ancova blunderbuss with a haphazard assortment of shot, pebbles, keys, chain links, and what-have-you, and closing our eyes and pulling the trigger, is not the most effective way of hitting the target. Every covariate costs a degree of freedom, so it is best to include only the most relevant. Moreover, since this is really a form of regression, inferences become slipperier as the variables increase. More often than not, there is more than a single covariate, however.

One way to minimize the strain of thinking about multiple covariates is to imagine that, before our analysis of covariance, some helpful soul had combined all our covariates into a single composite, so that what we have been calling x_1 is really a composite. If we ignore for the moment the question of how this

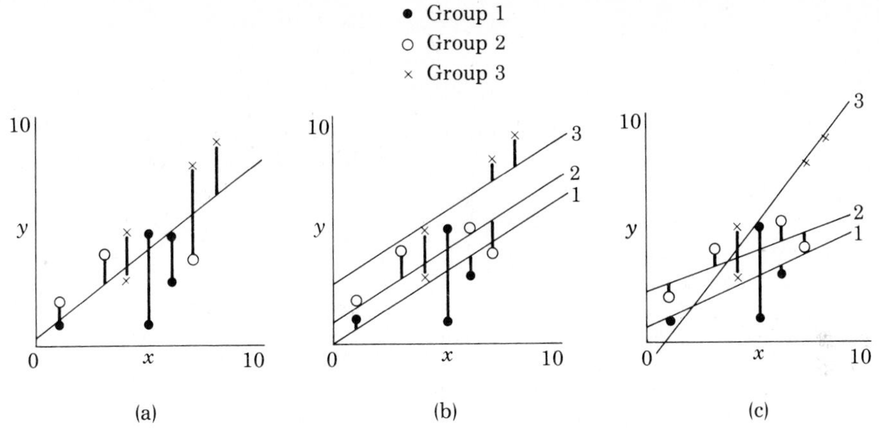

Figure 12-3 Fit of different models in analysis of covariance using data from Exhibit 12-2. Part (a) shows errors resulting from fitting a single regression line. Part (b) shows those from fitting three lines with equal slopes, and part (c) shows a different line for each group.

				Regression Only		*Different Intercepts*		*Different Regressions*	
i	h	x	y	$\hat{y}$	$y-\hat{y}$	$\hat{y}'$	$y-\hat{y}'$	$\hat{y}''$	$y-\hat{y}''$
1	1	1	1	1.267	−.267	.495	.505	1.068	−.068
2	1	5	5	4.360	.640	2.963	2.037	2.831	2.169
3	1	5	1	4.360	−3.360	2.963	−1.963	2.831	−1.831
4	1	6	3	5.133	−2.133	3.580	−.580	3.271	−.271
Mean		4.25	2.50	3.780	−1.280	2.500	0	2.500	0
1	2	7	4	5.907	−1.907	5.446	−1.446	4.747	−.747
2	2	3	4	2.813	1.187	2.979	1.021	3.297	.703
3	2	1	2	1.267	.733	1.745	.255	2.571	−.571
4	2	6	5	5.133	−.133	4.830	.170	4.385	.615
Mean		4.25	3.75	3.780	−.030	3.750	0	3.750	0
1	3	4	5	3.587	1.413	5.170	−.170	4.020	.980
2	3	8	9	6.680	2.320	7.638	1.362	9.118	−.118
3	3	7	8	5.907	2.093	7.021	.979	7.843	.157
4	3	4	3	3.587	−.587	5.170	−2.170	4.020	−1.020
Mean		5.75	6.25	4.940	1.310	6.250	0	6.250	0
Grand Mean		4.75	4.17	4.167	0	4.750	0	4.750	0
Sum of Squares					34.025		19.373		11.931

Exhibit 12-2(a) Analysis of covariance with single covariate shows the covariate x and dependent variable y for three groups of four observations. Then the predicted values of y based on three different models are shown, along with the corresponding errors. The $\hat{y}$ predictors are based on a single regression equation; $\hat{y}'$ predictors are based on regressions with different intercepts but the same slope. Finally, the $\hat{y}''$ predictors are based on different slopes as well as different intercepts.

1. Regression Only

$$b_1 = \frac{\sum_i \sum_h x_{ih} y_{ih} - \bar{x}_1 \sum_i \sum_h y_{ih}}{\sum_i \sum_h x_{1ih}^2 - \bar{x}_{1..} \sum_i \sum_h x_{ih}} \qquad b_0 = \bar{y}_{..} - b_1 \bar{x}_{..}$$

$$= .7733 \qquad b_0 = .4933$$

2. Different Intercepts

$$b_1 = \frac{\sum_h \left(\sum_i x_{1ih} y_{ih} - \bar{x}_{1 \cdot h} \sum_i y_{ih} \right)}{\sum_h \left(\sum_i x_{1ih}^2 - \bar{x}_{1 \cdot h} \sum_i x_{1ih} \right)} \qquad b_{0h} = \bar{y}_{\cdot h} - b \bar{x}_{\cdot h}$$

$$= .6169 \qquad b_{01} = -.1218$$

$$b_{02} = 1.1282$$

$$b_{03} = 2.7028$$

3. Different Slopes and Intercepts

$$b_{1h} = \frac{\sum_i x_{1ih} y_{ih} - \bar{x}_{1ih} \sum_i y_{ih}}{\sum_i x_{1ih}^2 - \bar{x}_{1 \cdot h} \sum x_{1ih}} \qquad b_{0h} = \bar{y}_{\cdot h} - b_h \bar{x}_{\cdot h}$$

$$b_{11} = .4407 \qquad b_{01} = .6271$$

$$b_{12} = .3626 \qquad b_{02} = 2.2090$$

$$b_{13} = 1.2745 \qquad b_{03} = -1.0784$$

Exhibit 12-2(b) Computation of parameter estimates for the different models. Part 1 is the single slope and intercept used to derive $\hat{y}$. Part 2 is the information for the model with different intercepts and the same slope, and part 3 is for different slopes and intercepts.

	ss	*df*
1. Deviations from grand mean	66.667	11
2. Deviations from $\hat{y}_{ih} = b_1 x_{1ih} + b_0$	34.025	10
3. Deviations from $\hat{y}_{ih}' = b_1 x_{1ih} + b_{0h}$	19.373	8
4. Deviations from $\hat{y}_{ih}'' = b_{1h} x_{1ih} + b_{0h}$	11.931	6

Exhibit 12-2(c) Sums of squares for deviations from each model and the corresponding degrees of freedom. The latter is equal to n minus the number of parameters fitted, that is, $n - 1$, $n - p - 1$, $n - p - g$, and $n - gp - g$, respectively. The first model fits the mean; the second fits p slopes (one for each covariate); the third fits p slopes and g intercepts; and the fourth fits gp slopes and g intercepts. These values are 11, 10, 8, and 6, respectively. The deviations for the last three models are from Exhibit 12-2(a). Figure 12-1 shows the regression lines and errors in each model.

Source	*ss*	*New Parameters*	*ms*	*F*
1. Different Slopes and Intercepts	19.373 − 11.931 = 7.442	2	3.72	1.872
2. Different Intercepts Only	34.025 − 19.373 = 14.652	2	7.326	3.685
3. Regression Error	67.667 − 34.025 = 33.642	1	33.642	16.922
	11.931	6	1.988	

Exhibit 12-2(d) Tests of models. The first test shows the effect of allowing different slopes, which is not significant here. Under H_0: $\beta_1 = \beta_2 = \beta_3$, the difference in deviation sums of squares is treated as a reflection of error. The second test shows the effect of allowing different intercepts. This is significant, reflecting, for example, treatment effects that are additive. Thus, H_0: $\beta_{01} = \beta_{02} = \beta_{03}$ is rejected. The third test is only when *both* previous tests are accepted. In the present case, where the intercepts are different, the appropriate test of $\beta = 0$ would be to compare the deviations using regression and intercept differences (19.373) to the *anova* sum of squares within groups, since the latter amounts to fitting only the dummy variables. In that way we find out if there is regression within groups.

composite was constructed, we can use the same scheme for general ancova as we did for the single covariate case. This is satisfactory up to a point; beyond that point it is necessary to admit that there are multiple covariates.

This can be done by judicious substitution in the equations we have used so far. For example, Equation 12.1 can be expanded by replacing $b_1 x_{ih}$ by $\Sigma b_j x_{ihj}$, where j designates the different covariates. In Equations 12.3 and 12.5 we would make the same substitution while the rest of the equation stays the same. We can still use the same symbols as before, adding a j subscript to x_{ih}. The general formulas then become

$$\hat{y}_{ih} = \sum_j b_j x_{ihj} + b_o \qquad (12.10)$$

$$\hat{y}_{ih}' = \Sigma b_j \mathrm{x_{ihj}} + b_{oh} \qquad (12.11)$$

$$\hat{y}_{ih}'' = \Sigma b_{hj} \mathrm{x_{ihj}} + b_{oh} \qquad (12.12)$$

In the first equation, x_{ihj} stands for the score of individual i of group h on covariate j, and b_j stands for the weight of that variable in a common regression equation; b_o is the common intercept. In Equation 12.11, these terms have the same meaning, except that b_{oh} refers to the intercept in group h. Finally, in the last equation, b_{hj} allows for different weights (slopes) in the different groups.

The regression analysis form generalizes readily to multiple control variables. Suppose there are p control variables. Then there is an $n \times (p + 1)$ X_{+r} matrix corresponding to that in Exhibit 12-1, reflecting just the prediction of y from these control variables. Thus, using weights $\mathbf{b}_{+r}$, $\hat{y} = (b_1, b_2, \ldots, b_p, b_0)'$. Then the generalization of the middle X_{+i} matrix is formed by adding $g - 1$ dummy variables to these multiple covariates. Now, $\hat{y}'$ is obtained using the weight vector

$$\mathbf{b}_+ = (b_1, b_2, \ldots, b_p, b_{01} - b_{0g}, b_{02} - b_{0g}, \ldots, b_{0g})$$

That is, $\mathbf{b}_{+i}$ has $p + g$ entries. Finally, the predictor matrix corresponding to different slopes can be generated, X_{+s}. Each of the p predictors is spread out into g versions. In each, the entries are zero except for the members of one group. This makes gp predictors of this kind, together with the $g - 1$ dummy codes and the intercept. These are used with the corresponding regression weights vector:

$$\mathbf{b}_{+s} = (b_{11}, b_{12}, \ldots, b_{g1}, b_{12}, b_{22}, \ldots, b_{g2}, \ldots, b_{1p},$$
$$b_{2p}, \ldots, b_{gp}, b_{01} - b_{0g}, b_{02} - b_{0g}, \ldots, b_{0g})'$$

The X_{+s} matrix is thus $n \times (g + 1)p$. The presence of more control variables multiplies the number of variables involved but does not change anything qualitatively.

There seems to be a difference between this ancova hierarchical regression analysis and the hierarchical multiple regression. In the latter case, the variables used in the models are subsets of each other. In ancova, this seems not to be literally true. However, since the variables in the smaller set are a linear combination (in this case, a simple sum) of the variables in the larger, the effect is the same as using nested sets of predictors.

The sums of squares of deviations from the various models require no new concepts or equations. One merely realizes that the three sums of squares in Equations 12.7, 12.8, and 12.9 have their predictions generated by the general forms of the Equations 12.10, 12.11, and 12.12.

Statistical Inference

So far our discussion has been at the descriptive level. However, the investigator's interest ordinarily lies in deciding which of the models is more appropriate in the population. The appropriate analysis is simple to identify when we look at the questions in terms of multiple regression. In all, there are $(p + 1)g - 1$ predictors, pg within-group control variables, and $g - 1$ dummy variables, so

Table 12-1 Sums of Squares and Degrees of Freedom in Analysis of Covariance

Null Hypothesis	*Prediction*	*Error*
1. $\beta_j = 0$ (for all j)	$\hat{y}_{ih} = \sum_j b_j x_{ihj} + b_o$	$\sum_{i,h} (y_{ih} - \hat{y}_{ih})^2 = ss_{dev_1}$
2. $\beta_{o1} = \beta_{o2} = \cdots = \beta_{og}$	$\hat{y}_{ih}' = \sum_j b_j x_{ihj} + b_{h0}$	$\sum_{i,h} (y_{ih} - \hat{y}_{ih}')^2 = ss_{dev_2}$
3. $\beta_{j1} = \beta_{j2} = \cdots = \beta_{jg}$ (for all j)	$\hat{y}_{ih}'' = \sum_j b_{hj} x_{ihj} + b_{h0}$	$\sum_{i,h} (y_{ih} - \hat{y}_{ih})^2 = ss_{dev_3}$

Note: p is the number of covariates and g is the number of groups.

there are $n - pg - g$ df for $ss_{dev\,3}$, but $n - p$(covariates) $- g$ (groups) for $ss_{dev\,2}$ because of the p covariates. The hypothesis of different slopes is tested by comparing the corresponding ss_{dev} since $\hat{y}''$ used $[(p + 1)g - 1] - (n - p - g)$ more parameters than $\hat{y}'$. Thus, applying Equation 9.33, we have the following regression analysis:

$$F = \frac{(ss_{dev_2} - ss_{dev_3})/p(g - 1)}{ss_{dev_3}/(n - pg - g)} \tag{12.13}$$

Equivalently, the corresponding squared multiple correlations could be compared.

If this hypothesis of equal slopes is rejected, then the full model is necessary, and ordinarily no further analysis is considered justified. If the slopes are different, it makes little sense to look at the intercepts. This conclusion is modified somewhat in the discussion later on.

If the test of different slopes is not significant, then it is legitimate to test the intercepts. This too is done by the F test. The F contrasts the errors using $g + p - 1$ predictors, that is, $g - 1$ dummy variables and the p covariates, and the p covariates alone. Again, Equation 9.33 becomes:

$$F = \frac{(ss_{dev_1} - ss_{dev_2})/(g - 1)}{ss_{dev_2}/(n - g - p)} \tag{12.14}$$

If this hypothesis is rejected, then the hypothesis of equal intercepts (or the equivalent hypothesis of equal means on y adjusted for differences on x) is rejected.

The various models, their parameters, degrees of freedom, and sums of squares are presented in Table 12-1.

Adjusted Means

Applications of ancova are most useful when the slopes are found to be, or even assumed to be, equal. In that case, it can be considered an analysis of variance of "adjusted group means" rather than a test of equality of intercepts. The error

df Used	*F Numerator*	*F Denominator*
$p + 1$	$\frac{(ss_T - ss_{dev_1})}{p}$	$\frac{ss_{dev_1}}{n - p - 1}$
$p + g$	$\frac{(ss_{dev1} - ss_{dev_2})}{g - 1}$	$\frac{ss_{dev_2}}{n - p - g}$
$gp + g$	$\frac{(ss_{dev2} - ss_{dev_3})}{p(g - 1)}$	$\frac{ss_{dev_3}}{n - gp - g}$

mean square is defined as ss_{dev_2}. The adjusted means are

$$\hat{\mu}_h = \bar{y}_{.h} - \sum_j b_j(\bar{x}_{.hj} - \bar{x}_{..j}) \tag{12.15}$$

This expression shows that the adjusted means are the obtained group means "adjusted" for differences on the covariates. It differs from the intercepts themselves by the addition of the term $\Sigma_j b_j x_{..j}$. Thus it examines the separation of the lines at the overall mean of the covariates.

Contrast Codes and Ancova

There is no need to limit the form of the variables representing the groups to the dummy code form. We can also use some contrast codes in place of the dummy codes. They will be exactly the same transformations of the dummy codes as in the anova case. Thus the second set of predictors may be in either dummy code or contrast form.

We may evaluate the contribution of any contrast or effect to the prediction or explanation of y. Let x_c be some contrast variable, for example, control-groups-versus-experimental. In the ancova context, the prediction of y may be made using *all* the contrasts *and* covariates and using all *but* this contrast. The difference in the residual sum of squares in the two cases is the contribution of this contrast, which takes the covariates into consideration as well as the other contrasts. In this way, we can evaluate the contribution of the effects in a 2^k factorial design since there is one contrast per effect. With more than one degree of freedom for an effect, the process is similar. We compute the ss_{res} using all the $p + g - 1$ predictors, again omitting the ones which define a particular effect such as the two contrasts that define an A main effect which has three levels. The reduction in ss_{res} due to the inclusion of these contrasts is the amount attributable to that effect, taking the covariates into consideration.

The significance of the contribution of any contrast or effect is evaluated as we would expect. Any contrast may be tested for significance by testing its regression coefficient. For a multilevel effect, we make the usual F test to evaluate the significance of the contribution of any set of variables to the prediction equation. This means dividing the reduction in ss_{res} by the degrees of freedom, and testing it against the final ms_{res}. In this way, the regression approach furnishes a straightforward procedure for doing ancova.

Magnitude of Computations

To anyone who has wrinkled his or her brain over the complexities of ancova as described in the analysis of variance context, the preceding discussion may seem like a cursory treatment. However, this is all there is to it, as long as the computations are being done by a computer using a regression model. The difficulties in ancova arise mainly from the computational complexities; we may lose sight of what is going on conceptually during the desperate attempt to keep track of the computations when done by hand.

The computational details of ancova will not be treated here. While it is feasible to carry out the computations by hand with one or two covariates, the process is quite tedious and not feasible with more covariates. Thus it is assumed that the computations will be carried out on a computer.

Extensions of Ancova

It may be obvious that ancova generalizes to factorial and other designs and to the testing of various contrasts. Interpretations are obviously complex except in the equal-slopes case, so we will only discuss that case. Viewing ancova as a regression problem, the dummy codes are transformed into contrast codes, just as in anova. Thus the various tests of contrasts (regression coefficients) and effects would proceed in a manner similar to that of regression analysis in anova.

Since the computations are rarely carried out in this literal multiple regression way, the approach to testing effects and contrasts should be spelled out. Recall that the transformation of dummy codes into contrasts did not literally need to be done; the same thing is accomplished by applying the inverse of the transformation to the group means.

Clearly the same thing applies in ancova. The difference is that the adjusted means, or equivalently, the intercepts, form the basis for the analysis rather than the group means themselves. When the group sizes are equal, one could straightforwardly perform analyses of variance of these intercepts. Where they are not equal, then exactly the same procedures apply as were used in the anova.

In summary, then, the methods of ancova readily extend to the case of designed studies and contrast variables.

DESIGN AND INFERENCE IN ANCOVA

Failures of Assumptions

The ideas of ancova apply best where there is a more-or-less continuous dependent variable, some qualitative or discrete independent variables, and some more or less continuous control variables. It is in this context that the description emphasizing the slopes and intercepts of regression lines applies most clearly. Furthermore, normality of errors and homogeneity of variance assumptions underlie the probability statements. The analyses can be applied where the variables have other properties, such as when some of the control variables are dichotomous, or even when the dependent variable is dichotomous. No one has yet built into a computer program a feature that checks the values of a variable and prints out "Shame! Shame! Shame!" when the computer discovers that a variable that is supposed to be continuous is not. The problem is that we are uneasy about just what the results might mean in those cases, both at the descriptive level and when it comes to inferential statements. Here, as elsewhere, statistical analysis is based on "assumptions" or what might be called idealizations of the real-life characteristics of data. The data never conform exactly to those idealizations, even under the best of circumstances. We are sometimes assured that the conclusions drawn from an analysis are statistically valid unless the data depart from the ideal "too far," but the definition of "too far" proves to be complex and dependent on characteristics of the situation that are difficult to assess.

Applying ancova methods which are designed for continuous, normally distributed variables, to variables that are dichotomous or radically skewed presents a case in which the data clearly depart from the model which underlies the analysis. But if that is the kind of data we have, what are we to do? In the vast majority of instances, most investigators will analyze the data anyway and make inferences in much the same way as if the data were ideal. (If we are advising someone else, however, we may protect ourselves by asking to be left out of the acknowledgment footnote.)

Assuming we are doing the analysis to find out what is in the data—a far from universal motivation—there is a certain amount we can do to guard against complete fatuity. Probably the most useful tool is the scatterplot. The most useful ones plot predicted values of y against the obtained values separately for the different groups. These plots are less commonly provided by computer programs, than are the error deviations, which are nearly as useful. In either case, the clearest results are obtained when the data from different groups are plotted on the same graph using different symbols. In this way, we can see just what the differences are between the groups and gain some confidence that the descriptive statements we are making are consistent with the data.

Intact and Randomized Groups

Reference has been made earlier to the differences in interpretation and purposes of ancova as a function of whether the independent variables represent characteristics that define intact groups or whether they are manipulated variables. In the latter case, the main function of the control variables is to sharpen the analysis by reducing error variance. In the former case, the control variables may show instead that the observed group differences are not the result of differences among the groups on the control variables.

In both cases, ancova may be regarded as a form of partial correlation analysis. We want to look at the relationship between the dependent variable and the independent variables when the effect of the control variable is removed. Therefore, all the cautions and concerns that were expressed in the context of partial correlation are applicable here.

The larger problems occur when we are dealing with intact groups. Here, the entire study is basically correlational, even though some of the variables are qualitative. In this case ancova is usually done to see whether there is a change in the ordinal relationships among group means on some dependent variable when the control variables are taken into account.

There is a problem in that a discrepancy almost always exists between the manifest variable analyzed and the latent variable that it represents. We do not completely control for the latent variable "social class," for example, when we use parents' reported education as a covariate. If two groups differ in their average social class, then they still differ in their average social class when we partial out parent's education. The degree to which this occurs is an inverse function of the accuracy with which the manifest variable corresponds to the latent one. With rare exceptions, this discrepancy is substantial with respect to almost all subject characteristics used in behavioral science research. Therefore, we can rarely have confidence that intact groups have been completely "equated" on relevant control variables.

It is unfortunate that many issues of direct social consequence, such as the effects of different educational treatments, may involve exactly these difficulties. It is barely possible to imagine a study that would empirically resolve such issues, much less be able to carry out such a study in practice.

In spite of these drawbacks, it is still desirable to include the control variables since they may shed considerable light on results, even though they are not perfect measures of what they are intended to measure. Suppose the group differences studied in an analysis of variance are found to be significant and large enough to be of practical importance. Then it is incumbent upon the investigators to see if these differences are accounted for by plausible covariates. If a fairly complete collection of covariates only slightly reduces the amount of variance that is accounted for, then it seems unlikely that better measures of the important variables would have had much more of an effect. If however, the covariates account for the major part of the group differences, then we may well suspect that the remainder might disappear given more precise and uncontaminated measures of the true variables.

Dichotomized Subject Variables Versus Covariates

Experimental psychologists often treat a subject characteristic as dichotomous when it is more or less continuous. This is done by taking a variable such as age, or some measure of ability or personality, and dividing the subjects in half on the basis of their scores, into one group of "highs" and another of "lows." Most often the split occurs at the mean or median although sometimes those subjects who score near the middle are discarded.

The problem with this procedure is that there remains substantial variance on the variable *within* those groups, but the groups are then treated as if they were homogeneous with respect to it. Treating the dichotomized variable as a covariate rather than a dichotomy will usually result in a more efficient design. It will also give a more reasonable estimate of the variance attributable to that variable.

The effect of dichotomizing covariates is best interpreted by specific consideration of some statistical effects. Suppose we substitute a dichotomous variable in place of a continuous covariate. The dichotomous variable should be as closely related to the continuous variable as possible. If the dichotomy is at the mean and the variable is approximately normally distributed, then the correlation between the continuous variable and the dichotomy derived from it will be about .8. This is a reasonably high correlation, although far from perfect. Thus, while conclusions derived using the dichotomy are likely to be similar to those derived from the original variable, they may sometimes differ from them in important respects. The practice of dichotomization probably owes much of its popularity to the simplification of the calculations that result. Given modern computing facilities, this is no longer important, so the frequency of the practice should be reduced, although many investigators—and advisors—may be reluctant to change their ways.

Like any other human activity, the effectiveness of the actual practice of data analysis depends on assimilating real-life experience. The most that can be done here is to alert the reader to the most common issues and sources of difficulty. The most important determinants of correct practice are likely to be the attitudes and abilities of the investigator. If one has a clear understanding of what a particular method does and is able to perceive the degree to which it matches up with our experimental questions, then the outcome and its interpretation are likely to be close to the optimum. The other main ingredient is a dispassionate stance with respect to the relationships in the data. Without this, prejudices and preconceptions may keep the investigator from getting the most relevant data or blind him or her to what is or is not there.

SUMMARY

Analysis of covariance is a member of the family of general linear models, of which multiple regression is the prototype. In ancova the ideal case is one in which there is a continuous dependent variable, one or more qualitative independent variables that divide the observations into nonoverlapping subgroups, and one or more continuous control variables or covariates. In effect, we examine the regression of y on the dependent variable within *groups,* and compare the estimates of the regression parameters obtained from different groups. There are two major differences between using ancova and anova on the same dependent variable. First, we can use ancova to answer the question as to whether some observed group differences on y can actually be explained in terms of corresponding differences on the control variables. Alternatively, or at the same time, the ancova can serve to *sharpen* differences among groups by reducing residual variation. To a large extent, we have the former purpose in mind when dealing with intact groups defined by natural variables and the latter purpose in mind when the groups are defined in terms of experimental manipulations.

In most applications, the ancova can be described as attempting to see if the regressions of y on the x's have different intercepts in different groups. This is equivalent to asking if there are consistent differences in the means of y for different groups given fixed values for the x's. Such a question is relatively meaningless unless the slopes of the regression are the same in the different groups; otherwise, the regression lines cross. Therefore, as a preliminary to the test of differences in intercepts, it is important to test for equality of slopes of the regression lines among the different groups.

COMPUTER APPLICATIONS

BMDP

Both P1V and P2V, described in Chapter 11, allow for covariates, although their specifications are different. In P1V, covariates are listed in the DESIGN paragraph using the statement INDEPENDENTS ARE followed by the list of covariates. The actual independent variable is specified in the GROUPING statement. Then P1V performs the complete analysis, including the information for testing equality of slopes.

The output of an ancova using P1V is quite complete. It shows the complete analysis of variance tables, the tests for equality of slopes, equality of intercepts (listed as EQUALITY OF ADJUSTED CELL MEANS), and of zero slopes.

It also prints—either by default or as options requested in a PRINT paragraph—a number of useful statistics. These include the within-group and overall regression coefficients; the group means, adjusted group means, and their standard errors; a matrix of t's for the adjusted group means and the associated alpha levels; and t values for contrasts on the adjusted group means. As an option, the program prints within-group correlation matrices for all variables named in the VARIABLES statement; total, between, and within-group variance-covariance matrices for the dependent, independent, and control variables; and the sampling correlation matrix for the regression coefficients of the covariates.

Thus, the investigator is provided with sufficient information for a thorough understanding of what is going on in the data.

The P2V program is similar in operation but not identical. The factors in the design are listed as GROUPING ARE in the DESIGN paragraph, and the covariates are listed as COVARIATES ARE followed by the list of their names. The output is less extensive than in P1V, and equality of slopes is not tested. Equality of slopes could be tested, however, using P1V first (treating the design as one-way) or even using P1V alone and performing the factorial analysis by specifying the appropriate contrasts to define the effects. This can only be done if the number of groups is not too great. It should be noted that P2V is suitable to do ancova on a design in which one or more factors is a repeated measures factor, as well as in latin squares, fractional factorials, and so on.

It is also possible to do ancova with P4V, although this program introduces a number of complexities, and furnishes less information than P1V. Applications to simple factorial ancova require little more than for P2V. However, familiarity with P4V will provide its user with a very general purpose program similar to SAS's GLM.

SAS

In SAS, ancova is a simple extension of the anova in the GLM procedure. One simply lists the covariates along with the anova effects in the model statement. For example,

MODEL Y = X1 X2 A B A*B

will do the analysis of a factorial design assuming common slopes. Where X1 and X2 are covariates, the TYPE III SS for the A, B, and the AB will be the ancova effects. The TYPE IV effects for X1 and X2 will test within-group regression, whereas their TYPE I effects will be the test of regression assuming no group differences at all. The possibility of different slopes would have to be investigated through a separate analysis. Here, the statement would be

MODEL Y = A*X1 A*X2 B*X1 B*X2 A*B*X1 A*B*X2 A B A*B

The difference in the sums of squares accounted for by this model and the previous one would then be used to perform the test of equality of slopes, which would be done by hand.

SPSS[x]

Covariates are handled in either ANOVA or MANOVA in much the same way as in SPSS. As noted in the previous chapter, in ANOVA, one merely specifies the covariates at the end of the ANOVA statement following WITH. The effects of covariates are always entered first by the program, no matter what option is used in the case of an unbalanced design. ANOVA does not directly allow the testing of equal slopes, although we can construct the corresponding contrasts.

In MANOVA the same format is used in denoting covariates via a WITH specification. The subcommand CONTIN allows the testing of parallel regression lines (that is, equal slopes). CONTIN specifies that a variable is to be treated as continuous rather than categorical. Used in a DESIGN subcommand it allows a covariate to be included among the factors. For example

DESIGN = CONTIN(X1), A

is one way to specify that X1 is a covariate and that there is a single factor A. Testing equal slopes could be done by using

DESIGN = CONTIN(X1), A, CONTIN(X1) BY A

The latter part of the subcommand has the effect of fitting a separate slope for each level of A and the former part fits a common slope. Thus the two residual sums of squares can be compared.

PROBLEMS

1. Below are the data for two groups and a single covariate. Perform the complete ancova analysis, following the methods used in Exhibit 12-2.

	$h = 1$				$h = 2$			
$\mathbf{x}_1' =$	[5	4	6	3	4	6	8	2]
$\mathbf{y}' =$	[4	2	3	5	6	9	7	2]

2. Using the data in Problem 1, construct the three X_+ matrices corresponding to each of the three models as shown in Exhibit 12-1 and find each $X_+'X_+$ and $X_+\mathbf{y}$ matrix. Find the inverses of the $X_+'X_+$ matrices and the $\mathbf{b}_+$ vectors. (If no computer facilities are available, you may find the inverses by peeking at the answers.)

3. Using the vectors of regression weights, find the three ss_{reg} and ss_{dev} and compare them to your answers in Problem 1.

4. Suppose there is a study with three conditions, four observations per condition, and two covariates. Construct the X_+ matrices that correspond to the three models.

5. Input the data of Problem 1 to a standard package program that performs ancova, and do the ancova. (Not all programs will test equality of slopes without special instructions.)

ANSWERS

1. Homogeneous regression: $b_1 = .7225$; $b_0 = 1.3039$; $ss_t = 43.50$; $ss_{regl} = 13.422$; $ss_{dev1} = 30.078$; Equal slopes, different intercepts: $b_{1w} = .400$; $b_{01} = .620$, $b_{02} = 2.80$; $ss_{dev2} = 20.76$; $ss_{reg2} = 22.74$. Different slopes and intercepts: $b_{11} = -.4000$; $b_{12} = .6000$; $b_{01} = 5.3000$; $b_{02} = 1.50$; $ss_{dev3} = 14.00$; $ss_{reg3} = 29.50$; F for different slopes $= 1.93$, $df = 1{,}4$; t for different intercepts $= 2.24$, $df = 1{,}5$; F for regression $= 2.67$, $df = 1{,}6$

2.

$$\mathbf{X}_{+j} = \begin{bmatrix} 5 & 1 \\ 4 & 1 \\ 6 & 1 \\ 3 & 1 \\ 4 & 1 \\ 6 & 1 \\ 8 & 1 \\ 2 & 1 \end{bmatrix} \qquad \mathbf{X}_{+i} = \begin{bmatrix} 5 & 1 & 1 \\ 4 & 1 & 1 \\ 6 & 1 & 1 \\ 3 & 1 & 1 \\ 4 & 0 & 1 \\ 6 & 0 & 1 \\ 8 & 0 & 1 \\ 2 & 0 & 1 \end{bmatrix} \qquad \mathbf{X}_{+s} = \begin{bmatrix} 5 & 0 & 1 & 1 \\ 4 & 0 & 1 & 1 \\ 6 & 0 & 1 & 1 \\ 3 & 0 & 1 & 1 \\ 0 & 4 & 0 & 1 \\ 0 & 6 & 0 & 1 \\ 0 & 8 & 0 & 1 \\ 0 & 2 & 0 & 1 \end{bmatrix}$$

$$(\mathbf{X}_{+r}'\mathbf{X}_{+r})^{-1} = \begin{bmatrix} .03922 & -.1863 \\ -.1863 & 1.0098 \end{bmatrix} \qquad \mathbf{b}_r = \begin{bmatrix} .7255 \\ 1.3039 \end{bmatrix}$$

$$(\mathbf{X}_{+i}'\mathbf{X}_{+i})^{-1} = \begin{bmatrix} .04 & .02 & -.20 \\ .02 & .51 & -.35 \\ -.20 & -.35 & 1.25 \end{bmatrix} \qquad \mathbf{b}_i = \begin{bmatrix} .64 \\ -2.18 \\ 2.80 \end{bmatrix}$$

$$(\mathbf{X}_{+s}'\mathbf{X}_{+s})^{-1} = \begin{bmatrix} .1143 & 0 & -.4857 & 0 \\ 0 & .0500 & .2500 & -.25 \\ -.4857 & .2500 & 3.8143 & -1.50 \\ 0 & -.2500 & -1.5000 & 1.50 \end{bmatrix} \qquad \mathbf{b}_s = \begin{bmatrix} -.4 \\ .9 \\ 3.7 \\ 1.5 \end{bmatrix}$$

3. The sums of squares are the same as from Problem 1.

4.

$$\mathbf{X}_{+r} = \begin{bmatrix} x_{111} & x_{112} & 1 \\ x_{211} & x_{212} & 1 \\ x_{311} & x_{312} & 1 \\ x_{411} & x_{412} & 1 \\ x_{121} & x_{122} & 1 \\ x_{221} & x_{222} & 1 \\ x_{321} & x_{322} & 1 \\ x_{421} & x_{422} & 1 \\ x_{131} & x_{132} & 1 \\ x_{231} & x_{232} & 1 \\ x_{331} & x_{332} & 1 \\ x_{431} & x_{432} & 1 \end{bmatrix} \qquad \mathbf{X}_{+i} = \begin{bmatrix} x_{111} & x_{112} & 1 & 0 & 1 \\ x_{211} & x_{212} & 1 & 0 & 1 \\ x_{311} & x_{312} & 1 & 0 & 1 \\ x_{411} & x_{412} & 1 & 0 & 1 \\ x_{121} & x_{122} & 0 & 1 & 1 \\ x_{221} & x_{222} & 0 & 1 & 1 \\ x_{321} & x_{322} & 0 & 1 & 1 \\ x_{421} & x_{422} & 0 & 1 & 1 \\ x_{131} & x_{132} & 0 & 0 & 1 \\ x_{231} & x_{232} & 0 & 0 & 1 \\ x_{331} & x_{332} & 0 & 0 & 1 \\ x_{431} & x_{432} & 0 & 0 & 1 \end{bmatrix}$$

$$
\mathbf{X}_{+rs} = \begin{bmatrix}
x_{111} & 0 & 0 & x_{112} & 0 & 0 & 1 & 0 & 1 \\
x_{211} & 0 & 0 & x_{212} & 0 & 0 & 1 & 0 & 1 \\
x_{311} & 0 & 0 & x_{312} & 0 & 0 & 1 & 0 & 1 \\
x_{411} & 0 & 0 & x_{412} & 0 & 0 & 1 & 0 & 1 \\
0 & x_{121} & 0 & 0 & x_{122} & 0 & 0 & 1 & 1 \\
0 & x_{221} & 0 & 0 & x_{222} & 0 & 0 & 1 & 1 \\
0 & x_{321} & 0 & 0 & x_{322} & 0 & 0 & 1 & 1 \\
0 & x_{421} & 0 & 0 & x_{422} & 0 & 0 & 1 & 1 \\
0 & 0 & x_{131} & 0 & 0 & x_{132} & 0 & 0 & 1 \\
0 & 0 & x_{231} & 0 & 0 & x_{232} & 0 & 0 & 1 \\
0 & 0 & x_{331} & 0 & 0 & x_{332} & 0 & 0 & 1 \\
0 & 0 & x_{432} & 0 & 0 & x_{432} & 0 & 0 & 1
\end{bmatrix}
$$

READINGS

Pedhazur, E. J. (1982) *Multiple regression in behavioral research: Explanation and prediction.* New York: Holt, Rinehart, and Winston, pp. 493–550.

Overall, J. E., and Klett, C. J. (1972) *Applied multivariate analysis.* New York: McGraw-Hill, pp. 430–438.

Tatsuoka, M. M. (1975) *The general linear model: A "new" trend in analysis of variance.* Champaign, Illinois: Institute for Personality Research, pp. 49–60.

13

Components and Principal Components of Variables

LINEAR COMBINATIONS AS SUMMARIES

Linear Combinations

A concept that provides some unification for this book is that of forming new variables from linear combinations of observed variables. We applied this concept in our study of multiple regression, where we sought the linear combination of predictors that allowed us to come as close as possible to the criterion. Similarly, we applied it in partial correlation analysis, where the partial correlations are correlations between linear combinations.

In Chapters 10 and 11, we found that different ways of doing an anova result from taking different linear composites of the arbitrary group-identification variables. Ancova, which is a combination of anova with a form of partial regression analysis, also involves studying the results of forming certain kinds of linear combinations. In all of these cases, the weights are defined primarily on the basis of empirical relationships in order to optimize the correlation or to minimize some form of residual sum of squares or variance.

We also form linear combinations on a priori bases. A test constructor uses the sum (that is, all weights equal 1.0) of scores on a number of test items as the individual's overall score on the test. Similarly, the psychophysicist uses the average of k judgments (that is, all weights equal $1/k$) of brightness of lights to represent perceived brightness; the graduate psychology department uses "twice verbal plus quantitative" as the index on which to select graduate students; or the basketball coach uses the sum of height and fingertip-to-fingertip arm span as an index of "effective height." In some of these cases, the weights are used as actual or rough approximations to regression weights for predicting some external criterion. In others, the linear combinations are used as *summaries* of the observations in some sense. In this chapter, the latter purpose will be considered, although we will explicitly construct the index so as to optimize the degree to which it represents a set of variables.

Two Types of Summaries

Weighted combination summaries fall into two closely related but subtly different classes. In many instances, the summary is used as a measure of some latent variable. Each manifest variable is taken to be a fallible indicator of a latent variable, and the sum or other weighted average of several such indicators is theoretically a better indicator of the latent variable than any individual indicator. The second type of summary uses the composite not so much as an index of a latent variable but as a summary description of the scores on the *manifest* variables themselves. The idea is that we could *reconstruct* with acceptable accuracy the scores on the several observed variables from the scores on the single summary variables.

Clearly, we might like to have the same summary variable serve both of these purposes, but it turns out that in many cases we cannot perform both types of summaries at the same time, at least not equally well. Even though the methodology for each summary looks very much alike, there are differences. These differences sometimes can affect conclusions from a study, even though they often seem trivial. Therefore, it is necessary to know the difference between summaries, if only to show off your sophistication to your peers.

If the composites are made of simple a priori weighting schemes, there is little need for elaborate discussion of analysis methods. However, there is very often a need to make the composites have certain optimal properties, properties in which the specific nature of the empirical data must be taken into account. Methodology for doing this has been developed over the years, starting in the early 1900s with the work of Spearman (1904) and continuing into the present day. In fact, there are so many variations in methodology that it is impossible to cover them all. All that can be done here is to present the central ideas and the most commonly available methods.

A *common factor analysis* or *classical factor analysis* is an approach that attempts to explain the relationship between observed variables in terms of latent variables, and conceivably uses the manifest variables to measure the latent variables. *Component analysis* uses a composite of the manifest variables as a summary of those variables. The latter analysis is perhaps easier to understand since it is more concrete, so we will consider it first.

PRINCIPAL COMPONENTS

Least Squares Fit to the Score Matrix

Throughout this chapter, it is assumed that the primary data consist of the scores of a set of individuals or other entities on a set of variables, such that there is a score for every individual on every variable. Thus, we start with an $n \times p$

matrix X, just as in multiple regression, but there is no dependent variable singled out for a special role. X is a deviation score matrix; ordinarily it should be thought of as a *standard score* matrix so that the variables receive equal weight in determining the solution. We would leave the variables in raw deviation form if we desired that those with larger variances should have greater influence on the components. The description below refers to the covariance matrix for generality, but in most applications this will actually be a correlation matrix.

The purpose of the principal components analysis is to find a composite variable z_1, where

$$\mathbf{z}_1 = \mathrm{X}\mathbf{w}_1 \tag{13.1}$$

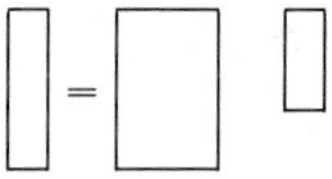

and $\mathbf{w}_1$ is a vector of weights w_j, so that $\mathbf{z}_1$, the composite variable, "represents" X as nearly as possible. The first question is to decide what we mean by "represent." The definition that is adopted is the least squares principle. Let

$$\mathbf{z}_1\mathbf{v}_1' = \hat{\mathrm{X}}_1 \tag{13.2}$$

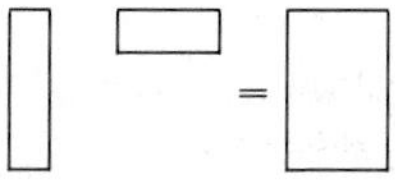

where $\mathbf{v}_1$ is a vector of scaling factors used to adjust $\mathbf{z}_1$ to make it fit the different variables x_j. Define the error matrix as

$$\mathrm{E}_1 = \mathrm{X} - \hat{\mathrm{X}}_1 \tag{13.3}$$

Then our least squares criterion is that we would like to minimize the sum of squared error scores:

$$\phi = \sum_j \sum_i e_{ij}^2 \tag{13.4}$$

It looks as if there are two things that must be found, $\mathbf{w}_1$—to provide $\mathbf{z}_1$—and $\mathbf{v}_1$, to scale $\mathbf{z}$ appropriately for the reconstruction of the various columns of X. In one of those surprising results that serve to make mathematics interesting, it turns out that when we find the $\mathbf{v}_1$ and the $\mathbf{w}_1$ that make ϕ as small as it can be made, their elements are proportional to each other. The reason for this and how we might go about finding $\mathbf{w}_1$ will be postponed until later.

A related problem is one of securing a composite $\mathbf{z}_1$ that has maximum variance. You may think this problem has no answer. We can make the variance of $\mathbf{z}_1$ larger and larger just by using larger and larger numbers in $\mathbf{w}_1$. That is true enough, so this means that the problem must be specified in a way that somehow limits the size of numbers that can be used in $\mathbf{w}_1$. The conventional limitation is

$$\sum_j w_{j1}^2 = 1.0 \tag{13.5}$$

Thus we want to find the linear composite of the x_j that has the largest possible variance subject to the restriction of Equation 13.5.

The variance of z can be expressed as

$$\lambda_1 = \sum_i z_{i1}^2/(n - 1) \tag{13.6}$$

Now there is another interesting convergence. We find that the $\mathbf{w}_1$ that maximizes λ_1 is the same as the $\mathbf{w}_1$ that minimizes ϕ. Maximizing the variance of the component $\mathbf{z}_1$ under the restriction of Equation 13.5 is the *same* as finding the component that gives the least squares fit to X when it is weighted by that same component, that is, $\mathbf{v}_1$.

EIGENVALUES AND EIGENVECTORS

A Little Mathematics

The vector that served as a solution to our problem of maximizing variance—and previously turned up as both halves of the solution to finding a component that gave a least squares fit to X—has a long history in mathematics under a variety of names. It is called an *eigenvector,* and its corresponding λ is called an *eigenvalue.* These terms are far from universal, however. In the earlier statistical literature of behavioral sciences they were called "characteristic" vectors and roots or "latent" vectors and roots. (Occasionally they were called "eigenroots" instead of eigenvalues, but the latter term was considered to be an illiteracy.) Recently, we begin to encounter the terms "singular" vectors and "singular" values or even "stationary" vectors and values. All these terms refer to the same concept in matrix algebra, one that turns out to have wide application in statistics. We will use the eigenvector-eigenvalue terminology.

While we have been focusing on a single eigenvalue and eigenvector, it turns out that every $p \times p$ matrix A has associated with it p different eigenvectors and p eigenvalues, particular values going with particular vectors. Some of the eigenvalues may be equal, and some may be zero—a very important possibility. The property that makes a vector an eigenvector of a particular matrix is one of the properties mentioned in our introduction. This property may be stated formally as follows. Let A be a p by p matrix and $\mathbf{w}$ a p-element vector. If it is true that

$$\mathrm{A}\mathbf{w} = \lambda\mathbf{w} \tag{13.7}$$

for some scalar λ, then $\mathbf{w}$ is an eigenvector of A and λ is the corresponding eigenvalue. That is, an eigenvector of a matrix is a vector such that when we multiply the matrix by the vector we get the vector back again except that it has been multiplied by a particular constant, called the eigenvalue.

For example, if we multiplied the following matrix A by the indicated vector $\mathbf{w}$ we would get the following product:

$$\begin{bmatrix} 6 & 2 & 1 \\ 2 & 6 & 1 \\ 1 & 1 & 7 \end{bmatrix} \begin{bmatrix} 1 \\ 1 \\ -2 \end{bmatrix} = \begin{bmatrix} 6 \\ 6 \\ -12 \end{bmatrix}$$

$$\text{A} \qquad \mathbf{w} \;=\; \lambda\mathbf{w}$$

Multiplying this A by this **w** gives $\lambda\mathbf{w}$, where here $\lambda = 6.0$. Therefore, **w** is an eigenvector of A and the corresponding eigenvalue is 6.0.

Equation 13.7 is often expressed in a different form where $\lambda\mathbf{w}$ is subtracted from both sides, and λ is expressed as λI, a multiple of the identity matrix. Then **w** can be factored out to give

$$(\text{A} - \lambda\text{I})\mathbf{w} = 0 \tag{13.8}$$

Both equation 13.7 and 13.8 may be called the "characteristic equation" of a matrix. As mentioned earlier, the number of eigenvalues and eigenvectors equals the order of the matrix. This means that the matrix used as an example on page 297 has two other sets of eigenvalues and eigenvectors as follows:

$$\begin{bmatrix} 6 & 2 & 1 \\ 2 & 6 & 1 \\ 1 & 1 & 7 \end{bmatrix} \begin{bmatrix} 1 \\ -1 \\ 0 \end{bmatrix} = \begin{bmatrix} 4 \\ -4 \\ 0 \end{bmatrix}$$

and

$$\begin{bmatrix} 6 & 2 & 1 \\ 2 & 6 & 1 \\ 1 & 1 & 7 \end{bmatrix} \begin{bmatrix} 1 \\ 1 \\ 1 \end{bmatrix} = \begin{bmatrix} 9 \\ 9 \\ 9 \end{bmatrix}$$

The other eigenvectors for this matrix are $(1, -1, 0)'$ with an eigenvalue of 4, and $(1, 1, 1)'$ with an eigenvalue of 9, respectively. For each of these combinations of **w** with λ, the characteristic Equation 13.7 is true.

A further aspect of eigenvectors—an indeterminacy—is that an eigenvector is defined only up to multiplication by a constant. For example, (2, 2, 2) could be used as the third eigenvector, making the product A**w** (18, 18, 18), but the eigenvalue is still 9.0 times the eigenvector (2, 2, 2). In fact, we could use $(-2, -2, -2)$, and then the product would be $(-18, -18, -18)$, which is also an eigenvector, with the same eigenvalue of 9.0. The eigenvector is thus unique only up to multiplication by a constant. The principle is quite general and may be stated formally:

> If **w** is an eigenvector of a matrix A, then $k\mathbf{w}$ is also an eigenvector of the matrix, associated with the same eigenvalue.

This principle is easy to prove as a theorem through the use of the properties of multiplication by a scalar. It is useful to remove this indeterminacy in the eigenvector by placing a restraint on it. In statistical applications, this restraint is that the sum of squares of the eigenvector's elements must equal 1.0, and from here on we assume that all eigenvectors have been normalized in this fashion.

PRINCIPAL COMPONENTS

Eigenvectors, Eigenvalues, and Principal Components

We will now describe the connection between eigenvalues and eigenvectors and our statistical problems of finding optimum composites. If λ_1 is the largest eigenvalue of the covariance matrix S, and $\mathbf{w}_1$ is the corresponding eigenvector, then the $\mathbf{z}_1$ which is equal to $X\mathbf{w}_1$ has a larger variance than any other composite that satisfies the normalization constraint of Equation 13.5. Moreover, λ_1 is the variance of this composite. It is also true that the residual matrix $E_1 = X - (X\mathbf{w}_1)\mathbf{w}_1'$ has the smallest sum of squares that can be found using any such composite. Note that it is also $\mathbf{w}_1$ that acts as the vector of scaling constants that we previously have symbolized as $\mathbf{v}_1$.

A geometric interpretation can be made of $\mathbf{w}_1$ and $\mathbf{z}_1$. Here, the matrix X defines the location of n points in terms of the p coordinate axes corresponding to each variable. The criterion we are trying to maximize is equivalent to trying to find the direction through the cloud of points in this p-dimensional space such that when all the points are projected down onto a line in that direction, the sum of squared projections is the maximum possible. We might say we are looking for the "longest" direction in the space. When this direction is found, it will be true that the cosines of the angles that the line makes with the axes are equal to the entries in $\mathbf{w}_1$. These entries are sometimes called "direction cosines." This geometrical interpretation is illustrated for $p = 2$ in Figure 13-1.

The eigenvector $\mathbf{w}_1$ of S thus provides the weights that produce a $\mathbf{z}$ with maximum variance; this variance is the corresponding eigenvalue of S. The elements of $\mathbf{w}$ in normalized form—that is, under the restrictions of Equation 13.5—are also the cosines of the angles between it and the axes defined by the variables.

Now we'll consider the question of trying to reproduce X as closely as possible. Again, let $\mathbf{w}_1$ be the eigenvector of S corresponding to the largest eigenvalue, and $\mathbf{z}_1 = X\mathbf{w}_1$. Then the $\hat{X}_1$ formed by

$$\hat{X}_1 = \mathbf{z}_1\mathbf{w}_1' \qquad (13.9)$$

is the one that minimizes the least squares criterion defined by Equation 13.4.

So much for the largest eigenvalue and its eigenvector. What about the other eigenvalues and vectors? We might well want more than one component, either because more than one score or a closer approximation to X was desired. In the former case, we would want a second z score with a variance that was as large as possible, subject to the restriction that it was *uncorrelated* with (orthogonal to) the first. The component score that has this property is one that uses the eigenvector corresponding to the *second* largest eigenvalue of S as the weights. This eigenvalue is then the variance on that component. We can continue this process finding each eigenvector in turn in order of magnitude. Each eigenvector

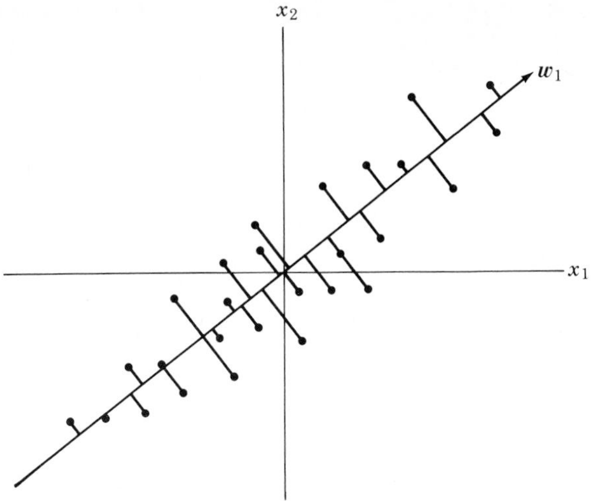

Figure 13-1 Illustration of an eigenvector. The points represent the positions of persons on variables x_1 and x_2, that is, the rows of an $n \times 2$ matrix X. The diagonal line corresponds to an eigenvector of the matrix X. It represents the direction in the space such that the sum of squares of the projections of the points down onto it is largest. Here, the value of the eigenvector's direction cosines are approximately (.7809, .6247)'.

will give the composite score that has largest variance, subject to the restriction that each composite is uncorrelated with all the others.

This process works the same way if our desire is to reproduce X as closely as possible. Given the error matrix E_1 (Equation 13.2), we want another $\mathbf{z}$ and another $\mathbf{w}'$, $\mathbf{z}_2$ and $\mathbf{w}_2'$ such that when they are multiplied together, $\hat{E}_1 = \mathbf{z}_2\mathbf{w}_2'$, they come as close as possible to these errors. We could do this by finding the first eigenvector of $E_1'E_1(n - 1)^{-1}$ but it turns out that this eigenvector is the same as that corresponding to the second largest eigenvalue* of S. This eigenvector $\mathbf{w}_2$, used as a weight vector on E_1, gives a second $\mathbf{z}$. When this $\mathbf{z}$ is used with $\mathbf{w}_2'$ to produce an $\hat{E}$, we are reproducing E_1 as accurately as possible in the least squares sense. The second-order errors could then be calculated. $E_2 = E_1 - \hat{E}_1$. This can correspondingly be best approximated using $\hat{E}_2 = (E_2\mathbf{w}_3)\mathbf{w}_3'$, where $\mathbf{w}_3$ is the third eigenvector of S. The process can be continued approximating the mth order error matrix using the $(m + 1)$ eigenvector, until n equals the rank of X, at which point E_m will be zero. Each time, E_m will be the least squares approximation to E_m.

The point of this discussion is that the successive eigenvectors furnish successive levels of approximation to X. If we arrange the first r eigenvectors in a matrix W_r, then $Z_r = XW_r$, and $\hat{X}_r = Z_rW_r'$ will give the best approximation to X

*To save space, we will assume that all the eigenvalues and their corresponding eigenvectors are known and that both are arranged in order of size of eigenvalue. Hence the first, second, and final eigenvector will correspond to the largest, next largest, and final eigenvalue.

using r components. The Z_r are the *principal components* of X. These are related to, but not the same as, the principal components factors of S.

Note also that X is usually a matrix of standard scores rather than of deviation scores which means all variables receive equal weight in the analysis. In the standard score case, S would be a correlation matrix R, and the relevant eigenvalues and eigenvectors would be those of the correlation matrix.

Geometrically, the involvement of succeeding eigenvectors and eigenvalues also proceeds in a stepwise fashion. Just as the first eigenvector projects a vector into the space defined by the variables such that the sum of squared projections of the points (observations) on it is as large as possible, the second eigenvector projects a vector *orthogonal* to the first such that the sum of squared projections on it is as large as possible. The third eigenvector, then, is subject to the restriction of orthogonality to the first two eigenvectors, and so on. In each case, the elements of the eigenvector are equal to the direction cosines in the space for that direction. The eigenvalues of S give that sum of squared projections divided by $n - 1$.

It may help to visualize a three-dimensional swarm of points shaped like a football that has been somewhat flattened. The direction of the long axis of the swarm corresponds to the direction cosines of the first eigenvector. Looking at the swarm from this direction—that is, with the point of the football in our direction—we would see a flattened oval. The direction corresponding to the second eigenvector is at right angles to the first (the length of the football) and in the direction of the long axis of this oval cross section. The third eigenvector is orthogonal to the first two—it is the short axis of the oval. This illustrates a general principle: The pth (last) eigenvector is always determined by the preceding $p - 1$ since it must be orthogonal to it.

The two-dimensional case is not very interesting, because as soon as the first eigenvector is defined, the other one must be at right angles to it. This is illustrated in Figure 13-2.

The eigenvectors then define the directions in the space in which the swarm is longest. The size of the eigenvalues tells how big the configuration is for each of these dimensions. For example, if the configuration were shaped like an oval meat platter, the first eigenvector would run in the direction of the long axis of the platter, parallel to the table. The second eigenvector would be at right angles to the first, in the direction of the short axis of the platter, parallel to the table. The third eigenvector would correspond to the depth of the platter, at right angles to the table. The first eigenvalue would be perhaps twice the size of the second, reflecting the length of the platter relative to its width. The second eigenvalue would be several times as large as the third because the platter is relatively shallow. This analogy is rather rough, because we are actually dealing with a set of points rather than a solid object, but the image of some dimensions of the configuration being larger than others is useful. A photograph of the distribution of stars in a distant galaxy would provide a more direct analogy.

Properties of Principal Components

The score components defined by eigenvectors have many nice properties, some of which have been mentioned above. These properties will now be stated in a more formal way, along with some others.

For now, let X be a score matrix in deviation form. This represents the more common situation in which principal components are used. The properties

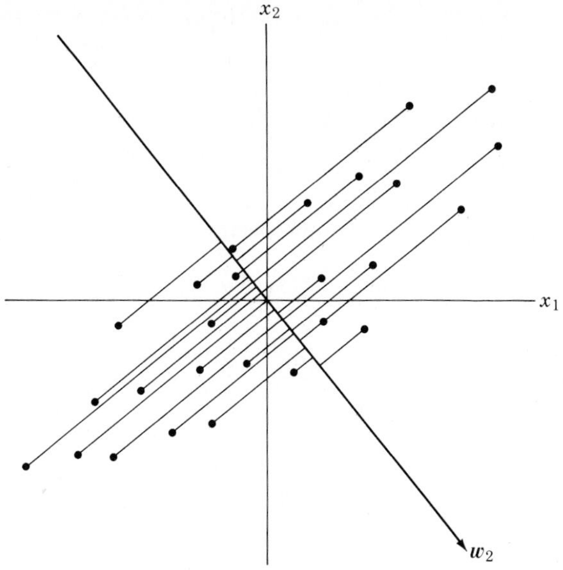

Figure 13-2 Illustration of second eigenvector using the points in Figure 13-1. The projections of the points down onto this vector are also shown. The sum of squares of these points will be much smaller since this direction corresponds to the short axis of the "ellipse" that is the cloud of points. This eigenvector $\mathbf{w}_2$ equals (.6247, −.7809). Either or both $\mathbf{w}_1$ and $\mathbf{w}_2$ can be multiplied by −1 without affecting the vector's properties.

described below need to be modified to accommodate the case where X is a raw score matrix.

Some general properties of eigenvalues and eigenvectors of symmetric matrices are as follows:

1. If S is a square symmetric matrix of order p, then $SW = \Lambda W$, where W is the matrix of eigenvectors (columnwise) and Λ is a diagonal matrix of eigenvalues. All the eigenvalues and eigenvectors are real numbers. (If S is not symmetric, then some eigenvalues may be imaginary, and their eigenvectors may be complex numbers. No more will be said about such horrifying possibilities.)

2. If S is the product of a matrix and its transpose, then all its eigenvalues are either positive or zero. If all are positive, it is called a *positive definite* matrix. If some are zero, then it is called positive *semi*definite. A covariance matrix is therefore positive semidefinite or positive definite.

3. The rank of a matrix is equal to the number of its nonzero eigenvalues. Finding the eigenvalues is then one method of finding the rank of a matrix, although a cumbersome one. It is useful here to remember that the rank of a product can be no more than the smallest rank of any of the factors.

4. The eigenvectors of a symmetric matrix are mutually orthogonal. That is, if W is a $p \times r$ matrix containing r of the normalized eigenvectors of some matrix, and $r \leqq p$, then

$$\mathrm{W'W} = \mathrm{I}_r \tag{13.10}$$

Note, however, that $\mathrm{WW'} \neq \mathrm{I}_p$ unless $r = p$, in which case the equality *does* hold. The orthogonality property is useful in a variety of situations because it simplifies a number of relationships among various matrices.

5. The nonzero eigenvalues of X′X and XX′ are the same.

In order to present the next principles it is necessary to introduce a subscript for W, Z, X, and E. The subscript r will indicate that r eigenvectors, components, and so on, are involved. Thus, W_r with elements w_{jm} will be a p by r matrix containing the first r eigenvectors of S; Z_r will be the corresponding n by r matrix of components, $\mathrm{Z}_r = \mathrm{XW}_r$, where Z_r has elements z_{im}. In addition, Λ_r is an r by r diagonal matrix containing the first r eigenvalues. From here, we define $\hat{\mathrm{X}}_r$:

$$\hat{\mathrm{X}}_r = \mathrm{Z}_r\mathrm{W}_r' \tag{13.11}$$

That is, $\hat{\mathrm{X}}_r$ is an approximation of X using the first r components. Note that $\hat{\mathrm{X}}_r$ is an $n \times p$ matrix, but that its elements have three subscripts including one to denote how many components are included, $\hat{x}_{ijr}$. We also use the corresponding equation

$$\mathrm{E}_r = \mathrm{X} - \hat{\mathrm{X}}_r \tag{13.12}$$

An element of the n by p matrix E_r needs three subscripts for the same reason, e_{ijr}.

Now we are ready to state two additional principles.

6. The components of a matrix are uncorrelated. That is,

$$\frac{1}{n-1}\mathrm{Z}_r'\mathrm{Z}_r = \Lambda_r \tag{13.13}$$

7. If r equals the rank of X, the errors are reduced to zero: $\mathrm{E}_q = 0$. This is obviously true since $\mathrm{X} = \hat{\mathrm{X}}_q$. Ordinarily, q will be p or $n - 1$, whichever is smaller, unless something in the data leads to a lower rank.

8. We can express the covariance matrix S in terms of the eigenvalues and eigenvectors:

$$\mathrm{S} = \mathrm{W}\Lambda\mathrm{W}' \tag{13.14}$$

Now we will consider the variance-covariance matrices of the $\hat{x}$'s and e's.

9. Because of the orthogonality principle for eigenvectors, it is true that

$$S = S_{\hat{x}\hat{x}} + S_{ee} \tag{13.15}$$

These covariance matrices can also be expressed in terms of the eigenvalues and eigenvectors.

10. The $S_{\hat{x}\hat{x}}$ involves the first r eigenvectors, where

$$S_{\hat{x}\hat{x}} = W_r \Lambda_r W_r' \tag{13.16}$$

11. The value of S_{ee} depends on the $p - r$ remaining eigenvectors, W_s and Λ_s:

$$S_{ee} = W_s \Lambda_s W_s' \tag{13.17}$$

The sum of the diagonal elements of any square matrix is called its *trace*, abbreviated $tr(S)$. In the case of the covariance matrix S, $tr(S)$ represents the sum of the variances of the variables. Similarly, $tr(S_{\hat{x}\hat{x}})$ represents the sum of the variances of the reproduced $\hat{x}$'s and $tr(S_{ee})$ is the sum of variances of the errors.

There are several other principles that involve the eigenvalues. One is that the sum of the diagonal entries of a symmetric matrix is the sum of its eigenvalues. For example,

$$tr(S) = \sum_{m=1}^{p} \lambda_m \tag{13.18}$$

12. This means that the total variance of the x's is the sum of the eigenvalues of S:

$$\sum_{j=1}^{p} s_j^2 = \sum_{m=1}^{p} \lambda_m \tag{13.19}$$

13. Also, the total variance of the $\hat{x}$'s, using r components, is the sum of the r largest eigenvalues:

$$tr(S_{\hat{x}\hat{x}}) = \sum_{m=1}^{r} \lambda_m \tag{13.20}$$

14. Finally, the total variance of the errors is the sum of the $p - r$ smallest eigenvalues:

$$tr(S_{ee}) = \sum_{m=r+1}^{p} \lambda_m \tag{13.21}$$

This means that the total variance of the errors in reproducing X from $\hat{X}_r$ is equal to the sum of the $p - r$ smallest eigenvalues. This allows us to express the accuracy with which X is reproduced by $\hat{X}_r$ in terms of a proportion that is analogous to a squared multiple correlation. This proportion is the ratio of the r largest eigenvalues to the sum of the eigenvalues, or equivalently to the sum of the variables' variances. If it is the correlation matrix that is analyzed, the variances are all unity, so this sum is the number of variables.

These principles form the basis for what is usually called "principal components analysis." They may seem different from the type of analysis as it is usually carried out at countless computer centers across the country. The differences are not as major as they may seem at first glance, however. The connections are fairly readily made, and will be described in the next section. First, however, it may be useful to trace the example given in Exhibit 13-1. The exhibit shows how a data matrix is successively approximated by the components and how they are derived from the eigenvectors of S.

Components Loadings

A slightly different formulation is needed to find the numbers implied when we speak of "factor loadings" in the principal components sense. This method requires the standardization of the *component* scores so that they have unit variance. Remember that in the section on eigenvalues, eigenvectors, and principal components (p. 239), $\mathbf{w}_1$ appeared in two places. It appeared as a weight vector applied to X to give the component $\mathbf{z}_1$ and as a way of scaling this component to reproduce the individual observed variables. The variance of this z_1 was λ_1. When we moved on to multiple components, the individual λ's were the variances of the different components; λ_m was the variance of component m.

In most applications, it is desirable to think of *standardized* component scores with a mean of zero and a variance of unity instead of a variance equal to λ_m. It is easy to obtain such standard scores using Equation 13.11. All that is necessary is to divide each of the columns of Z_r by the corresponding $\lambda^{1/2}$ as follows:

$$Y_r = Z_r \Lambda_r^{-1/2} \tag{13.22}$$

In order to preserve equality of Equation 13.11, we must also multiply $\mathbf{w}_r$ by $\lambda^{1/2}$, yielding

$$\hat{X}_r = (Z_r \Lambda_r^{-1/2}) (\Lambda_r^{1/2} W_r') \tag{13.23}$$

We can also let

$$F_r' = \Lambda_r^{1/2} W_r' \tag{13.24}$$

These are the numbers that are referred to in computer output as "principal components factor loadings." They represent the weights to apply to the standardized components scores in order to reproduce the variables. If the variables are standardized, making S_{xx} a correlation matrix R_{xx}, then the F_r matrix is also the correlations between the variables and the components. This duality results from the fact noted in Chapter 9: Standardized regression coefficients are the same as correlations when the predictors are uncorrelated. Here, the "predictors" are the components.

Now we have a different form of Equation 13.11, one expressed in terms of scores on standardized components and the principal components factor loadings:

$$X = YF' \tag{13.25}$$

$$
\begin{array}{c}
\quad j \\
\begin{array}{c|ccc}
i & 1 & 2 & 3 \\
1 & -.2 & .4 & -.9 \\
2 & -.2 & .5 & -.1 \\
3 & 1.9 & .8 & .8 \\
4 & -.4 & -.1 & .3 \\
5 & -1.5 & -1.9 & 1.1 \\
6 & .4 & .3 & -1.1
\end{array} \\
\mathrm{X}
\end{array}
\qquad
\begin{array}{c}
\quad j \\
\begin{array}{cc|ccc}
 & & 1 & 2 & 3 \\
 & 1 & 1.252 & .870 & -.098 \\
j & 2 & .870 & .952 & -.444 \\
 & 3 & -.098 & -.444 & .794
\end{array} \\
\mathrm{S} \\
tr(\mathrm{S}) = 2.998
\end{array}
\qquad
\begin{array}{c}
\quad m \\
\begin{array}{cc|ccc}
 & & 1 & 2 & 3 \\
 & 1 & .708 & .468 & .529 \\
j & 2 & .650 & -.138 & -.747 \\
 & 3 & -.276 & -.873 & -.401
\end{array} \\
\mathrm{W}
\end{array}
$$

$$
\begin{array}{c}
\quad m \\
\begin{array}{c|ccc}
m & 1 & 2 & 3 \\
1 & 2.089 & 0 & 0 \\
2 & 0 & .812 & 0 \\
3 & 0 & 0 & .097
\end{array} \\
\Lambda \\
\sum_{m=1}^{3} \lambda_m = 2.998
\end{array}
\qquad
\begin{array}{c}
\quad m \\
\begin{array}{c|ccc}
i & 1 & 2 & 3 \\
1 & .367 & -.935 & -.043 \\
2 & .211 & -.250 & -.439 \\
3 & 1.644 & 1.476 & .086 \\
4 & -.431 & .089 & -.257 \\
5 & -2.601 & .521 & .184 \\
6 & .782 & -.815 & .430
\end{array} \\
\mathrm{Z} = \mathrm{XW}
\end{array}
\qquad
(n-1)^{-1}\,\mathrm{Z'Z} = \begin{bmatrix} 2.089 & 0 & 0 \\ 0 & .812 & 0 \\ 0 & 0 & .097 \end{bmatrix}
$$

$$(n-1)^{-1}\,\mathrm{Z'Z} = \Lambda$$

$$
\begin{array}{c}
\begin{array}{c|ccc}
i & 1 & 2 & 3 \\
1 & .260 & 239 & -.101 \\
2 & .149 & .137 & -.058 \\
3 & 1.164 & 1.068 & -.454 \\
4 & -.305 & -.280 & .119 \\
5 & -1.841 & -1.690 & .719 \\
6 & .554 & .508 & -.216
\end{array} \\
\hat{\mathrm{X}}_1 = \mathrm{Z}_1\mathrm{W}_1'
\end{array}
\qquad
\begin{array}{c}
\begin{array}{c|ccc}
i & 1 & 2 & 3 \\
1 & -.460 & .161 & -.799 \\
2 & -.349 & .363 & -.042 \\
3 & .736 & -.268 & 1.254 \\
4 & -.095 & .180 & .181 \\
5 & .341 & -.210 & .381 \\
6 & -.154 & -.208 & -.884
\end{array} \\
\mathrm{E}_1 = \mathrm{X} - \hat{\mathrm{X}}_1
\end{array}
\qquad
\begin{array}{c}
\mathrm{S}_{\hat{x}\hat{x}} = \begin{bmatrix} 1.047 & .961 & -.409 \\ .961 & .882 & -.375 \\ -.409 & -.375 & .160 \end{bmatrix} \\
tr(\mathrm{S}_{\hat{x}\hat{x}}) = 2.089 = \lambda_1 \\
\mathrm{S}_{\mathrm{EE}} = \begin{bmatrix} -.205 & -.091 & .311 \\ -.091 & .070 & -.069 \\ .311 & -.069 & .634 \end{bmatrix} \\
tr(\mathrm{S}_{\mathrm{EE}}) = .909 = \lambda_2 + \lambda_3
\end{array}
$$

Exhibit 13-1 Principal components. In the upper panel are X, S, and W, the eigenvectors of S. In the middle is the diagonal matrix of eigenvalues Λ and the unstandardized components Z. Note that $tr(\mathrm{S}) = \Sigma\lambda_m$. On the right is the variance-covariance matrix of the z's, showing that the variances are λ's and the covariances are zero. The lower panel shows $\hat{\mathrm{X}}_1 = \mathrm{Z}_1\mathrm{W}_1'$ and $\mathrm{E}_1 = \mathrm{X} - \hat{\mathrm{X}}_1$. On the right are $\mathrm{S}_{\hat{x}\hat{x}}$ and S_{ee} for $r = 1$, showing that $tr(\mathrm{S}_{\hat{x}\hat{x}}) = \lambda_1$ and $tr(\mathrm{S}_{ee}) = \lambda_2 + \lambda_3$.

This equation is typically presented as forming the basis of a factor analytic interpretation of principal components analysis. Similarly,

$$\hat{X}_r = Y_r F_r' \tag{13.26}$$

The model that underlies components analysis is frequently stated as

$$x_{ij} = \sum_{m=1}^{r} y_{im} f_{mj} + e_{ij} \tag{13.27}$$

This follows from Equations 13.25 and 13.26 when we remember that $X = X_r + E_r$. Therefore, the matrix form of Equation 13.26 is

$$X = Y_r F_r' + E_r \tag{13.28}$$

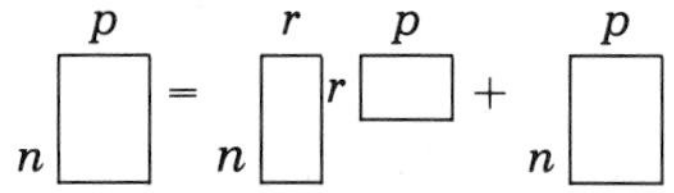

These equations describe the principal components model as one that explains an observed score x_{ij} as a weighted function of the standardized component scores y_{im}, where the *weights are the components "factor-loadings"* f_{mj}. In addition, there is a matrix of error scores e_{ij}.

This reformulation into standardized component scores allows reinterpretation of $S_{\hat{x}\hat{x}}$ and S_{ee} from Equations 13.16 and 13.17:

$$S_{\hat{x}\hat{x}} = \frac{1}{n-1} (\hat{X}_r)' (\hat{X}_r)$$

Substituting for $\hat{X}_r$ in terms of Equation 13.26 yields

$$S_{\hat{x}\hat{x}} = \frac{1}{n-1} (F_r' Y_r') (Y_r F_r) \tag{13.29}$$

Because the y are uncorrelated,

$$S_{\hat{x}\hat{x}} = F_r \left(\frac{1}{n-1}\right) (Y_r' Y_r) F_r = F_r I F_r' = F_r F_r' \tag{13.30}$$

At the beginning of this chapter we spoke of two types of factor analysis—components and common factor. The components approach focuses on approximating the score matrix with a small number of components, whereas the common factor approach emphasizes reproducing the correlation matrix. Sometimes an investigator may want to know how well the *components* account for the correlation matrix. This is not entirely consistent because reproducing the correlation matrix of principal components is not the objective. Nevertheless, because it *is* done, we will discuss it. (This type of goodness of fit can be illustrated in a general way using Equations 13.16 to 13.18.)

The principle of least-squares reproduction of matrices by means of eigenvectors, and the principle that the goodness of fit is shown by eigenvalues are quite general principles. For *any* matrix A, the eigenvectors of A′A show how to form a least-squares composite to approximate A, and the eigenvalues of A′A show how well A will be approximated. It is S we are trying to approximate here, not X, but the principles still hold. It turns out that

$$\hat{S}_r = S_{\hat{x}\hat{x}} = F_r F_r' \tag{13.31}$$

is the best approximation of S, just as $\hat{X}_r$ was the best approximation to X. Furthermore, the *squared* eigenvalues of S show how well it is approximated by S_r.

The reasoning for this is as follows. Remember that the trace of a matrix that is itself a product of a matrix and its transpose is the sum of the squared elements of the matrix. For example, $tr(X'X)$ will be equal to the sum of squares of all the x's. Remember also that the sum of the diagonal entries of a symmetric matrix is equal to the sum of its eigenvalues. We should also remember that eigenvectors are orthonormal, that is, $W'W = I$. If we remember all these things, then the connection is fairly straightforward. First, because S is symmetric, $S'S = SS$. Then in Equation 13.14 we found $S = W\Lambda W'$, so

$$SS = (W\Lambda W')\,(W\Lambda W') = W\Lambda(WW')\Lambda W' = W\Lambda^2 W'$$

This is interesting in itself because it shows that the eigenvectors of S^2 are the same as the eigenvectors of S, and its eigenvalues are the squares of the eigenvalues of S. The relevance here is that the $tr(S^2)$ is the sum of squared elements of S, *and* that it is equal to $\Sigma\lambda_m^2$. Applying similar reasoning to S_{ee} will show that

$$tr(S_{ee}^2) = \sum_{m=r+1}^{p} \lambda_m^2 \tag{13.32}$$

Thus we need not actually find S_{ee} to see how well r components fit the covariance matrix. We simply find the ratio

$$\frac{\sum_{m=1}^{r} \lambda_m^2}{\sum_{m=1}^{p} \lambda_m^2}$$

This reasoning will be elaborated in Chapter 15 when we describe the common factor model. Some of these relationships are illustrated in Exhibit 13-2.

So the principal components factors have a number of elegant properties. However, there are many different types of factors besides these. The other factors share some but not all of these properties, which leads to some confusion in the minds of users. Principal components factors can also be devilishly hard to interpret substantively, so that it is rare that they constitute the final form of a factor analysis.

In almost all applications of principal components, a further step is taken. This is called "rotation" of the components. The most common form of rotation is called *varimax*. The object of such procedures is to reorient the axes in the

	m				*m*				*j*		
j	1	2	3	*i*	1	2	3	J	1	2	3
1	1.023	.421	.165	1	.254	−1.037	−.138	1	1.047	.961	−.409
2	.939	−.125	−.233	2	.146	−.277	−1.408	2	.961	.882	−.375
3	−.400	.787	−.125	3	1.138	1.639	.275	3	−.409	−.375	.160
				4	−.298	.098	−.825				
				5	1.799	.579	.591				
				6	.541	.904	1.376				
	$F = W\Lambda^{1/2}$				$Y = Z\Lambda^{-1/2}$				$F_1F_1' = \hat{S}_1 = S_{\hat{x}\hat{x}}$		

Exhibit 13-2 Principal components factors F and standardized scores Y for the data from Exhibit 13-1. It can be verified that $X = YF'$ and $\hat{X}_r = Y_rF_r$. Note also that $\hat{S}_1 = F_1F_1'$ is the same as $S_{\hat{x}\hat{x}}$ from Exhibit 13-1. Finally, the sum of squared entries in $\hat{S}_1$ is $4.363 = \lambda_1^2$. It can also be found that the sum of squared entries in $S - \hat{S}_1$ is $\lambda_2^2 + \lambda_3^2$.

principal components space so that the components will be more easily interpreted, and, we hope, correspond more nearly to the reality underlying the variables. The methods of rotation—transformation is a more accurate term—of components and other types of factors are presented in some detail in Chapter 14.

Empirical Example

The principles of accounting for the observed data and for the correlation matrix may be illustrated further with the eight-variable empirical matrix given in Table 13-1. As in most applications, the objective is to reproduce the standard scores, not the deviation scores, so the eigenvalues and vectors are those of R, not S. R represents the intercorrelations among eight physical measurements of the human body taken from Heath (1952). Table 13-2 gives the eight principal components factors that would completely reproduce the data. For present purposes, the data are considered in standard score form, although it is arguable that the raw or deviation-score data are more relevant because the units in which the measurements are made have real interpretation in this particular case. The sum of the variances of the standard score variables is 8.0, and it is reasonable to see how the various factors account for this total amount of variance.

The factor loadings are given in Table 13-2, with the eigenvalues and percentages of variance at the bottom of the table. The first eigenvalue is 4.033, accounting for slightly more than half the total variance of the measures. The second is 1.776, 22.2 percent, showing that the first two eigenvalues account for 72.6 percent. The next largest eigenvalue accounts for 8.1 percent, and so on, with each subsequent eigenvalue accounting for less and less, given the nature of the eigenvalue-eigenvector process.

As far as reproducing the correlation matrix is concerned, the squared eigenvalues show that the first factor accounts for very nearly 80 percent of the squared entries and the second accounts for another 15.5 percent, leaving 4.8 percent for the remaining six factors.

Table 13-1 Intercorrelations among Eight Physical Measurements*

	1	2	3	4	5	6	7	8
1. Stature	1.00	.82	.74	.76	.67	.27	.21	.31
2. Hip Height	.82	1.00	.72	.74	.46	.16	.11	.19
3. Lower Leg Length	.74	.72	1.00	.65	.43	.22	.14	.22
4. Arm Length	.76	.74	.65	1.00	.44	.26	.25	.34
5. Sitting Height	.67	.46	.43	.44	1.00	.26	.24	.29
6. Knee Girth	.27	.16	.22	.26	.26	1.00	.59	.57
7. Elbow Girth	.21	.11	.14	.25	.24	.59	1.00	.58
8. Wrist Girth	.31	.19	.22	.34	.29	.57	.58	1.00

*From Heath, 1952, pp. 96–97.

The entries in the principal components factor matrix in Table 13-2 deserve some comment because they typify components results in several ways. The first factor has all large and positive loadings. This is almost always the case except when R has an appreciable mixture of positive and negative correlations. The signs of all the loadings could just as well have come out negative because if **w** is an eigenvector, so is $-\mathbf{w}$. This also means that when the signs of all the loadings are negative, there is no harm in reversing *all* the signs. Some principal component programs do that automatically.

The second factor in Table 13-2 has positive loadings for the first five variables and negative loadings for the last three, reflecting the grouping of the variables into the sets with relatively high correlations within groups and lower correlations between sets. (Variable 5 is between the two.) The relatively higher loadings for the first four variables on the first factor is reflected in part by the fact that there are four variables in the first group and only three in the second and by the fact that the correlations among the first group are somewhat higher. Thus, the first principal axis goes closer to these four variables. It is the second group that has the higher loadings on the second component because now there is more

Table 13-2 Principal Components of Physical Measurements

Variables	*I*	*II*	*III*	*IV*	*V*	*VI*	*VII*	*VIII*
1	.906	.270	.094	−.001	.003	−.037	.089	.297
2	.817	.396	−.167	−.003	−.047	−.094	.336	−.151
3	.788	.314	−.207	−.184	.062	.422	−.141	−.040
4	.836	.195	−.233	.174	−.079	−.294	−.289	−.045
5	.688	.059	.713	−.005	−.001	.007	−.063	−.102
6	.502	−.678	−.066	−.460	.204	−.176	−.005	−.009
7	.453	−.730	−.038	.031	−.497	.107	.030	.008
8	.541	−.643	−.057	.411	.327	.108	.054	−.011
λ	4.033	1.776	.652	.445	.409	.329	.232	.125
% Variance	50.4	22.2	8.2	5.6	5.1	4.1	2.9	1.6
Cum %	50.4	72.6	80.8	86.3	91.4	95.5	98.4	100.0
λ^2	16.265	3.154	.425	.198	.167	.108	.054	.016
% of Squared Elements	79.8	15.4	2.1	1.0	0.8	0.5	0.3	0.1
Cum %	79.8	95.3	97.3	98.3	99.1	99.7	99.9	100.0

residual variance on them. The numerical process of calculating the eigenvectors and eigenvalues is more or less responsible for the way the loadings look at this stage.

The third factor in Table 13-2 is mainly accounting for variable 5, which was not very well represented by either of the first two factors. This result is common because the smaller factors frequently represent instances where the process focuses on the variance of an individual variable that had relatively little variance on earlier factors. This is less true of the fourth factor and later factors.

The number of components needed to furnish an adequate description depends on the accuracy with which we want to account for the variance. If 70 percent is adequate, then two components are enough; if 80 percent is required, then three are necessary. It will take five components—more than half the number of variables—to account for 90 percent of the variance. Those who require all the variance will have to take as many components as there are variables. These results are also fairly typical. If anything, they represent an instance where the variance is relatively well accounted for by a relatively small number of components.

As far as accounting for the correlation matrix is concerned, much the same picture emerges. Squaring the eigenvalues increases their spread, however, making it appear that a given number of factors is doing a more complete job of reproducing the correlations than it reproduces the variance. This is primarily a result of squaring the entries. If we look at the last line of Table 13-2, which gives the rms deviations—that is, the square root of the average squared entry in the residual matrix—then the rms are seen to decline gradually.

The preceding paragraph should be read with the caveat that there are many factor theorists who feel that one should not even think of components as accounting for the correlations; they feel that this is the job of the common factor model. It is presented here nevertheless, both to illustrate the extent to which the correlations are reproduced and to reinforce the idea that accounting for the variances of the variables themselves and accounting for the intercorrelations are two separate, although related, processes.

THE NUMBER OF COMPONENTS

Technical Difficulties

The reader may have noted the absence of any mention of inferential matters—tests of hypotheses, significance of loadings, and so on—in this chapter. The reason is that applications of the usual armory of inferential concepts has proved to be very difficult with components analysis and other traditional forms of factor analysis. Great progress has been made recently in some regards, but not with the components procedures described here. There is very little of a formal nature that can be said to aid the inferential process in the methods described in this chapter. The procedures and tests listed in some sources are too limited in application to be of general interest.

Over the years of using these methods a series of informal rules has evolved, sometimes bolstered by some theoretical reasoning. Some of these rules of thumb are more sound than others. The following sections will attempt to describe them, along with some dangers and pitfalls that should be avoided.

Number of Components

The first question asked of a factor analysis is almost invariably, "How many factors are there?" The answer to this question should always be "lots." This is certainly the case with the components type of analysis, where it is rare indeed to get a really close approximation to p variables with many fewer than p components. Thus, "How many factors are there?" is the wrong question. Two others should be substituted for it: "How many components are necessary for an adequate description of the data matrix?" and "For how many factors can we get statistically stable estimates of their loadings?" Rough answers to these questions—or at least some concepts to guide our decisions—are usually available.

If one is performing an analysis for the purpose of describing or summarizing the score matrix, then it would seem that the major criterion should be the proportion of variance that is accounted for by r components. If the proportion is not satisfactory with r, then more components should be taken. What proportion is "satisfactory" depends on the standards of the investigator: 70, 80, 90, 95, 99, 99.9. In many applications, a number of components are necessary to account for a reasonable amount of variance, perhaps more than half the number of variables. If so, the data are telling us something. They are telling us that *they cannot be summarized very well with a few components* and that we should stick with the data themselves.

Often, though, we do not enter the principal components analysis with a components orientation. Rather, we may have a common factors purpose; our analysis is aimed at understanding the correlations among the variables. This is not really the purpose of the principal components process. Nonetheless, this is a common practice, and the results of a rotated principal components analysis are at least as easily interpreted as those of a common factor analysis of the same data. Therefore, an investigator may do a principal components analysis for the purpose of trying to find a reasonably small but reasonably well-defined set of underlying dimensions with which to understand the consistent aspects of the variables.

Eigenvalues Greater than Unity

It is regrettable that the most frequently employed procedure for deciding on the number of relevant components or factors is undoubtedly the "eigenvalues greater than unity" rule. This is probably because it is the default rule in most of the common statistical packages. It is regrettable because almost all the evidence and expert opinion indicates that it gives an inappropriate number.

The rule has three bases. First, the user should not pay attention to components that account for less variance than a single variable. However, if one or a few components account for most of the variance, there may be small but consistent and meaningful additional components, so the rule may be wrong.

The second basis of the default rule is that a component should define a composite that has positive reliability in the internal consistency sense (Kaiser, 1961). It has recently been demonstrated (Cliff, 1985) that the basis for the rule is erroneous. The "eigenvalues greater than unity" rule does not tell how many components would have positive internal consistency. Table 13-3 is a case in

Table 13-3 Hypothetical Correlation Matrix

	1	2	3	4	5	6
1	1.0	.8	.8	.6	.6	.6
2	.8	1.0	.8	.6	.6	.6
3	.8	.8	1.0	.6	.6	.6
4	.6	.6	.6	1.0	.8	.8
5	.6	.6	.6	.8	1.0	.8
6	.6	.6	.6	.8	.8	1.0

point. Clearly there are two groups of variables in this table—1, 2, and 3 versus 4, 5, and 6. This suggests that two components should be retained, but the first eigenvalue is 4.4 and the second is only .8, meaning there is only one component according to the rule. Any reasonable estimate of the reliability of the individual variables will lead to a positive reliability for the second component. When the correlations among the variables are very low, a number of eigenvalues will be just above unity, whereas the number of components that represent anything consistent in the data may be considerably less, even none.

The third basis for the rule is Guttman's lower bound theorem (Guttman, 1950). He proved that in the *population* correlation matrix, there must be at least as many consistent factors as there are eigenvalues greater than unity. However, in *samples,* there may be many *more* eigenvalues greater than unity. Thus this rule of thumb is not valid on these grounds, either. A number of simulation studies have been done over the years in which correlation matrices are constructed on the basis of a predetermined structure. The conclusions from almost all of these studies (for example, Zwick and Velicer, 1984) is that eigenvalues greater than unity is a poor rule for deciding on the appropriate number of components.

Another procedure used for deciding on the number of components to retain is an inferential procedure called Bartlett's test for sphericity. It is an inferential procedure for testing the equality of the $p - r$ smallest eigenvalues. It is employed for successive values of r from one up to the point where the null hypothesis is accepted. The number of components is the number for which the null hypothesis is rejected. Simulations have shown this procedure works fairly well (Velicer, 1982). There are, however, circumstances where it would not be expected to work. If the sample size is small, Bartlett's test may accept too few components; if the sample size is large and there is variability in the amount of variance that the variables share with each other, the test may accept too many. The procedure does seem to have sufficient utility to consider the evidence from it's own, as well as other sources.

We might hesitate to admit to such informality, but the number of factors question is usually decided by a combination of eyeballing the eigenvalues and trying to interpret the factors. Consider the interpretational approach first. Generally, the investigator rotates several different numbers of components and attempts to interpret the resulting factors and chooses the solution that seems to yield the largest number of most interpretable factors. In defense of this, it should be pointed out that the first factors are rarely changed much by rotating a few additional factors. Also, if we regard the principal components process as purely exploratory, then this approach may be suitable. The problem is its dependence on the ingenuity of the investigator, and also the extent to which it allows for self-confirmation—that is, it allows us to find what we expect or want to find. This approach, is completely lacking in the objectivity that is the cornerstone of scientific endeavor. Thus it is one of those statistical procedures that we think is fine for ourselves, but of questionable value in the hands of others.

The process of examining the eigenvalues to decide on the number of factors is a timeworn method. There are a number of terms for this process. "Rootstaring" is one of the older terms used; it originates from the days when we called eigenvalues characteristic or latent roots. A more dignified term that carries an air of formality and statistical expertise is the "scree test." To perform this test, we arrange the eigenvalues in order, examine the magnitudes of the adjacent ones, and look for a "break" in the differences between them. This is aided usually by a graph plotting the eigenvalues as a function of their ordinal positions. Figure 13-3 is an example, using data from Table 13-2.

Figure 13-3 illustrates some typical characteristics, as well as some that are not so typical. The plot necessarily decreases from left to right, and the largest differences are usually at the beginning. In fact, with a 40-variable study, the first eigenvalue may be four or five times as large as the next. The right-hand side of the plot, for the last half or more of the eigenvalues, usually shows a fairly uniform gradual decrease. Looking from right to left, we mentally or literally draw a line through these points, looking leftward for the first point that seems to be well above this line. This jump is taken as evidence that the corresponding factor represents something real, whereas the smooth curve at the right reflects components composed of some kind of error. (This is called the *scree* test, based on a rather fanciful analogy to the scree one sees at the base of a mountain—the dirt that rolls down forming a gradual slope up to the steep rise of the mountain.

From Figure 13-3, we would conclude rather confidently that the last five factors represented something random. There might be some question about the third eigenvalue, which sticks up a bit above the line. However, because it is less than 1.0, it represents less variance than a single variable. Moreover, examination of this eigenvector shows that it has one large loading for variable 5 and small loadings for the rest of the variables. Thus, it is primarily there to account for variable 5, which is the variable least represented, remember, by the two larger

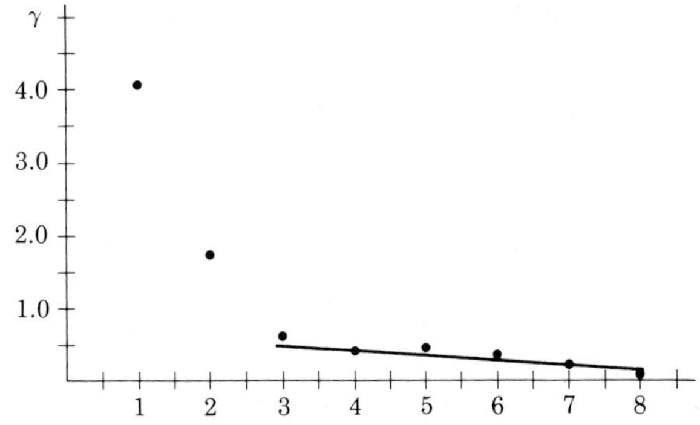

Figure 13-3 Eigenvalues for Table 13-2 plotted against order. The line at the right has been fitted by eye to the last six points.

factors. On several grounds, then, one might conclude that this third factor is not worth including with the first two factors.

This type of a decision process on the number of factors to rotate is perhaps better suited to the principal *factors* method, which will be discussed in Chapter 15. The plot with the eigenvalues of the correlation matrix itself tends to be irregular even under the best of circumstances, making the decision somewhat ambiguous. Often there is no discernible break; the eigenvalues decrease gradually, falling along a smooth curve. Or sometimes there are several breaks in the curve, and the investigator is hard put to decide which of these is the most appropriate.

Looking for a discontinuity in the pattern of eigenvalues is not just a modern analogue of examining the entrails of the sacrifice; it has some statistical justification also. For example, consider the population correlation matrix and its eigenvalues and eigenvectors. Suppose we really knew these values, and used the largest ones to define a certain number of components factors, rotated or not. This *population* factor solution is the ideal we are trying to achieve. On the other hand, all we really have is the *sample* correlation matrix and we wish to use it to define the factors. We would like the sample factors to represent the population factors as much as possible. It turns out that looking for a break in the eigenvalues helps us do this.

The reason for this is a little hard to explain. Every p by p matrix of eigenvectors W is an orthonormal transformation T of any other such matrix W*. This must be true because W and W* are themselves orthonormal. An element t_{pq} of this transformation matrix is the cosine of the angle between vector p of the one matrix W and vector q of the other W*. If T is the identity matrix, then the two sets of eigenvectors are identical.

Suppose W is the complete matrix of population eigenvectors and W* is the same for the sample. It turns out that the cosine t_{pq} between vector q of the sample and vector p of the population depends on the difference in the corresponding population eigenvalues (Cliff, 1970). This means that the extent to which the r largest *sample* principal components factors are influenced by the $p - r$ smaller *population* principal components depends on the differences between the first r population eigenvalues and the remainder. Because we do not know what these differences are, we fall back on assuming that they correspond to the differences in the sample eigenvalues. This suggests that the rule of looking for a break in the eigenvalues and using it to define the number of factors is a sensible rule. If we do this, the sample components are likely to resemble the corresponding population components because they are less confounded by the population components farther down in the order.

This discovery provides some support for the practice of rotating one or two more components than we think are really there. We do this in the hope that, by rotating $r + 1$ components, and then discarding the last one or two, we can disentangle the sample factors to some extent from the minor population influences. This is likely to be especially useful when the separation between eigenvalues is not a sharp one.

A useful but neglected method of determining the number of factors that are well-defined is the split-sample approach. In this approach, the total sample is divided into two halves, and the analyses are carried out independently on each half. Then the factors are compared, usually by correlating the loadings on the factors from the two samples. The number of factors to be interpreted is the number that shows good levels of correspondence—say, with correlations between the patterns of loadings above .70. This may be carried out on either the rotated or unrotated loadings. In the latter case, the rotations would be carried out subsequently.

Probably the best program to carry out this procedure is a factor-matching or Procrustes program (see Chapter 14). This procedure takes the factor matrices from the two samples and rotates them simultaneously to the "best matching position" such that the first pair of factors match as well as possible, the next pair match as well as possible (subject to the restriction that they are orthogonal to the first ones), and so on. The details for this procedure will be presented in Chapter 14.

NONPRINCIPAL COMPONENTS

General Process

For about half the history of factor analysis, the precomputer era, the principal components method was impossible to carry out for any but the smallest problems. The factors were extracted one at a time from the correlation matrix, with matrices of "residual" correlations computed after each factor. "Extracted," with all its connotations of effort, pain, and resistance, is the appropriate word here.

A variety of methods performed such extractions, including centroid, group, multiple group, diagonal, bifactor, and so on. The favorite method—at least of those who did not actually have to do the calculations—was the centroid method. Generally speaking, it was the most nearly successful at extracting the maximum amount of variance with each factor, and was thought to be fairly close to the principal components.

Recall that principal components are defined in terms of uncorrelated weighted composites of the observed variables. Other forms of component analysis also lead to uncorrelated composites, but in the other forms simple weights are used to define standardized, linear composites, and the correlations of the variables with these composites. This is done first with a single composite; this composite is then partialed out of the correlations among the variables. A second composite is formed in a similar way, using different weights, and the correlations with the second composite are found. It is partialed out too, and the process continues until the extractor cannot stand it any longer or until the residual covariances are deemed small enough to ignore.

The process of partialing out successive composites before computing a new composite guaranteed that the composites were uncorrelated. Thus the correlations of the variables with the composite were the same as regression weights for reproducing the variable from the composite.

The different methods had different names, essentially because they used weights of different kinds. The centroid method used weights of ± 1.0. The grouping method started by grouping the variables together that had high correlations with each other. Then it used weights of 1.0 for variables in the group and zero for the others, different factors using different variables to define the group. The diagonal method was similar except that it used single variables instead of groups. A thorough discussion of these methods can be found in Mulaik (1972).

In addition to the principal components method of defining the procedure that is almost universally employed today, there are a wide variety of other procedures that can be used to define the initial orthogonal factor loadings. In all cases, we are using linear combinations of variables to define an orthonormal basis for the vector space of scores. It is only the ways in which this is done that vary among the methods.

The utility of these ideas is partly historical. In the "olden days" of factor analysis, these methods were important, but the connections among them were often spelled out only in the technical literature. The connections are of some use conceptually in that they broaden our perspective on what factor loadings represent. Currently, we usually define factors starting from principal components. Although this has some degree of optimality, it is not the only process possible. Particularly if we want to define the clusters of variables that define factors on an a priori basis, a good deal may be said for going back to the grouping method.

SUMMARY

Summarizing Data Matrices

Once again we have taken a relatively narrow problem and expanded it to cover a great deal of ground. We began with the notion that, for a variety of purposes, it would be nice to have a single variable that was as accurate a summary of the p variables in the data matrix as possible. As usual, we chose the least squares criterion to specify what we meant by "summarizing as closely as possible." The solution to this problem turned out to involve the curious and ubiquitous functions of eigenvalues and eigenvectors, defined by Equation 13.7.

The eigenvalues and eigenvectors of a matrix define the "principal components" of the matrix. Mostly, we work with the eigenvalues and vectors of square, symmetric matrices, $X'X$, S_{xx}, or R_{xx}. There are as many eigenvectors and eigenvalues of a matrix as the order of the matrix, although some of the eigenvalues may be zero. The number of nonzero eigenvalues is the rank of the matrix.

Eigenvalues and eigenvectors have many convenient and useful properties. Not only can they be used to compose least squares summaries of the data matrix, but the eigenvalues furnish a numerical description of the adequacy of the summary (see Equation 13.24). Furthermore, eigenvectors are orthonormal, a fact that can be used to simplify many algebraic relationships. Thus the principle components of a matrix are very useful in summarizing the matrix of observations.

In most statistical applications, the attention focuses on the correlation matrix—or more rarely, on the covariance matrix—rather than on the data matrix. However, the correlation matrix is a convenience, and one should not lose sight of the data matrix that lies behind the symmetric matrices derived from it.

Although the principal components method is virtually the only components method currently in use, this chapter considers its predecessors such as the centroid, grouping, multiple-group, and diagonal methods. Consideration of these earlier methods broadens our perspective on the components process.

PROBLEMS

The following data and preliminary results are from a principal components analysis:

R_{xx} =

	1	2	3	4	5	6	7	8	9
	1.000	.486	.536	.309	.166	.124	−.059	−.039	.024
		1.000	.392	.234	.075	.161	.185	.143	.134
			1.000	.158	−.056	.012	−.038	−.108	−.040
				1.000	.579	.685	.263	.285	.193
					1.000	.557	.310	.352	.303
						1.000	.116	.183	.113
							1.000	.577	.592
								1.000	.689
									1.000

Eigenvalues:

m	1	2	3	4	5	6	7	8	9
λ_m	3.0712	2.0187	1.4481	.5980	.5177	.4484	.3531	.3135	.2313

First five components:

F_5 =

1	.31479	−.76151	.20990	−.22044	.35842
2	.40684	−.53799	.40576	.57967	.11693
3	.13955	−.72911	.35732	−.31888	−.40348
4	.74985	−.25426	−.40897	−.01070	−.16483
5	.72274	.03717	−.41874	−.19924	.22810
6	.62591	−.16678	−.61076	.17954	−.13917
7	.64116	.39303	.36076	.08582	−.30499
8	.69077	.44516	.29783	−.03247	.09491
9	.64998	.41389	.42080	−.17675	.11037

1. Using data from the preceeding analysis, plot the eigenvalues against order of eigenvalues. Calculate the cumulative proportions of variance accounted for 1, 2, . . . , 9 components. What number of components seems reasonable to account for these variables?

2. For each variable in the preceeding analysis, calculate the proportion of variance accounted for by three components and five components.

3. Convert the first component in the preceeding analysis into an eigenvector, $\mathbf{w}_1$. Multiply the first two rows of R_{xx} by w_1 and verify that the result is the first two entries in $\mathbf{w}_1$, multiplied by λ_1.

4. The second two eigenvectors of R are at right.
   (a) Suppose a person i has scores of $x_i = (1, 1, 1, 0, 0, 0, 0, 0, 0)$ on the nine measures. Use W to find her scores on the first three components.
   (b) Use these component scores to "reproduce" her $\mathbf{x}_i$. How does the sum of squared deviations, $\Sigma_j(x_{ij} - \hat{x}_{ij})^2$, compare with what would be expected from the eigenvalues?

$\mathbf{w}_2$	$\mathbf{w}_3$
−.53597	.17443
−.37865	.33719
−.51317	.29710
−.17895	−.33985
.02616	−.34797
−.11738	−.50754
.27662	.29979
.31331	.24750
.29131	.34968

5. Input the correlation matrix of a statistical package and do a principal components analysis of it. Compare the results to those presented here. (There may be very small discrepancies due to differences in computing routines. Remember also that all the signs on a component may be reversed.) If possible, save these results on the system for later use.

ANSWERS

1. Cumulative proportions of variance: .341, .566, .726, .793, .850, .900, .939, .974, 1.000. Three components seem reasonable due to a "break" in the plot and small proportions of variance added by the later components.

2. 3: .723, .620, .679, .794, .699, .793, .696, .764, .771
   5: .900, .969, .943, .821, .791, .844, .796, .774, .814

3. $\mathbf{w}_1' = (.1796, .2322, .0796, .4279, .4124, .3572, .3659, .3942, .3709)$
   $.552 \div 3.0712 = .1797; .713 \div 3.0712 = .2322$

4. (a) $\mathbf{z}_i = (.4914, -1.4278, .8087)$; (b) $\hat{\mathbf{x}}_i = (.9945, .9274, 1.0121, .1957, -.1162, -.0673, .0273, -.0535, .0491)$; $\Sigma_j (x_{ij} - \hat{x}_{ij})^2 = .0678$ $\Sigma_4^9\lambda_r = 2.4620$, so the errors here are very small compared to the average subject.

READINGS

Cooley, W. W., and P. R. Lohnes (1971). *Multivariate data analysis.* New York: Wiley, pp. 96–128.

Gorsuch, R. L. (1974). *Factor analysis.* Philadelphia: Saunders, pp. 12–32.

Green, P. E., and J. D. Carroll. (1976). *Mathematical tools for applied multivariate analysis.* New York: Academic Press, pp. 194–240.

Morrison, D. F. (1976). *Multivariate statistical methods* (2nd ed.). New York: McGraw-Hill, pp. 266–301.

Mulaik, S. A. (1972). *The foundations of factor analysis.* New York: McGraw-Hill, pp. 95–131, 173–185.

14

Transformation of Component Loadings and Computation of Component Scores

IMPORTANCE OF TRANSFORMATION

According to the component model stated in Equation 13.26 or 13.27, the components loadings in F_r are regression coefficients for reconstructing the observed variables from the standardized components. If the loadings are derived from the correlation matrix, as is usually the case, then they also represent correlations between variables and components since the components are uncorrelated. The main purpose of such an analysis is typically that of trying to understand the composition of the variables. Unfortunately, the coefficients in the F_r matrix are usually difficult to understand or reconcile with our expectations. The mixture of positive and negative signs, necessitated by the orthogonality of eigenvectors, is particularly hard to understand in most domains. If the investigator has taken some pains to construct the battery of measures according to a predetermined plan, the contents of F_r depart radically from the plan. Furthermore, if two or more samples are analyzed, the resulting component loadings often differ substantially, with the exception of the first component.

Therefore, virtually all matrices of principal components loadings are transformed in a way that is intended to make them more interpretable, and, we hope, closer to reality. Just as different transformations of the score matrix can be used in anova, transformations of the original solution are employed in principal components and other forms of factor analysis. A small minority of data analysts disagrees with this practice, maintaining that the directions of maximum variance in the space are important to preserve, but virtually all investigators feel it necessary to resort to transformation.

In most factor analysis literature, the transformation process is called the "rotation of factors" (or some similar term), although the term "rotation" should logically only be used when the transformation is known to be orthogonal. Sometimes it is not orthogonal, so the more general term is transformation. When the principal components are rotated according to some criterion or method, it means that a new set of loadings has been computed:

$$A_r = F_r T \tag{14.1}$$

The transformation T may or may not be orthogonal; the context or method used should make that clear.

GENERAL PRINCIPLES IN TRANSFORMATION

Balancing the Books

Because the loading matrix A_r is different from F_r, it must apply to a different set of scores. Indeed, because the object of components analysis is to find a small number of components that nearly reproduce X, we want the altered system to do this just as well. To preserve the equality $\hat{X}_r = Y_r F_r'$ after F_r has been transformed, it is necessary to transform Y_r also. This is an inverse transformation that counteracts, so to speak, the effect of T. We express it as

$$P_r = Y_r (T')^{-1}$$

This ensures that $\hat{X}_r = Y_r A_r'$ just as $\hat{X}_r = Y_r F_r'$. Substituting $T'F_r'$ for A_r' and $Y_r(T')^{-1}$ for P_r gives

$$\hat{X}_r = [Y_r (T')^{-1}](T'F_r') \tag{14.2}$$

The equality in Equation 13.26 is preserved because the transformation and its inverse cancel each other out. Therefore, the transformed components reproduce the observed scores to exactly the same extent that the original component scores did.

Although the extent to which the transformed components account for the data stays the same, the *way* in which they do this may be different. The effect of transforming components depends to a considerable extent on whether or not T is an orthonormal transformation. Remember the orthonormality means that $T'T = TT' = I$. Using this kind of a transformation has much less complex effects on how the vectors are expressed than if T is not orthonormal. We will discuss the orthonormal transformation first.

Orthogonal Rotation

If T is orthonormal, then the transformation process is referred to as "orthogonal rotation." This is because an orthonormal transformation can be visualized as a rigid rotation of the coordinate axes representing the basis vectors of the space.

Not only is the transformation process easy to visualize, the bookkeeping in Equation 14.2 is made simple. If $TT' = T'T = I$, then $T^{-1} = T'$ and $(T')^{-1} = T$. If in general we let the transformed component scores be

$$P_r = Y_r(T')^{-1} \tag{14.3}$$

this means that in the case of orthogonal rotation,

$$P_r = Y_rT \tag{14.4}$$

Simple substitution will show that because the y's are uncorrelated standard scores, the p's are also:

$$R_p = \frac{1}{n-1} P_r'P_r \tag{14.5}$$

This will be an identity matrix if $T'T = I$. The correlation matrix will be reproduced in exactly the same way by A_r as it was by F_r:

$$\hat{R}_r = A_rA_r' = (F_rT)(T'F_r') = F_r(TT')F_r' = F_rIF_r' \tag{14.6}$$

If T is not an orthogonal rotation, the relationships are similar, although they are expressed in less simple or convenient ways, which will be taken up later.

The proportion of each variable's variance that is accounted for by the transformed components is not changed by the transformation. This is illustrated in the fact that the $\hat{X}_r$ is the same. It can also be confirmed by considering the diagonal of R_r in Equation 14.6. What does change is the way in which the variance is distributed over the transformed components. Using the principal components, we accounted for as much variance as possible by the largest component, and then by the next largest, and so on. When the components are transformed, the variance across the factors evens out when all the variables are considered. However, as we will see, the transformation is usually designed to concentrate as much of an individual variable's variances as possible on one or a few components, and different variables' variance on different components. The details of this process are considered next.

CHOOSING THE TRANSFORMATION

Simple Structure

Factor loadings represent two things. On the one hand, they represent a set of coefficients, analogous to regression coefficients, for constructing the *variables* from the *factors*. Because the factors are so far uncorrelated, and we are assuming everything to be in standard score form, this means that the loadings also represent the correlations of the variables with the factors.

Most applications of factor analysis are based on the belief that only one or a few components are involved in most variables. This belief has been formalized in factor analysis under the concept called *simple structure*. With few exceptions, we expect the influences of the underlying processes on the observed variables to be positive, not negative. This expectation is denoted in the factor analytic literature as the *positive manifold* concept. The vast majority of rotational schemes in factor and components analysis are based on both simple structure and positive manifold—particularly the former.

The idea of simple structure suggests that we should rotate the principal components so that most of the loadings for each variable are quite small and its variance is concentrated on only a few factors—preferably one. Conversely, as we consider each factor, there should be major loadings on only a minority of variables, except when there are only two or three factors. With few exceptions, the loadings should be positive unless they are near zero, in which case there may be a number of loadings close to zero in both the positive and negative directions.

Hand Rotation

During the first twenty years of its existence, multiple factor analysis relied on a combination of visual inspection and hand calculation as the basis for choosing the transformation that carried the original set of loadings into its final position. Today the process is almost always carried out by a computer algorithm. Doing the calculations by hand did provide a certain feel for rotation. However, the labor involved in a large problem—involving perhaps 50 variables and a dozen factors—might take hundreds of person-hours. With this in mind, we will use a relatively simple example to demonstrate this method.

Consider the factor loading on the first two components reproduced in Table 14-1. These loadings display a pattern that is typical of a set of variables where there are two well-defined factors. Note that all the loadings on the first component are substantial and positive, whereas those on the second are about half positive and half negative. We might be satisfied with interpreting the first loading factor as a general size factor, but the second factor poses some difficulties. It would be difficult to imagine a physical growth process that had the effect of making people longer as it made them thinner. In addition, we know that it would be possible to reverse the signs of all the variables on either or both factors without altering the way in which we reproduce either the score or the

Table 14-1 Principal Components Factors

	Factor Loadings		
Variables	*1*	*2*	*% Variance*
1. stature	.906	.270	89.6
2. hip height	.817	.396	82.5
3. lower leg length	.788	.314	72.0
4. arm length	.836	.195	73.7
5. sitting height	.688	.059	47.7
6. knee girth	.502	−.678	71.1
7. elbow girth	.453	−.730	73.8
8. wrist girth	.541	−.643	70.6
$\Sigma_i f_{im}^2$	4.033	1.776	72.6

correlation matrices. This means that the second factor could be shown as one that was making people wider as it was making them shorter. Thus, these factors are not satisfactory as explanatory concepts in their current forms.

The loadings are plotted in Figure 14-1. Note that there are two groups of points. This suggests transforming the basis of the space so that the points lie near the coordinate axes. New positions for the axes are suggested by the dashed lines in the figure (note the orthogonality) and plots using these transformed axes are shown in Figure 14-2. Notice that the axes are not able to pass directly through the clusters of points as we might hope in an ideal case, but they come close.

The transformation matrix that corresponds to this rotation is given in Table 14-2, along with the loadings of the variables on the transformed axes. In this case, the transformation was determined by placing a transparent drafting triangle over the plot with the point of the right angle at the origin and rotating it until the most satisfactory position had been reached. The transformation is

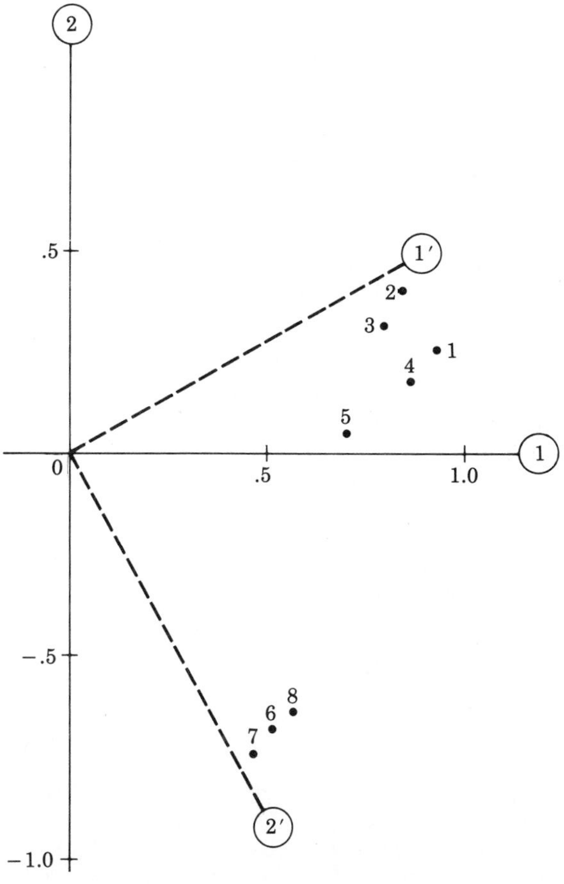

Figure 14-1 Plot of loadings of the variables on the first two principal components. The dotted lines indicate the intended positions of the new axes.

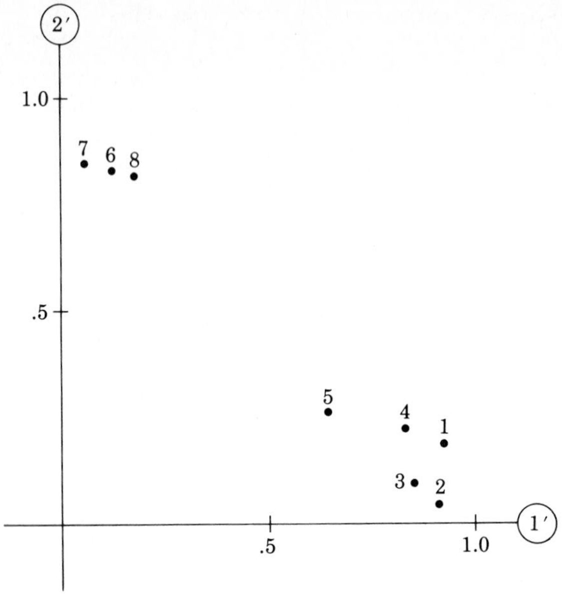

Figure 14-2 Plot of loadings on the orthogonally transformed factors. Note that variables 6, 7, and 8 now appear at the top of the figure because the direction of the second axis has been reversed.

determined as a function of the angles that the new axes made with the original axes, as follows:

$$T = \begin{bmatrix} \cos_{11'} & \cos_{12'} \\ \cos_{21'} & \cos_{22'} \end{bmatrix} \qquad (14.7)$$

Thus the elements of the transformation are the cosines of the angles that the new axes, designated 1′ and 2′, make with the old ones, 1 and 2. In many sources, we see this transformation as

$$\begin{bmatrix} \cos\phi & -\sin\phi \\ \sin\phi & \cos\phi \end{bmatrix} \qquad (14.8)$$

where ϕ is the "angle of rotation." Although this form is equivalent, it does not generalize to more than two components as Equation 14.7 does.

Note that what happened here was a change of basis. The points representing the variables may be thought of as vectors. The loadings tell how to make up these vectors as a linear function of the basis vectors, in this case, the factors. Thus, in the original component matrix, stature is defined as .906 of factor 1 and

Table 14-2 Transformation Matrix T and Rotated Components Matrix A

Old Components	*New Components* 1′	2′
1	.876	.482
2	.482	−.876

Variables	*New Loadings* 1′	2′	*% Variance*
1	.924	.200	89.4
2	.907	.047	82.5
3	.842	.105	72.0
4	.826	.232	73.6
5	.631	.280	47.7
6	.113	.836	71.2
7	.045	.858	73.8
8	.164	.824	70.6
Σa_{jm}^2	3.508	2.299	72.6

.270 of factor 2 (see Table 14-1). The transformation represents a change of basis. The new basis vectors are defined in terms of the original basis vectors according to the coefficients of the transformation matrix. In Table 14-2, the new first component is .876 of the original first component plus .482 of the original second component; the new second component is .482 of the original first component and −.876 of the original second component. Note that the latter is negative. This reflects the fact that the ends of the second component have been reversed so that the large loadings will end up positive. If the dotted lines in Figure 14-1 indicate the positions of the new basis vectors, it can be seen that these new basis vectors are defined in terms of the original ones according to these coefficients. The new axis 1′ is about .88 of the old 1 plus about .48 of the old 2; the new axis 2′ is about .48 of the old axis 1 plus about .88 of the *negative* end of the old second axis.

The transformation $TT' = T'T = I$ is orthogonal and may be verified by noting that $.876^2 + .482^2 = .9997$ and that the sum of products between rows is zero, as is the sum of products between columns. The sum of squared loadings on the two new components are now 3.508 and 2.299, respectively. The overall sum is still 5.807, the same as for the sum of the first two eigenvalues, within rounding error, verifying that the total amount of variance is still the same, as it must be, because the transformation is orthogonal. This is true not only for the total, but for every variable individually. Note, however, that the larger component accounts for slightly less than the original first principal component, and that the reverse is true for the second component. Remember that the first principal component accounted for as much variance as possible. Therefore, any other direction in the space must account for less than this. Note also that the sum of products between components is no longer zero in the rotated case. This will always be true of rotated factors; $F'F$ is diagonal only when the factors are the principal components; A′A is therefore not diagonal, except in the unlikely case where eigenvalues are equal.*

The loadings of the variables on the factors of the physical variables in Table 14-1 are much easier to interpret in Table 14-2. Rather clearly, there seems to be a "length" and a "girth" factor. The major part of the variance of the variables associated with bone length is now attributable to one component, whereas the variance of the variables associated with circumference measurements is attributable to the other component. One height variable, sitting height, has a moderate loading on component 2′. That is not too surprising since this variable is likely to reflect the length of one's spinal column and the girth of one's posterior. Overall, the transformed components seem much more readily interpretable than the untransformed components. When we obtain the corresponding component scores, it becomes apparent that those with high scores on component 1′ are tall and those with low scores are short, but that some have thick limbs whereas others have slender limbs. On the other hand, the high scores on 2′ represent those with heavy arms and legs (in an absolute sense, rather than a relative sense), whereas those with a low score have slender wrists and ankles. A well-built basketball forward would score very high on the first component and at least fairly high on the second. A person as tall and heavy as Wilt Chamberlain, for example, would be near the extreme on the second component as well as the first.

Hand rotation involving more than two components or factors is a complex process. It requires successive approximations, rotating pairs of factors into new locations, and then looking back to see how rotating 1 with 2 and 3 with 4 affects the relation between, say, 2′ and 3′ and how this justifies rotating 2′ and 3′ with each other. Hand rotation is not only tedious; it also calls upon our ability to visualize complex spatial relationships.

Orthogonal Analytic Rotation: Varimax

"Analytic" transformation or rotation methods employ a numerical rather than an intuitive or spatial definition of the simple structure concept. This numerical definition of degree of simplicity of structure is then taken as an "objective" function, a numerical quantity that reflects the degree of accomplishing the objective, that is, the simplicity of the structure in a given matrix of factor loadings. Using numerical methods, we try to find the position for the axes that gives the maximum value for the criterion or objective function. The calculations involved are so horrendous that hardly anyone, with the possible exception of the inventors of the methods, has ever carried them out by hand. All that we will do here is sketch the rationale for several of the major alternative methods and try to understand what their differences are, concentrating on the most widely used and most often satisfactory method, *varimax*.

The varimax method starts from a particular definition of the simplicity of the loadings on a *factor*.† In an ideal case of simple structure, the distribution of

---

*A common confusion arises out of this terminology. We say that a transformation or matrix T is orthogonal if T′T is a diagonal matrix. When the term "orthogonal factors" is used to refer to a matrix A of factor loadings it means only that the factors are orthogonal rotations of orthogonal components, not that A′A is diagonal. For example, the sum of products of columns in Table 14-2 is .952.

†To avoid awkward phraseology, we will use the term "factor" to refer to both rotated components and the common factor loadings described in Chapter 15.

loadings on a factor would be *bimodal*—some high loadings and many near zero loadings, with nothing in between. The variance for such a distribution is larger than for any other distribution that has a similar mean and is confined to the interval between zero and one. Henry Kaiser, the inventor of varimax, decided that this logic could form the basis of a rotation criterion when used in reverse. A transformation that maximizes the variance of factor loadings should also maximize the simplicity of the structure—hence the name varimax.

This idea had to be modified before it could be used as the basis for a successful analytic rotation method. One major modification was the decision to work with the squares of the loadings rather than the loadings themselves. This allows us to ignore the signs of the loadings and also leads to some technical simplifications in the process of defining the procedure for maximizing the varimax criterion. In any event, the varimax criterion for a given factor m, v_m^2, is defined as

$$v_m^2 = \frac{1}{p} \sum_j [a_{jm}^2 - (\bar{a}_{.m})]^2 \tag{14.9}$$

where $\bar{a}_{.m}$ is the average squared loading on factor m, not the squared average loading. Thus v_m^2 is the variance of the squared loadings on the factor. The overall criterion v^2 is the sum of these variances across factors:

$$v^2 = \frac{1}{p} \sum_j \sum_m (a_{jm}^2 - \bar{a}_{.m})^2 \tag{14.10}$$

Using the varimax criterion is equivalent to finding the orthogonal transformation matrix T that gives the highest value for v^2 as defined across the matrix of loadings A = FT.

Very often, varimax rotation is carried out on variables whose loadings have been "normalized," that is, multiplied by a conversion factor of $(\Sigma_m a_{jm}^2)^{-1/2}$ so that their sums of squared normalized loadings are all 1.0. This is done for technical reasons and is "invisible" to the user of computer packages since normalization is reversed before the rotated loadings are printed out.

Other Analytic Rotation Methods

The major early alternative to varimax is a method called *quartimax*. It was actually discovered independently by several sets of investigators who used several different rationales. Here, we will consider only one of the rationales, beginning with the idea that the factors should be transformed in such a way as to simplify the loadings of each *variable* across each *factor*. Analogous to the varimax idea is the quartimax notion that the maximum simplicity of loadings for variable j will occur when all of its variance is concentrated on a single factor. The sum of a variable's squared loadings stays constant under orthogonal transformation of factors, but as more and more of this variance is concentrated on a single factor, the sum of fourth powers of the loadings increases. For example, if there were four factors and a variable had a loading of .5 on each, the sum of squares would be 1.0 whereas the sum of fourth powers would be only .25. However, if we concentrated all the variance on one factor—that is, if we had one loading of 1.0 and three of zero—then the sum of fourth powers would also be 1.0. The quartimax criterion, then, tries to maximize the sum of fourth powers of loadings for all of the variables on all of the factors simultaneously, hence the name quartimax. The value of T that does this is sought by analytical methods in the belief that it will lead to simplification in the pattern of loadings, which it does, by and large.

Mathematically, varimax and quartimax represent two points on a continuum of possible rotational criteria leading to the possibility of intermediate points on the continuum, any of which might be a suitable definition of simple structure. One popular intermediate method is called *equamax*. This method is more like quartimax when the number of factors is a small proportion of the number of variables and more like varimax when the number of factors is relatively large.

These three analytic rotation methods are available in computer packages for orthogonal analytic transformation. More often than not, varimax gives the more satisfactory results, but there is little harm in sampling one or more of the other methods in these days of nearly cost-free computing and picking whichever method looks best.

Table 14-3 shows the effect of applying varimax and quartimax to the loadings in Table 14-2. (It makes no difference whether we start from these or the unrotated loadings.) The results for each method are virtually identical here. This is generally true of cases in which the number of factors is small and the simple structure is clearly defined. In other cases, the results of varimax and quartimax will differ, particularly when the simple structure is not so clearly defined.

Rotation to Targets

An alternative to rotating to maximize simple structure is to rotate the principal components to conform as closely as possible to a "target" matrix. The target is a matrix of hypothetical loadings for the variables. It is based on theoretical expectations or previous experience with the same set of variables. The rotation is used to make the obtained factor matrix conform to the theoretical one as nearly as possible. Thus, it is a kind of confirmatory or hypothesis-testing procedure, although this is only true in an informal sense, not in the sense of formal statistical hypothesis testing.

To do this, the investigator prepares a hypothesis or target matrix, consisting of the loadings the variables are expected to have. This need only be an approximation, using 1.0 where the loadings are expected to be high and 0 where they are expected to be low. If we anticipated the obtained structure in Table 14-1 on theoretical grounds, or from inspection of the unrotated loadings, the

Table 14-3 Orthogonal Analytic Rotations

	Varimax		*Quartimax*	
	Factor 1	*Factor 2*	*Factor 1*	*Factor 2*
1	.930	.168	.938	.122
2	.908	.016	.908	−.029
3	.845	.076	.848	.034
4	.833	.204	.843	.162
5	.640	.258	.652	.226
6	.141	.832	.183	.824
7	.074	.856	.117	.851
8	.192	.818	.233	.807

loadings on the first factor of the target matrix would be 1.0 for the first four variables and 0 for the rest, and 0 for the first five variables and 1.0 for the rest on the second factor (variable 5 just rides along). A slight refinement would be to scale the target loadings so that the sum of squared loadings for each variable in the target matrix is the same as the sum of squared loadings on the unrotated factors. The unrotated matrix must have at least as many factors as the target matrix. A computer program is used to find the transformation matrix that will transform the obtained factors and make the sum of squared differences between the rotated and target loadings as small as possible. Unfortunately, this procedure has not been incorporated into any of the most widely available computer packages as of yet. It is, however, relatively simple (Cliff, 1966).

Orthogonal rotation to congruence is not widely available in packages, although it is easy to program if we have access to a package for matrix operations such as SAS PROC MATRIX, Gauss, or Speakeasy. To do this, let F_r be the matrix of factor loadings to be rotated and A_r be the target or hypothesis. Experience has shown that both should be the same order. Find the eigenvectors W_F of $(F_r'A_r)(A_r'F_r)$, and W_A of $(A_r'F_r)(F_r'A_r)$, where each column is an eigenvector. Then $\hat{A}_r = F_r W_F W_A'$.

This program can also be used to match factors. Simply multiply $M_f = F_r W_f$ and $M_a = A_r W_a$ to match corresponding factors as closely as possible. The degree of correlation between corresponding factors can be used to decide on the number of reliable components. For example, it can be used to assess the consistency between results of different studies that use the same variables, or to estimate the number of reliable components by splitting a sample in half and analyzing each half.

Rotation to target is sometimes carried out in an iterative fashion. The fit may be found to be fairly good except for a few major discrepancies between target and rotated matrix; a few variables depart substantially from expectations. Since the procedure is trying to fit all the data at the same time, the presence of these major errors "pulls" the solution away somewhat from the loadings that could be fitted fairly accurately. Sometimes the investigator may decide to adjust the target to take account of this, to make the target more like the rotated matrix, particularly in respect to the largest discrepancies. The adjusted target matrix becomes the target and the process is carried out again. The investigator who employs this strategy must recognize that the procedure is no longer one of hypothesis testing, and that a substantial part of the rigor of the procedure has been lost due to tailoring the "hypothesis" matrix to fit the data.

The user of rotation-to-target methods must also be aware that there is substantial opportunity for spurious approximate confirmation of the target. The transformation matrix contains r^2 elements that are being adjusted to maximize the fit to the *rm* elements of the target matrix. Even though the restriction to orthogonality means that only $1/2r(r - 1)$ of the elements of T are independently free to vary, rotation to target is an extremely powerful wrench to apply to the data. Although it may succeed in putting the solution in the correct position, it may also do so arbitrarily, in effect producing an apparent fit by twisting things radically out of shape. It appears that the procedure is likely to give spurious confirmatory results unless the number of variables in a large problem is at least five and preferably ten times the number of factors. When the target is adjusted in a post hoc basis, these considerations are even stronger.

For this reason, the name sometimes given to this kind of procedure—*Procrustean rotation*—is particularly apt. Procrustes was the leader of a band of robbers in Greek mythology. He was in the habit of putting his victims in a bed—whether they fit in this bed or not. If they were too long for it, he performed radical surgery on their legs to improve the fit. If they were too short, he stretched their limbs so that they became the right length. One of the most important fundamentals of multivariate analysis is respect for data. The investigator should therefore try to ensure that his rotation to a target is not literally Procrustean. When tempted by the very human desire to confirm our expectations, it may help to remember Procrustes's fate: The hero Theseus "fitted" Procrustes to his *own* bed as Procrustes had fitted others.

Rotation and Theory

Remember that the goal of factor analysis is finding out what major variables underlie a larger set of manifest ones. The concept of simple structure—that only a few of the underlying factors are important in determining any given manifest variable—is really a hypothesis, a belief about the nature of the variables involved. As such, it may be true or false in any given instance. Indeed, in most cases there are many influences on a variable, such as a test score, rather than only a few influences, as simple structure requires. Thus, if there is some "true" underlying factor structure for a collection of variables, application of the simple structure principle in rotation may give results that are misleading. In a single analysis with a single set of data, we never know.

Factor or component analysis is not well suited to a single analysis of a single set of data, although it is often applied in this way. It is much more likely to give meaningful results, however, when used in a programmatic fashion. Exploratory analysis of a single set of data should be followed by the gathering of new data, using measures that have been refined or developed on the basis of the results of the first study. The process continues until a reasonable closure is obtained concerning the nature of the dimensions underlying a given area. It is only in these latter stages that the concept of simple structure is likely to be a very close approximation to the true state of affairs. In the earlier stages, the results of a simple structure-seeking rotation should be considered advisory, rather than as a revelation of the true state of affairs. More will be said about the tactical and strategic matters in factor analytic research, after more of the major fundamental procedures in factor analysis have been considered.

CORRELATED FACTORS

Motivation for Correlated Factors

Restricting the transformation of factor space to the orthogonal case has the advantage of keeping the mathematics simple, but there are frequently cases where the data do not seem to support the position that the factors are uncorrelated. Consider the example of the physical measurements in Table 14-1. This case is unusual in that the points representing the variables are quite close to the axes, reflecting the fact that the correlations between length and girth measures are near zero. However, even here the length variables have small but mostly positive loadings on the girth factor, and vice versa. We may attempt some interpretation of these small loadings, but the interpretation will be ambiguous. Shall we say that the girth measurements depend to some extent on tallness and that bone length depends to some extent on a factor leading to girth? Shall we change the rotation slightly, eliminating one set of cross-loadings, perhaps those of the length variables on the girth factor? Here the orthogonality constraint necessarily increases the other set of positive loadings. The interpretation here might be that girth measurement depends to some degree on the same growth factor that leads to a general lengthening of bones, whereas the reverse is not true. Perhaps.

Instead, we could imagine two distinct but slightly *correlated* growth factors, one for height and one for girth. This can be accomplished by letting the factors become oblique to each other, so that the corresponding factor scores are now correlated, in this case positively.

The nature of what happens under oblique rotation is the subject of considerable confusion. We might think that the new factor axes run through the clusters of points so that the angle between the factor axes becomes less than 90 degrees. Not so. It is the factor *scores* that are going to become positively correlated. This means that the axes representing the factors in the variables plot actually move *away* from each other in situations like the present one. If we wish to visualize the axes in the factor space, imagine rotating one axis so that all but one set of variables is as nearly *orthogonal* as possible to that axis. Thus, in Figure 14-3 we rotate the horizontal axis representing factor 1′ *down* and clockwise, so that variable 6's projection down on it is reduced to near zero; as a result, the loading of 7 becomes negative, and is reduced. In the same way, axis 2′ is rotated a bit to the left (counter-clockwise), to make the loading of variable 3 near zero, reducing 1, 4, and 5, and making 2 slightly negative.

To many, this procedure seems to be the reverse of what should be happening intuitively. The paradox may be resolved by concentrating on the correct interpretation of the factor loading matrix. This matrix is a set of coefficients for

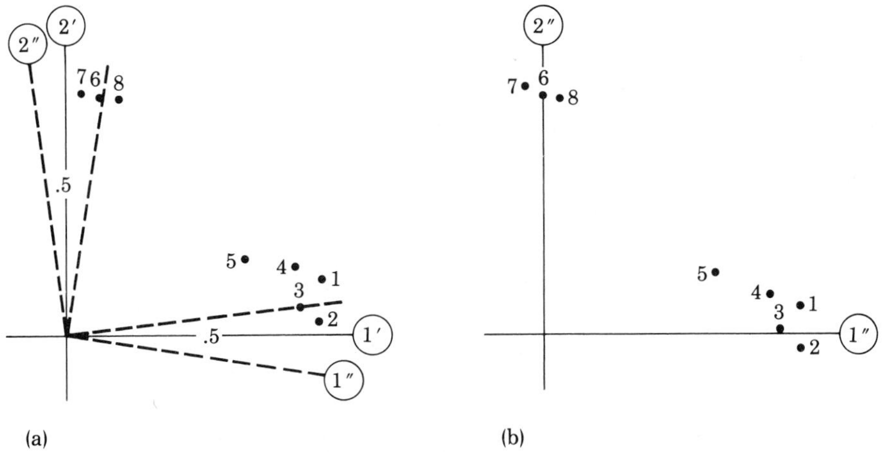

Figure 14-3 (a) Approximate intended positions of the new factor axes. (b) Oblique factor loadings plotted as if the axes were orthogonal (see Table 14-4).

constructing *scores* on observed variables out of *scores* on factor variables, in this case, rotated principal components. Remember that there are always two transformations involved: a transformation of the factor score matrix and a transformation of the factor loading or coefficient matrix. These must be inverses of each other so that the books stay balanced. In the orthogonal case, one transformation is the same as the other, only it needs to be transposed in order to match corresponding indices. In the oblique situation this simplicity breaks down, so a little more attention must be paid to details.

Next we are going to transform the factor *scores* in order to simplify the coefficients that need to be applied to them to construct the variables. The question is, How do we use the factor plots of the variables to transform the factor scores of the persons in order to obtain the desired effect? For all practical purposes, it seems logical to change the orientation of the axis in the variable space so that the variables that do *not* represent that factor project down near zero on that axis. The actual process of defining the transformation is less straightforward because of the complication introduced by our desire to keep the factor scores in standard score metric—that is, with variance 1.0. If the factors axes are adjusted in the way described above, then a final correction is necessary to make the factors scores into standard scores. We'll skip the details of this adjustment process, assuming the computer will take care of this calculation anyway. The results for these data are shown in Table 14-4, where the oblique transformation matrix has been applied to the orthogonal rotated loading matrix A to give the oblique loading matrix A_c. Here, the degree of obliquity is so slight that it makes only a little change in the loadings. Still, the loading pattern is slightly simpler, making the following generalization slightly more precise: *The explained variance in the three girth measurements is entirely due to the girth factor.* In cases where the degree of obliquity is greater, the change is more substantial.

Table 14-4 Oblique Transformation Matrix and Oblique Factor Loadings A_c for Figure 14-4

	Oblique	
Orthogonal	1″	2″
1′	1.022	−.113
2′	−.143	1.026

T

Variable	1″	2″
1	.915	.101
2	.920	−.054
3	.845	.013
4	.811	.145
5	.605	.216
6	−.004	.845
7	−.077	.875
8	.049	.827

A_c

The effects of oblique transformation on the algebra used to describe factor analysis are substantial, as is its effect on the computations involved. Equation 14.2 may be used when the factors are orthogonal or oblique, but in the oblique case subsequent equations must be altered if the various identities are to be preserved. If we want

$$P_r A_r' = \hat{X}_r \qquad (14.11)$$

to represent the same approximation to the score matrix that we had from the principal components (which we do, since the latter is the best possible approximation to X itself), then it is necessary to use Equation 14.3,

$$P_r = Y_r (T')^{-1}$$

in order to preserve this identity. When the transformation is orthogonal, $(T')^{-1} = T$. This is not true when the transformation is oblique.

Presumably, we want to stay in standard score form so that each column has a variance of 1.0. In this case, some care must be taken in constructing T, because the columns of T^{-1} must preserve this property. In general, it does not work to simply use the form for T as, say, we would read it off the plot of Figure 14-3. This transformation must be corrected to retain unit variance in P_r by working on T^{-1}. We'll leave the details of this process to those who write the programs.

If the factors are correlated, we would like to know what these correlations are. It is not necessary to literally compute the correlations among the factors from the factor scores themselves. Because Y_r is orthogonal with unit variance, premultiplying P_r by its transpose and dividing by $n - 1$ gives

$$\frac{1}{n-1} P_r' P_r = \frac{1}{n-1} T^{-1} Y_r' Y_r T'^{-1} \qquad (14.12)$$

$$R_p = T^{-1} T'^{-1} \qquad (14.13)$$

as the matrix of correlations among the factors R_p. This matrix is also necessary in order to reproduce the correlation matrix in exactly the same way with the oblique factors as with the orthogonal ones. When the factors are oblique, since $F = AT^{-1}$, we have

$$\hat{R}_r = A_r R_p A_r' \qquad (14.14)$$

instead of the simpler relation $\hat{R}_r = A_r A_r'$. The latter relation occurs when the factors are uncorrelated so that $R_p = I$. The factor correlation matrix for the example is shown in Table 14-5.

Table 14-5 Factor Intercorrelation Matrix

	1″	*2″*
1″	1.000	.246
2″	.246	1.000

In the oblique case it is also necessary to distinguish between the matrix of factor coefficients, a_{jm}, which indicates the degree of involvement of the *factors* in the *variables,* and the matrix of *correlations* between the factors and the variables. In the orthogonal case these are the same, whereas in the oblique case they are different. In factor analytic literature, the matrix of factor coefficients is the one that is typically reported when a factor study is described; it is called the *pattern* matrix. The matrix of factor-variable correlations is called the *structure.*

Since P_r is in standard score form, this set of correlations, like any other, is derived simply from it and the variables' standard score matrix by $(n-1)^{-1}X'P_r$. A little judicious substitution then shows

$$S_r = F_r T'^{-1} \quad (14.15)$$

Here, S_r is the $p \times r$ matrix of correlations between factors and variables, F_r is an orthogonal component loading matrix, and T is the transformation matrix that takes that set of loadings into the oblique pattern matrix. Applying these concepts to the example gives Table 14-6 as the structure matrix.

The pattern and structure matrices are connected the same way the regression weights and the validities are connected in multiple regression. Recall that $\mathbf{b}^* = R_{xx}^{-1}\mathbf{r}_{xy}$ is the solution for the regression weights. In the same way, $S_r R_p^{-1} = A_p$. The difference is that the factors are being used to "predict" several different variables rather than just one. (Also, the matrices are presented as transposes of each other in the two contexts.) Remember that in the regression context, when two positively correlated predictors are put into a regression equation, the regression weights of both predictors move closer to zero than their respective validities, and the difference between the regression weights is exaggerated relative to the difference in their validities. A similar effect can be observed in each row of the example. Here, the structure correlations of each variable consist of one high value and one low value. The pattern weights are one value that is not quite so high and one that is near zero.

Oblique Analytic Rotation

Oblique transformation methods, like orthogonal methods, arose in the pre-computer era of intuitive rotation by eye. As with the orthogonal methods, a wide

Table 14-6 Factor Structure Matrix: Correlations between Factors and Variables

Variable	*1″*	*2″*
1	.940	.327
2	.907	.173
3	.848	.221
4	.846	.345
5	.657	.365
6	.204	.844
7	.138	.856
8	.253	.839

variety of analytical oblique transformation methods have been suggested. Indeed, since there are various *degrees* of obliquity, there is an even wider set of possibilities in the oblique case. This text will not go into all of these—nor will it even do justice to the full gamut of theoretical possibilities. However, in the following section, we will consider some of the most widely used and seemingly satisfactory alternatives.

The conceptual complexities of oblique transformation impose themselves quite forcefully in the context of oblique analytic rotation. In our description of the orthogonal analytic methods we mentioned that several investigators had reached the quartimin criterion via different routes. In the oblique case, these different starting assumptions—variations in the quantitative definition of simple structure—lead to *different* endpoints. Although we will not trace these routes here, we will note one starting definition of simple structure, called the *covarimin* criterion. This definition tries to minimize the between-factor covariances of squared factor loadings:

$$c = \sum_{m,m'} (p\Sigma a_{jm}^2 a_{jm'}^2 - \gamma \Sigma a_{jm}^2 \Sigma a_{jm'}^2) \tag{14.16}$$

The factor structure of a battery of measures is simplest when each measure has only one high loading and the rest of the loadings are zero. Therefore, the sum of products of loadings between factors should be near zero if the simple structure criterion is nearly satisfied. We use squared loadings instead of the actual loadings to avoid various anomolies, like the situation where the criterion is made small by having positive and negative elements of the sum cancel each other out.

This is called the *general oblimin* criterion because of the presence of γ as a weight on the right-hand term. Different values of this parameter allow the investigator to control the degree of obliquity in the transformation. Setting the parameter equal to zero gives a highly oblique solution; setting it equal to 1.0 makes it only slightly oblique. Intermediate values have intermediate effects. In analysis, we might try various possible values to see which value gives the most satisfactory—that is, simplest—results.

There are two other variations in oblimin depending on how the computer procedure is constructed. The solutions originally used did not place the restriction of unit variance directly on the factor scores. Instead, a restriction was placed on the transformation matrix. Imposing the restriction of unit variance on the factor scores rather than on the transformation is a more difficult mathematical problem to solve, and it was not solved until later. When the restriction is imposed correctly, the method is called *direct* oblimin.

Several variations on the oblique rotation procedures have been employed on our physical measurements variables of Table 14-1. For this simple example there is little variation in the solutions. However, where there are more factors, where the simple structure is not clear, or where the apparent degree of obliquity is not so small, there will be more variation in the results.

Oblique Procrustes

If we have a hypothesis about the factor structure of a set of variables, oblique Procrustean solutions are just as possible as orthogonal ones. In fact, Procrustean solutions are easier in one sense and have been around since the early 1940s. However, most of the available solutions are inexact in the sense that they do not preserve the standard score aspect of the factor scores. This presents the same theoretical difficulty as direct versus indirect oblimin. Even so, the consequences of this are often relatively minor; the solution is not far from the correct one, so little harm is usually done in employing whatever oblique Procrustes solution comes to hand. If possible, a novice should seek technical advice on this matter before employing oblique Procrustes.

One of the more useful applications of both the oblique and orthogonal Procrustean methods occurs in the final stages of transformation. When we recognize a potential pattern, that pattern is used as a target, and a Procrustean transformation is used to attempt to reach that target as nearly as possible. This use of the data to suggest the pattern removes all the confirmatory aspects of the Procrustean procedures, but the process is useful.

This procedure is incorporated into the oblique rotation method called *Promax,* the only oblique method currently available in the SAS system. The major difference between oblique- and orthogonal simple-structure solutions often lies in the fact that the orthogonal solution shows a pattern of large and small-to-medium loadings whereas the oblique solution for the same data shows large (or largish) and near-zero-to-small loadings. Because loadings are in the ± 1 range, raising all of them to a power (say, cubing them), will make the large ones largish ($.9^3 = .729$) but make the smallish ones even smaller ($.2^3 = .008$). So, the difference between the original loading and a matrix of their cubes or fifth powers is much like that between orthogonal and oblique loadings. (The power must be odd to preserve the signs.) This suggests the following procedure: rotate orthogonally by, for example, varimax, cube all the loadings, and use the cube, fifth, or even, seventh powers of the loadings as the target in an oblique Procrustean transformation. This is exactly what the oblique rotation procedure called promax does.

Oblique Solutions Reconsidered

The idea of correlated factors is a generalization of the ideas in orthogonal factoring. Put another way, orthogonal factors are a special case of correlated ones. From either point of view, the possibility of correlation represents an intellectual complication. Indeed, until fairly recently a number of authoritative sources did not clearly present all the correct aspects of the differences. Thus, there is something to be said for using orthogonal factoring until the data clearly indicate otherwise. In most of the problems to which factor analysis is applied, however, it seems highly plausible that, if factors exist in any real sense, they are likely to be correlated rather than not.

As far as handling factor concepts is concerned, it is important to bear in mind that the elements in the matrix of loadings represent regression weights, not correlations. When the factors are orthogonal, then they are also equal to the correlations between the factors and the variables, although that is not what they represent in the factor model.

In interpreting a factor study, the investigator must interpret the factor intercorrelations as well as the loadings. The loadings are often difficult enough to manage; interpreting the factor intercorrelations themselves may be even more difficult. If there are more than two or three factors, we can factor the factor correlations to yield "second order" factors. This is usually what is meant by a "hierarchical factor analysis." This method introduces further intellectual complexities, not the least of which is that the factor correlations are the outcome of decisions made by the investigator, rather than directly representing empirical relations.

There is an inherent ambiguity in factor problems in that an infinite number of factor loading matrices reproduce a given correlation matrix or data matrix equally well. The investigator chooses a particular strategy to transform the principal components. There is no guarantee that the strategy is correct. In fact, its most direct aim is to simplifying the ideas of the investigator, and simple interpretations are not always the correct ones. This uncertainty applies to all forms of factor analysis, in spite of the fact that some forms are referred to as "confirmatory." This applies to the question of oblique versus orthogonal factors in another way, as illustrated in our discussion of the orthogonal rotation of the example. When we interpret the orthogonal loadings by saying that the variables are complex, we mean that all the factors are appreciably involved in all the variables. But in fact some variables—perhaps only one—are more involved than others in each case. Oblique transformation may sharply reduce the extent to which this happens. Still, this does not mean that the simpler pattern of loadings is correct. This same general phenomenon can occur in a more systematic fashion.

Consider the following four patterns of loadings. The first three are orthogonal and the fourth is oblique.

$$\begin{bmatrix} 1.0 & 0 \\ 1.0 & 0 \\ .7 & .7 \\ .7 & .7 \end{bmatrix} \quad \begin{bmatrix} .7 & .7 \\ .7 & .7 \\ 0 & 1.0 \\ 0 & 1.0 \end{bmatrix} \quad \begin{bmatrix} .93 & .37 \\ .93 & .37 \\ .37 & .94 \\ .37 & .94 \end{bmatrix} \quad \begin{bmatrix} 1.0 & 0 \\ 1.0 & 0 \\ 0 & 1.0 \\ 0 & 1.0 \end{bmatrix}$$

All of these loadings are equally good as far as the data are concerned, but they have quite different interpretations: We may prefer the oblique loading out of a desire for simplicity, but be aware that the injudicious use of Occam's razor may remove important facial features as well as unwanted stubble.

Substantive considerations should determine which set of transformed loadings should be accepted as the "real" ones. Deciding whether to opt for oblique or orthogonal rotation is part of this general principle. In almost all cases where the interfactor correlations are going to be small, it is better to stick with orthogonal rotation. Otherwise, there is no substitute for knowing what you are doing, and knowing the variables that you are studying.

SCORES ON COMPONENTS

Scores on principal components and transformed principal components are relatively simple to obtain because they are linear combinations of the observed variables. Let V be a $p \times r$ matrix of weights to produce scores on the desired r components. Then

$$V = A_r(A_r'A_r)^{-1} \tag{14.17}$$

has entries v_{jm}. These are the weights that are applied to standard score variables

j to produce components m. Equation 14.17 can be used when A_r is the original component matrix F_r and whether A_r is orthogonal or oblique. For raw scores, V must be premultiplied by the reciprocals of the standard deviations D_s^{-1}:

$$P_r = XD_s^{-1}V \tag{14.18}$$

If X is in standard scores, D_s^{-1} may be omitted.

Equation 14.18 produces the P_r that, weighted by A_r', reproduces $\hat{X}_r$. Sometimes simplified weights are used instead of V. The variables with high loadings on a factor are given weights of 1.0 and the others weights of 0. This is a simple way to construct composite variables and may have attractive simplicities of interpretation. However, such approximate components will not be the same as the exact ones given by Equation 14.18. They usually correlate highly with their corresponding columns of P_r, but it typically turns out that these components are considerably more highly correlated with each other than R_p would indicate. It should be noted that these correlations are likely to be appreciable when the factors are supposed to be orthogonal. The covariances among such approximate components can be calculated using the methods described in Chapter 4.

It should be noted, however, that these simple methods (Equations 14.17 and 14.18) of deriving component scores do not apply directly in the common factor model. This issue will be discussed in more detail in Chapter 15.

INTERPRETING LOADINGS

Ask a factor-analytic investigator whether his loadings are statistically significant and his expression will go blank or shifty-eyed. In the latter case, the investigator probably feels that he ought to be able to answer but knows he cannot. In the former case, it probably never occurred to him that such concepts applied. Of course, the question of statistical significance is framed in a deceptively straightforward fashion. The main source of ambiguity in defining factor loadings is the possibility of rotation. If h_j^2 is the sum of squared loadings of variable j on all the factors, then its loading on any one of these factors can be made to vary over the full range $\pm\, h_j$.

Almost nothing of an inferential nature can be said about component loadings; there are no standard errors, significance tests, or related matters. What can be said is so full of restrictions that it does not merit description here. Typically, when a factor analyst writes about the results of a study certain loadings are singled out as "significant." Such a description is uniformly based on some simple numerical rule which recognizes .30 as a cutoff; those loading at or above this value are "significant" and those below it are "nonsignificant." Why .30? Because if corresponds to about 10 percent of the variance of the variable, and many people feel that a construct that does not correspond to at least 10 percent of a variable's variance is not significant to it. Obviously, this is just a rule of thumb. Sometimes .20 is used as the cutoff, and sometimes it is raised to .40. In any case, this cutoff is being used in a purely descriptive sense.

It is clear from this discussion that we cannot say much of an inferential nature in principal components types of factor analysis. Although some investigators rely heavily on the factor-matching strategy, this is far from universal. The same situation applies to traditional methods of common factor analysis. In recent years, some rigorously inferential methods have evolved, but they too have their problems. In any event, these methods change the nature of the factor analytic concepts.

Design Considerations

Designing a factor study requires thought, just like any other type of study. The methods described in this chapter are often applied when very little thought has been given to the analysis when the data were gathered. This is not necessarily wrong however, if the data are primarily being used for some other purpose, but the conclusions from such an adventitious study are rarely more than suggestive. There may be great gains from a little planning in deciding on the variables to be studied.

The majority of thought in components factor analysis should be devoted to providing enough of the right variables to give a good definition to the emerging components. Typically, this means at least three and preferably four or five times as many variables as important components. If the variables are individual test or questionnaire items, this ratio should be increased to eight or ten due to the fact that error and specific variance forms a greater proportion of the variance in that case. However, throwing in irrelevant or complex variables just to meet the quota is likely to be self-defeating because they add additional factors.

Something to be aware of in the selection of variables for inclusion in a factor analysis is *induced* rather than natural covariation. The variables need to be operationally independent of each other. For example, use of several linear composites of the same observed variables should be avoided. Similarly, variables that are nonlinear functions of each other—such as a time score and its reciprocal, a rate score—should not be included in the same analysis. Another type of variable that should be avoided is a two-part test problem where a subject cannot get the second part correct unless he has succeeded on the first part. The trouble with these variables is that they induce "artifactors"—factors that are the outcome of the induced relations between the variables. These factors would not be much of a problem if we could confidently identify them and toss them aside. The difficulty is that they become confused with the natural relations of the variables, making it difficult to determine just what part of a factor is the result of true relations among variables and what part is the result of artificial relations. Thus, before the analysis is carried out, the list of variables should be examined carefully in an effort to eliminate such dependencies.

The sample size should be sufficient to give good stability to the correlations. It is rarely profitable to do an analysis with a sample below 50, and with one below 100 we may not expect very good results with more than 15 or 20 variables. With 40 or so variables, a group of 150 persons is about the minimum, although 500 is preferable. If we do a split-sample analysis, these minima are even more important. These guidelines become more stringent for common factors analysis (see Chapter 15).

Of course, not all factor analyses are carried out on samples. Some are carried out on a significant portion of a population. This is often the case when the units of observation are not individuals. For example, we may have data on a number of variables for all the countries of Europe or the 50 largest American manufacturing corporations. Components-type factor analyses are frequently carried out on *averages* or even on individuals' ratings of a number of objects or stimuli of some sort on a number of scales, such as a semantic differential. In these cases, the rows of the matrix are the only units that we are interested in. The inferential problems are avoided, or at least shifted to a different domain. In these contexts we are justified in carrying out components analyses on quite small data matrices, even on matrices where the number of columns is nearly equal to the number of rows.

Exploratory and Confirmatory Analyses

To use a sweeping generality, most factor analyses—even most of those that are published—furnish relatively little in the way of useful information. To some extent, this is a result of using procedures that are questionable on technical grounds, samples that are too small, and so on. The primary reason, however, is that factor analysis is treated as a one-shot procedure rather than as a tool in programmatic research. A distinction is often made between *exploratory* and *confirmatory* factor analyses. In essence, these terms correspond to whether the investigator is exploring the content of a battery of measures or confirming his expectations about its underlying dimensions. The downbeat nature of the opening statement in this paragraph is largely based on the observation that most studies stop at the exploratory level—an early stage of it at that—whereas we really need to go beyond this stage to confirm the apparent nature of the variables.

Like most other dichotomies, the exploratory-confirmatory analysis actually masks a continuum. There are several other levels on this continuum. The lowest—most exploratory—level is the accidental study. Here, the battery of measures is collected for other reasons or is somehow available. In any event, the investigator did not deliberately go out to gather the data with the explicit purpose of doing a factor analysis. This type of study has become increasingly common as computerized data files are put to this purpose. Even within this type of study there is variation, ranging from the most hodgepodge of variables to the reasonably coherent. Also contained within the exploratory range are the batteries that are put together deliberately by the investigator with the idea of finding out what the main variables underlying some domain might be—whether autonomic nervous system responses, political attitudes, or intellectual ability. Relatively little of a solid nature is learned if the analyses stop at any of these exploratory levels.

It is important to attain as high a level of confirmation as possible. This takes a range of forms, from just getting a similar batch of data and seeing if the same factors emerge to a highly formalized process that includes the construction of *new* measures whose factor content is predicted and tested by the elegant new confirmatory procedures described in Chapter 15. Between these two extremes lie intermediate levels that include a winnowing-and-replacement of variables to reduce the number that have complex structures, or the addition of variables to provide better definition of factors that have only a couple of high loadings. In the confirmatory process, it is often necessary to include representatives of previously-established constructs so that one may establish that the supposed factors in the battery are indeed different from those already known. Ideally, factor-analytic research follows the spiral pattern of activities that characterizes other aspects of the scientific endeavor: one gathers data, ruminates on the results, uses the conclusions to construct new apparatus, gathers more data, and so on. Too often factor analysts are content to stop at the first stage.

SUMMARY

Transforming Principal Components

Except in those instances where a single principal component is sufficient for the investigator's purposes, it is necessary to alter the loadings on the components rather than accept them in their eigenvector-eigenvalue form. This is because statistical optimality rarely corresponds to conceptual simplicity. Thus, after the investigator accepts the r largest principal components as sufficient to furnish an acceptably accurate description of the data matrix, he subjects them to a linear transformation.

Such transformation is carried out to aid interpretation. The statistical properties of the components, taken as a group, are preserved during transformation because the scores are transformed on the components in such a way as to compensate for the transformation of the loadings of the variables.

In all types of factor analysis interest focuses on the matrix of loadings, referred to here as F_r for the original components and A_r for the loadings after transformation. These loadings represent weights for regressing the *observed* variables on the components. The purpose of rotating the principal components is to try to ensure that the components correspond to some sort of underlying reality.

The most common principle applied in rotation is simple structure. Here, the components are rotated to simplify the loading pattern by making it have, insofar as possible, a few large loadings, ideally only one for each variable, and a majority of small loadings that account for negligible amounts of variance. Rotating for simple structure can be carried out using a laborious graphical process or analytic methods. These methods are derived from numerical definitions of the simplicity of loading patterns. Although there are many such methods, the favorite method by far is varimax.

The major alternative to transforming to simple structure is to bring the loadings as closely as possible into conformation with a theoretical or anticipated pattern. This is called Procrustean transformation.

The principal components factors themselves are uncorrelated, and an important distinction should be made between those types of transformations that preserve this orthogonality and those that allow the factors to become correlated. As long as the components are orthogonal, the loadings correspond to correlations with the components in addition to being regression weights. However, once the components are allowed to become oblique, they lose their identity as correlations. Furthermore, there has been some difficulty in finding fully satisfactory analytic rotation methods. Nonetheless, oblique rotation may sometimes supply enough additional simplicity in interpretation to justify its use.

Scores on Components

For the most part, components analyses are carried out with the goal of understanding the variables in the battery. However, when the need arises, procedures for getting scores on the components are quite straightforward. Sometimes, it seems plausible to use simple weights instead of the literal factor score weights, but some care must be taken then because of correlations that are likely to arise among such approximate factor scores. We should evaluate the effect of using the simple weights in any given instance, rather than just assuming that approximate factors correspond to the intended ones.

Inference and Design

Due to the complexities involved, and due to the fact that components analysis is a multi-stage, partially subjective process, there is almost a complete lack of formal aids to the inferential process in components loadings. A split-sample approach is the best guide for judging the sampling stability of loadings and factors. For the most part, references to the significance of factor loadings simply reflect their descriptive magnitudes. The value of .30 is frequently taken as the minimum below which one pays little attention to the loadings.

In factor analysis, as well as elsewhere, the empirical data should provide the basis for sound inferences and conclusions, and a factor study should be planned carefully. Sufficient observations should provide statistically stable results and enough well-structured variables to provide unambiguous definitions of the factors. A factor study should not be carried out with fewer than 50 persons or observational units, and then only if the number of variables is no more than 12 or 15. The results of such a study would be very rough, however; 200 persons is a more useful minimum.

Including the right variables in the study is the most important consideration, and it can rarely be done with a one-shot approach. Rather, we should plan on a more long-term, programmatic approach where there are several sequential studies, and where the variables and factors are subjected to a series of refinements.

COMPUTER APPLICATIONS

SPSS^x

As the default option, the FACTOR procedure in SPSS^x performs principal components factoring with varimax rotation. The statement FACTOR VARIABLES= followed by a list of the variables to be analyzed, is the basic instruction. The number of factors rotated is the number of eigenvalues greater than unity. FACTOR must come in the first column, but additional commands used to modify or extend the analysis must be indented. The number of factors to be rotated can be controlled by the CRITERIA= subcommand. Following this command with MINEIGEN (minimum eigenvalue) will lead to rotating the number of factors to eigenvalues greater than the specified level. CRITERIA=FACTORS (number of factors) will rotate just that many factors.

Different types of rotation can be specified by using the ROTATION subcommand. The options are VARIMAX, EQUAMAX, QUARTIMAX, and OBLIMIN. The degree of obliquity can be controlled in OBLIMIN by specifying CRITERIA=DELTA (obliquity parameter).

Different subsets of the variables can be analyzed by the command ANALYSIS= followed by a list of the variables to be analyzed. Changes in the default options must come before the subcommand that controls that option. Several different numbers of factors can be extracted by using several different CRITERIA= options. Each of these can be rotated several different ways by following each CRITERIA= with several different ROTATION= options. Two programming points deserve mention. Beginning with the FACTOR command, all commands and subcommands must end with a slash (/). Default options for a particular subcommand remain in effect until they are superceded by a command, whereupon the new option remains in effect. The user can return to the default by following a subcommand by DEFAULT.

There are a variety of PRINT= and PLOT= options. Particularly useful are the CORRELATION and REPR options, which cause the program to print the original and reproduced correlation matrices. Using PLOT one can plot the eigenvalues for the scree plot by using PLOT=EIGEN, but the line printer output usually does not provide good enough resolution for the decision on the number of factors. The pairwise plots of factor loadings on factors can also be requested.

The output needs a little care in reading. The INITIAL STATISTICS list the COMMUNALITIES of the variables and the successive eigenvalues. These input communalities are the squared multiple correlations of each variable with all the others, and are irrelevant to a principal components analysis. We will discuss communalities extensively in Chapter 15. The eigenvalues of the correlation matrix are listed in the right half of the table. Note that to save space the programmers have listed communalities for variables $j = 1, 2, \ldots, p$ and eigenvalues for factors $m = 1, 2, \ldots, r, \ldots, p$ in parallel columns.

After this initial information, the program prints extraction information as FINAL STATISTICS. This output consists of the output communalities, which are the sums of squared factor loadings, and the eigenvalues corresponding to the number of factors rotated. This is followed by the unrotated factor loadings and then the rotated factor loadings and the transformation matrix. For obliquely rotated factors, the structure matrix S is printed, as well as the pattern matrix A and the factor correlation matrix R_p.

If both the original and residual correlation matrices are requested, they are printed together as upper (residual) and lower (original) triangles of the same matrix. The diagonal is the output communality.

SAS

The SAS system offers a similar range of possibilities in its PROC FACTOR. Its METHOD = PRIN option, the default, gives the principal components loadings, corresponding to our description of $SPSS^x$. The user specifies a number with NFACT =, or through use of MINEIGEN=.

SAS offers the same orthogonal rotation methods, but there is no default; one must be specified. The only possibility for oblique rotation is Promax, the Procrustean rotation to a powered version of the orthogonal solution. This requires the prespecification of a choice of orthogonal rotation methods to define the target, using the PREROTATE parameter. Factor scores may also be requested, and the range of input and output options is large.

BMDP

As in other procedures, BMDP offers the most options as a factoring program. The details of the problem are specified under two subheadings or "paragraphs" in the jargon of this package. The first is the FACTOR paragraph that specifies the method of extracting factors, and the second is the ROTATE paragraph that gives the method and details of rotation.

For principal components analysis, only three cards are of concern under FACTOR. First we specify METHOD = PCA, rather than one of the other options that refer to various forms of common factor analysis. We also may specify NUMBER = to give the maximum number of factors to be rotated and/or CONSTANT = to specify the minimum acceptable eigenvalue. The other cards in this paragraph specify various details relevant to common factor analysis which are not relevant here.

There are a number of possible things to detail in the ROTATE paragraph. Here, METHOD = selects from among VMAX (varimax), QRMAX (quartimax), EQMAX (equamax), and ORTHOG (generalized orthogonal rotation using some value for the gamma parameter other than that implicit in the three listed previously). The oblique options include DQUART (direct quartimin, a special case of the direct oblimin family), DOBLI (direct oblimin, requiring the specification of gamma), ORTHOB (orthoblique, suggested by Kaiser (1970) but not discussed here), and NONE. GAMMA = specifies the value of gamma for ORTHOG or DOBLI. MAXIT and CONSTANT can be used to control the length of the iteration process. Rotation may be carried out on the obtained factor loadings or the matrix normalized so that the sum of squared loadings is 1.0 for each variable using NORMAL or NONORMAL, with NORMAL as the default.

PROBLEMS

1. Transform the first three components factors from the Chapter 13 problems using the following transformation:

$$T = \begin{bmatrix} .2859 & .6967 & .6580 \\ -.8373 & -.1522 & .5250 \\ .4659 & -.7011 & .5399 \end{bmatrix}$$

   Which solution exhibits better simple structure? Which variables have patterns that depart most from a good simple structure pattern?

2. Show that T is orthonormal. Verify that the transformation does not affect the variance accounted for in each variable. Which of the transformed factors accounts for the most variance?

3. What will be the scores of the person from Chapter 13, Problem 4, on these rotated components? Use them to reproduce x_i and compare to the results of Chapter 13, Problem 4.

4. Input the unrotated three factors to a computer program and rotate them by several standard procedures. The rotated components from Problem 1 may be input as well.

(There should be no effect of using different input matrices.) It is also instructive to rotate 2, 4, and 5 factors to see the differences.

5. Plot the loadings of the first two components, decide on an appropriate orthogonal rotation by eye, and compute the rotated loadings. (Further rotations involving the third component would be necessary to approximate the transformed solution from Problem 1.)

6. Below are the first two components of the correlation matrix for six variables. Plot the components and decide on axes for an oblique rotation, making the loading on one factor about zero for variable 3 and variable 6.

$$F' = \begin{bmatrix} .712 & .743 & .617 & .679 & .742 & .695 \\ -.526 & -.353 & -.649 & .466 & .452 & .555 \end{bmatrix}$$

ANSWERS

1. $$A = \begin{bmatrix} .8254 & .1881 & -.0793 \\ .7558 & .0808 & .2043 \\ .8169 & -.0425 & -.0979 \\ .2367 & .8478 & .1391 \\ -.0196 & .7914 & .2690 \\ .0340 & .8897 & -.0055 \\ .0223 & .1340 & .8230 \\ -.0365 & .2047 & .8490 \\ .0353 & .0948 & .8721 \end{bmatrix}$$

2. $$T'T = \begin{bmatrix} .9999 & 0 & 0 \\ 0 & .9999 & .0001 \\ 0 & .0001 & 1.0001 \end{bmatrix}$$

$$\sum_m a_{jm}^2 = .7230, .6196, .6786, .7942, .6991, .7922, .6958, .7641, .7709$$

$$\sum_j a_{jm}^2 = 1.9805, 2.2489, 2.308$$

3. $\mathbf{p}_i = [1.713 \quad -.007 \quad .010]$
$\hat{\mathbf{x}}_i = [1.415 \quad 1.296 \quad 1.398 \quad .401 \quad -.037 \quad .052 \quad .046 \quad -.055 \quad .069]$
(Differences from Chapter 13 are very small.)

5.

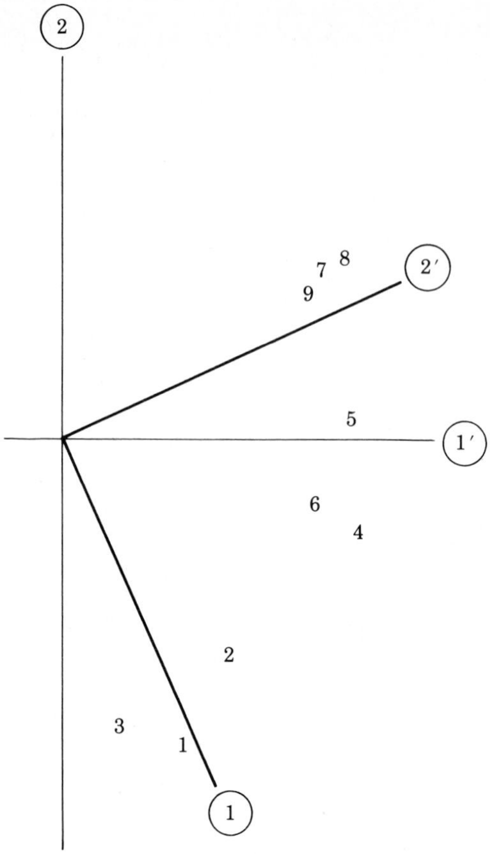

$$T = \begin{bmatrix} .425 & .905 & 0 \\ -.905 & .425 & 0 \\ 0 & 0 & 1.0 \end{bmatrix}$$

$$A = \begin{bmatrix} & 1' & 2' & 3' \\ & .823 & -.039 & \text{same} \\ & .660 & .140 & \text{as} \\ & .719 & -.184 & \text{before} \\ & .549 & .571 & \\ & .341 & .670 & \\ & .428 & .520 & \\ & -.083 & .747 & \\ & -.109 & .814 & \\ & -.098 & .764 & \end{bmatrix}$$

Transformations may be slightly different. Also, factors 1 and 2 may be reversed.

6.

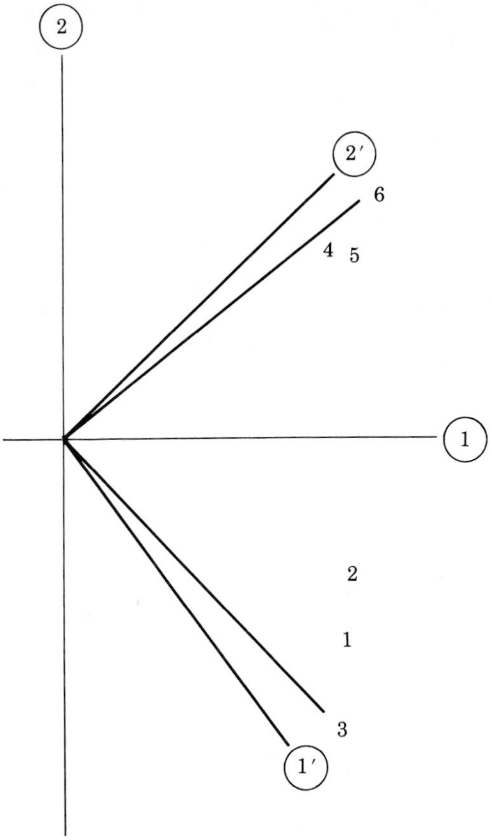

READINGS

Gorsuch, R. L. (1974). *Factor analysis,* 2nd. ed. Philadelphia: Saunders, pp. 175–234.

Harman, H. H. (1976). *Modern factor analysis,* 3rd. ed. Chicago: University of Chicago, pp. 247–390.

Kim, J-O, and C. W. Mueller (1978b). *Factor analysis: statistical methods and practical issues.* Beverly Hills, CA: Sage, pp. 29–40.

Levine, M. S. (1977). *Canonical analysis and factor comparison.* Beverly Hills, CA: Sage, pp. 37–55.

Mulaik, S. A. (1972). *The foundations of factor analysis.* New York: McGraw-Hill, pp. 217–336.

Overall, J. E., and C. J. Klett (1972). *Applied multivariate analysis.* New York: McGraw-Hill, pp. 118–35, 157–67.

15

The Common Factor Model

COMMON VERSUS SPECIFIC VARIANCE

Explaining the Correlations

The components model seeks to explain the *data* or score matrix with summary variables. The model does its best to account for as much of the variance as possible, with relatively few components, although p components always account for all the variance in p variables. The common factor model *starts* with the assumption that not all the variance can be accounted for. Instead, it focuses on the attempt to explain the correlations between variables through their common or shared variance.

The common factor model assumes that the variation in scores of individuals on a variable has two sources. On the one hand, there are *common factors;* these are latent (that is, unobserved) variables that influence scores on more than one measure in the battery. In addition there are influences that are specific to any particular measure, and affect no other measure. These influences include both pure error variance and variance that is reliable but specific to a single measure. These latter types can never be disentangled from each other in the present data, so they are lumped together into a single latent factor that is unique to just one measure. The different observed variables involve the different common factors to varying extents. The similarity in their pattern and degree of involvement with the common factors determines the correlation between them.

For example, in the ability domain, we explain the moderately high correlation between a vocabulary and a reading test by positing that both involve an underlying factor of *verbal* ability. The correlation between the two is less than perfect for two reasons. First, there are specific abilities that are called upon by a reading test that are not relevant to vocabulary, and vice versa. Second, there is error of measurement in both scores.

Similarly, in the attitude domain, responses to one item on feelings toward nuclear power and another on preservation of endangered species are not only dependent on where the respondent stands on an underlying environmental factor that influences both, but also on considerations specific to each item and on the vagaries of the moment. In consumer behavior, for example, liking for two TV commercials will depend on variables underlying both (perhaps they both involve athletes) and also on the individual's reactions to aspects that are specific to one or the other commercial.

In components analysis, the variables are taken as they are. The components try their best to explain all the variance, without considering whether it reflects processes that are general to all the measures or specific to only one, or whether they reflect plain error. The procedures of common factor analysis, on the other hand, try to separate out that portion of the variance that is unique to a measure, and then to explain the remaining variance. In other words, they try to account for the correlations in the correlation matrix, ignoring the diagonal entries, or just letting them ride along.

As everywhere else, there is a certain amount of ideology in factor analysis. Some authorities insist that components analysis is the only suitable approach, and that the common-factors methods just superimpose a lot of extraneous mumbo jumbo, dealing with fundamentally unmeasurable things, the common factors. Feelings are, if anything, even stronger on the other side. Militant common-factorists insist that components analysis is at best a common factor analysis with some error added and at worst an unrecognizable hodgepodge of things from which nothing can be determined. Some even insist that the term "factor analysis" must not be used when a components analysis is performed.

This book comes out forthrightly in the middle on this issue for several reasons. The first is pragmatic in that the choice of common factors or components methods often makes virtually no difference to the conclusions of a study. Second, where it does make a difference, there are cases in which one method makes more sense, and other cases in which another method does. For this reason, it is helpful to know about both common factors and component methods. Finally, because the purpose of this book is to inform, we will discuss both methods. The components model was introduced in Chapter 13 due to its more concrete nature. In this chapter, we now turn to the common factor model.

The Common Factor Model

In the common factor model, the fundamental equation represents the score of a given individual i on a given test as a function of the scores p_{im} on the r common factors, weighted by the loadings a_{jm} on these factors, plus a unique score u_{ij}:

$$x_{ij} = \sum_{m}^{r} p_{im} a_{jm} + u_{ij} \tag{15.1}$$

or

$$\mathrm{X} = \mathrm{PA'} + \mathrm{U} \tag{15.2}$$

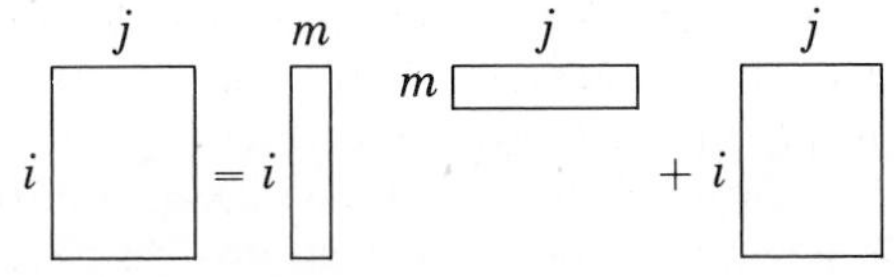

This model is very similar in form to Equations 13.27 or 13.28, but this similarity is deceptive. As in Equation 13.27, there is a term representing the score of the individual on the common factor (p_{im}) and the loading (regression weight) of the variable on that factor (a_{jm}). The difference is the last term. In 13.27, e_{ij} simply represented an error or deviation term reflecting the inadequacy of the components in reproducing the scores exactly. Here, u_{ij} represents the score of the individual on *an additional factor,* the factor unique to variable j. One of these factors exists for every variable. The presence of these unique factors, along with the properties imposed on them, is what distinguishes the common factor model.

The common factor scores p_{im} may be defined as standard scores, and for convenience in the present discussion, the x_{ij} may be in standard score form also. The distinguishing property of the u's is that, by definition, they are uncorrelated with each other and with the p's:

$$(n - 1)^{-1}\mathrm{U'U} = \mathrm{S}_{uu} \tag{15.3}$$

a diagonal matrix, and

$$(n - 1)^{-1}\mathrm{P'U} = 0 \tag{15.4}$$

By substitution, the correlation matrix R is

$$\mathrm{R} = (n - 1)^{-1}\mathrm{X'X} = (n - 1)^{-1}(\mathrm{PA'} + \mathrm{U})'(\mathrm{PA'} + \mathrm{U}) \tag{15.5}$$

or

$$\mathrm{R} = (n - 1)^{-1}(\mathrm{AP'PA'} + \mathrm{AP'U} + \mathrm{U'PA'} + \mathrm{U'U}) \tag{15.6}$$

The zero correlations between common and unique factors mean that the two middle terms are zero:

$$\mathrm{R} = \mathrm{AR}_p\mathrm{A'} + \mathrm{S}_{uu} \tag{15.7}$$

The zero correlations between unique factors mean that S_{uu} is a diagonal matrix whose entries are s_u^2, the unique-score variances. Following tradition, we will use the symbol u_j^2 to represent the variance of the scores on the unique factor of variable j. (This risks some confusion, since u_{ij} was the score on this factor, but we will have little need to refer to these scores.) In any event, we arrive at the equation

$$\mathrm{R} = \mathrm{AR}_p\mathrm{A'} + \mathrm{U}^2 \tag{15.8}$$

where U^2 is a diagonal matrix containing the u_j^2. As with the components model, common factors may be either orthogonal or oblique. If they are orthogonal, then $\mathrm{R}_p = \mathrm{I}$ and may be ignored.

This equation has several consequences. First, the correlations among the variables are accounted for by the common factor loadings. Assuming uncorrelated factors, as we may for the initial unrotated ones, the model implies through

Equation 15.8, that

$$r_{jk} = \sum_m a_{jm} a_{km} \tag{15.9}$$

The variance of a standard score variable is then expressed as

$$s_j^2 = 1.0 = \sum_m a_{jm}^2 + u_j^2 \tag{15.10}$$

The sum of squared common factor loadings, called the *communality* of the variable, is usually referred to as h_j^2. It is the proportion of the variable's variance that is attributable to common factors,

$$h_j^2 = \sum_m a_{jm}^2 \tag{15.11}$$

or the complement of the unique variance

$$h_j^2 = 1 - u_j^2 \tag{15.12}$$

The concept of communality is important in common factor analysis and estimating the communalities is one of the major features that distinguishes common factor analytic methods from their counterparts in component analysis. The theoretical upper limit for the communality is the reliability of the variable, since measurement error is assumed to be uncorrelated with everything else. Furthermore, the communality of a variable depends on the context in which it is analyzed. It can be low if the variable shares little with the other variables in the battery, or it can reach its upper limit if its companions cover the whole spectrum of its reliable variance.

By now the concept of "communality" should be fairly clear, but operationally this term refers to three different things. One is what might be called the "true communality," the actual proportion of a variable's variance that is due to factors common to other variables in the current battery in the population. The symbol h_j^2 will refer to this abstract quantity. Communality also refers to the *estimates* used in the diagonal of the correlation matrix from which the factors are extracted. We will refer to these as the estimated or input communality, $\hat{h}_j^2$. Finally, after the sample factor loadings have been determined, we have the sum of squared loadings of each variable on those factors. This is referred to as the reproduced or output communalities, $h_j'^2$. It is hoped that these distinctions among qualities will not cause confusion, although it is necessary to have them in mind.

If the number of common factors is small, which is the usual assumption, then $R - U^2$ is of reduced rank. Much of the early theoretical work on factor analysis centered around finding the communalities that would exactly reduce the rank of the correlation matrix to some small number, say $p/2$ at most. Today this is recognized in general as an impossibility. While numbers can always be found to reduce the rank to about $p/2$ when they are subtracted from the diagonal, this usually requires that the communalities be made either negative or greater than 1.0. (The latter is called a "Heywood case" after its discoverer.) Either of these eventualities conflict with our idea of variance because in the former case, common variance is negative, and in the latter, uniqueness is negative. Thus we now look on the problem as one of finding the communalities that "almost" reduce the rank to some fraction of p, preferably quite a bit less than $p/2$. That is, common factor analysis tries to find acceptable communality estimates such that when substituted in the diagonal of R, the matrix can be accounted for very closely using only a few factors. Equivalently, we can think of searching for values for the u_j^2 that have the same effect when subtracted from the main diagonal.

As it was originally formulated, common factor analysis tried to estimate the communalities, substituting them in the diagonal of R, and then proceeding to extract and rotate factors, using the same methods that were discussed under the components methods. Recent advances in statistical sophistication and computational methods have led to a different approach, although direct extensions of the older methods are still used. Because these methods are still in wide use, and because they aid in the understanding of the principles involved, we will focus on them first, before discussing the more sophisticated methods such as maximum likelihood factor analysis.

THE PRINCIPAL FACTORS METHOD

Estimating Communalities

In the principal factors method (PFA), we first substitute some initial estimate of each variable's communality for the diagonal of the correlation matrix and then apply the principal components method to this *reduced* correlation matrix. A variety of methods have been used for finding these communality estimates. Currently, the most widespread default option of several computer programs involves using the squared multiple correlation of each variable with all the other variables as its input communality estimate $\hat{h}_j^2$. This estimate is used because it can be shown that in the population—that is, ignoring sampling problems—the squared multiple correlation of a variable with all others, P_j^2, approaches the true communality as the number of *variables* increases without limit. Since we always deal with a finite set of variables, P_j^2 is referred to as a *lower bound* estimate of the communality h_j^2. Note, however, that in a sample, R_j^2 is an overestimate of P_j^2. None of the common computer programs attempt to correct R_j^2 for bias, according to the methods of Chapter 9.

Other communality estimates are used with some degree of frequency. Probably the next most common involves using a variable's highest correlation with another variable as its estimated communality. The rationale here is that if two variables had the *same* loadings on all factors then

$$\sum_m a_{jm}a_{km} = \sum_m a_{jm}^2$$

This will never exactly be the case, however, so the highest r is also considered to be a lower bound for the communality.

However, consider two variables whose profiles of factor loadings are similar, but where the loadings of one variable are systematically lower than the other's. Here the correlation between the two variables will be *intermediate* in value to the two communalities—that is, higher than that of the lower-communality variable. The highest r_{jk} has one clear advantage over the squared multiple correlation as a communality estimate: it can be employed even when the matrix is singular. It is probably a fairly reasonable estimate in most cases, even though it seems likely to suffer from the problems of any extreme value estimator—an upward bias and a considerable degree of sampling fluctuation.

The triad method was actually the first one suggested for use as a communality estimate. Common factor analysis originates in Spearman's theory that there is only a single common ability underlying all tests of mental ability, and that this single common factor accounts for all the correlations among such measures. This reduces Equation 15.9 to a single term, $r_{jk} = a_j a_k$, and Equation 15.11 to $h_j^2 = a_j^2$. This means we can solve any set of correlations among three variables for their communalities. For example,

$$\hat{h}_1^2 = \frac{r_{12}r_{13}}{r_{23}} \tag{15.13}$$

because

$$\frac{r_{12}r_{13}}{r_{23}} = \frac{(a_1a_2)\,(a_1a_3)}{a_2a_3} = a_1^2 \tag{15.14}$$

This method may be used when we are willing to find for every variable two other variables that can be assumed to have very similar profiles of factor loadings. We might use this method routinely, finding for each variable the two others that correlated most highly with it using Equation 15.14. However, this procedure is rarely used.

Computing Loadings

Whichever method of communality estimation is used, the resulting coefficient is substituted for the 1.0 in the diagonal of the correlation matrix, and then the eigenvalues and eigenvectors of the matrix are computed as in the principal components method. If we use R_c to refer to the correlation matrix with communalities in the diagonal, then

$$\mathrm{R}_c = \mathrm{R} - \hat{\mathrm{U}}^2 \tag{15.15}$$

If we use all p eigenvectors and eigenvalues, then R_c is exactly reproduced:

$$\mathrm{R}_c = \mathrm{FF}' \tag{15.16}$$

and if the r largest eigenvalues are selected we have

$$\hat{\mathrm{R}}_{cr} = \mathrm{F}_r\mathrm{F}_r' \tag{15.17}$$

The F_r are then transformed to simple structure or some other theoretically plausible solution by either an orthogonal or oblique transformation. The discussion of rotational methodology in Chapter 14 applies equally well here, and indeed most of it was developed in the common factors framework.

The discussion of the properties of eigenvalues and eigenvectors in Chapter 14 also applies here *except* that it is no longer possible to refer to reproducing the *score* matrix in any direct sense, and there are no longer factor scores that can be directly obtained. Procedures for *estimating* the factor scores are a separate topic and will be discussed later.

The F_r matrix furnishes the best r-component least-squares approximation to R_c as it did with components, although this F_r differs from the one that would be derived from R itself. Also, some of the eigenvalues may be negative which makes the corresponding factors imaginary in the mathematical sense. (Skeptics say that all common factors are purely imaginary anyway.) $F_r'F_r$ is also still diagonal. The eigenvalues still correspond to variance accounted for by the corresponding factor, but in a more abstract sense than with the components factors since there is no longer any directly computable matrix of factor scores.

It is necessary, however, to exercise a certain amount of care in referring to aspects of the solution. Dividing the sum of the r largest eigenvalues by p does give the proportion of variance that is accounted for by the r largest common factors. Sometimes, however, we see references to the proportion of *common* variance that is accounted for by a certain number of factors. This is a much less solid concept. Presumably it refers to

$$\sum_{m=1}^{r} \lambda_m^2 / \Sigma \hat{h}_j^2 \tag{15.18}$$

so it reflects primarily the accuracy of the sum of the communality estimates. It is not hard for this ratio to go over 1.0 because of the presence of negative eigenvalues.

Remember that the goal of a common factor analysis is to account for the correlations among the variables. Rather inconsistently, the evaluative focus often taken by investigators is on how much *variance* the common factors account for, reflected by the size of the largest eigenvalues. Remember that the *squared* eigenvalues reflect the degree to which the sum of the squared elements of the correlation matrix is accounted for.

Equation 13.32 expressed the sum of squared residual correlations as a sum of squared eigenvalues. We will repeat it here in a different form:

$$\sum_j \sum_k (r_{jk} - \hat{r}_{jkr})^2 = \sum_{r+1}^{p} \lambda_m^2 \tag{15.19}$$

This equation is an overall indicator of the goodness of fit of the factors to the correlations. Perhaps the most meaningful ratio in a principal factors analysis is

$$g_r = \sum_{m=1}^{r} \lambda_m^2 \Big/ \sum_{m=1}^{p} \lambda_m^2 \tag{15.20}$$

This ratio is the proportion of the total squared entries in the R_c, the correlation matrix with communality estimates, that is accounted for by the first r principal factors.

Another way of looking at goodness of fit is to ask what the typical residual will be. A common way of doing this is to use a root-mean-square deviation. Recall from Equation 15.19 that the sum of the squared $p - r$ smallest eigenvalues is the sum of squared residuals. Dividing this by the number of elements in the correlation matrix, p^2, gives the average squared deviation, and taking the square root gives a root-mean-square deviation:

$$d_r = \left(\frac{\sum_{j,k} (r_{jk} - \hat{r}_{jkr})^2}{p^2} \right)^{1/2} = \left(\frac{\sum_{r+1}^{p} \lambda_m^2}{p^2} \right)^{1/2} \tag{15.21}$$

Equation 15.21 is an expression for the "average" or typical deviation of an observed correlation from the value reproduced using r common factors.

Iterated Principal Factors

For various reasons, the initial communality estimates may be inaccurate, so it may be desirable to adjust them on the basis of the obtained factor loadings. Here, we make a decision on the appropriate number of factors, r, and compute the reproduced or output communalities:

$$h_j'^2 = \sum_{m=1}^{r} f_{jm}^2 \tag{15.22}$$

substituting $h_j'^2$ or something close to it for the input communalities $\hat{h}_j^2$. Then the principal factors are found again. These may be used to make a second adjustment, and so on, until someone gets tired or until the amount of change from iteration to iteration is considered negligible. This process, called *iterated principal factor,* is automated in many computer programs for factor analysis, such as the PAF option in SPSS. In general, little is usually gained after the first one or two iterations. Also, as a general rule, the pattern of loadings on the larger principal factors does not change much as a result of changes in the input communalities when there are more than a few variables. Thus extensive iteration is not necessarily desirable.

The procedure for iterating communalities described above is an informal one in that there is no mathematical proof that it does any good, although the general belief is that it should do at least some good. In contrast to this ad hoc situation, there are several formal procedures for finding communality estimates. The major varities are the unweighted least squares procedure, originally called the MINRES (short for *min*imum *res*idual), the weighted least squares procedure, and the unrestricted- and restricted-maximum likelihood methods, of which LISREL is the most widely used. These will be described later in this chapter.

Example

We can use the data on the eight physical measurements in Table 13-1 to illustrate the principal factors approach. Visual inspection of the correlations in Table 13-1 and some rather informal use of Equation 15.13 leads to the values in the first row of the initial communality estimates in Table 15-1. Applying the principal factors method gives the eigenvalues in the first row of Table 15-2. As in the principal components analysis, it appears that two factors are sufficient because the remaining eigenvalues are quite small and fit is good, as indicated in the first row of Table 15-3. Comparing the output communalities with the input values suggests that some small adjustments are in order, so the diagonals can be replaced by the values in the second row of Table 15-1 with the resulting

Table 15-1 Input Communality Estimates for Each Trial Solution

	1	*2*	*3*	*4*	*5*	*6*	*7*	*8*
Initial Estimates	.840	.800	.650	.690	.380	.600	.600	.600
2nd 2-factor	.880	.800	.635	.675	.370	.583	.600	.585
1st 3-factor	.930	.800	.635	.675	.430	.583	.600	.585
2nd 3-factor	.965	.815	.635	.690	.465	.583	.600	.585

eigenvalues in the second row of Table 15-2. The fit is slightly better as shown in the second row of Table 15-3. Further comparison of the second output communalities to the second input values, together with the very small result of the previous correction, leads to the conclusion that further adjustment would have negligible effects.

The second d_r values in Table 15-3 indicate that the typical residual correlation is in the neighborhood of $\pm.026$. The complete residual correlation matrix is given in Table 15-4. Surprisingly, most of the residual correlations are considerably smaller than .026, although a few are considerably larger—notably $r_{15} - r_{152}$, which is .108 or four times the rms value. This suggests that there is another very small factor involving variables 1 and 5 almost exclusively. Such two-variable factors are called *doublets*. They are a bit of a nuisance because they have an unsettling effect on the communalities of the variables involved. The problem is that a single residual correlation can be reproduced as the product of a large loading and a small loading, or two middle-sized loadings, with consequent uncertainty as to what the communalities of the variables are. In this case, the other residual correlations are so tiny that they do little to help define matters.

Nevertheless, it seems worthwhile to go after this third little factor. The third eigenvector from the two-factor analysis (not shown here) has fairly high values for variables 1 and 5 and small ones for the rest. The communalities for these two variables can be increased by a modest amount, as shown in the third line of Table 15.1, since there seems to be more common variance for these two variables. The other communalities can be left unchanged, however, since their residuals do not suggest that they have much additional common variance. A three-factor solution using these communalities yields the results shown in the third lines of Tables 15-2 and 15-3. The effect is rather small in absolute terms, but the sum of squared residuals is apparently reduced by a factor of four. Thus the third d_r value is about half the second. Examination of the eigenvalues shows that we have increased the third one and made the last (most negative) value nearer to zero, while the other values remain unchanged.

Inspection of the residuals themselves shows that $r_{15} - \hat{r}_{153}$ was still the largest of the residuals at about .05, suggesting that we have not gone far enough to

Table 15-2 Eigenvalues from Each Trial Solution

	1	*2*	*3*	*4*	*5*	*6*	*7*	*8*
Initial Solution	3.7182	1.4077	.1480	.0628	.0152	−.0170	−.0333	−.1417
2nd 2-factor	3.7193	1.4004	.1496	.0473	.0083	−.0251	−.0446	−.1271
1st 3-factor	3.7368	1.4025	.1873	.0474	.0085	−.0251	−.0425	−.0770
2nd 3-factor	3.7489	1.4064	.2109	.0458	.0120	−.0200	−.0322	−.0588

Table 15-3 Accuracy Indices

	g_r	d_r
Initial Solution	.9970	.0273
2nd 2-factor	.9972	.0261
1st 3-factor	.9994	.0129
2nd 3-factor	.9996	.0106

completely account for this correlation. We can increase the communalities of 1 and 5 again, and the communalities of a few other variables very slightly, as indicated in the last line of Table 15-3. We obtain the principal factors again, with the result shown in the last lines of Tables 15-2 and 15-3. The root-mean-square residual d_r is now down to .01! The common factor model is indeed accounting for these correlations. The complete residuals are given in Table 15-5, and inspection of it suggests that we may still be underestimating the communalities of 1 and 5 by about .02. However, because further fussing is probably not productive, the process ends here.

The effect of adjusting communalities on the factor loadings themselves bears examination. All four sets of loadings are given in Table 15-6. Note that the first two factors in all four solutions are remarkably similar. The largest difference in the whole table is .028, which occurs between the first-factor loadings of height when we compare the first two-factor solution to the second three-factor solutions. The other loadings change very little.

None of these changes should affect the interpretation of the factors. Indeed, the first two factors resemble very closely the first two principal components (see Table 13-1), and their rotated versions correspond to the bone-length and limb-girth factors that were the results of rotating the components (see tables 14-2 and 15-7). The main effect of the principal factors method has been to permit the emergence of a third, very small factor. In Table 15-7, only two variables end up with appreciable loadings on this factor, and these variables suggest a factor of torso length. Note that the output communality of height is now .984, showing that almost all its variance is attributable to the three factors, whereas sitting height still has a large amount of unique variance, and the other variables are intermediate.

The above example is treated in some detail in order to illustrate the process of using the principal factors method to estimate factor loadings. The process may be a little misleading as a representation of the typical data and the operation of

Table 15-4 Second Two-Factor Residuals

	1	*2*	*3*	*4*	*5*	*6*	*7*	*8*
1	−.003							
2	−.008	.002						
3	−.007	.012	.000					
4	−.010	.031	.008	.006				
5	.108	−.042	−.031	−.051	.002			
6	−.005	.004	.026	−.021	−.007	.008		
7	−.012	.009	−.008	.012	.001	.004	−.001	
8	−.006	−.007	−.009	.025	.000	−.006	−.004	.006

Table 15-5 Second Three-Factor Residuals

	1	*2*	*3*	*4*	*5*	*6*	*7*	*8*
1	−.020							
2	.001	−.007						
3	.002	.003	−.002					
4	.008	.005	−.006	−.006				
5	.019	−.000	−.003	−.007	−.018			
6	.001	.002	.024	−.025	−.004	.007		
7	−.004	.005	−.011	.007	.005	.004	−.001	
8	−.001	−.009	−.011	.021	.003	−.006	−.004	.006

fully computerized methods. One rarely finds a structure that is so precise a representation of the correlations, and the sample of over 4,000 is unusually large. Furthermore, the initial communality estimates were perhaps better than the typical automated estimates. Nevertheless, the basic ideas concerning how the factors reproduce the correlations and how the squared eigenvalues reflect the accuracy with which this is done remain valid. Investigators rarely go to the trouble of examining the residuals themselves, and this is much more difficult to do meaningfully with 80 variables than it is with 8. Nevertheless, using a less than fully automated approach may provide the user with greater insights into the data, particularly at the early stages of experience with these methods.

EIGENVALUES AND THE NUMBER OF FACTORS

Users of the principal factors method and other heuristically based procedures derived from the traditional methods of factor analysis spend much time and thought trying to decide how many factors exist in a given battery. Surely the answer to that question is always that there are many, many systematic influences on scores, some of them large, some of them minute, some of them confined to a single measure, but many of them common to two or more. Thus, the question of how many factors there are is immaterial; the real question is how many factors have appreciable influence and have loadings that can be reliably estimated.

Table 15-6 Factor Loadings for the Four Solutions

	1st 2-Factor		*2nd 2-Factor*		*1st 3-Factor*			*2nd 3-Factor*		
Variable	I	II	I	II	I	II	III	I	II	III
1	.909	−.239	.919	−.242	.929	−.248	.203	.937	−.249	.213
2	.816	−.364	.816	−.362	.814	−.358	−.151	.816	−.358	−.164
3	.755	−.255	.752	−.250	.750	−.247	−.110	.749	−.243	−.114
4	.808	−.151	.805	−.146	.803	−.143	−.157	.804	−.141	−.170
5	.608	−.021	.606	−.019	.615	−.018	.290	.620	−.016	.313
6	.462	.608	.459	.604	.457	.605	−.025	.455	.607	−.026
7	.416	.651	.415	.655	.413	.656	−.029	.411	.657	−.030
8	.500	.579	.497	.577	.493	.578	−.025	.493	.580	−.028

Table 15-7 Rotated Common Factors

	I	*II*	*III*	$h_i'^2$
1	.896	.123	.408	.984
2	.905	−.032	.028	.821
3	.792	.051	.053	.633
4	.817	.165	−.003	.695
5	.503	.224	.423	.482
6	.200	.732	−.002	.576
7	.143	.762	−.019	.601
8	.245	.720	.006	.578

The sizes of the obtained eigenvalues provide suggestive although not definitive answers to this question. First, recall the principles that were discussed at some length in Chapter 13. The sphericity test does not apply well to the common factors case, although the rest of the discussion in Chapter 13 does. It has even been found through Monte Carlo studies (Cliff, 1970) that the principles concerning sampling effects on eigenvectors apply pretty well in the common factors case, at least when there really is a fairly clear definition of how many large factors there are. The second consideration is the size of the squared residual eigenvalues, which can be divided by p^2 to get the average squared residual correlation. If this is about $1/n$ or so, there presumably is mostly sampling variance left.

Another thing should be mentioned. The number of common factors that can be reliably determined is influenced by the communalities of the variables and therefore by their reliabilities. If the total proportion of common variance in a battery is quite low, it stands to reason that it will be more difficult to separate this common variance from the unique variance and from the instability resulting from pure sampling variation. Therefore, we should not try to identify as many factors in a battery of 80 individual test items as we would out of 80 well-defined and reliable variables.

Finally, the split-sample-factor-matching procedure is probably even more useful and valid in the common-factor case than it is in the components.

One procedure that is not recommended is the widely-used practice of rotating all the factors that correspond to principal *components* eigenvalues greater than 1.0. The reasons for this are discussed in Chapter 13. Where intercorrelations are low, the procedure tends to select too *high* a number of factors, because then nearly half the PCA eigenvalues will be greater than, but close to, 1.0. On the other hand, when communalities are high, the procedure tends to select too small a number. The above example is a case in point; the third eigenvalue is only about .6 (Table 15-2), even though the third factor is quite convincing.

LEAST SQUARES AND MAXIMUM LIKELIHOOD FACTORING

Explicit Solutions for Communalities

The methods described so far have evolved out of the experience of investigators over the years. Their relationship to the common factors model is heuristic or intuitive; they put no great strain on the computing facilities of even the 1960s.

Several methods that require explicit mathematical routes to solve for the common factor loadings have also been developed. They have strong conceptual advantages, although they tax the computing facilities available to the average social scientist in the 1980s. These methods are also somewhat fragile. However, they will and should come into wider use as the cost of computing continues to fall. The level of mathematical sophistication necessary to understand these methods is considerably higher than has been required so far, so we will only try to sketch the general ideas involved.

The major problem in applying the common factor model has always been that of estimating the communalities, or equivalently, the unique variances. If the communalities are known, then principal factors provide least squares estimates of the factor loadings, and other types of estimation could also be applied. The methods that we will describe here make communality estimation part of the overall problem. They seek to estimate the communalities simultaneously with estimating the factor loadings, rather than separating the two aspects as we have done so far.

MINRES or Unweighted Least Squares

The procedure developed by the late H. H. Harman, called MINRES, (Harmon and Jones, 1966) is closest in spirit to the principal factors method. The basic idea is a rather direct one: We seek factor loadings that give a least squares fit to the *off-diagonal* part of the correlation matrix. Undoubtedly several persons had this same idea at various times, but it turned out to be difficult to find a fully satisfactory solution to the problem stated in this way. Indeed, the final aspects of the procedure were developed only within the past few years.

MINRES proceeds in an iterative fashion, starting from a principal *components* solution, assuming a particular number of factors—that is, taking the first r principal components. It then works on the set of loadings for each variable j in turn, solving for adjustments to *all* r of that variable's loadings so as to reduce the variable's residual correlations.

The main difference in this process, as compared to the principal factors method, is that it does not make use of the communalities at all, but directly fits the off-diagonal entries. However, there is a connection between the principal factors method and MINRES. Principal factors applied to a correlation matrix which has the output MINRES communalities in the diagonal will give back the MINRES factors. Thus, MINRES can be regarded primarily as a communality-estimating procedure.

Associated with MINRES is a test of the number of factors. As with the significance test associated with the maximum likelihood methods, discussed below, this test has a particular form. In essence, it compares the sample correlation matrix R to $F_rF_r' + U^2$ by means of a χ^2 test. If the χ^2 test is significant, it means that F_r and U^2 are *rejected* as being insufficient to account for the correlation matrix. Essentially, the residuals are too large. If the hypothesis is rejected, then presumably it means that there are too few factors in F_r, so we would try again with more factors. Some philosophical and practical issues surrounding this test and a similar one associated with the maximum likelihood methods will be discussed later.

The test associated with MINRES is based on an assumption of multivariate normality for the variables factored. It is a large-sample test. As usual, no one really knows how large a sample should be, although we should have both $n - p$

and $(p - r)^2$ in the neighborhood of 50 or more for the test to be valid. The robustness of this test in the face of non-normality is also unknown, but its validity may be acceptable as long as none of the variables are dichotomous, bimodal, or j-shaped.

Methods that are essentially the same as MINRES are widely available as unweighted least squares analysis.

Maximum Likelihood Factor Analysis

So far, we have based the methods of factor analysis on trying to fit the data in various ways or by various criteria. These methods are essentially descriptive. In this section we turn to a family of methods that is more explicitly based on inferential considerations.

The principle of maximum likelihood is widely used in theoretical statistics. Recall that a theoretical probability function states a probability or probability density as a particular mathematical function of one or more random variables. The function involves parameters, and it is these parameters that the sample statistics are used to estimate. A question central to theoretical statistics is, "How should these estimates be made?" Maximum likelihood methods are one approach to estimation. Using these methods, the sample of data consists of a particular set of values for the random variables. This particular set is fairly likely to occur if *some* values for the parameters are assumed and less likely to occur for other values. For example, a sample consisting of values of 6, 7, and 8 for x has a fairly high probability of occurring if we assume that the population has a mean about 7.0 and a standard deviation about .5. It is much less likely to occur, however, if we assume that the mean is 20 and the standard deviation is 5. The principle of maximum likelihood says that we should use as parameter estimates those values for the parameter estimates that make the obtained sample data *most likely to occur.*

In the case of factor analysis, the parameters involved are those in the factor equation—the loadings, uniquenesses and correlations between the factors. Assuming multivariate normality, all the information about these parameters is in the sample covariance matrix; it is a sufficient statistic. It is possible, although complex, to formulate the *likelihood* of the obtained covariance matrix as a function of the factor parameters. Certain values of the factor parameters maximize the likelihood that the sample covariance matrix will occur. Methods for iteratively solving the equations that relate the maximum for the likelihood have been evolved, and they constitute "maximum likelihood factor analysis."

There are two general types of maximum likelihood factor analysis; restricted and unrestricted. The latter is similar in spirit to traditional forms of factor extraction such as principal factors, and the expectation is that such factors would be rotated. We will refer to it as UML, for "unrestricted maximum likelihood." In the restricted form, some elements of the factor equation—that is, some loadings, uniquenesses, and/or factor intercorrelations—are specified by the investigator, and the remaining terms are estimated from the data. This will be discussed in more detail in a later section. Indeed, the methods that evolved from it (LISREL and its competitors) can be used for many purposes beyond common factor analysis.

The unrestricted maximum likelihood (UML) and unweighted least squares (ULS) approaches are only two of several that are widely available for solving for the parameters in the common factor model. Examination of the options in various general-purpose statistical packages reveals a list of others, including generalized least squares, alpha factor analysis, image analysis, and perhaps others.

Generalized least squares (GLS) is similar in philosophy to ULS and UML. In ULS, the analytical procedure provides a least squares fit to the off-diagonal entries of R, all discrepancies being weighted equally. In GLS, the discrepancies are weighted differently. That is, some errors in reproducing R are treated as more important than others. This weighting is a function of R^{-1} (or S^{-1} if the analysis uses the covariance matrix).

These three approaches—ULS, UML, and GLS—are all fitting the common factor model. When we say we are "fitting a model" it means we are estimating some parameters in such a way as to minimize or maximize some criterion function. The most familiar criterion is the least squares; we know that ULS is fitting the terms in the factor equation so as to minimize the sum of squared deviations in $R - \hat{R}_r$. In the same sense, GLS is minimizing the sum of squared elements in $R^{-1}\hat{R}_r - I$. Clearly, this difference is zero if $\hat{R}_r = R$. On the other hand, the UML procedure minimizes the rather technical function $\log(\det\hat{R}_r/\det R) + \mathrm{tr}(R\hat{R}_r^{-1})$, where "det" means "determinant of." The determinant may be recalled as a complex function of the elements of a matrix; it is also the product of all the eigenvalues of the matrix. If the model were to fit exactly, that is, if $\hat{R}_r = R$, then all three types of solutions would be the same. As a general rule, a solution that minimizes one of the functions comes close to minimizing either of the others. However, this is not the same as saying that the solutions (factor loadings, etc.) that minimize the criteria are nearly the same. Often they are, but depending on which method is used, they may also differ substantially.

Image factor analysis is one of the many ideas in measurement and data analysis introduced by Louis Guttman. It analyzes those parts of the observed scores that are predictable from the other measures in the battery. These parts are called their images, hence the name. In effect, the variances and covariances analyzed are those among the $\hat{x}_j$, where $\hat{x}_j$ is predicted from all the other variables. In this way the communality problem is circumvented. Some aspects of image analysis are interesting to theorists, although they do not seem to offer any practical advantages in data analysis. Hence there is no strong motivation to discuss it further.

The situation is similar with respect to alpha factor analysis. This analysis is of conceptual interest to theorists as a way of circumventing the problem of estimating communalities while connecting the components and common factor models. However, it has little to offer the data analyst that is particularly advantageous.

Choosing an Appropriate Method

Given a correlation matrix that we want to factor analyze, we can consider six major options for finding factor loadings that will be subsequently rotated. These are principal components, principal factor analysis with and without iterating communalities, ULS, GLS, and UML. Which should be used?

The list above is pretty much in order of sophistication, from rough-and-ready principal components to the finely tuned maximum likelihood. Which method is most appropriate depends mostly on the quality of the data. If data-summarization-variance-reproduction is the object, however, then principal components is the choice.

Assuming that our objective is to understand the relationships among the variables, our choice of method will depend on how much refinement has gone into the measures. The more refined the battery, the more refined the analysis. For a grab-bag of variables that were gathered for other purposes, principal components may still be the choice even for common factors analysis. The alternative is principal factors with just one iteration of communalities. That is, $\hat{h}_j^2$ is replaced by $h_j'^2$ just once. Principal factors is preferable to principal components when the factors are fairly well-defined, particularly when there is considerable variation in communalities.

Any method that uses more refined communality estimates should require that the common factor model be a reasonably accurate description of the data. If it really is possible to reproduce the correlations fairly well with just a few common factors, then iterated principal factors or ULS, GLS, or UML may be used. For any of these methods to work well, at least three substantial loadings are needed for *each* common factor. Otherwise, there is likely to be trouble with improper communality estimates of 1.0 or greater. Such results mean that some variable has no unique variance, according to the solution, and this is very rarely a plausible description. If we assume that there really are at least some true common factors, finding communality estimates of 1.0 means that one or more of the common factors have been twisted over toward the unique factor of those variables having communalities of 1.0. In a sense, uniquenesses of zero contradict the common factor model.

Remember that Equation 15.1 is a model for the data, a hypothetical description. It is virtually certain to be false in detail, but with good data it may represent a reasonable approximation as a description. The times when the model is reasonably accurate are almost always those where the measures being factored have been subjected to considerable selection and modification.

As to which of the methods to use with reasonably good data, the situation is a little paradoxical. When the data fits the model very well, it makes little difference what we do; all the solutions are likely to be very similar whether we use iterated principal factors, ULS, GLS, or UML. When the data are not quite so good, then there seems to be a greater likelihood that improper solutions will occur in order of sophistication of methods. ULS seems likely to lead to fewer problems than GLS, which in turn has fewer than UML. If there are no problems, then we would lean toward the better inferential status of UML. If computer time or capacity is at all a consideration, UML is much more demanding than the others.

These data-fitting methods are extremely powerful. Using one is a little like telling a special-purpose robot, "Og, fit the factor model." The robot replies, "Okay. Og fit model," and it does so in the most literal-minded fashion. Whether it is a good idea or not, orders are carried out.

Goodness of Fit Statistics

The more statistically sophisticated common factor model fitting methods provide one or more statistics that indicate how well the model fits. The most common one is a chi-square test. It assesses the deviations of the covariance (correlation) matrix from the common factor model with the obtained factor loadings and unique variances. In the maximum likelihood case, it is called a *likelihood ratio test.* If the test is significant, it means that the model is rejected. If not, it is not rejected. Ordinarily, rejection means that more factors are needed to account for the data. However, adding factors may mean that one of these zero-uniqueness situations will arise, indicating that the model is not really appropriate.

Such tests are notoriously sensitive to the sample size. With a small sample, say fewer than 100 observations, it may well happen that a model with a small number of factors is not rejected by the statistic. If the sample is much larger, but from the same population, say 1,000 cases, the model with the same number of factors may well be resoundingly rejected. What is going on?

The answer is that very likely the small number of factors model *is* wrong for the data. With the smaller sample, the test's *power* is not great enough to reject the model. Failure to reject means that the sample R could have occurred where the population correlation matrix P equals $AA' + U^2$, where A has just r columns. With the larger sample, we *can* tell that such a model is false because the sample correlations must then fall within much narrower boundaries if the model is true. The nonsignificant statistic in the small-sample case is also telling us that attempting to fit more factors is unwise; additional factors will be largely determined by sampling errors in the data.

There is another side to the coin, as with all hypothesis-testing procedures. With a large sample, a hypothesis may be rejected for deviations that are essentially trivial in an absolute sense. We will see that this is the case with the physical measurements data. This issue becomes particularly salient with the restricted maximum likelihood methods that are discussed in the next section.

Some common-factor-fitting procedures provide additional statistical information. Some of this information will be discussed later in this chapter along with restricted maximum likelihood methods.

Example

We can apply maximum likelihood factor analysis to the data factored earlier, trying both two and three factors. The results of this analysis are given in Table 15-8. The solutions are similar to the ones obtained earlier. For the most part, the differences can be reconciled by rotating the solutions slightly.

The goodness of fit statistics are given below the loadings. In both cases, the deviations are highly significant, according to the χ^2 statistic. However, fitting the third factor causes a 92 percent reduction in χ^2; fitting 6 more parameters reduces the χ^2 by over 800. This supports the plausibility of the third factor. The fact that the three-factor χ^2 is still highly significant is, to a large extent, the result of a very large sample of over 4,000. The root-mean-square residuals, d_r, are also included; these are similar to the ones obtained by the more informal procedure for the same numbers of factors: d_r is less than .01 for three factors. The significant χ^2, however, indicates that even these small deviations cannot be dismissed as random when the sample is this large.

In a very large sample, the information provided by the third factor is highly reliable statistically, even though the actual amount of information, as indicated by the sizes of the loadings, is quite small. A fourth factor would be much smaller still, and therefore we would consider its information to be of little value, even though it might be reliable. Furthermore, eight variables are not enough for a clear definition of four factors.

Table 15-8 Maximum Likelihood Factors

	Two-Factor Solution				Three-Factor Solution			
	1	2	h_i^2		1	2	3	h_i^2
i								
1	0.944	−0.077	.897		.984	.027	.062	.972
2	0.860	−0.204	.781		.848	.151	.284	.822
3	0.782	−0.113	.625		.764	.058	.212	.632
4	0.817	−0.001	.667		.791	.058	.273	.703
5	0.641	0.079	.418		.665	.121	.316	.557
6	0.342	0.670	.566		.296	.689	.040	.564
7	0.289	0.724	.607		.238	.742	.054	.611
8	0.384	0.655	.576		.337	.679	.051	.577
χ^2	911.35			χ^2	71.99			
df	13			*df*	7			
d_r	.0268			d_r	.0080			

CONFIRMATORY FACTOR ANALYSIS

The Exploratory-Confirmatory Continuum

Factor analytic studies are often characterized as being either "exploratory" or "confirmatory," but this dichotomy is really a continuum. Using a dichotomy to substitute for a continuum is no more valid here than it is in statistical applications. There is a continuous range of possible circumstances in factor analytic studies, ranging from gropingly exploratory to precisely confirmatory. Almost all the original machinery of factor analysis tended toward the exploratory application, and with some exceptions it is only recently that much in the way of rigorous confirmatory methods have been available.

The most exploratory stage in factor analysis is that in which the investigator has a fortuitous set of measures on a fortuitous sample of cases. At the other end of the continuum is the rare case in which we are willing to make precise specifications of not only the number of factors but the values of the loadings of each variable on each factor and even the correlations between the factors. In between these we have all possible gradations.

We have already encountered some aids for confirmatory approaches. The most important is perhaps the rotation-to-target procedure in which we take the obtained matrix and fit it to a hypothetical one. There are also related methods for matching the factors from two groups. In addition, the maximum likelihood method and MINRES method allow tests of the hypothesis that a particular number of factors is present. All of these procedures are rather ad hoc, however, so it is desirable to have more rigorous ones.

Restricted Maximum Likelihood Methods

The most exciting recent development in factor analysis, ranking with the development of rotational methods as the most important of all conceptual improvements, is the implementation of methods for testing a specified structure for the factors underlying a given matrix, or set of matrices. These methods allow for almost unlimited degrees and forms of specification of the structure of a matrix. Generally, they fall under the name of restricted maximum likelihood factor analysis (RMLFA). The currently most widely used RMLFA program is called LISREL, by Karl Jöreskog and Dag Sörbom (1982). Two others are Peter Bentler's EQS (Bentler, 1985) and Roderick McDonald's COSAN (McDonald, 1980).

An entire book could be devoted to the full ramification of these methods and their implementation in various research strategies. In the interest of space, we will simply sketch the ways in which some of the more common and simpler purposes in common factor analysis can be formulated using RMLFA.

The factor model is very general: it allows for any number of factors, any possible factor intercorrelations, and any possible values for uniquenesses. It can always be made to fit any correlation or covariance matrix, provided enough factors are allowed. In RMLFA, the investigator can be more specific about the form of the factor model being fit, and in that way can test hypotheses about the underlying structure for the data. In the basic equation that expresses the correlation matrix in terms of the common factor model (Equation 15.8), there are three different matrices—A, R_p, and U^2. We make the factor model more specific by specifying entries in these matrices, and by specifying the number of common factors. For p variables and r factors, there are pr entries in A, p^2 entries in U^2, and r^2 entries in R_p. These are the parameters of the common factor model. We need to estimate these parameters from the $p(p + 1)/2$ entries in the correlation or covariance matrix. The system is obviously indeterminate in the general form since there are more things to estimate than there are things to estimate them with; therefore, some restrictions are in order. Some of these restrictions are obvious; others are conventional, and/or automatic. For example, R_p should be symmetric with unit diagonal, leaving only $r(r - 1)/2$ elements that are seemingly free to vary. By convention, U^2 is a diagonal matrix, so only p of its entries are free (although we will see that there are occasions where this restriction can be relaxed slightly).

Not all the remaining parameters can vary independently, for various reasons. First, the possibility of transformation of factors exists, just as it does in other forms. This means that $r(r - 1)/2$ of the factor loadings can be set arbitrarily, leaving only $pr - r(r - 1)/2$ of the loadings determined from the data. Furthermore, the correlations in R_p do not really represent additional degrees of freedom separate from A, since we have seen that a correlation matrix can be fit equally well by correlated and uncorrelated factors. The total number of *free* parameters to estimate from the $p(p + 1)/2$ entries in the covariance matrix remains at $pr - r(r - 1)/2$ loadings plus p uniquenesses.

Restricted MLFA differs from the unrestricted version by "using up" some or even all of these free parameters by specifying some terms of the model.

Many restrictions can be made. Perhaps the simplest thing to do is to specify the values of certain loadings. This is done most obviously as a means of specifying the nature of a factor by requiring certain loadings to be zero. This is often easier than stating what the nonzero loadings should be.

For example, in the case of the eight body measurement variables, we could hypothesize a bone-length factor by stating that the three girth measurements have zero loadings on factor 1, and hypothesize a girth factor by stating that the first four variables have zero loadings on factor 2.

In the present case, we think the factors may be slightly correlated, but since we are not sure of the correlations' exact value, we probably would allow this inter-factor correlation to be a free parameter. To illustrate another form of restriction, we could require that the three main girth variables, 6, 7, and 8, have equal loadings on the girth factor. This illustrates a linear constraint, reminiscent of some things that happen in anova; there are actually two constraints here:

$$a_{62} - a_{72} = 0$$

and

$$a_{72} - a_{82} = 0$$

The third equality is redundant with the other two. Thus we have imposed a total of nine constraints on the model by specifying that seven loadings be zero and that three loadings be equal.

If we specify these constraints to an appropriate RMLFA program, it will fit the model to the correlation matrix as well as it can, *subject to the restriction that the restraints be imposed exactly.* The complete model is shown in Exhibit 15-1. Those entries with an asterisk are the specified parameters, which, in the present case, are all zeros; those marked with a dagger are the parameters constrained to be equal. Note that the subscripts for these parameters are all the same to show that they are all to be equal to a_{62}. If there were more than one equality constraint, a different subscript could be used for each set, referring to one of the members of the set that are equal.

The fit of the model to the data is typically expressed in terms of the likelihood ratio chi-square statistic. Based on the hypothesis that the model does account for the population covariance matrix, this statistic is asymptotically distributed as a chi-square with degrees of freedom that depend on the number of variables, the number of factors, and the number of restraints. Hence

$$df = p(p+1)/2 - v \qquad (15.23)$$

$$\mathrm{R} = \begin{bmatrix} a_{11} & 0^* \\ a_{21} & 0^* \\ a_{42} & 0^* \\ a_{41} & 0^* \\ a_{51} & a_{52} \\ 0^* & a_{62}\dagger \\ 0^* & a_{62}\dagger \\ 0^* & a_{62}\dagger \end{bmatrix} \begin{bmatrix} 1.0^* & r_{12} \\ r_{12} & 1.0^* \end{bmatrix} \mathrm{A}' + \begin{bmatrix} u_1^2 & & & & & & & \\ & u_2^2 & & & & & & \\ & & u_3^2 & & & & 0^* & \\ & & & u_4^2 & & & & \\ & 0^* & & & u_5^2 & & & \\ & & & & & u_6^2 & & \\ & & & & & & u_7^2 & \\ & & & & & & & u_8^2 \end{bmatrix}$$

Exhibit 15-1 Hypothesized model for the two-factor case.

*Parameter value fixed.
†Parameter values set equal.

where v is the number of parameters that are estimated—free loadings, uniquenesses, and factor correlations. The number of different parameters given are in Exhibit 15-1, including the five free loadings on factor 1, the two different loadings on factor 2 (a_{52} and a_{62}), the correlation between the two factors, and the eight uniquenesses (16 in all). This leaves 20 degrees of freedom for estimating these parameters from the 36 elements of the "covariance" matrix.

If we fit the model to the data, we obtain the results shown in Exhibit 15-2. Included in the table is the poorness-of-fit chi-square. The resultant value indicates that the likelihood is very small that the sample covariance matrix arose from a population whose covariance matrix is the one specified by Exhibit 15-1 with the parameter values found in Exhibit 15-2. Note that by specifying certain loadings, the χ^2 has been increased from 911.35 to 1235.94, with 13 versus 20 df. We can interpret this large difference as grounds for rejecting the hypothesis that the loadings can be fixed as indicated. Given the overall poor fit, we could try to add a torso-length factor. However, if we merely stated that the three girth variables have zero loadings there would be an indeterminacy in the pattern because this is exactly the same pattern that was specified for the length factor. Therefore, the solution would not be defined. To remedy this, we would further specify that one or two of the length variables—say numbers 2 and 4—are zero also, making a total of five specified zero loadings on the factor. Since there would be three factors, there would be three factor intercorrelations to be treated as free parameters. An alternative might be to require that the factor intercorrelations be equal, but there is no particular reason to do that here.

The second thing to consider is the loadings that have been set at precisely zero. Recall that the loadings in the rotated traditional common factor analysis of

	A: 1	A: 2	u_j^{2*} (U²)
1	0.955	0.0	.088
2	0.865	0.0	.252
3	0.785	0.0	.384
4	0.807	0.0	.349
5	0.591	0.141	.574
6	0.0	0.762	.420
7	0.0	0.762	.420
8	0.0	0.762	.420

$$R_r = \begin{bmatrix} 1.000 & .347 \\ .347 & 1.000 \end{bmatrix}$$

$\chi^2 = 1{,}235.94$
$df = 20$

Exhibit 15-2 Empirical matrix fit to the data.

*U² is presented as a column to save space.

these data were small, but not exactly zero. Thus part of the lack of fit is due undoubtedly to this very stringent substantive requirement that has been imposed.

In hypothesis-testing contexts, the investigator's reaction to resounding rejection of the model can range from intense chagrin—even depression—to lighthearted acceptance. Whichever is the case, we would be inclined to want to try to find out more specifically where the model failed.

LISREL, the most widely used current confirmatory factor analysis program, provides an almost bewildering variety of information in this regard. In addition to the matrix of residual covariances, which can be inspected in the usual way, LISREL shows "normalized residuals." This normalization has the effect of adjusting for the magnitude of the variances of the different variables and of the sample size. In addition, LISREL provides a set of "modification indices" for those parameters that have been fixed in the model. A large value for the parameter's modification index reflects its hunger to be free—or at least its discomfort with the role it has been forced into. Included among the fixed parameters are the covariances between uniquenesses that have been set equal to zero. Large values for these parameters, considered together with the residuals, may suggest the nature of additional factors. For example, if there are only isolated large off-diagonal modification indices, this suggests a "doublet" factor, although doublet factors are poorly determined, as we have seen. An alternative is to allow occasional off-diagonal entries in U^2 to be free, not fixed at zero.

It should be emphasized that we cannot interpret the chi-square from our example as representing a genuine test of a "hypothesis" that the correlation matrix had the form indicated by Exhibit 15-1. Exhibit 15-1 was suggested by the data, indeed by the previous analyses of these correlations. We cannot test a hypothesis on the same data that suggested it! The enormous chi-square makes the issue moot in the present case, although it may arise in analyses of less exact data. One frequently sees a nonsignificant chi-square statistic reported for a RMLFA as if it tested the hypothesis presented by the target matrix. However, it is often clear that the "model" was suggested by the data, indeed that the final model results from a number of adjustments of successive analyses.

Whenever the fitted model is in any way suggested by the data, the inferential aspects of the chi-square statistic are lost. All that is being tested then is the investigator's ability to translate the appearance of a correlation matrix into the corresponding factor equation. The problem is made particularly acute when use is made of the hypothesis-adjusting aids provided by the modern programs. This allows the investigator to pick out only those adjustments to the previously fitted (and apparently rejectable) model that will make the most difference in fit. In doing this, we are capitalizing to an unknown extent on chance fluctuations. The situation is similar to the necessity for the Scheffé correction for post hoc hypotheses about regression weights or mean differences. One reason that the Scheffé correction is so large is to prevent the "hypothesis" from becoming exactly those regression coefficients or mean differences that happened to occur in the sample. The same situation applies in RMLFA, but in an enormously more complex way, since there are so many other things that can be adjusted. Here, there is no Scheffé correction or Tukey test to give us honest tests of hypotheses. Unless the hypotheses used to generate the restrictions placed on A, R_p, and U are genuinely independent of the data analyzed, the probabilities associated with the chi-square are just as invalid as those for the F tests in variable-selection multiple regression. It is regrettable, and bodes ill for scientific progress in the behavioral sciences, that this elementary principle is often ignored by investigators of all degrees of renown.

One difficulty that occurs with RMLFA and not elsewhere is the problem of making a solution determinate. One must, of course, be sure that the number of parameters to be estimated does not exceed $p(p + 1)/2$, but beyond that there are subtle interrelationships among the parameters that make it difficult to ensure that a given problem will successfully run with a given set of free and fixed parameters.

Goodness of Fit Indices

There are two points of view on goodness-of-fit tests. The more Puritan view says that if the test is significant, the model is rejected as representing falsehood, and no more should be done with it. A more forgiving view is that a rejected model may not be all bad—in fact, it may be a close approximation to the truth, departing only in irrelevant details. *Failure* to reject a model may merely mean that the data are so weak they would be hardly able to reject a bizarre distortion of the truth. Clearly, we need some means of assessing degree of departure. There have been several proposals, but we will concentrate on the method used in the LISREL package. The maximum likelihood procedure provides a goodness of fit (GFI) index where

$$\text{GFI} = 1 - \frac{tr(\hat{\Sigma}^{-1}\text{S} - \text{I})^2}{tr(\hat{\Sigma}^{-1}\text{S})^2} \tag{15.24}$$

and $\hat{\Sigma}$ is the fitted matrix. The product in the ratio is the quantity that the method is minimizing. If the fit is perfect, then $\hat{\Sigma} = \text{S}$, making the numerator zero. Recall that the trace of a symmetric matrix times itself is the sum of squared entries in the matrix. Thus, the index represents the closeness with which the fit approaches perfection.

Since the fit improves if more degrees of freedom are allowed—that is, if there are fewer restrictions or more factors—there is an adjusted index that takes this into consideration, called AGFI:

$$\text{AGFI} = 1 - p(p + 1)/2d\ (1 - \text{GFI}) \tag{15.25}$$

where d is the number of degrees of freedom in the model.

Special Models for Covariance Matrices

So far in our discussion we have considered the application of programs such as LISREL, EQS, and COSAN to the problems of factor analysis. The reader may well be aware that these highly sophisticated and flexible procedures can be used to test a wide variety of structures for the covariance matrix. For example, we can test whether the same factor structure exists in two or more groups, or whether it is the same longitudinally, or whether two batteries of measures are parallel. These procedures can also be used to test whether the variables can be considered a simplex, or whether they conform to the pattern of a multitrait multimethod matrix, although there is some controversy about the latter. In addition, these methods can be used to test path-analytic models, which are a kind of sequential multiple regression. This type of analysis can be carried out using latent (that is, factor) variables as well as observed ones. Unfortunately, it would take an entire book nearly the size of this one to adequately cover these topics, so they are not dealt with here.

There is a dearth of general sources to suggest on these matters. Many treatments are highly technical or very uncritical of these procedures. Perhaps the scarcity will be remedied in the near future. One should certainly apply the concerns expressed for the inferential aspects of such model-fitting with equal force to these other models. Perhaps even more importantly, the attempts to use such model-fitting or path analyses to "verify" causal relations seem to be misplaced in many cases (Cliff, 1983).

Additional Considerations in RMLFA

The applications of RMLFA are of great potential utility, but it is important not to forget the basic tenets of scientific inference in applying them. For example, if a given model can be fit to a correlation matrix with a nonsignificant chi-square, this does not prove that the model is correct, only that it cannot be rejected. Many other models will fit equally well. If $R = AR_rA' + U^2$ for some A, R_r, and U, then it is always possible to insert transformations: $R = (AT)(T^{-1}R_rT^{-1\prime})(T'A') + U^2$ that give a different set of factor loadings AT and intercorrelations $(T^{-1}R_rT^{-1\prime})$. If U is permitted to be nondiagonal, then there is additional freedom involving it as well. At least some of the alternative models may have plausible substantive interpretations. This, of course, is another manifestation of the fundamental indeterminacy of the factor model. An advantage that the RMLFA does provide, however, is that it may be possible to show that one or more competing models do *not* fit the data.

We must be particularly careful about over-interpreting the data where causal inferences are made. In almost all applications, the data are fundamentally and irretrievably correlational. Even if the variables are observed at different times, the inference that one causes the other is very rarely justified without an extreme amount of careful research. "Post hoc, ego propter hoc" is one of the most widely recognized of inferential fallacies, and it remains a fallacy no matter how sophisticated the computer program.

One final point must be noted. There is a tendency to treat a maximum likelihood solution as if it were markedly better than any other. This is frequently not the case, however, even by the likelihood ratio criterion itself. In Chaper 9 we found that R^2 could be surprisingly insensitive to large variations in the b's, indicating that while **b** was the "best" solution, there was a wide range of alternatives that were nearly as good. In the same way, while the obtained solution does maximize the likelihood statistic, a wide range of values for the parameters may yield likelihoods that are only slightly smaller. Thus, one does not take the parameter estimates as revealed truth, but rather as suggested values. In this sense, split sample analysis in RMLFA is a good antidote to excessive faith in mere numbers.

Utility of RMLFA

The idea of restrictions on the parameters in the maximum likelihood estimation methods has proven extremely useful in many areas of research, and will continue to do so in the future. The great benefit that RMLFA provides is that it allows the estimation of parameters of the factor model and the evaluation of theories, something that was extremely difficult using the traditional methods of factor extraction and rotation.

The method has its drawbacks as well, however. Some of these are practical. For example, the methods have voracious appetites for computer time and memory, with consequent restrictions on the size of problem that can be studied. More important is the mesmerizing effect that their use seems to have on investigators, paralyzing their critical faculties. An attempt has been made here to point out some of the most obvious instances where the general principles of statistical inquiry need to be followed rather than ignored.

RMLFA is only rarely the most appropriate approach to factor analysis. To make the analysis restricted, loadings have to be specified exactly and it is rare that an investigator can firmly believe that such exactness is true. Except in small samples, the data will ordinarily indicate the values are not *exactly* true. However, they may be more or less accurate. In such cases, the more traditional approach of factor-extraction-followed-by-rotation may work better. The simple structure rotation may benefit from some adjustment through a targeting procedure. There is little doubt that these methods are the current new wave. They are rather complex, however, and their treatment here has been brief. Readers who intend to use these methods are urged to consult other sources such as Long (1983) and McDonald (1985).

FACTOR SCORES IN THE COMMON FACTOR CASE

Nature of the Problem

The concept of factor scores is rather straightforward in the components case. The correlations of the variables with the component scores are the same as the entries in the factor structure matrix; the pattern matrix contains the weights to use in estimating the observed scores from the components. In short, there is an identity between the factors in the model and the obtained factor scores.

The same is not true of the factor scores in the common factors model. Too many things are required of the factor scores in this model. They must both reproduce the variables in the manner indicated by the factor loadings and communalities *and* be uncorrelated with the uniquenesses. Although this is possible for hypothetical constructs, it is more than can be asked of actual numbers printed on computer paper. Therefore, what we get in the common factors case are not factor scores but factor score *estimates.*

Furthermore, the estimates cannot have all the properties we would like them to have; it is necessary to choose from among these properties. It would be nice if the factor score estimates correlated with the observed variables in the way indicated by the structure matrix. It would also be nice if applying the weights in the pattern matrix to the factor scores gave predictions whose correlations with the observed variables were exactly the same as the communalities. Most of all, it would be nice if the factor score estimates correlated as highly as possible with the "true" factor scores, if we should somehow come to know them—perhaps in the afterlife, or through the aid of a psychic medium. Finally, it would be nice if the intercorrelations of the common factor score estimates (FSEs) were the same as the entries in R_r, and if they correlated zero with the unique factors. Unfortunately, we cannot have all of these things at the same time; we have to choose which properties we prefer.

Least Squares Estimates of Factor Scores

In the context of linear models, the FSEs are going to be linear combinations of the observed variables. Traditionally, we would opt for the FSEs that would be the least squares estimate of the "actual" factor scores. But how can we find least squares estimates of the factor scores without knowing them? If we know them, who needs the predictions? The answer is that even though we do not know the scores, we do know all the things necessary to solve the regression equations. After all, that is what we do to get least squares estimates. As usual, we would find the estimated factor score matrix $\hat{P}_r$ by applying a regression coefficient matrix B_r to the observed scores; we could find the regression coefficient matrix from the product of the inverse of the correlation matrix of the predictors, which here are the observed variables, and those variable correlations with the variables to be predicted, in this case, the factors. In equations, we have

$$\hat{P}_r = XB_r \tag{15.26}$$

and, assuming X is in standard scores,

$$B_r^* = R_{xx}^{-1}R_{xp} \tag{15.27}$$

The correlations between the observed scores and the factors are precisely the entries in the structure loading matrix S_r. In this context, $R_{xp} = S_r$, the structure matrix. Thus we can find the coefficients that can be applied to get least squares estimates of the factor scores even though the scores themselves are unknown. This least squares estimate is virtually always used in packaged computer programs, and it seems a very reasonable one to use. However, several other possible factor score estimation methods can be used, depending on which properties are desired. See Harman (1976) for a good discussion of these methods.

We can find the squared multiple correlations of the FSEs with theoretical factors by the usual method: the sum of products of weights and validities. This may be done simultaneously for all factors with the additional benefit of providing the covariances between all the factors and FSEs as follows:

$$\hat{S}_{\hat{p}p} = B_r'S_r \tag{15.28}$$

where $\hat{S}_{\hat{p}p}$ is the covariance matrix of the FSEs with the "true" factor scores. The squared multiples of the estimated and "true" factor scores are in the diagonal of this matrix. Because of the method we are using to get $\hat{P}$, the estimated covariances between true and estimated factors will be the same as those between the estimated factors.

As mentioned in our discussion of components, we can also use simple weighting schemes to get approximate factor scores in the common factors case. In this case, however, it is even more important to check back and see how closely the properties of the approximate factor scores resemble those of the factors. We need to find the intercorrelations of the approximate scores along with their correlations with the observed variables themselves and check them against S_r and R_r to see that the approximate FSEs have the properties that the factors do themselves.

Factor Score Indeterminacy

These factor score multiple *R*'s are never 1.0, and indeed they can be appreciably lower. This fact has led to a considerable amount of controversy. Specifically, if the FSEs are found to have a certain correlation with some other variable, then the true factor scores could have a quite different correlation with that variable. Remember that if we know r_{12} and r_{23}, there is considerable latitude for r_{23} in most cases. Suppose we find the correlation of some variable x with an FSE. Suppose we call this correlation $r_{x\hat{p}}$, and would like to know how close it will be to r_{xp}, the correlation with the true factor. In this case, the inequality is

$$r_{p\hat{p}}r_{x\hat{p}} - (1 - r_{p\hat{p}})^{1/2}\,(1 - r_{y\hat{p}})^{1/2} \leqq r_{xp} \leqq r_{pp}r_{x\hat{p}} + (1 - r_{p\hat{p}})\,(1 - r_{x\hat{p}})^{1/2} \quad (15.29)$$

Here, $r_{p\hat{p}}$ is the multiple correlation of the predicted and true factors, that is, the square root of a diagonal entry in Equation 15.27. The limits are quite broad unless $r_{p\hat{p}}$ is quite high, higher than it often is in practice. This conclusion has been very disturbing to some investigators, others regard it as just another one of the vicissitudes of factor analytic research.

Neither hand-wringing nor shoulder-shrugging are very effective ways of coping mechanisms, so it would be preferable to try to do something about maximizing the degree of factor score determinacy. One way to increase multiple correlations, of course, is to increase the zero-order correlations on which they are based and increase the number of predictors—in this case, in the structure matrix. Basically, this means that the loadings should be as high as possible; scores on a factor defined by three or four loadings around .40 are very poorly determined. If this is impractical, increasing the number of variables that define a factor can substitute to some degree for the size of the loadings—but see McDonald and Mulaik (1979) for a limitation here as well.

Another dimension of indeterminacy is introduced by the sensitivity of a factor solution to the composition of the battery that is analyzed. Adding new variables to a battery may cause the loadings of variables on a factor to break up into loadings on several factors that are more specialized as the "old" variables share an additional portion of their variance with some of the "new" variables but not with others. The process of factor differentiation will reverse if variables are deleted from a battery; specialized factors will coalesce into a more general one. This change in loading patterns is likely to happen unless the original communalities approach the reliabilities of the variables.

For example, consider a battery consisting of several measures of verbal, reasoning, and arithmetic probabilities, yielding three correlated ability factors. If the measures are reduced to one of each kind, there will be only a single general ability factor. However, if additional, specialized tests are added to the original battery, the verbal factor will probably break up into reading, vocabulary, and so on, as will the other factors. Let the measures be factorially complex rather than simple, or let some factors be more poorly represented than others, and the situation will become quite confusing. In judging a factor result we need not only be confident that there are an adequate number of variables defining each factor but that the communalities of the variables are fairly close to their reliabilities. Otherwise, we have an unclear understanding of the factor structure of the variables and an inadequate set of FSEs since the latter will change as the definitions of the factors change.

Indeterminacy is of particular concern in the supposedly sophisticated types of analyses using RMLFA in an attempt to describe the functional and causal relations between latent factor variables. Essentially, these analyses are based on the correlations between estimated factor scores, which are often very tenuously defined. They are highly subject to revision on the basis of changes in the collections of variables analyzed, as we have described here. Thus, the conclusions drawn from such studies should be appropriately tentative, because they may not be as firm as one would like.

SUMMARY

The common factor model differs from the components model in that there is no attempt to account for all the variance in a variable. There is an explicit provision in Equation 15.1 for unexplained, unique variance. Some of this variance is measurement error and some of it may be reliable but specific to the measure. The assumption that the common and unique factors are uncorrelated and that unique factors are uncorrelated with each other leads to what is called the fundamental theorem of factor analysis (Equation 15.8). This theorem shows that the correlations between the variables are functions of the common factor loadings, so the basic problem in factor analysis is to solve for the loadings from the correlations.

This would be simple if the variables' communalities ($h_j^2 = 1 - u_j^2$) were known, but of course they are not. In the precomputer years of factor analysis, a number of ways of estimating these communalities were developed. The traditional methods of common factor analysis consist of substituting the estimates into the diagonals of the correlation matrix and applying the eigenvector-eigenvalue method to this "reduced" correlation matrix R_c. This is called the principal factor method. The squared eigenvalues in the principal factors analysis are the major indicators of the goodness-of-fit of a given number of factors to the correlation matrix.

Various methods have evolved for improving the estimates of the communalities. The most straightforward is the use of "output" communalities—that is, sum of squared loadings—to replace the "input" communalities and refactor. This is the principal axis method with iteration.

More sophisticated than this approach are the unweighted least squares or MINRES, generalized least squares, and maximum likelihood methods for estimating common factor loadings. The last two make estimation of uniquenesses an explicit part of the analysis and have an explicit basis in statistical estimation theory. MINRES bypasses the communality problem by fitting the factor loadings to the off-diagonal correlations. Using these methods, it is possible to test the hypothesis that the correlations can be explained by a particular number of factors.

Methods for confirmatory factor analysis are an important recent development. They allow us not only to test a hypothesis concerning the number of factors, but also to test a variety of specific aspects of the factor equation. The most common program for such analyses is LISREL, although COSAN (McDonald 1978, 1980) and EQS (Bentler 1985) are alternatives. Care must be taken in making inferences from these programs since it is easy to let post hoc considerations influence the nature of the model that is tested.

Factor score estimation is somewhat complex empirically in the case of the common factor model in that there are no factor score matrices that have all the properties of the factor scores in the model. The compromise that is most frequently taken is to use the least-squares regression estimates if the factor scores are needed (Equations 15.26 and 15.27). The basic indeterminacy of the factor scores corresponds to an indeterminacy in the definition of factors. The nature of a factor may change if the variables that load on it are analyzed in the context of a different set of variables. Even under the best of circumstances, this is an imperfect correlation between estimated and "true" factor scores; this correlation is low insofar as the loadings are low and the variables are few.

Transformation of factors was not discussed in the common factor context because the methods are exactly the same as in the components case. We simply take the output from a common factors solution and transform it by whatever method we deem appropriate, according to the discussion in the previous chapter. Rotation is presumably unnecessary in RMLFA since the specification process determines an orientation for the factors. In some applications, however, it may be useful to perform a further rotation.

COMPUTER APPLICATIONS

SPSS^x and SPSS

The procedures for common factor analysis are quite similar by the two SPSS systems except that the older SPSS version does not offer the newer methods like ULS, GLS, or UML, and the command structure for SPSS^x is not as clearly described in the manual as the old version was. The probability of having SPSS^x FACTOR do what you want it to do is increased if you realize that the system reads the control cards sequentially and carries out an analysis step using the parameters in force at the time. Therefore, we must make all the method specifications before the commands that they control. Otherwise, execution of some step takes place *before* our intended modifications to the system parameters are carried out.

The PAF option is the SPSS^x version of iterated principal factors. It starts with the squared multiple correlations as initial input communalities (or alternative values supplied by the user through the "DIAGONAL" subcommand) and replaces them as many times as is indicated by the ITERATE keyword (default is 25), which is part of the CRITERIA subcommand, for example, CRITERIA-ITERATE (10). The other control is the ECONVERGE parameter, which specifies the smallest change in communality that permits continued iteration (default is .001). The default values allow more iterations than are likely to be useful, however. Remember that small changes in communality lead to even smaller changes in loadings.

Iteration of communalities requires the specification of the number of factors. This is done using the minimum acceptable eigenvalue (default = 1.0), although this decision is made from the eigenvalues in the *principal components analysis*. One can also use NFACTORS= or MINEIGEN=. In combination with a two-stage analysis this is likely to be the more satisfactory way of specifying the number of factors. One uses the results of an initial analysis to decide on the number of factors and then does a reanalysis with that specification.

The SPSS^x output can be confusing at first. Regardless of what extraction method is specified, it first performs a principal components analysis, and the first section of output is the same as that described in Chapter 14. Under COMMUNALITY, it lists the input communalities $\hat{h}_2^2$ in a column. Opposite to this are the eigenvalues of the *principal components* analysis of the data. After straightforwardly presenting the unrotated factor loadings from the extraction method specified in the "EXTRACTION" command, it again lists COMMUNALITY, but this is the *output* communality, $h_y'^2$. To the right of this is another set of eigenvalues. These are the eigenvalues that correspond to the retained factors only (after iteration of communalities, if called for) and the "PCT OF VAR" for them is the sum of the first r eigenvalues divided by the sum of all the eigenvalues. This is a rather useless ratio, as noted here in the text, due to the negative eigenvalues.

Rather frustratingly, FACTOR will not provide the eigenvalues that are necessary to assess fit in the common factor case. It is possible, however, to circumvent this deficiency by using the eigenvalues from the principal *components* analysis, which are provided. If for the moment, we call the eigenvalues from the principal components analysis λ_{pc}, and those from the common factor analysis λ_{cf}, then

$$\Sigma\lambda_{cf}^2 = \Sigma\lambda_{pc}^2 - p + \Sigma(h_j')^2$$

where all summations run from 1 to p. The eigenvalues for the retained factors are provided, so the summary statistics d_r and g_r can be calculated by subtracting $\Sigma_1^r\lambda_m^2$ from $\Sigma\lambda_{cf}^2$ as defined above.

Alternative EXTRACTION subcommands can choose unweighted least squares, that is, MINRES, with EXTRACTION = ULS, or unrestricted maximum likelihood with =ML, image with =IMAGE, alpha with =ALPHA, or generalized least squares with =GLS. More than one extraction method can be chosen. The number of iterations for these methods is specified with the =ITERATE aspect of CRITERIA=.

SPSS^x calculates FSEs by the usual regression method as the default option. Two other methods are also available that are not discussed here. The same rotational methods and controls apply here as in the components case (see Chapter 14).

BMDP

BMDP offers a somewhat better selection of alternatives and output for the common factor analysis. In its METHOD= statement it offers PCA (principal components analysis, the default), PFA (principal factor analysis with iterations, similar to SPSS^x's PAF), MLFA (unrestricted maximum likelihood factor analysis), and others.

The parameters that control the process are ITERATE= (the number of iterations for PFA or MLFA); EPS (epsilon, the amount of change in communality for PFA, like the CONVERGE parameter of SPSS^x); COMMUNALITY= (the way communality is to be estimated for PFA); UNALT (unaltered, that is, 1.0 or the variances); SMCS (squared multiple correlations); and MAXROW (the maximum correlation of the variable). The word may be followed by a specified list of communality values. Additional control words are NUMBER=, which gives the maximum number of factors to be rotated, and CONSTANT= which specifies the minimum eigenvalue for a factor to be rotated.

The output from the BMDP factor analysis program is quite straightforward and unambiguous.

SAS

The alternatives and specifications in SAS are similar to those in BMDP. The METHOD= statement offers several factoring methods. PRIN is the principal components method or principal factors method, depending on the specification of the diagonal elements. PRINIT is principal factors with iterations, like SPSS^x's PAF or BMD-P's PFA. IMAGE and ALPHA have the obvious referents and ML is the SAS version of unrestricted maximum likelihood methodology.

The number of factors can be limited by NFACT=, in which the user specifies the number, MINEGEN=, which specifies the minimum eigenvalue corresponding to a factor to be rotated, or by PORTION=, which specifies the ratio of the sum of the eigenvalues of the rotated factors to the sum of all the eigenvalues (sum of input communalities). These are relevant primarily to the PRIN and PRINIT methods, and as with these other programs a two-stage analysis is often necessary to decide on the number of factors. Start by using MINEGIN= .5 for the first analysis, then examine the results and set NFACTORS= at whatever number seems appropriate in the second.

There are two control words that refer to the iteration process. MAXITER= is used for the maximum number of iterations, like ITERATE= in SPSS^x and BMDP, whereas CONVERGE= has the same function as STOPFACT= and EPS=.

The input communalities are controlled by the PRIORS statement. Numerical values following PRIORS are the input communalities. Alternatively, PRIORS SMC specifies squared multiple correlations and PRIORS MAX uses the highest correlation.

The output from SAS is quite self-explanatory and well-organized. The only part that seems likely to cause confusion is the extensive output associated with the factor scores; this is inadequately explained in the manual.

LISREL

The LISREL system is actually much more complex than simply providing a RMLFA procedure. We use only a few of its features. The version considered here is LISREL V (Joreskog and Sorbom, 1982). Later versions can be expected to differ in details, although they can be expected to take the form of increased flexibility, convenience, and so on. Reading about the full LISREL model can be confusing since it contains many more matrices than we have considered and their notation is different from ours.

LISREL allows for the analysis of a set of dependent variables Y in terms of a set of independent variables X where each set is resolved into a factor structure.

In RMLFA applications, there is only one set of variables. Furthermore, in LISREL the factor equation is generalized. Conventionally, it is assumed here that, in RMLFA, the Y set is present. The least general form of the factor equation that we can consider is in their notation:

$$\Sigma = \Lambda_y B^{-1} \Psi B^{-1} \Lambda_y' + \Theta_\epsilon$$

where Σ is the variable covariance matrix; Λ_y (lambda) is the pattern matrix that we have called A; Ψ (psi) is the factor covariance matrix that we have designated R_p and restricted to being a correlation matrix; Θ_ϵ (theta epsilon) is the unique factors covariance matrix that we have called U^2; and B (beta) is a matrix of regression weights that in RMLFA applications is specified to be the identity matrix, so it has no effect. The only reason for mentioning B is that it must be specified in the control statements.

The different symbols and names for these various matrices in different contexts along with the nature of their default values in LISREL are given in Table 15-9. As with other systems, we must specify the nature of the input, what to do with it, and what to output. These are fairly clearly specified in the manual, although the explanations are rather scattered, so we will recapitulate the ones that are necessary for RMLFA.

The program begins with a dataparameters card (DA) that specifies the number of groups (NG) in case data for several groups is to be analyzed in parallel; (default = 1); the number of input variables (NI); the number of observations (NO); and the type of matrix to be analyzed, for which the options are MM (moment matrix, that is, uncorrected for means), CM (covariance), or KM "Korrelation" matrix (default = CM). Then the data are read, along with identifying variable names (labels, LA).

Our major concern is with how the model is specified. This starts with a model (MO) card that specifies the number of variables (NY) and factors (NE) and that BETA = IDENTITY. Now comes the tricky part, specifying which elements are fixed, free, or equal. Here, one must be aware of the default natures of three parameter matrices, which have their default properties unless otherwise specified. For example, TE, the uniqueness matrix is diagonal with fixed (that is, user-specified) entries unless otherwise stated. The LY loading matrix is fixed and "full;" and Ψ, the factor covariance matrix, is diagonal and fixed. The character of a matrix can be changed as a whole in the MO statement. For

Table 15-9 Explanation of LISREL Notation

Text Symbol	*LISREL Symbol*	*Content*	*Input Name*	*Output Name*	*Default Nature**
A	y	Factor Loadings	LY	LAMBDA Y	Full, Fixed
R_p	Ψ	Factor Intercorrelations (covariances)	PSI	PSI	Diagonal, Fixed
U^2	Θ_ϵ	Uniquenesses	TE	THETA EPS	Diagonal, Fixed

*Full means a matrix that is not necessarily symmetric or diagonal. The alternatives to *fixed* are *free* and *equal*.

example,

MO NY = 8 NE = 2 PSI = SYM, FREE,

TE = DIAG, FREE, BETA = IDENTITY

makes Ψ into a covariance matrix and frees the elements of Θ_ϵ to be estimated, as would be done in an RMLFA problem. Having defined a matrix as free or fixed, either through accepting the default or in the MO statement, one can now go on to *reverse* the character of some of its entries. For example,

FREE LY(1,1) LY(2,1) LY(3,1)

allows the first three loadings on factor one to be free, assuming that the MO statement has left the rest of it fixed. Similarly, after allowing the factor covariances to be "SYM,FREE" we can convert these covariances into a correlation matrix by requiring the diagonal to be 1.0:

FIX PSI(1,1) PSI(2,2)

This will allow us to specify that the diagonal of PSI is 1.0 for a two-factor solution.

The final question is how to specify the fixed entries. Fixing entries is actually combined with furnishing the program with starting values for the free elements of the parameter matrices. LISREL is an iterative program; it improves the trial solution, then looks at how well it has done and repeats the process until the solution is "good enough" or time runs out. Therefore, it needs somewhere to start, so the user provides starting values for LY (that is, A), PSI (R_p), and TE (U^2). The values of these fixed parameters are included in these matrices of starting values. For example, we would provide the starting values for the two-factor solution to the eight physical measurements in Table 14-1 by freeing the appropriate entries of A, corresponding to the unstarred entries of Table 15-9:

FREE LY(1,1) LY(2,1) LY(3,1) LY(4,1) LY(5,1)

LY(5,2) LY(6,2) LY(7,2) LY(8.2)

Now we need to specify the fixed values, which are zero here, and also must provide starting values for the free entries. This is done with a MATRIX statement:

MA LY

* (The asterisk specifies that the data is being input in "free field" rather than some other format)

.9 0 .8 0 .8 0 .8 0 .5 .3 0 .8 0 .8 0 .8

This specifies where the three zeros fall on factor I and the four on factor 2, as well as providing approximate estimates of the nonzero loadings. We also need to specify any factor correlations (PSI) and the uniquenesses (TE):

```
MA TE
*
.2, .4, .4, .4, .5, .4, .4, .4
```

Finally, if there are any parameters that are constrained to be equal, these are specified with *equal cards,* which have each set of equalities on (its own) card. In our example, we have only one parameter involving the last three loadings on the second factor:

```
EQ LY(6,2) LY(7,2) LY(8,2)
```

The above description may seem rather formidable. Even this level of detail, however, is not likely to be sufficient to permit the user who is unfamiliar with LISREL to run a program successfully without expert advice. However, given such advice or prior experience with LISREL, it should help us to translate our plans into data analysis.

PROBLEMS

1. What communality estimates seem appropriate for the data in Table 13-3? Insert these in the diagonal using the eigenvectors

$$w' = \begin{bmatrix} .4082 & .4082 & .4082 & .4082 & .4082 & .4082 \\ .4082 & .4082 & .4082 & -.4082 & -.4082 & -.4082 \end{bmatrix}$$

to obtain common factor loadings. Compute the residuals. (Problems 2 through 5 require the use of a computer package.)

2. Using the intercorrelations of the first six variables of the nine-variable matrix from the problems of Chapter 14, find the squared multiple correlations R_j^2 for each variable with the other five and substitute them as communality estimates, h_j^2. Verify that the following eigenvalues result: 1.9899, 1.0818, −.0466, −.0854, −.1744, −.2290. Compute d_r and g_r for $r = 1, 2, 3$.

3. Take the two-factor solution from Problem 2 to re-estimate the communalities and find a new two-factor solution. Find the eigenvalues, g_r, d_r for $r = 1, 2, 3$ and the factor matrix for the appropriate number of factors.

4. Assume $n = 100$ and submit the Problem 2 data to an UMLFA program, testing one and two-factor solutions.

5. Given below is an orthogonally rotated two-factor solution for the data in Problem 2. Compute the factor score coefficients B_2 and the factor covariance matrix $S_{\hat{p}p}$.

$$A = \begin{bmatrix} .1760 & .1433 & -.0148 & .7986 & .6925 & .7812 \\ .7276 & .5801 & .6727 & .2149 & -.0113 & .0261 \end{bmatrix}$$

What are the correlations between the estimated and "actual" factor scores? What is the correlation between the two estimated factor scores?

6. For the data above, suppose approximate factor scores are computed by weighting the last three variables 1.0 and the rest 0 (factor 1) and the reverse for factor two. Find the covariances and correlations of these approximate factors with the true factors and with each other.

ANSWERS

1. $h_j^2 = .8$ for all j

$$\mathbf{F}' = \begin{bmatrix} .8367 & .8367 & .8367 & .8367 & .8367 & .8367 \\ .3162 & .3162 & .3162 & -.3162 & -.3162 & -.3162 \end{bmatrix}$$

All residuals = 0

2. R_j^2: .422, .276, .337, .571, .409, .521; For r = 1, 2, 3, respectively, g_r = .758, .982, .982. d_r = .1874, .0512, .0506

3. λ = 2.0791, 1.174, .0361, .0006, −.0837, −.1344
g_r = .755, .995, .995. d_r = .1975, .0271, .0264; somewhat better

$$\mathbf{F}' = \begin{bmatrix} .5334 & .4279 & .3381 & .8093 & .5987 & .6956 \\ -.5433 & -.4316 & -.5950 & .2292 & .3713 & .3850 \end{bmatrix}$$

5. $$\mathbf{B}_2 = \begin{bmatrix} -.0003 & -.0038 & -.0711 & .4265 & .2446 & .3543 \\ .4347 & .2408 & .3306 & .0692 & -.0832 & -.0716 \end{bmatrix}$$

$$\mathbf{S}_{p\hat{p}} = \begin{bmatrix} .7873 & .0478 \\ .0478 & .6922 \end{bmatrix} \quad r_{p_1\hat{p}_1} = .887;\ r_{p_2\hat{p}_2} = .832 \quad r_{\hat{p}_1\hat{p}_2} = .0648$$

6. Correlations:

	p_2	$\hat{p}_1$	$\hat{p}_2$
p_1	0	.882	.089
p_2		.126	.820
$\hat{p}_1$			.190

READINGS

Gorsuch, R. L. (1983). *Factor analysis,* 2nd ed. Philadelphia: Saunders, pp. 94–174.

Harman, H. H. (1976). *Modern factors analysis,* 3rd ed. Chicago: University of Chicago, pp. 3–32, 113–246, 363–87.

Kim, J. O., and C. W. Mueller (1978a). *Introduction to factor analysis.* Beverly Hills, CA: Sage pp. 7–79.

Kim, J. O., and C. W. Mueller (1978b). *Factor analysis: Statistical methods and practical Issues.* Beverly Hills, CA: Sage, pp. 7–28, 41–79.

Long, J. S. (1983a). *Confirmatory factor analysis.* Beverly Hills, CA: Sage.

McDonald, R. P., and S. A. Mulaik (1979). Determinacy of common factors: A non–technical review. *Psychological Bulletin* 86:297–306.

McDonald, R. P. (1985). *Factor analysis and related methods.* Hillsdale, NJ: Erlbaum, pp. 28–81; 96–107.

Mulaik, S. A. (1972). *The foundations of factor analysis.* New York: McGraw-Hill, pp. 95–102, 132–172, 186–216, 361–401.

Overall, J. E., and C. J. Klett (1972). *Applied multivariate analysis.* New York: McGraw-Hill.

16

Discriminant Analysis and Multivariate Analysis of Variance

THE EMPIRICAL CONTEXT

The Simplest Case

One of the first statistical techniques that we learn is the *t* test for independent means. In it, the independent variable represents some distinction between two groups, and the dependent variable is some continuous measure. The independent variable could be an experimental manipulation, or it could be a natural distinction such as sex, country of origin, or age group. Occasionally, the roles of the variables are reversed, and we may wish to predict from some continuous measure *to* a dichotomy. For example, from scores on a personality trait test we may wish to predict success versus failure as a manager. To assess the accuracy of the prediction, we traditionally use the point-biserial correlation. In a perceptual experiment, it is equally reasonable to predict success in identification from length of exposure of the stimulus, but this approach is rarely taken here.

In the context of experimental research, we have learned about the extension of the dichotomy (*t* test) to a polytomy* (one-way anova), and about a multiple dichotomy or polytomy (factorial designs) in which there are multiple dichotomous or polytomous predictors and a continuous dependent variable. The reverse situation also occurs—that is, a single dichotomous independent variable sometimes occurs with multiple continuous dependent variables. For example, in an

---

*We sometimes see the term "polychotomy," but "polytomy" is considered to be more correct etymologically.

EEG experiment with two experimental conditions, there are several placements of electrodes and numerous frequency bands, and each provides a dependent variable. In a clinical treatment, there may also be multiple dependent measures that reflect different aspects of the outcome. Similarly, the prediction of job success-failure could be attempted with multiple measures of the individuals. The classical case of dichotomy-predicting is one in which we measure multiple aspects of two species of organisms and use the measures to "predict" species membership. Thus, we frequently have occasion to assess the statistical effect of a dichotomy on multiple dependent measures, or to use multiple continuous measures to predict or explain a dichotomy. In both cases the dichotomy may be generalized to a polytomy. Which of the two concerns is primary depends on the field of research and the research questions.

The single-dichotomy-multiple-dependent-variable context is the simplest case of *multivariate analysis of variance—manova,* for short. The multiple-measure-predicting-group-membership case is the simplest type of *discriminant function analysis.*

In a manova context, the investigator typically has two types of concerns. Historically, the main concern has been how to assess alpha levels in the case of multiple dependent measures, although this is perhaps not the most important concern. The experimentalist who uses multiple dependent measures may be concerned that finding a significant difference on some single dependent variable, or a subset of them, would be dismissed as a chance finding since multiple tests obviously increase analysiswise alpha levels. Thus, a procedure for controlling the experimentwise alpha level is needed when multiple dependent variables are used. Equally important is the question of which variables are most affected by the manipulation: What does this independent variable affect? Manova, and the analyses that accompany it, are suited for both purposes.

The concerns in discriminant analysis parallel those of manova, although the emphasis is slightly different. First, there is the question of whether we can discriminate between the groups, and if we can, how well. Second, there is the question of what combination of variables makes the best discrimination. This combination is called the *canonical variate.*

We already have a way of doing both manova and discriminant analyses in the two-group case, although it is not obvious. They can be done using multiple regression with the dichotomy as the dependent variable. This is perhaps best illustrated in the discriminant case, although this is not precisely the way in which a discriminant analysis problem is typically formulated. However, the idea is that if we wish to predict which member of the dichotomy an observation belongs to, we can simply code the dichotomy as a 1–0 variable and use it as the dependent variable in a multiple regression analysis. The regression weights will give us the weights that best discriminate, the squared multiple correlation will tell us how good the discrimination is, and its significance will allow us to judge the probability that discrimination took place by capitalizing on chance differences. All of these answers need to be surrounded by a stockade of if's, and's, but's, however's, and other caveats, but that is the general idea.

The same type of analysis answers the manova questions. It helps here to remember that the t ratio is a direct function of the point-biserial correlation. Thus the mean differences on the dependent variables correspond to the validity correlations of a multiple regression. We can assess the overall multivariate significance with the following questions: If we called the optimum possible combination of dependent variables a composite dependent variable, what would be the degree of difference on it, and to what extent can a chance explanation be ruled out? The obvious candidate for the combination rule is to use the regression weights to form the composite dependent variable and then the multiple regression formulas to assess the significance of the separation of the two groups on it. This case is a little harder conceptually than the preceding case in which we started out thinking of the dichotomy as the independent variable. Here, we act as if we are trying to predict the independent variable. However, the formal properties of the analysis are the same, regardless of which way we look at things. So multiple regression is again the way in which the analysis could take place.

The mathematical formulations for manova and discriminant analysis are really quite different looking. The multivariate generalization of the t test is actually called Hotelling's T^2. However, both manova and discriminant analysis are directly related to multiple regression against a dichotomous prediction, so we will begin with the familiar and emphasize the similarities of the various methods.

The General Cases

Obviously, we would like to go beyond the two-group case in both manova and discriminant analysis. An investigator can easily have more than two experimental conditions or more than two groups to discriminate. If so, we can think of having p continuous variables and $g - 1$ membership variables that are dummy-coded. Multigroup manova and discriminant analysis constitute our first encounter with what some statisticians consider the true multi*variate* techniques. By contrast, all of the preceding chapters of this text deal with multi*variable* methods.

In manova, the empirical organization is the same, whether we have two groups or many groups. But here there is more room for complication, since with several groups there may be more than one possible kind of difference, or the groups may be organized into a factorial design or some other structure. The same holds true of discriminant analysis. If there are more than two groups, there is obviously more than one discrimination to be made, and indeed there may be more than one way to make an overall discrimination, both ways being nearly as good. As usual, the multivariate aspect of the empirical situation places a heavier cognitive burden on the investigator. Initially the investigator must decide just what he or she is after in the analysis—or, for that matter, in the study as a whole.

The statistical methodology for the manova and discriminant cases are basically the same—although they can be approached in somewhat different ways and the emphases are somewhat different.

DISCRIMINANT ANALYSIS

Two-Group Discriminant Analysis

Suppose that the admissions committee of a graduate program in a psychology department is concerned about the large fraction of students that perform satisfactorily but never finish their degrees, and decides to see if it is possible to identify such individuals upon admission. The committee decides to focus on two predictor variables—the GRE Quantitative score, and the number of units of undergraduate psychology courses taken. These variables were selected for some good a priori reasons. The GRE-Q is important because the psychology program requires a number of statistics courses that might be less of a chore for the higher aptitude students. The number of units is important because this might reflect familiarity with and commitment to the field. Note that the two variables must not be selected from a battery of others on the basis of promising aspects of the *data*. That would be peeking, which throws off the probabilities in the significance tests.

First, the committee examines the records for students who entered the program over the last ten years to find out which students received a degree and which students received no degree within that period, or are known to have left the program. (Students who entered the program within the last five years and are still in good standing are not included in the study.) Thus there are two groups: Successful Completions (Wins) and No Degrees (Losses). The former is composed of three subgroups: students who finished in three years, four years, or five-or-more years. (If this example strikes too close to home, or seems naive or implausible, feel free to change the details.) These latter three groups are lumped together in this first analysis. The program is a relatively small one: the committee finds 30 Wins and 20 Losses [see Exhibit 16-1(a)].

As always, the first step is to plot the data. The initial plot reveals three outliers (we are trying to be realistic here). Re-examination shows that these constituted two errors in recording the data and one case where the student had not had a transcript. These errors and omissions are corrected, and the data are plotted again, with the results shown in Figure 16-1.

The bivariate plot suggests that there is an appreciable separation between the means of the two groups, although there is considerable overlap. It is instructive to take a clear plastic ruler and try to find a line of demarcation that most purely separates the two sets of points; this separating line has been estimated by eye in the figure. Univariate frequency distributions are entered on the margins of the figure to show the degree of separation of the individual variables. Another line, which represents a vector in the space, has been drawn at right angles to the fitted line of demarcation and through the origin. The frequency distribution on this line has been drawn also. This shows what the distribution would be if this vector were used as an axis. Note that the overlap in the distributions is somewhat less along this line than in either of the univariate distributions.

This line, like any other line in the space, is a linear composite of the two axes in the space. In this case, the space represents the observed variables; it is a linear composite of the observed variables composed in an attempt to minimize the overlap in the distributions. As such, this new axis represents an approximate *discriminant* or *canonical variate,* and the weights define how to make it up out of the observed variables. That is,

$$z_{ih} = w_1 x_{ih1} + w_2 x_{ih2} \tag{16.1}$$

where, z_{ih} is the canonical score of individual i of group h; x_{ih1} is i's score on the first observed variable, and x_{ih2} is i's score on the second. The w's are the weights that define the canonical variate in terms of the two observed variables.

Fitting the line of demarcation by eye in terms of minimizing overlap is a rather informal procedure. This procedure is similar in spirit to our little story that

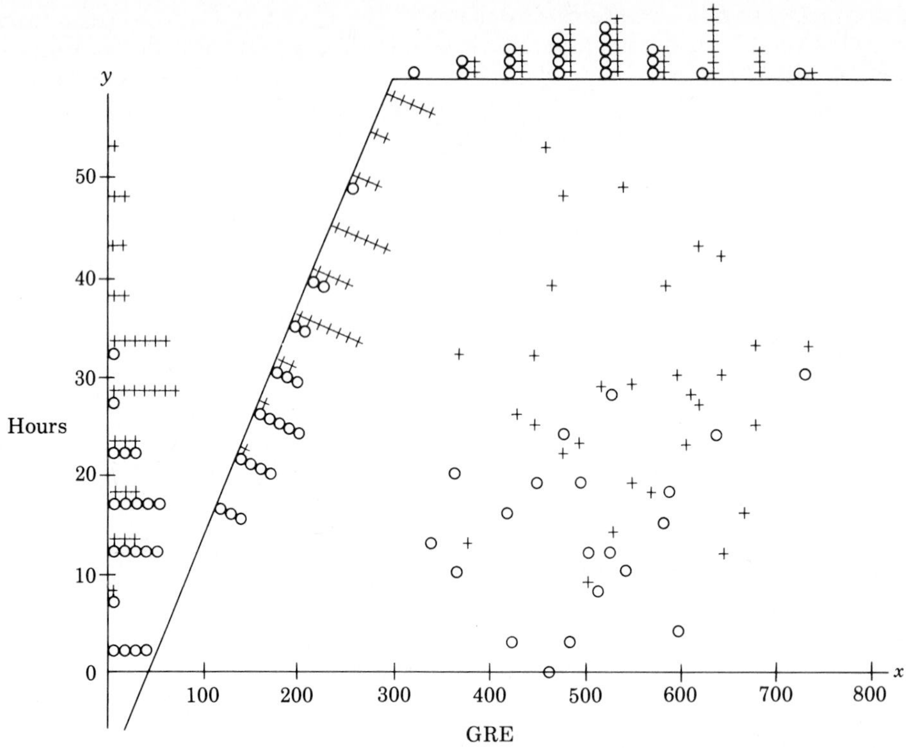

Figure 16-1 Bivariate distribution of data for Exhibit 16-1. The plus indicates successful completion; zero indicates no degree. The line through the origin is a composite variable fit by eye to separate the two groups as completely as possible. Also shown are univariate distributions.

introduced initial regression in Chapter 6. The actual process of discriminant analysis is more formal. It starts with an "objective function" that defines what is meant by "separation" between the groups, and expresses the separation in terms of the definition of the discriminant variate.

As usual, the process has a descriptive aspect, couched in terms of the observed data. The process also has an inferential aspect, referring to the populations from which the data are assumed to be drawn. We will approach the problem through the descriptive side first and then connect it to the inferential side.

Discriminant analysis defines a weighted combination of scores like those in Equation 16.1 such that the ratio of the among-groups mean square to the within-groups mean square on it is as large as possible. Thus we are trying to find the weighted combination of observed variables such that a generalized F ratio is as large as possible. It does not matter whether we maximize the ratio of among/within mean squares, among/within sums of squares, or among/total sums of squares or mean squares. None of these choices affect the nature of the discriminant variable that we end up with, but in discriminant analysis, the problem is usually defined in terms of the ratio of among sums of squares to within sums of squares, and sometimes as the ratio of among to total sums of squares. (Note that here we are linguistically correct in using "among" rather than "between." For some reason, the latter term did not become entrenched as it did in anova.)

We seek a variable z that is a weighted combination of x's:

$$z = \Sigma w_j x_j \tag{16.2}$$

so that the ratio for z,

$$v = \frac{ss_a}{ss_w} \tag{16.3}$$

is as large as possible. There is an indeterminancy in the w's since, if we use a particular set of w's, the resulting ratio is not affected if we use another set that is the same as the first except all are multiplied by the same constant. This indeterminancy can be removed by requiring that

$$\Sigma w_j^2 = 1.0 \tag{16.4}$$

which is similar to the requirement imposed on eigenvectors in principal components.

To find the vector $\mathbf{w}_1$ that has these two properties maximizing Equation 16.3 and conforming to Equation 16.4, we need to define multivariate extensions of the sums of squares on the observed variables. That is, we need *matrices* of the sums of squares of each variable and of the sums of products between variables. These matrices are defined below.

Let us go back to the score matrix X on the continuous predictor variables where X is n by p, and where the total sample size is

$$n = \Sigma n_h \tag{16.5}$$

as in our discussion of anova. For convenience

$$1'\mathrm{X} = 0 \tag{16.6}$$

where x's are expressed as deviations from the overall means for each variable. Then for each observation x_{ihj},

$$x_{ihj} = (x_{ihj} - \overline{x}_{.hj}) + \overline{x}_{.hj} \tag{16.7}$$

Thus, in matrix notation,

$$\mathrm{X} = \mathrm{X}_w + \mathrm{X}_a \tag{16.8}$$

where X_w contains the within-group deviations on each variable and X_a is an $n \times p$ matrix of the group means, as in the last term of 16.7. (This last term represents the deviations of group means from the grand means; the latter is set to zero since we are dealing with deviations.)

The total sum of squares and products is X′X, analogous to the anova total sum of squares:

$$T_t = X'X \tag{16.9}$$

Because $X = X_a + X_w$ and $X_a'X_w = 0$, the following is a generalization of $ss_t = ss_b + ss_w$:

$$T_t = T_a + T_w \tag{16.10}$$

where

$$T_w = X_w'X_w \tag{16.11}$$

Equation 16.11 is the matrix of sums of squares and products within-groups; its diagonal is the within-group sum of squares for each variable; the off-diagonal terms are sums of products of within-group deviations for different variables. Dividing by the degrees of freedom $n - g$ (g groups), gives a matrix of within-groups variances and covariances:

$$S_w = T_w(n - g)^{-1} \tag{16.12}$$

Correspondingly,

$$T_a = X_a'X_a \tag{16.13}$$

Here, the degrees of freedom are $g - 1$, so

$$S_a = T_a(g - 1)^{-1} \tag{16.14}$$

If all the groups are really samples from the same population, then both S_w and S_a are estimates of the population variance-covariance matrix Σ, just as ms_w and ms_b are estimates of σ_e^2 in anova.

The ratio of among-to within sums of squares on the discriminant variate, can now be expressed in terms of T_a and T_w:

$$v = \frac{\mathbf{w}_1'T_a\mathbf{w}_1}{\mathbf{w}_1'T_w\mathbf{w}_1} \tag{16.15}$$

We wish to pick $\mathbf{w}_1$ in order to maximize the ratio of among to within sums of squares (Equation 16.15). It turns out that this is an eigenvalue-eigenvector problem. The weights vector $\mathbf{w}_1$ is the solution to the characteristic equation

$$T_w^{-1}T_a\mathbf{w}_1 = v\mathbf{w}_1 \tag{16.16}$$

This is a new type of eigenvalue problem in that $T_w^{-1}T_a$ is not a symmetric matrix.

However this poses very little difficulty since we will not have to worry about computing it, except in very simple cases.

In the two-group case, there is only one solution. Consideration of X_a quickly shows that its rank is one. Alternatively, if we think of the multivariate space of the observed variables, the means of the groups each defines a point in that space. When there are only two groups, they form a line, that is, a one-dimensional space. If there are more groups, the means clearly define a $g - 1$ space; except that the dimensionality is limited by the number of variables since the rank of T_w will ordinarily be its order. For the moment, we are concerned only with the single eigenvalue v and the corresponding vector $\mathbf{w}_1$ for the two-group case.

Example

Exhibit 16-1 shows the calculations in some detail. Part (a) shows the raw data, along with the means of the variables. Part (b) shows the X matrix in deviation form, along with the corresponding X_a and X_w. Part (c) shows the sums of squares and products matrices using Equations 16.9, 16.10, 16.11, and 16.13. The two within-groups variance-covariance matrices are shown in part (d) so that they can be inspected for homogeneity. We will return to this issue later, but for the moment we will accept the matrices as homogenous.

The T_w^{-1} and $T_w^{-1}T_a$ matrices are in part (e) and the characteristic equation and eigenvalues are given in part (f), together with the eigenvector that corresponds to the larger eigenvalue. Note that one eigenvalue is zero since $g - 1 = 1$ so that S_a is rank one. In the eigenvector, the weight for the GRE is low relative to that for Hours. This is partly to compensate for the different variances for the variables. In this case it primarily reflects the fact that the differentiation is almost completely on the Hours variable. (Note that our fit by eye in Figure 16-1 was off in this regard.) This is analogous to a two-predictor multiple regression in which one regression weight is near zero. Indeed, if we use X as a predictor matrix and have a dichotomous criterion that records group membership, the resulting regression weights will be proportional to the elements of $\mathbf{w}_1$. This is done for some other data in Problem 4, page 427.

In Exhibit 16-1(g) we have computed the sample means and common standard deviation on the canonical variate using the weight vector $\mathbf{w}_1$. The results allow us to conclude that, expressed in terms of the common standard deviation $\hat{\sigma}$, the means for the groups are separated by a bit less than 1.5 standard deviations. The distribution on the canonical variate is shown in Figure 16-2.

Statistical Significance

The null hypothesis for the Exhibit 16-1 data is that the means of both groups are the same on all variables, $H_0: \boldsymbol{\mu}_1 = \boldsymbol{\mu}_2$, where $\boldsymbol{\mu}_1$ and $\boldsymbol{\mu}_2$ are the vectors of means for the two groups. This hypothesis can be tested in several different ways, all of which are essentially equivalent in the two-group case but somewhat different in the case of multiple groups. For the moment, we will be concerned with only one—Bock's version of the largest-root criterion (Bock, 1975)—although we will return to alternative methods later. ("Root" here refers to eigenvalue.)

Bock's procedure converts the largest root v_1 into a generalized F ratio, F_0, and then makes use of special tables (see Appendix Table 6). The largest root v_1 is a ratio of sums of squares. In the univariate case the ratio of sums of squares could be converted into the ratio of mean squares—that is, an F—by multiplying it by

Raw Data Wins GRE	Raw Data Wins Hours	Raw Data Losses GRE	Raw Data Losses Hours
462	39	727	30
643	12	511	8
595	30	493	19
617	43	526	28
548	19	501	12
364	32	586	18
374	13	364	10
605	23	421	3
428	26	417	16
581	39	461	0
608	28	362	20
675	33	482	3
446	32	634	24
501	9	596	4
539	49	580	15
642	30	514	12
455	53	539	10
640	42	447	19
730	33	338	13
475	48	477	24
665	16		
516	29		
568	18		
618	27		
490	23		
677	25		
548	29		
475	22		
527	14		
444	25		

Means	GRE	Hours
Wins	548.53	28.70
Losses	498.80	14.40
Total	528.64	22.98

Exhibit 16-1(a) Raw data and means on the two variables.

$$
\begin{array}{cc} \text{GRE} & \text{Hours} \end{array}
$$

$$
\begin{bmatrix} -66.64 & 16.02 \\ 114.36 & -10.98 \\ 66.36 & 7.02 \\ \vdots & \vdots \\ 198.36 & 7.02 \\ -17.64 & -14.98 \\ -35.64 & -3.98 \\ \vdots & \vdots \end{bmatrix} = \begin{bmatrix} 19.89 & 5.72 \\ 19.89 & 5.72 \\ 19.89 & 5.72 \\ \vdots & \vdots \\ -29.84 & -8.58 \\ -29.84 & -8.58 \\ -29.84 & -8.58 \\ \vdots & \vdots \end{bmatrix} + \begin{bmatrix} -86.53 & 10.30 \\ 94.47 & -16.70 \\ 46.47 & 1.30 \\ \vdots & \vdots \\ 228.20 & 15.60 \\ 12.20 & -6.40 \\ -5.80 & 4.60 \\ \vdots & \vdots \end{bmatrix}
$$

$$
\mathrm{X} \quad = \quad \mathrm{X}_a \quad + \quad \mathrm{X}_w
$$

Exhibit 16-1(b) Score matrix X, in deviations from overall means, for first three cases in each group.

$$\begin{bmatrix} 466{,}486.0 & 13{,}503.1 \\ 13{,}503.1 & 7{,}503.0 \end{bmatrix} = \begin{bmatrix} 29{,}677.0 & 8{,}533.7 \\ 8{,}533.7 & 2{,}453.9 \end{bmatrix} + \begin{bmatrix} 436{,}809.0 & 4{,}969.4 \\ 4{,}969.4 & 5{,}049.1 \end{bmatrix}$$

$$T_t \qquad = \qquad T_a \qquad + \qquad T_w$$

Exhibit 16-1(c) Total, among, and within sums-of-squares-and-products matrices.

$$\underset{\text{Wins}}{\overset{\text{GRE} \quad \text{Hours}}{\begin{bmatrix} 9{,}322.6 & 281.72 \\ 281.72 & 72.15 \end{bmatrix}}} \quad \underset{\text{Losses}}{\overset{\text{GRE} \quad \text{Hours}}{\begin{bmatrix} 8{,}954.5 & -13.21 \\ -13.21 & 126.84 \end{bmatrix}}} \quad \underset{\text{Pooled}}{\overset{\text{GRE} \quad \text{Hours}}{\begin{bmatrix} 9{,}100.2 & 103.53 \\ 103.53 & 105.19 \end{bmatrix}}}$$

Exhibit 16-1(d) Within-group and pooled within-group covariance matrices

$$\underset{T_w^{-1}}{10^{-6} \times \begin{bmatrix} 2.3153 & -2.2787 \\ -2.2787 & 200.2978 \end{bmatrix}} \qquad \underset{T_w^{-1}T_a}{\begin{bmatrix} .049265 & .014166 \\ 1.641656 & .47207 \end{bmatrix}}$$

Exhibit 16-1(e) Inverse of T_w and product $T_w^{-1}T_a$. Note that elements of T_w^{-1} are to be multiplied by 10^{-6}.

$$\begin{aligned}
\text{Characteristic equation:} &\quad (.049265 - v)\,(.47207 - v) - (.014166)\,(1.6416) = 0 \\
\text{Solutions:} &\quad v_1 = .52133;\ v_2 = 0 \\
\text{First eigenvector:} &\quad \mathbf{w}_1 = (.03000, .99955) \\
\text{Sums of squares:} &\quad \mathbf{w}_1{}'T_a\mathbf{w}_1 = 2990.1;\ \mathbf{w}_1{}'T_w\mathbf{w}_1 = 5735.7 \\
\text{Ratio:} &\quad 2990.1/5735.7 = .52133 = v_1
\end{aligned}$$

Exhibit 16-1(f) Characteristic equation, eigenvalues v_1 and v_2, eigenvector corresponding to v_1, and within groups sum of squares $w_1'TW_1$ and among groups sum of squares $W_1'T_aW_1$.

$$\begin{aligned}
\bar{z}_1 &= (.03000)\,(19.89) + (.99955)\,(5.72) = 6.314 \\
\bar{z}_2 &= (.03000)\,(-29.84) + (.99955)\,(-8.58) = -9.471 \\
\hat{s} &= \sqrt{5735.7/48} = 10.931 \\
\Delta &= (\bar{z}_1 - \bar{z}_2)/\hat{S} = [6.314 - (-9.471)]/10.931 = 1.4441
\end{aligned}$$

Exhibit 16-1(g) Means on the discriminant variate $\bar{z}_1$ and $\bar{z}_2$ and estimated within-group standard deviation, $[w_1'\, T_w w_1/(n - g)]^{1/2}$. Delta is then the separation of the means expressed in terms of the within-group standard deviations.

Generalized F:

$$r = |2 - 1 - 2| + 1 = 2.$$

$$s = \min(1, 2) = 1.$$

$$t = (50 - 2 - 2 + 1) = 47.$$

$$F_0 = \frac{47}{2}.52133 = 12.251 \quad p < .01.$$

Individual F's:

$$\text{GRE: } F = \frac{29{,}677/1}{436{,}809/48} = 3.21 \qquad \text{Hours: } F = \frac{2{,}453.9/1}{5{,}049.1/48} = 23.328$$

$$p > .05 \qquad\qquad p < .01$$

Exhibit 16-1(h) Computation of individual and generalized F ratios. The parameters r, s, and t are computed from Equations 16.17, 16.18, and 16.19, respectively. Data for the numerators and denominators of the univariate F ratios are derived from diagonal entries of T_a and T_w of 16-1(c), respectively.

the ratio of the denominator and numerator degrees of freedom. In F_0, we do essentially the same thing although we need to employ the multivariate analogues of the degrees of freedom. The tables are then entered using these two numbers and a third number that represents the number of eigenvalues.

The generalized degrees of freedom are designated r and t and the rank of the matrix (that is, the number of nonzero eigenvalues) is s. These variables are defined as follows:

$$r = |g - 1 - p| + 1 \tag{16.17}$$

$$s = \text{minimum of } (g - 1, p) \tag{16.18}$$

$$t = n - p - (g - 1) \tag{16.19}$$

Then F_0 is defined as

$$F_0 = \frac{t}{r} v_1 \tag{16.20}$$

It can be seen that r reduces to $g - 1$ when $p = 1$ and t then becomes $n - g$. In the multivariate case, t needs to be corrected because additional parameters—the values of elements of the weight vector—are estimated from the data.

In Exhibit 16-1, $F_0 = 12.251$, which is highly significant, well beyond the tabled value for $\alpha = .01$ (see Appendix Table 6). This means we can, with confidence—in fact, with cheerful nonchalance—reject the H_0 stating that the means of the two groups are the same on both variables. It is now legitimate, using the protection levels concept, to test the individual observed variables by means of univariate F

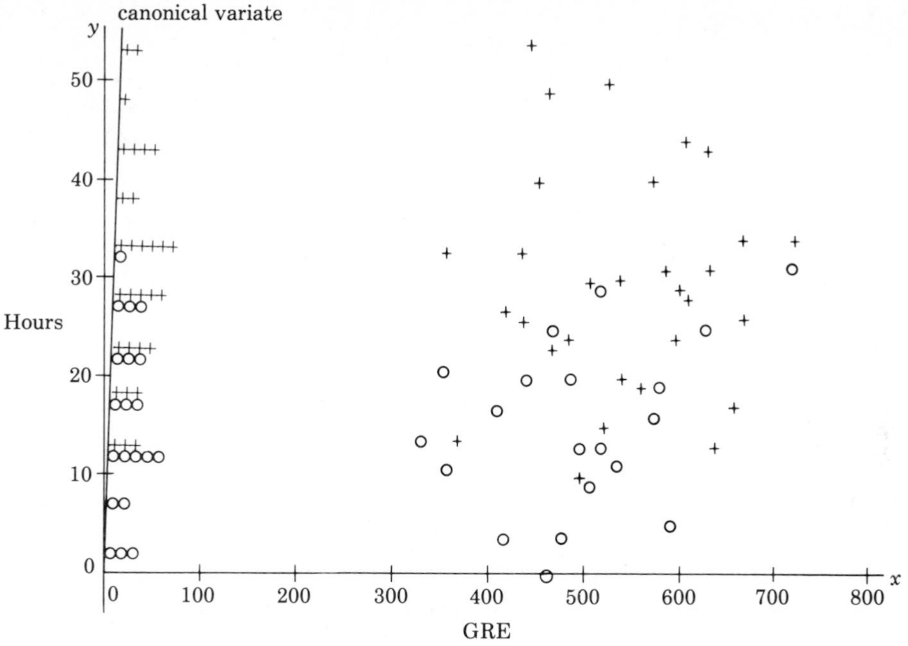

Figure 16-2 Distribution on canonical variate.

tests in order to see if there are clear differences on both variables or only one. This is analogous to testing individual validity correlations (not regression coefficients) in the regression case after finding an overall significant F ratio. The same concepts and cautions apply here, if shaky inferences are to be avoided.

These calculations, along with those for F_0, are carried out in the part (h) of Exhibit 16-1. Note that the F for Hours is highly significant, whereas that for the GRE is not.

The data for Exhibit 16-1 were constructed by random sampling from a population with specific values for the parameters. It may be interesting to compare the sample results to the population parameters just to remind ourselves that the relationship between samples and populations is only approximate. In the population, the means for the Wins were 550 and 30, whereas for the Losses they were 500 and 20. In both groups the within-groups variances were 10,000 and 100, respectively, with zero correlation between the two variables. The population discriminant weights were therefore .24262 and .97012, respectively. Thus our sample weights correspond fairly closely to these, although the difference between the two weights is exaggerated in the sample, and although we have actually accepted the null hypothesis that the one for the GRE is zero.

Multiple Discriminants

So far our discussion has emphasized finding a single best-discriminating function, although we have allowed for the multiple-function possibility in the multiple-group case. In this section, further consideration will be given to the more general case and the meaning of the multiple eigenvalues and their relative sizes.

Geometrically the means of g groups measured on p variables form a figure with g vertices whose dimensionality is $g - 1$ or p, whichever is smaller, in p-dimensional space defined by the variables. The number of nonzero eigenvalues represents the dimensionality of the space, the parameter s of Equation 16.18. For example, if there are three groups, and at least two variables, the means of the groups form a triangle. The overall size of the triangle will reflect the degree of separation of the groups. Assuming the variables are standardized in terms of the within-group standard deviations, the relative size of the two eigenvalues indicates whether the triangle is highly oblique—that is, with the three points nearly in a line (large ratio between eigenvalues)—as opposed to being equilateral, reflecting equal eigenvalues. Strictly speaking, the correlation between the variables should be taken into consideration here as well, but this is the general idea. In the highly oblique triangle, the single large eigenvalue means that there is essentially only one way to take a combination of the variables to differentiate the groups. In the equal-root case, one combination is almost as good as another, and two separate differentiations may be considered simultaneously. These ideas generalize to more groups and more variables. The size of the eigenvalues reflect the overall shape of the s-dimensional figure with g vertices that are formed by the points representing the means. This configuration is perhaps most meaningfully revealed by a plot or plots of the means for the groups expressed in terms of the discriminant variates.

Thus, in the multiple-group case, there is more than one type of discrimination possible. The first eigenvector represents the combination of variables that leads to an absolute maximum for the ratio in Equation 16.15. The succeeding eigenvectors represent smaller maxima also, although they are subject to the restriction of orthogonality from the first. In a sense, we are finding the principal components of the space of the means, although the basis has been transformed to orthonormality with respect to the within-groups vector space. Sometimes it helps to rotate the axes of the space to make an interpretation of the group differences. This will be considered further in our discussion of canonical correlation analysis in Chapter 17.

To some, the major question in discriminant analysis is whether it is possible to discriminate the groups at all. This viewpoint focuses on the overall degree of differentiation rather than on the nature of the different discriminant variates. While it has some validity, it seems more likely that the individual canonical variates and the ways in which they differentiate the groups are also important.

In the multiple discriminant case, several questions may be of interest as far as overall significance testing is concerned, and several different approaches can be taken, one of which is the largest root criterion, expressed as F_0. As always, the approach that we take will depend on the research question, and how we expect the data to operate.

The first consideration is whether there is any differentiation at all, or whether the main interest is in the number and nature of dimensions along which the groups can be reliably differentiated. The second consideration is whether we *expect* only a single dimension of differentiation or more than one.

If our concern is with the overall question of differentiation, then there are two possible tests: Wilks' Λ and the largest root criterion. The test that we use will depend on whether we expect only a single dimension or more than one. If only one is expected, then we use the largest root, presumably through the F_0

procedure (but several alternatives are used in computer programs). Otherwise, we use Wilks' Λ (Wilks, 1932), where Λ is the ratio of the determinants of T_w and $T_a + T_w$:

$$\Lambda = \frac{|T_w|}{|T_a + T_w|} \tag{16.21}$$

Under the null hypothesis $\mu - \mu_g = 0$, the quantity $-\log_e \Lambda[(n - 1 - 1/2\,(g + p)]$ has a distribution of approximately χ^2 with $p(g - 1)$ degrees of freedom. Because the determinant of a matrix is equal to the product of its eigenvalues, Λ reflects the comparison of the eigenvalues of T_w to those of $T_a + T_w$. In fact, it can be expressed in terms of the v_j, as described later in the chapter. The test depends on all the eigenvalues, so it will tend to reject H_0 if there are several fairly large eigenvalues as well as when there is a single very large eigenvalue. There are two test criteria that are closely related to Λ. One is called Bartlett's chi-square approximation, which gives an adjusted value of Λ whose distribution is even closer to χ^2 (Bartlett, 1947). A second approximation makes use of the F distribution and was suggested by Rao (1951). It employs a function of Λ that has the F distribution with degrees of freedom related to g and p. Neither of these tests change the essential nature of the process, and only rarely do they make any difference in the inferences. They are often given by computer programs for discriminant analysis, although we will not specify them here.

The largest root and Wilks' Λ test the overall null hypothesis. However, if we are interested in deciding on how *many* dimensions of differentiation are worth considering, then there are two analogous possibilities. One is to make use of the fact that Wilks' Λ is related to the product of the eigenvalues v_j. Here we would test the product of all of them first, then all but the largest, all but the two largest, and so on, concluding that there are only j significant dimensions as soon as the $j + 1$st is not significant. That is, we test

$$q_j = -m \sum_{k=j}^{s} \log_e \lambda_k \tag{16.22}$$

where

$$\lambda_k = \frac{1}{1 + v_k} \tag{16.23}$$

and

$$m = n - 1 - (g + p)/2 \tag{16.24}$$

We evaluate q_j as χ^2 with $(p - j + 1)(g - j)$ df. Failure to reach significance is taken as an indication that the remaining eigenvalues are zero in the population, which means they do not represent reliable dimensions of discrimination.

The alternative is to successively apply the largest root criterion to each eigenvalue in turn, adjusting the degrees of freedom accordingly. A Monte Carlo study shows that both procedures tend to give similar results, and that both are quite conservative, tending to give too few significant dimensions rather than too many. The q_j procedure (Equation 16.22), which tended to be somewhat less conservative, is preferable.

Thus we can make an omnibus test of the hypothesis that no discrimination is possible using either Wilks' Λ or the largest-root-criterion-generalized-F. We lean toward using the former test if we suspect diffuse discrimination along several dimensions. The latter test is used if we expect differences to be concentrated on only one. We decide on the number of dimensions of discrimination by successive applications of Wilks' Λ criterion to each root in order of size. One or more of these tests is part of the typical statistical package's discriminant procedure.

Example

Using the same data as before, we will illustrate multiple discriminant function analysis by subdividing the Wins into several subgroups on the basis of how long they took to finish their degrees. In the original data matrix [Exhibit 16-1(a)], the first ten students (rows) of the Wins group are those who finished in less than four years, while the next ten students finished in four or five years, and the last ten students in more than five years. These three subgroups are now designated I_1, I_2, and I_3, respectively, and their means and standard deviations are shown in Exhibit 16-2(a). Also shown are the T_a and T_w matrices that result from this finer breakdown of the grouping of the data of Exhibit 16-1. Because there are four groups and only two variables, the rank of T_a is determined by the variables, so there are two eigenvalues and vectors.

Part (b) of this exhibit shows the two eigenvalues and their corresponding eigenvectors. Note that one eigenvalue is somewhat larger than it was in the two-group analysis, whereas the other is very small. The first vector ($\mathbf{w}_1$) is quite similar to the one for the two-group analysis; both entries are positive and the entry for Hours is much larger. The second vector ($\mathbf{w}_2$) has one positive and one negative entry, which is virtually guaranteed since it must be orthogonal to the first vector, within groups. Remember that which entry is positive and which negative is arbitrary because if $\mathbf{w}_2$ is an eigenvector, $-\mathbf{w}_2$ must also be an eigenvector. The difference in the absolute magnitudes of the weights in $\mathbf{w}_2$ reflects the fact that the GRE has a larger variance.

In part (c) we see the within- and among-group sums of squares on the canonical variates, noting that the correlation between variates must be zero within groups. This must be true among groups also as one of the properties of the solution. The ratios of the within-groups and among-groups *ss* are found to equal the roots v_1 and v_2, as they should.

Next we examine the means on the canonical variates:

$$\overline{Z} = \overline{X}\,W \tag{16.25}$$

where W is p by s, $\overline{X}$ is g by p, and $\overline{Z}$ is g groups by s variates. In order to get a better feel for the relative sizes of the differences, these means can be divided by the within-group standard deviations; the resulting means are designated $\overline{Z}^*$. These quantities are shown in part (d) of the exhibit and illustrated in Figure 16-3.

Part (e) examines the question of significance. The question of overall differences is tested by means of the largest root criterion, where H_0 is resoundingly rejected. In part (f) the first test shown is also an overall test using Wilks' Λ.

	GRE	Hours
I_1	521.7	27.60
I_2	571.1	35.70
I_3	552.8	22.80
Loss	498.8	14.40
All	528.64	22.98

$$\underset{T_a}{\begin{bmatrix} 42{,}152 & 10{,}157 \\ 10{,}157 & 3{,}304.1 \end{bmatrix}} \qquad \underset{T_w}{\begin{bmatrix} 424{,}334 & 3{,}346.3 \\ 3{,}346.3 & 4{,}198.9 \end{bmatrix}}$$

Exhibit 16-2(a) Group means when the In group is subdivided into three subgroups in terms of length of time to finish, the first 10 subjects from group I now making up I_1, the next 10, I_2, and the last 10, I_3. This additional variation among groups leads to the new values given here for T_a and T_w.

$$\text{Eigenvalues of } T_w^{-1}T_a\text{:} \quad v_1 = .82884;\ v_2 = .024609$$
$$\text{Eigenvectors of } T_w^{-1}T_a\text{:} \mathbf{w}_1' = .02384, .99971$$
$$\mathbf{w}_2' = -.30279, .95304$$

Exhibit 16-2(b) The two eigenvalues of $T_w^{-1}T_a$ and their corresponding eigenvectors.

$$\begin{bmatrix} 3{,}810.4 & 0 \\ 0 & 1{,}003.7 \end{bmatrix} \qquad \begin{bmatrix} 4{,}597.3 & 0 \\ 0 & 40{,}787 \end{bmatrix} \qquad \begin{aligned} \frac{3{,}810.4}{4{,}597.3} &= .82883 = v_1 \\ \frac{1{,}003.7}{40{,}787} &= .024608 = v_2 \end{aligned}$$

Exhibit 16-2(c) $W'T_aW$ is the among-group sums of squares and products on the two discriminant variates and $W'T_wW$ is the corresponding within-group data. The latter must be diagonal. The ratios of the diagonal elements of $W'T_aW$ to those of $W'T_wW$ are the same as the eigenvalues v_1 and v_2.

	$\bar{X}'W_1$	$\bar{X}'W_2$	$\frac{1}{s_1}\bar{X}'W_1$	$\frac{1}{s_2}\bar{X}'W_2$
I_1	40.029	131.66	4.004	4.422
I_2	49.305	138.90	4.932	4.665
I_3	35.979	145.74	3.599	4.894
Loss	26.297	137.31	2.630	4.611
	$\bar{Z} = \bar{X}W$		$\bar{Z}^*$	

Exhibit 16-2(d) Group means on the discriminant variates where $\bar{Z} = \bar{X}W$. The same means are expressed in terms of the within-group standard deviations, where $\bar{z}^*_{gm} = \bar{z}_{gm}/s_m$, $s_m^2 = t_m/46$, and t_m is a diagonal element of $W'T_wW$. The $\bar{z}^*_{g1}$ are plotted against $\bar{z}^*_{g2}$ in Figure 16-3, which shows the large variation in $\bar{z}^*_{g1}$ as compared to $\bar{z}^*_{g2}$.

$$F_0 = \frac{t}{r} v_1 = \frac{45}{2} .82884 = 18.65; \alpha \ll .01$$

$$r = |g - 1 - p| + 1 = |4 - 1 - 2| + 1 = 2$$

$$s = \min(g - 1, p) = 2$$

$$t = n - g - p + 1 = 50 - 4 - 2 + 1 = 45$$

Exhibit 16-2(e) Largest root test applied to data in 16-2(d). Significance is assessed through Appendix Table 6.

j	v_j	$\log_e\left(\frac{1}{1+v_j}\right)$	$-\sum_{k=j}^{s} \log_e \frac{1}{1+v_k}$	q_j	df	α
1	.82884	−.60368	.62799	28.888	6	.001
2	.024609	−.024311	.024311	1.118	2	.70

$$m = n - 1 - \tfrac{1}{2}(g + p) = 50 - 1 - \tfrac{1}{2}(4 + 2) = 46$$

$$df = (p - j + 1)(g - j)$$

Exhibit 16-2(f) Use of Wilks' Λ as an overall test of discrimination and of the number of dimensions of discrimination. Both of the q_i are computed according to Equations 16.22, 16.23, and 16.24. The fact that q_i is highly significant indicates that the means of all groups are not equal on all variables, which is parallel to the largest root test in 16-2(e). The fact that q_i (see Equation 16.22) is not significant supports the conclusion that only one dimension of discrimination can be reliably defined.

$$F_{\text{GRE}} = \frac{42{,}152/3}{424{,}334/46} = 1.523 \qquad F_{\text{Hours}} = \frac{3{,}304.1/3}{4{,}198.9/46} = 12.066$$

Exhibit 16-2(g) Univariate F's for the two original variables.

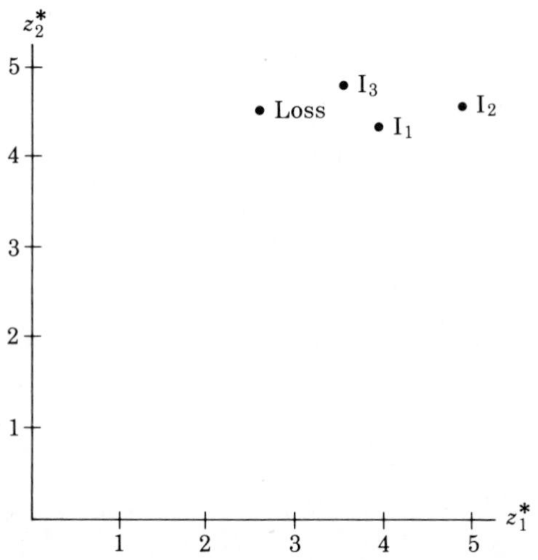

Figure 16-3 Standardized means on the two discriminant variates.

Rejecting H_0 is a rejection of the hypothesis that all the v_j are zero, which is equivalent to rejecting the hypothesis that there is no among-groups variation. The second test in part (f) concerns the hypothesis that all linear combinations of the group means orthogonal to the first are zero. That is, there is only one dimension of discrimination. This null hypothesis is accepted, since the chi-square of 1.118 is far below the value needed for significance at even the .50 level. Thus only the kind of discrimination corresponding to $\mathbf{w}_1$ is reliably estimated.

Having rejected the overall H_0, we can go back and make univariate F tests of the individual variables, shown in Exhibit 16-2(g). As in the two-group case, it can be seen that the F test for Hours is significant whereas the test for the GRE is not. Having found the significant univariate F, we could apply the protection levels logic one step further by testing differences between pairs of groups. This step is not shown here.

The means of the four groups on the two discriminant variates are standardized by the within-group variances [see Exhibit 16-2(d) and Figure 16-3]. Clearly, the differentiation in the horizontal direction (first variable) is much more than in the vertical (second variable).

Significance of Weights

Recall that in multiple regression there were two ways of measuring the importance of a variable in predicting a criterion. One was the significance of its zero-order correlation and the other was the significance of its regression weight. The zero-order correlation reflects its relevance to the dependent variable in isolation, whereas the regression weight reflects its importance when the influence of all the other variables is taken into account, the amount of prediction that it adds to all the other variables. The same sort of consideration applies with respect to discriminant analysis, although the logic is more difficult to follow because of the two-sided nature of the problem. The variables that discriminate best may depend on which subset of groups we are discriminating. Trying to keep all these aspects in mind is likely to produce headaches and irritability, so we begin with simplified cases.

First, we'll consider the two-group case. There is only one discriminant variate, and the weights for it are directly proportional to the regression weights that would result if the dichotomous group-membership variable were treated as the dependent variable in a regression problem. All the concepts and methods we learned concerning inferences and conclusions about regression weights and the importance of variables apply directly in this case.

Now suppose that there are more than two groups. It may help to ease into this case by considering what goes on when there is only a single dependent variable. Here, we would simply be doing a one-way anova, and all the things that were discussed about comparisons in one-way anova will apply. We may use single degree of freedom contrasts, or the concept of protection levels and make whatever comparisons seem to be of interest once an overall significant F is found. Or, if the comparisons have a post hoc flavor, we may fall back on a Scheffé approach.

Let us try to see how the same ideas would work in the case of multiple groups *and* multiple variables. First, we'll consider the question of the importance of an individual variable to the discrimination. Recall that one of several equivalent ways of looking at the test of significance of a regression coefficient is in terms of an F test of the significance of the improvement in R^2_{all} over $R^2_{\text{all but }j}$. A directly analogous procedure can be used in discriminant analysis. If a variable is irrelevant to the discrimination once all other variables have been taken into consideration, the ratio in Equation 16.15 will be about the same whether or not it is included in the battery. However, in practice, there must be *some* improvement in the ratio if an additional variable is included; the question is whether the improvement in the ratio could be expected by chance. If the improvement is large, then we conclude that the variable is relevant, meaning its weight is not zero in the population. In this way, we can assess the significance of the contribution of each variable. This is in effect what we do when we assess the significance of a weight. Here, the test is often called a *stepdown* test.

The ideas we discussed in stepwise multiple regression also apply to the variables in a discriminant analysis. That is, we can do a hierarchical analysis specifying the order in which variables or sets of variables are to be used in developing the discrimination. The significance of the improvement provided by each set of variables is assessed at each step in the analysis, and the process is cut off once a nonsignificant improvement is found so that analysiswise alpha is controlled. Similarly, discriminant analysis can take place in the variable selection mode, where the computer selects at each step the variable that will most increase the ratio in Equation 16.15. The extensive inferential caveats that were introduced with respect to variable selection in multiple regression apply here even more strongly since the search for a variable to improve discrimination also involves the opportunity to choose any comparisons whatsoever, including any subset of groups against any other. We will not go into the mechanics of these procedures here, assuming that some computer program will provide the information and that our task is only to interpret it. (Note that the term is *only* to interpret, not *merely* to interpret.)

We can think of adding new variables to the pool as a geometrical process. The generalized distances between the means of the groups increase as new variables are included, and the degree to which the ratio (Equation 16.15) increases reflects how much the average distance between means is increasing. Note, however, that the *shape* of the constellation of points may change also as new variables are added. The patterns of which groups are most different will alter—perhaps only slightly, or perhaps substantially. Thus not only the degree but the pattern of differentiation may change as new variables are added. This corresponds to the way in which the weights may change if new groups are added to the analysis or if one of the groups is subdivided. Thus, the specific way in which the groups are differentiated depends on the variables that are used and their roles. These roles change as new variables are added to different groups.

Generalized Confidence Intervals

The set of conclusions that we might formulate, or the set of hypotheses we might want to test, may be very broad in this context. Therefore, it is desirable to use one single system to investigate them all. For example, we would probably like to know wherein the differences lie, whether they involve several different contrasts on a single dependent variable, or differences on a number of combinations of variables. In all these cases, a "combination of variables" may be just one variable taken by itself.

The approach that we can take here is based on a system developed by S.N. Roy (1957). This system can be expressed in several different ways, but the simplest way would be a generalization of the approach we used to test all conceivable hypotheses about regression coefficients. In the discriminant analysis case, the focus is on the groups-by-variables matrix of means. Suppose **c** is a contrast vector, corresponding to one particular comparison among the groups. We would compose the entries in the same way as in the anova context, making sure that they represent the contrast in terms of the qualitative nature and are also orthogonal to the unit vector, that is, taking group sizes into account. For example, we would compare the three groups of Wins with the Losses by weighting the means by 1/3, 1/3, 1/3, −1, respectively. When this is done, the result would show the effect of the contrast with respect to each of the variables. That is, if we multiply the g by p matrix of means by a g by 1 vector we would get a 1 by p vector reflecting the contrast on each variable. We could then divide each of these by some term representing the standard error of the contrast on that variable and look up the result in some appropriate table to reach a conclusion about the contrast on that variable.

Note that the contrast's effect on each variable could also be weighted. This would reflect the effect of our contrast on a combination of variables. We can even formalize the process of "picking out one variable to look at" by using a vector with a 1.0 in the position corresponding to the variable being examined and zeros everywhere else. There are only two variables in the example, so here such a vector could be (1, 0) or (0, 1). Thus we can represent the whole process by pre- and post-multiplying a matrix by vectors:

$$\phi_{ca} = \mathbf{c}_1'\overline{\mathrm{X}}\mathbf{a}_1 \tag{16.26}$$

where $\mathbf{c}_1$ is the row vector for the contrasts, $\mathbf{a}_1$ is a vector for pulling out one variable to look at, and ϕ_{ca} is the effect of using the contrast represented by $\mathbf{c}_1$ on the variable represented by $\mathbf{a}_1$.

We could have various $\mathbf{a}_1$'s just as we can have various $\mathbf{c}_1$'s. So far, we have limited the choice among **a**'s to those with a single entry of 1.0 and the rest zero, but in principle, the choice is just as wide for $\mathbf{a}_1$ as it is for $\mathbf{c}_1$. The $\mathbf{a}_1$ vector just represents one way of making a linear combination of variables—we could investigate the effect of the contrast on *any* linear combination of the variables. If we took the weights corresponding to one or the other of the canonical variates, the contrast will be tested with respect to that canonical variate, rather than one of the observed variables.

The weights in $\mathbf{a}_1$ define a new dependent variable that is a linear combination of the observed variables. To find the within-groups sum of squares on that variable, we would pre- and post-multiply the T_w matrix by $\mathbf{a}_1$. Testing the significance of the contrast on the combination involves dividing the combination by its degrees of freedom and then dividing the effect by a factor that takes account of the coefficients used to define the contrast. The specifics of the latter will be dealt with later in this chapter.

The last piece of information needed is how to judge the significance of the contrast. The conclusion should use the concepts we have developed in this book for avoiding egregious capitalization on chance while still allowing reasonable power to test what is an a priori hypothesis. If the contrast truly is a planned one, and the combination of dependent variables truly is defined a priori, then we can treat the test defined by this contrast as if it were a univariate case.

This is rarely defensible in the case of discriminant analysis, where the clustering of groups together to define the contrast and the selection of what single variable or combination of variables to use is likely to be done on the basis of how the data come out. We must at least do the omnibus test, represented by Λ

or by F_0, and reject H_0. Then, as in the example, we can test the differences on each variable by means of the univariate F test and find the significance of each variable in the overall discrimination by the stepdown procedure.

What we need to consider now is the case in which the inferences are at least partially post hoc. Here we use a highly conservative procedure that involves the generalized F. It sets the probability at α that any one of every *conceivable* contrast among groups carried out on every *conceivable* combination of variables will lead to rejection of the corresponding null hypothesis when all the means are equal on all the variables in the population. Naturally, we would make the usual assumptions about homogeneity of variance-covariance matrices also.

One possible combination of variables is that given by the discriminant weights. One conceivable contrast among means is one that corresponds to groups' differences on that variate—that is, their deviations from the grand mean on that variate. The procedure has to make the probability α that the null hypothesis will be rejected if a sneaky investigator just "happens" to think of those weights and "happens" to think of that contrast. To protect ourselves against these investigators, we have to use the largest-root-generalized F statistic to define the critical value of a contrast-combination. Not only is the criterion used for this test, it is used for all other tests involving all other combinations and contrasts as well, even if they are largely a priori. That is what makes the procedure so conservative.

To assess the significance of a contrast, we need to take into account several things. First, we need to account for the amount of freedom involved in defining it. This depends on the number of groups and the number of variables, since the more there are of each, the greater the likelihood of finding *some* large distance in the space. Second, we need to take account of the accuracy of the estimation of the means and of the variances and covariances among the variables. We also have to consider the within-groups variance on the composite variable, and the magnitude of the coefficients used to define the contrast. This is a complex process, but it can be done if we take a deep breath before we start and go about it in a systematic way.

In the long run, the simplest approach to assessing significance is to use its parallel in the confidence interval. Recall that finding a confidence interval that does not contain zero is equivalent to finding that the corresponding parameter estimate was significantly different from zero. Thus we assess the significance of a particular contrast on a particular combination by computing the contrast and then constructing the confidence interval around it. If the interval contains zero, we accept null hypothesis.

The overall formula used is

$$\mathbf{c}_1'\overline{\mathrm{X}}\mathbf{a}_1 \pm \left[\frac{(n-g)r}{t} F_0(\mathbf{a}_1'\mathrm{S}_w\mathbf{a}_1)\left(\sum_{h=1}^{g} c_h^2/n_h\right)\right]^{1/2} \tag{16.27}$$

where r and t take their definitions from Equations 16.17 and 16.19, respectively, and F_0 is the generalized F with appropriate degrees of freedom and α level. The middle term involving $\mathbf{a}_1'\mathrm{S}_w\mathbf{a}_1$ computes the variance of the combination, and the final term involving c_h^2 takes account of the coefficients and group sizes. This final term may be translated into the variance of a linear combination of the dummy variables that define the group membership. The final term is in fact $\mathbf{c}_1'(\mathrm{X}'\mathrm{X}')^{-1}$ $\mathbf{c}_1$, where X is the complete n by g matrix of scores on dummy variables.

In an empirical case, then, we have a set of coefficients that defines the particular contrast and a second set of coefficients that defines the combination of variables that is used to assess the effect of the contrast. The contrast can be anything from the difference between a single pair of groups to the mean deviations on the first canonical variate. Likewise, the combination can be anything from the (1, 0, 0, . . .) vector that corresponds to a single dependent variable to the weights in $\mathbf{w}_1$ itself. The procedure here guarantees that, no matter which of these we use, the probability is α that one or more combinations do not contain zero when all the null hypotheses are true.

The cost in terms of statistical power is correspondingly large, becoming larger as the number of groups or variables increases. Thus, where defensible, we should use the less conservative protection-levels ideas. Otherwise, we may be in the position of the person who has so much life insurance that his children go without shoes. Nevertheless, when the discriminant analysis is undertaken with a purely exploratory motive, we are probably wise to employ the above procedure in deciding which groups really differ and what variables and combinations of variables they differ on.

One note of caution may be added. This procedure is difficult to use when assessing the independent contribution of individual variables. For example, if we wish to find out whether a variable has a nonzero weight in the population, we should use the procedure described in the previous section, in which the discrimination involving all variables is compared to the discrimination involving all but the specific variable. We cannot simply substitute zero for a variable's weight in $\mathbf{a}$ in order to find out whether the confidence interval contains zero. Such a test is conditional on the values for the weights being used for the other variables. This same reasoning was made with respect to regression weights.

Classification of Observations

So far we have described discriminant analysis as a technique for testing the hypothesis that several groups can be differentiated and for exploring the nature of that differentiation. This is the most common interest of those who use it in the social sciences. However, discriminant analysis may also be used for classifying observations. Textbook treatments on this subject usually include the example used by R. A. Fisher in his original presentation of the topic—the classification of several different species of iris.* The purpose of the weights in this example was to establish a score or vector of scores that would allow botanists to automatically classify a plant of unknown species into its correct niche, and thus to minimize the probability of misclassification.

Under the assumptions of multivariate normality and homogeneity of within-groups variance-covariance, this can be done using the discriminant weights that maximize the ratio in Equation 16.15. The canonical variates furnish the basis for the classification. In addition, we need a way of establishing cutoff values within the discriminant space such that an object on one side is classified into one group while an object on the other side is classified into another. The only other piece of information necessary to do this is the relative sizes of the groups, because this influences the probabilities on an a priori basis.

---

*R. A. Fisher (1936). The use of multiple measurements in taxomonic problems. *Annals of Eugenics, 9,* 238–49.

First, we'll consider the unidimensional case. Suppose that there are two equal-sized populations measured on a single variable. Theoretically, the probability distribution might look like that shown in Figure 16-4(a). Suppose that the two populations in Figure 16-4 are aardvarks and anteaters, and that the variable is snout length. As we move along the continuum from short snouts—a very relative term here—to longer ones, the odds given by the height of the curves are initially very much in favor of the animal with the corresponding snout-length being an aardvark because the aardvark curve is higher. Then at some point the two curves cross, and the odds favor the animal being an anteater as indicated by the higher curve. If the distributions are of equal size and the variances are equal, the point where the curves cross will come halfway between the two means. If, on the other hand, one species is much more common than the other, as indicated by the relative heights of the curves throughout their ranges, then the crossover point moves away from the mean of the larger curve, as shown in Figure 16-4(b). If the disparity in group sizes is large enough, the odds may favor one group

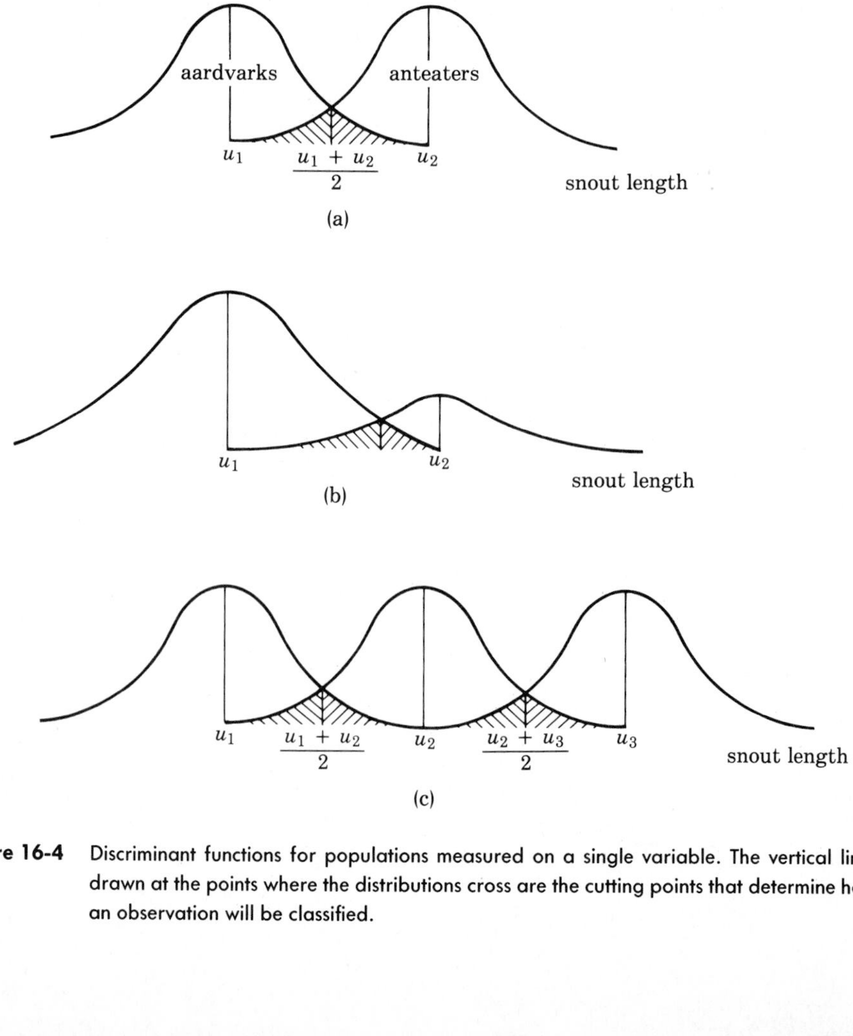

Figure 16-4 Discriminant functions for populations measured on a single variable. The vertical lines drawn at the points where the distributions cross are the cutting points that determine how an observation will be classified.

throughout the range. For example, aardvarks live in Africa and anteaters in South America—except for a few in zoos—so we should classify any animal found in Africa as an aardvark, no matter how long the snout.

For three species, there would be two crucial values corresponding to the two points at which the lowest and highest distributions cross the middle one, as shown in Figure 16-4(c). Again, the crossing points would depend on the sizes of the groups, but for equal-sized groups, the cutoff points are half-way between the adjacent means.

If there is more than one variable, the situation becomes somewhat more complicated, unless the variables are combined into a single canonical variate. That is, the continuum can be a single derived canonical variate rather than an observed variable. Once there is more than one canonical variate, however, the complication can no longer be dodged. Instead of dividing a single continuum by means of points on it, we are dividing a *space* into *regions*. It takes lines to divide a plane into regions, and in general, to divide up an s-dimensional space requires an $s - 1$ dimensional hyperplane. (If this sounds confusing, then just try to accept the general idea of dividing the space up into territories such that there is one territory for each type.) If the values of an observation fall into some territory, then it is more likely to belong to the type for that region than to any other type. The process is illustrated in Figure 16-5 for the set of four student groups

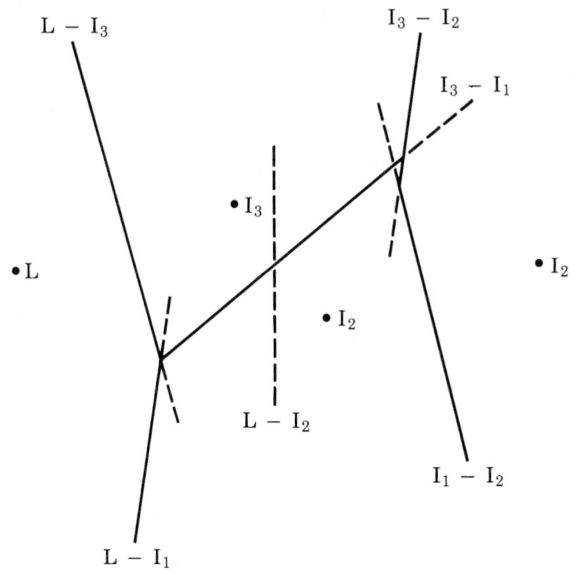

Figure 16-5 Regions belonging to the four student groups of Exhibit 16-2 in discriminant space, assuming equal populations. The lines are drawn by connecting the points corresponding to a pair of means and then making the perpendicular bisector of the line. The heavy lines define the boundaries; their dashed continuations move into areas where a third mean is closer than either. The line L-I_2 is entirely within the territories of I_1 and I_3 so it does not become part of any boundary.

described in Exhibit 16-2. Note that because the second eigenvalue is not significant, it may be counter-productive to use two dimensions, but we do so anyway for purposes of illustration. We will leave the general equation for the discriminant boundaries to the computer.

Modern computer programs include the option of classifying the observations in the sample according to the discriminant functions. That is, they "predict" which group the observations should belong to. This classification can then be compared to the actual known group membership, which is analogous to comparing predicted and actual scores on the dependent variable in multiple regression.

This classification system is also subject to the same sorts of biases. It takes advantage of every chance fluctuation in the data to draw the boundaries to its maximum advantage. Thus, depending on the number of variables and groups relative to the number of observations, the rate of successful classification in the sample can be misleadingly high. There may be an instance in which the classification is almost perfect in the sample, even though the function is not even significant with moderate numbers of variables and group sizes. In any event, if we intend to use the discriminant functions as a means of classifying *future* observations (predicting, for example, whether a graduate applicant will complete a degree within a reasonable time) then we are much more interested in the accuracy of those predictions. After all, with the current data, we do not need to predict what groups the individuals belong to. We already know what they are.

Thus we need a means of estimating how accurate the functions will be in classifying future observations. Most of the formulas for doing this are very cumbersome. More importantly, they do not take account of the failure of the assumptions. These assumptions are much more crucial when predicting *actual* classifications than they are when testing hypotheses about equality of mean vectors. They also seem likely to be violated. Since nearly every example based on real data seems to show clear evidence of heterogeneity of variance on the canonical variate, these formulas are not likely to provide very realistic results.

The alternative recommended here relies on the services of that patient workhorse known as cross-validation. A discriminant analysis should take place by dividing the groups in half, using one half to develop the discriminant functions. These functions are then used to predict the group membership of the other half. When we cross-classify by predicted and actual group membership, we can assess the effectiveness of the functions by the proportion of cases where the predicted and actual memberships coincide. Naturally, the double cross-validation strategy can be employed using a 2/3 versus 1/3 split, as well as other divisions. In assessing expected shrinkage, the general considerations in discriminant analysis are much the same as they are in multiple regression. The more variables and groups and the fewer cases per group, the greater the deterioration in accuracy of classification will be when we go from the sample used to develop the function to the cross-validation sample.

Some computer programs offer a form of the jackknife for use with discriminant analysis. The analysis is carried out n times, each time leaving out one of the observations. Group membership is predicted on the basis of the data from the $n - 1$ others. Then the comparison of the predicted and actual group membership for each observation can be carried out. This furnishes a realistic estimate of the accuracy of classification. Clever programming of the procedures avoids much of the apparent cost of inverting the T_w matrix n times.

Failure of Assumptions

At one or two points in the preceding discussion, we have alluded to the assumptions underlying discriminant analysis and the possible effects of their failure. The next few paragraphs will address what these assumptions are as well as their importance. First of all, we will distinguish between discriminant analysis used as a sample-descriptive technique and as an inferential technique.

We introduced discriminant analysis as a way of combining variables in such a way as to separate two groups most clearly and completely. This is not exactly what the analysis does, however. Instead, it finds a vector of weights that maximizes among-groups variance relative to within-groups. Although this has more or less the same effect, it is a somewhat different criterion from correct classification. It is sometimes possible to find a way of combining the variables that gives a lower probability of misclassification than the discriminant function does, even in the sample. For example, if the groups differ in their variances and covariances as well as their means, this is particularly likely to be the case. As with other techniques, we should always make a plot of the data to see what seems to be going on. Thus, even at the sample level, it is important to understand that the procedure does something quite specific, which may not always correspond precisely to what the investigator has in mind.

Such considerations have even more force in the case of inferences from the sample to the population, although the remedies are not so straightforward. Although we may be able to *see* how the data departs from the assumed forms, it is virtually impossible to state just how these deviations should affect our inferences and the confidence we have in them. A few things may be said about this, however.

Consider, for example, the effect of heterogeneity of variance-covariance matrices on inferences about means. Here, it seems desirable to adopt the multivariate generalization of the "unpooled" variance estimate, as in Equation 16.11. Instead of simply combining the T_w matrices from different groups, we can divide each matrix by its respective degrees of freedom before adding them together. When we multiply the result by the geometric means of the group sizes this produces an adjusted total T_w matrix. The procedure is directly analogous to that used in one-way anova to protect against heterogeneity. The details of this procedure will not be given here; they are not generally available in computer program packages, unfortunately.

The normality assumptions that underlie discriminant analysis are usually not considered to be very important, but there are limits here also. Often the variables that we consider using in a discriminant analysis are a mixed bag—perhaps some are dichotomous, or nearly so, or have modes at the end of the range. These variables depart so radically from the assumed normality, and their regressions are so likely to be nonlinear, that we can question the propriety of their inclusion. Here again, we are more likely than not to include them. (Who are we to propose moral rearmament in the statistical field?) If we really care about what we are doing, we'll be sure to have adequate sized samples—a minimum of ten in the smallest group—and cross-validate.

A well-known procedure that can be used to statistically decide about heterogeneity was introduced by Bartlett (1947) and refined by Box (1949). If several matrices are similar, then their average will be similar to them. This similarity will be reflected by the determinants of the average and the individual matrices.

Thus the generalized Bartlett's test for homogeneity of covariance matrices is based on the statistic

$$\chi^2 = -m[\Sigma(n_h - 1) \log_e |S_h| - (n - g) \log_e |S|] \tag{16.28}$$

where S_h is the covariance matrix within group h, and S is their average: $S = \Sigma(n_h - 1)S_h/(n - g)$. The constant m depends on the group sizes and the number of variables:

$$m = \left[1 - \left(\Sigma \frac{1}{n_h - 1} - \frac{1}{n - g}\right)\left(\frac{2p^2 + 3p - 1}{6(p + 1)(g - 1)}\right)\right] \tag{16.29}$$

Based on the assumption that all g covariance matrices are drawn from the same population, Equation 16.28 is distributed as χ^2 with $p(p + 1)(g - 1)/2$ df.

As noted in earlier discussions of homogeneity of variance, this test is sensitive when groups are large, but the inferential consequences of heterogeneity are greatest when they are small. If the null hypothesis is rejected when n is small—say 10 in the smallest group—then we have reason to worry about the validity of the test in Equations 16.20 or 16.21. If, on the other hand, homogeneity is accepted with large n—say 30 or more in the smallest group—then there is little cause for concern. Under other circumstances, the results are ambiguous in this regard.

Note, however, that the fact that covariance matrices are different under different conditions or in different groups may be of considerable interest in itself. Thus Bartlett's test is perhaps most useful when it focuses directly on the hypothesis it is testing: Are the covariance matrices equal? Component analysis or factor analysis of the matrices and subsequent comparison of the loadings may then be of considerable substantive interest.

Another aspect of assumption-failure is its effect on the probability of misclassification. The estimates that the computer furnishes are almost certainly based on assumptions that the populations from which the groups are drawn are multivariate normal and differ only in their mean vectors. Indeed, it is hard to see how such estimates can be made otherwise. Therefore, any substantial heterogeneity will make these estimates of classification accuracy erroneous. The only robust way of obtaining these estimates is through cross-validation. Morrison (1976, pp. 236–45) has a general discussion of classification probabilities.

MULTIVARIATE ANALYSIS OF VARIANCE

Applications and Motives

The difference between discriminant analysis and multivariate analysis of variance (manova) is largely based on the roles of the variables. In discriminant analysis, we are using the variables to differentiate the groups, whereas in manova we are concerned with whether the independent variables represented by group membership will have an effect on the continuous variables.

There seem to be two or three motives for doing a manova. One motive is genuinely multivariate; an investigator has several dependent variables and wishes to observe the effects of the treatments on them simultaneously. For example, she may record EEG waves from several sites and at several frequencies. Then in the experiment, she varies the conditions using different stimuli or drug treatments, and evaluates their effect on all the brainwaves. Alternatively, a market researcher asks a sample of consumers to rate several products on different descriptive scales and to question whether the products differ on the scales. An educational psychologist assesses the effect of a change in curriculum on the performance of students on a variety of tests and other variables. Or a social psychologist varies the conditions under which groups carry out a standard task in order to find out the effects on the frequency of several different kinds of communication within the group, and upon the feelings that the workers have about the task. In each of these cases, the experimenter is thinking about the dependent variables jointly, and evaluating a multivariate outcome.

In other applications of manova, the experimenter has a special interest in his dependent variables. He is interested in them individually, and if left to his own devices, would simply test them separately. He becomes concerned, however, about the effect of multiple effects on his experimentwise α level and decides to make an omnibus test of all the variables at once before going on to test them individually. For example, in an information processing experiment, the investigator wants to know whether the experimental conditions affect speed of response *and* number of errors. A sociologist wants to know if different social groups differ in their crime rates and in their frequencies of divorce and in their usage of consumer credit. A consumer psychologist wants to know if buyers of one product differ from buyers of another on each of several different socioeconomic characteristics. In each of these cases the investigator is mainly interested in multiple analyses of variance on the individual variables but wants to control the probability of rejecting one or more of the numerous null hypotheses when they are all true. Thus, this second purpose of manova is mainly one of controlling the experimentwise α level when the investigator has several dependent variables that are individually of interest.

Still a third application of manova may be distinguished, although it may be considered a special case of the first category. The most common type of example in this category is tracing the course of learning across trials under various conditions. Here, the investigator may be interested in finding differences in the learning curves for the different groups or conditions. It was formerly common to treat this as a "repeated measures analysis of variance," examining the main effect for groups, which would reflect differences in the overall levels of the curves for the different groups, and the groups-by-trials interaction, which would reflect differences in the shapes of the curves. This method poses some difficulty, however, because the investigator is assuming that the variance on each trial is the same and, even more bothersome, that the correlations between pairs of trials are all equal. The analysis requires not only that these quantities be equal across groups but also that they be equal within groups across trials and pairs of trials. Otherwise, the model we are testing is false, but false in a way that is different from simply saying the means are different. More and more, investigators are switching from repeated measures analysis of variance in this context and using manova instead. While the repeated measures analysis is more powerful if its assumptions are met, manova is more appropriate in most cases. Still, we need to be clear on how to interpret manova and how it corresponds with the various hypotheses.

These are the three major types of application of manova. We may wonder what all the fuss is about, since we could simply restate each of these applications as a discriminant analysis problem. Indeed, this is often possible, as noted earlier. There are two reasons for dealing with manova as a separate topic, however. One is simply a matter of familiarity or mental set. People who are used to thinking in analysis of variance terms naturally find the transition to manova easier than the seemingly new idea of discriminant analysis. The other reason is that it usually makes sense in the case of a more-than-one-way design to be able to break down the differences among cell means in the anova way. That is, it makes sense if we want to look at various main effects, simple, and higher order interactions. It is true that here, as well as in the univariate case, these various effects correspond to contrasts among groups. They can be formulated, as described at the end of the previous section. Indeed, these contrasts may well reflect more directly the question that the experimenter has in mind. Nevertheless, because of the familiarity of the anova models and the way in which they allow the breakdown of sums of squares in fairly routine ways, there is considerable utility in presenting the basic ideas of multivariate analysis of variance.

Sums of Squares and Products

We have already covered the breakdown of the total sum of squares in anova into various components. We have broken them down into within and between sums of squares in the simplest case, and into further subdivisions as the design becomes more complex. Null hypotheses can be tested by dividing these sums of squares by their appropriate degrees of freedom and then computing the appropriate F ratios. We can do the same thing in manova, breaking down a total sums-of-squares-and-products matrix into the corresponding sums-of-squares-and-products matrices. Hypotheses about various effects are tested by means of generalized F ratios or Wilks' Λ that correspond to the F test of each effect in anova.

We begin by consideration of the two-factor, fixed effects situation. Going from univariate to multivariate analysis of variance simply means replacing all the terms in Equation 11.1 by their vector counterparts in which the elements of the vector represent the various dependent variables

$$\mathbf{y}_{ijk} = \boldsymbol{\mu} + \boldsymbol{\alpha}_j + \boldsymbol{\beta}_k + \boldsymbol{\alpha\beta}_{jk} + \mathbf{e}_{ijk} \tag{16.30}$$

Thus $\mathbf{y}_{ijk}$ is a vector of observations on the p dependent variables,* $\boldsymbol{\alpha}_j$ (1 by p) is the vector of main effects for level j of factor A, $\boldsymbol{\beta}_k$ is the vector of main effects for level k of B, $\boldsymbol{\alpha\beta}_{jk}$ is the vector of interactions, and $\mathbf{e}_{ijk}$ is the vector of errors. The various null hypotheses say that, except for $\mathbf{e}_{ijk}$, these are null vectors. The homogeneity assumption about σ_e^2 becomes an assumption of homogeneity of covariance matrices Σ across groups.

*Note that here the y's are the *continuous* measures in contrast to the previous sections on discriminant analysis.

In the same way, the means defined by Equations 11.2 to 11.5 become mean vectors, and the model specification of Equations 11.6 to 11.8 now refers to null vectors. The way the means reflect the model parameters in Equations 11.9 to 11.12 and the way the means may be used to estimate these parameters in Equations 11.13 to 11.16 also describe vector relationships rather than simple relationships among scalars. If you find it hard to think about manova in this situation, then think about the univariate case, and make the translations only where they are necessary in setting up the problem initially and in interpreting the results.

We have already discussed the multivariate extensions of the sums of squares within and between (among) groups of Equations 11.17 and 11.18 from our discussion of discriminant analysis (Equations 16.11 and 16.13). These are exactly the same here, except that in the factorial manova T_a has three parts. These are the multivariate analogues of the parts in Equation 11.19 and their model parameters in Equations 11.21 to 11.23. The difference is that these are now sums of squares and products matrices, not just sums of squares. Thus they include cross-product terms between the dependent variables as well as the sums of squares on them.

We encountered the overall T_a matrix in the discriminant analysis case. Here we are showing that it may be broken down into its three components: A, B, and interaction. To illustrate these principles, the two dependent variables case will be presented explicitly here. In the equations below, the fourth subscript 1 or 2, refers to the variable. As in Chapter 11, $i = 1$ to r refers to replications, $j = 1$ to g_A refers to levels of the row variable, $k = 1$ to g_B refers to levels of the column variable. Thus

$$T_a = T_A + T_B + T_{AB} \tag{16.31}$$

is the matrix counterpart of Equation 11.20. In the case of two dependent variables the three components of the matrix are as follows:

$$T_A = \begin{bmatrix} g_B r \sum_k (\bar{y}_{.j.1} - \bar{y}_{...1})^2 & g_B r \sum_k (\bar{y}_{.j.1} - \bar{y}_{...1})(\bar{y}_{.j.2} - \bar{y}_{...2}) \\ g_B r \sum_k (\bar{y}_{.j.1} - \bar{y}_{...1})(\bar{y}_{.j.2} - \bar{y}_{...2}) & g_B r \sum_k (\bar{y}_{.j.2} - \bar{y}_{...2})^2 \end{bmatrix} \tag{16.32}$$

The T_B and T_{AB} terms are defined similarly except that they involve the appropriate mean differences. In a manova, then, there is a p by p matrix for each effect. Each diagonal term contains the sum of squares for that effect for one of the variables, whereas each off-diagonal term is the sum of products of the effect for the corresponding pair of variables.

Tests of Hypotheses

We assess significance of each effect as might be expected. We combine the ideas of how the significance of the differences in mean vectors is tested in the one-way analysis or the discriminant analysis with the procedures for testing the effects in a one-way analysis. That is, we use either the generalized F criterion or Wilks' Λ with the T matrix for the effect in the numerator and T_w in the denominator. Letting T_E stand for the sums-of-squares-and-products matrix of any of the effects, A, B, or AB, we find the eigenvectors that maximize the eigenvalues

$$v_E = \frac{\mathbf{w}_1'\mathrm{T}_E\mathbf{w}_1}{\mathbf{w}_1'\mathrm{T}_w\mathbf{w}_1} \tag{16.33}$$

and convert them to a generalized F using Equation 16.20. Note that $g - 1$ in the expressions for r, s, and t must be replaced by the appropriate degrees of freedom for the effect as

$$g_A - 1, g_B - 1, (g_A - 1)(g_B - 1)$$

or whatever else is appropriate in more complex designs. The same replacements are made in the Λ criterion in Equation 16.21. This criterion also may be employed when we expect that the effects are not concentrated on one single underlying dimension.

Once these omnibus tests have been made and the corresponding null hypothesis rejected, we may then take advantage of the protection levels concept and go on to make the corresponding univariate F tests. This protects against exceeding nominal α levels when the effect has no influence on any of the dependent variables. This is done with the understanding that the investigator is taking a comparisonwise stance with respect to the *effects* but an experimentwise stance with respect to the dependent variables. If there is experimentwise concern with α for the whole analysis, then we would begin with a test of the complete T_a, going on to make even the multivariate tests of the effects only if the overall null hypothesis could be rejected. Thus, while the multivariate nature of the experiment does complicate matters, the complications are straightforward generalizations of the univariate procedures.

Manova and Contrasts

Just as factorial forms of anova can be recast in terms of contrasts among means, the same can be done with manova. Manova can also be translated into a kind of generalized regression format using dummy variables, as discussed in the next chapter. For now, we will consider the contrasts interpretation of factorial manova.

Equations 16.26 and 16.27 can be used to illustrate this process. Recall that $\mathbf{c}_1$ is a vector that combines groups, whereas $\mathbf{a}_1$ is a vector that combines variables. The $\mathbf{c}_1$ vectors can be the orthogonal contrasts that define the effects in the factorial design. For example, in the two-by-two case, $\overline{\mathrm{X}}$ would be four-groups-by-p variables and the $\mathbf{c}_1$ for the row effect contrast would be (1, 1, -1, and -1)′; the column effect would be (1, -1, 1, and -1,)′ and the interaction would be (1, -1, -1, and 1)′. We would apply the contrasts one at a time as a priori contrasts; the F_0 is the ratio whose parameters correspond to the degrees of freedom for the effect.

If there are more than two levels on one or both factors, then a variety of different possible contrasts will span the space of the effect while staying orthogonal to the other effects. Various contrasts that are within the subspace defined by the contrast can be tested, as discussed in Chapter 11. For example, a main effect can be broken down into orthogonal polynomial contrasts or into nested contrasts against a control, and so on. In the same way, interactions can be broken down into the contrasts that correspond to the corresponding components of the independent variables, as described in Chapter 11. We do the same things here, making the necessary adaptations that the multivariate nature requires: We test the contrasts using F_0 instead of F and we remember to use the $\mathbf{a}_1$ vector to get the appropriate within-groups variance to use as the error term to guard against heterogeneity.

The only vector that provides a distinctly new aspect is $\mathbf{a}_1$. We have spoken rather casually of using "*an* $\mathbf{a}_1$," but when there is more than a single dependent variable, there is a whole space *full* of $\mathbf{a}_1$ shooting off into different directions like the display on an electronic space-warfare game. Even if there is only a single degree of freedom for an effect, there are different $\mathbf{a}$'s for different effects. In the testing of contrasts, the $\mathbf{a}$'s are quite directly given by the analysis since the T_E matrix is of rank one, and since the coefficients for any one of the $\mathbf{a}$'s are proportional to the product $\mathbf{c}_1'\overline{\mathrm{X}}$.

When there is more than a single degree of freedom for an effect, we must give a little more thought to just what a given test refers to. If our stance with respect to the dependent variables is that they are all representing essentially the same thing, so that we expect only a single significant eigenvalue, or have found that only one is significant by the procedure in Equations 16.22 to 16.24, then $\mathbf{a}_1$ would be the eigenvector corresponding to the largest root. Thus there is again no ambiguity about $\mathbf{a}_1$. However, if it seems that there is more than one significant eigenvalue for an effect, then unfortunately, we have some more thinking to do.

This means that there is more than a single relevant $\mathbf{a}_1$ vector for the effect. The specific natures of the differences attributable to an effect are numerous, to say the least, and there is no very well established strategy for how to deal with the ambiguity. When there is more than one significant eigenvalue for an effect—say there are four levels for the row variable and two of the eigenvalues are significant—this means that the profiles of means for different contrasts within the effect are different. The most ambiguous situation arises when there are multiple levels for an effect, multiple significant eigenvalues, and when we have little in the way of expectations. Then any contrast vectors $\mathbf{c}_1$ would be arbitrary. The presence of the multiple roots implies that the nature of the differences with respect to the effect are different on different variables and combinations of variables. Just how to interpret what is going on is pretty much left to the ingenuity of the investigator.

A little help in interpretation may be found in observing that the presence of multiple roots is rather like a higher order interaction. It is as if there were an interaction between the dependent variable and that main effect or interaction. The analogy is not quite complete but is useful for intuitive purposes. Perhaps we are going into unnecessary degrees of complication here. After all, only a minority of users of anova extend their analysis to the manova level, and even among these users it is quite rare to see evidence of concern about multiple roots.

Another issue that is rarely discussed in the context of manova is that the $\mathbf{a}_1$'s for the different effects will be different to a greater or less degree. That is, the combination of dependent variables that most completely shows an effect will be different for different effects. These differences may or may not be of interest to the experimenter whose interest is usually focused on the independent variables. Also, the difference between the largest eigenvector for the A effect and that for the AB interaction may be simply attributable to chance. In that regard, it may be useful to begin a factorial manova using an analysis that treats it as a one-way design with g_Ag_B groups and testing for the number of significant eigenvalues. If there is only one significant eigenvalue, then presumably the eigenvectors corresponding to the largest eigenvalue of an effect must be similar and their differences might be attributed to chance. Sometimes, however, there will be more than a single significant effect in the one-way part of the analysis. In that case, we need to examine the eigenvectors that define the canonical variables for the different effects and see if the results suggest that the different effects are having their effect on different combinations of the dependent variables. All of these considerations may inhibit the use of manova. This is perhaps not entirely bad, however, since it is often a lack of conviction concerning just what the dependent variable should be, rather than any scientific curiosity, that leads to manova. If we know just what the dependent variable should be, it is usually better to just go ahead and use anova instead.

Error Terms

One of the topics that requires extensive discussion in texts on anova is the choice of error terms in complex analyses of variance. Under the null hypotheses, the mean squares for various effects can have different compositions, depending on the nature of the independent variables. We will not deal with these issues here beyond saying that the error term in a manova corresponds directly to the error term in the corresponding anova. Fortunately, one can hardly imagine doing a manova without the services of a sophisticated computer program. This program will usually choose the correct error term without human intervention if the nature of the data is specified correctly.

Example

Exhibit 16-3 shows a manova in which there are two independent variables, labelled "Rows" and "Columns" having two and three levels, respectively. The example is an artificial one, based on the model shown in Table 16-1. Four observations were sampled for each cell, and all cells have a common within-cells covariance matrix shown in the lower part of the table.

The model indicates a relatively strong column effect on y_1, a weak one on y_2 that exactly parallels y_1, and a moderate effect of y_3 that is different from y_1. So we have linear effects on y_1 and y_2, and a quadratic effect on y_3. There are no row effects whatsoever, and a mild interaction on y_1 and no interaction on the other variables.

The various aspects of the analysis are shown in Exhibit 16-3. The Y matrix is

Table 16-1 Model Specification for Manova Example

	α_1	α_2	β_1	β_2	β_3	$\alpha\beta_{11}$	$\alpha\beta_{12}$	$\alpha\beta_{13}$	$\alpha\beta_{21}$	$\alpha\beta_{22}$	$\alpha\beta_{23}$
y_1	0	0	6	0	−6	2	2	−2	−2	0	2
y_2	0	0	2	0	−2	0	0	0	0	0	0
y_3	0	0	−2	4	−2	0	0	0	0	0	0

$$\Sigma = \begin{bmatrix} 33.0 & 16.5 & 0 \\ 16.5 & 16.5 & 0 \\ 0 & 0 & 16.5 \end{bmatrix}$$

	Column 1			Column 2			Column 3		
	y_1	y_2	y_3	y_1	y_2	y_3	y_1	y_2	y_3
Row 1	2.083	−.083	.208	−2.917	−5.083	7.208	−1.917	−4.983	−.792
	4.083	1.917	−.792	−.917	−2.083	−2.792	−8.917	−.083	−1.792
	4.083	2.917	−2.792	−.917	.917	−1.792	−6.917	.917	−3.792
	15.083	6.917	−3.792	−1.917	2.917	3.208	−14.917	−7.083	4.208
Row 2	15.083	9.9917	2.208	12.083	5.917	3.208	4.083	.917	1.208
	2.083	−2.083	−1.792	−.917	−6.083	7.208	−10.917	−5.083	−1.792
	−3.917	−3.083	−5.792	−.917	−1.083	−.792	9.083	6.917	−6.792
	−1.917	−2.083	.208	−2.917	1.917	4.208	−6.917	−4.083	2.208

Exhibit 16-3(a) Deviations from $\hat{\mu}$ for a 2-row by 3-column design in which there are four cases per cell and three dependent variables y_1, y_2, and y_3. $\hat{\mu} = (14.917, 8.083, 6.792)$. The Y matrix is thought of as a 24 × 3 matrix with the y's as columns of the matrix.

	Row 1			Row 2		
	Column 1	Column 2	Column 3	Column 1	Column 2	Column 3
y_1	6.333	−1.667	−8.167	2.833	1.833	−1.167
y_2	2.917	−0.833	−2.583	0.667	0.167	−0.333
y_3	−1.792	1.458	−0.542	−1.292	3.458	−1.292

Exhibit 16-3(b) Cell means of the data in 16-3(a). Note that the data have been rearranged so that the different *variables* are now in different rows of the table, whereas in Exhibit 16-3(a) they were in different columns.

	$\hat{\alpha}_1$	$\hat{\alpha}_2$	$\hat{\beta}_1$	$\hat{\beta}_2$	$\hat{\beta}_3$	$\widehat{\alpha\beta}_{11}$	$\widehat{\alpha\beta}_{12}$	$\widehat{\alpha\beta}_{13}$	$\widehat{\alpha\beta}_{21}$	$\widehat{\alpha\beta}_{22}$	$\widehat{\alpha\beta}_{23}$
y_1	−1.2	1.2	4.6	0.1	−4.7	2.9	−.6	−2.3	−2.9	.6	2.3
y_2	−0.2	0.2	1.8	−0.3	−1.5	1.3	−.3	−1.0	−1.3	.3	1.0
y_3	−0.3	0.3	−1.5	2.5	−0.9	0.0	−.7	0.7	0.0	.7	.7

Exhibit 16-3(c) Estimates of the effect vectors as derived from the means. This is done using the same process as in the univariate case, taking each dependent variable separately. For example: 11 = ½(6.333 + 2.833), and 131 = $\bar{x}_{131} - \bar{x}_{1.1} - \bar{x}_{.31} + \bar{x}_{...} = \bar{x}_{131}$ − 11 − 13 = −8.167 − (−1.167) − (−4.667) = 2.333.

$$\begin{bmatrix} 1305.8 & 626.17 & -93.417 \\ 626.17 & 451.83 & -107.58 \\ -93.417 & -107.58 & 309.96 \end{bmatrix} = \begin{bmatrix} 489.34 & 174.17 & -20.656 \\ 174.17 & 65.835 & -19.579 \\ -20.656 & -19.579 & 83.713 \end{bmatrix} + \begin{bmatrix} 816.48 & 451.99 & -72.762 \\ 451.99 & 386.00 & -88.004 \\ -72.762 & -88.004 & 226.25 \end{bmatrix}$$

$$\mathrm{T}_T = \mathrm{T}_a + \mathrm{T}_w$$

$$\begin{bmatrix} 489.34 & 174.17 & -20.656 \\ 174.17 & 65.835 & -19.579 \\ -20.656 & -19.579 & 83.713 \end{bmatrix} = \begin{bmatrix} 342.33 & 119.92 & -20.667 \\ 119.92 & 43.583 & -17.958 \\ -20.667 & -17.958 & 74.083 \end{bmatrix}$$

$$\mathrm{T}_a = \mathrm{T}_C$$

$$+ \begin{bmatrix} 32.685 & 4.6773 & 8.1783 \\ 4.6773 & .66934 & 1.1703 \\ 8.1783 & 1.1703 & 2.0463 \end{bmatrix} + \begin{bmatrix} 114.33 & 49.583 & -8.1667 \\ 49.583 & 21.583 & -2.7917 \\ -8.1667 & -2.7917 & 7.5833 \end{bmatrix}$$

$$\mathrm{T}_R \qquad \mathrm{T}_{RC}$$

Exhibit 16-3(d) Sums of cross-products matrices. T_T is Y′Y from 16-3(a). T_a is the total sums of cross-products among groups, based on the deviations of the cell means on each variable from their respective grand means. The entries in T_W are the sums of squares and cross-products of the deviations of the 24 observations from their cell means. T_C is an example of Equation 16.32 as is T_R. Each may be thought of as the result of multiplying a matrix of effects by its transpose. The two row effects are repeated 12 times, once for each of the 3 columns × 4 replications; the column effects are similarly repeated 8 times and the interaction 4. The sum of T_R, T_C, and T_{RC} are T_a, as in 16.31.

Source	v_1	v_2	v_3	r	s	t	F_0	α
$T_w^{-1}T_a$	.74568	.37909	.01297	3	3	16	3.977	>.05
$T_w^{-1}T_R$	.08672	0	0	3	1	16	.463	>.05
$T_w^{-1}T_C$	.52660	.33422	0	2	2	16	4.213	>.05
$T_w^{-1}T_{RC}$	.15281	.03738	0	2	2	16	1.222	>.05

Exhibit 16-3(e) Eigenvalues of the four effect matrices and tests of the null hypothesis by means of the largest root test. All the null hypotheses are accepted. If a completely experimentwise stance is taken with respect to α, the other tests would not have been made after T_a was found to be not significant. The acceptance of the null hypotheses here is an example of a Type II error since the data were artificially constructed by a model containing all three effects. Evidently, the group sizes were too small for them to be detectable.

Source	j	v_j	$\log_e\left(\frac{1}{1+v_j}\right)$	$-\sum_{k=j}^{s}\log_e\left(\frac{1}{1+v_k}\right)$	m	q_u	df	α
All								
	1	.74568	−.55714	.89145	18.5	16.492	15	$.5 > \alpha > .3$
	2	.37909	−.32142	.33431	18.5	—	8	not tested
	3	.01297	−0.1289	.01289	18.5	—	3	not tested
Rows								
	1	.08672	−.08316	.08316	20.5	1.705	3	$.7 > \alpha > .5$
Columns								
	1	.52660	−.42304	.71139	20.0	14.228	6	$.05 > \alpha > .02$
	2	.33422	−.28834	.28834	20.0	5.767	2	$.10 > \alpha > .05$
Interaction								
	1	.15281	−.14220	.17889	20.0	3.578	6	$.8 > \alpha > .7$
	2	.03738	−.03670	.03670	20.0	.734	2	not tested

Exhibit 16-3(f) Wilkes' Λ test of H_0. By this criterion, the column effect is found to be significant. The inconsistency with the largest root results, where the column effect approached significance, is presumably attributable to the presence of a second, moderately large root.

		y_1			y_2			y_3		
Source	df	ms	F	α	ms	F	α	ms	F	α
All	5	97.868	2.158	>.05	13.167	<1	>.05	16.743	1.332	>.05
Rows	1	32.685	<1	>.05	.669	<1	>.05	2.406	<1	>.05
Columns	2	171.167	3.774	>.05	21.792	1.016	>.05	37.042	2.947	>.05
Interaction	2	57.167	1.260	>.05	10.792	<1	>.05	3.792	<1	>.05
Within	18	45.360			21.444			12.569		

Exhibit 16-3(h) Univariate F tests. The only significant effect is columns on y_1, although columns on y_3 approaches the .05 level. The "total among groups" also approaches significance.

shown in 16-3(a) in a rearranged form to save space, and the overall mean vector has been subtracted from each entry. The mean vectors for each of the six cells are shown in 16-3(b), and from these we have derived the estimates of the effects in 16-3(c). These have been rounded to one decimal, and are fairly similar to the population values shown in Table 16-1. The various calculations are carried through in detail for the remainder of the exhibit. We will not trace them through in detail in this discussion, but only refer to results. In practice, some of the detailed tests would not be carried out when the omnibus tests are not significant, although they are given here for illustrative purposes.

There is a general lack of significant effects in this example, in spite of several null hypotheses being false. This is attributable to the cell sizes, which are rather small, and the effects, which are only moderate.

The eigenvalues associated with the effects shown in Exhibit 16-3(e), reflect the input model quite well, even though they are of questionable significance, at best, as shown in Exhibit 16-3(f). Of the two smallest eigenvalues, one is associated with the Row effect, where the null hypothesis is true, and one with the second eigenvalue of the interaction.

We saw that the cell means reflected the underlying model in spite of the lack of significance, and the same is true of the eigenvalues. The two largest eigenvalues are the ones associated with the column effect. These correspond to the true effects that are parallel on y_1 and y_2 but different on y_3. The next largest eigenvalue was v_1 for the interaction, where a single small true effect had been set up, as reflected by the small second eigenvalue. The single row eigenvalue is small also, which corresponds to the fact that the null hypothesis was true there.

The eigenvectors of the column effect (which are not shown here) correspond to the underlying model also. The first eigenvector had large weights for y_1 and y_2, and the model included effects on both, but here the weights were opposite in sign, whereas the true effects were parallel. The reason for the opposite sign is that the true effect on y_2 is small, relative to its correlation with y_1. This is similar to the suppressor effect in regression. The second eigenvector showed a large weight only for y_3, which corresponds to the fact that the model included an effect on y_3 that was orthogonal to the effect on y_1 and y_2, and is uncorrelated with the other two variables. Thus, when we know the model, we can see its effects in the data, even when the noise is relatively loud. However, inferring the true model from the data would be much more difficult in the absence of the clairvoyance that was possible using a known model.

MULTIVARIATE ANALYSIS OF COVARIANCE

Univariate Ancova Reviewed

The analysis of covariance, remember, is equivalent to an analysis of variance of the deviations of the dependent variable from the predictions provided by the control variables. This is the simplest way to conceive of ancova in the multivariate case. The only difference is that now we have multiple dependent variables. Presumably there is a separate regression equation for predicting each from the control variables, although the specific nature of that equation need not concern us as long as such an equation is possible. In our discussion of ancova, we distinguished two cases. In the simpler one, there was a matrix of scores on the control variables X bordered by a column of 1's, called X_+ and a single dependent variable arranged as a vector $\mathbf{y}$. Then

$$\mathbf{b} = (X_+'X_+)^{-1}X_+'\mathbf{y}$$

and

$$\mathbf{e} = \mathbf{y} - X\mathbf{b}_+$$

Thus anova is the anova of the vector of deviations $\mathbf{e}$.

The same thing occurs in the multivariate case. All that changes is that $\mathbf{y}$, $\mathbf{b}$, and $\mathbf{e}$ now have p columns and become p-column matrices Y, B, and E. The equations look the same in multivariate and univariate cases except that boldface, lowercase letters representing vectors are replaced by capitals.

Thus the multivariate analysis of covariance (mancova) corresponds to the manova of the E matrix. Any sort of structure in the data matrix, such as a factorial design, is simply used to define the way in which the among-groups sums of squares and products matrix will be divided into components.

Another way to think about ancova is to extend the X matrix to include dummy variables reflecting group membership, giving the matrix $c + g$ columns, one for each of the c control variables, one for the constant, and one for each of the $g - 1$ dummy variables. Later, these may also be transformed so as to represent the relevant contrast structure for the groups. Now there is a new matrix B called $\tilde{B}$ with $c + g$ entries. We apply $\tilde{B}$ to the extended form of X with $c + g$ columns to give a new set of predicted dependent variables. The difference between the predicted Y matrix and the actual one is a new set of deviations, $\tilde{E}$. If there are no group differences, then $E'E$ and $\tilde{E}'\tilde{E}$ differ only by the chance effects of fitting extra constants.

Homogeneity of Regression

In discussing ancova we had to deal with the issue of equality of the slopes of the regression equations in the various groups. After all, differences in intercepts may mean little if the regression lines cross. This was done as a form of extended regression by introducing the idea of having g separate variables to represent *each* control variable. Each of the entries of these variables was zero, except for the n_h entries that were the scores of the individuals in that group on the one control variable. Thus, the number of control variables was extended to gc, making the total number of variables $gc + g$. The deviations using these predictors were then compared to deviations using just the $g + c$, assuming homogeneity of slopes. If there was no significant decrease in errors using the larger predictor set, then we could assume homogeneity and go on to do the standard ancova and test equality of intercepts.

The same procedure is used in the mancova case. The difference here is that the matrices T_w and $T_w{}^*$ are compared, where T_w is the sums of squares and products of deviants using the $g + c$ predictors and $T_w{}^*$ is that using the $gc + g$. Rejection of the null hypothesis means that there is a significant reduction in deviations when the slopes are assumed to be different, so the comparison of intercepts has little meaning.

One further aspect of homogeneity is the question of whether the deviations from the regression line are the same in the different groups. This is done by comparing the T_{wh} matrices across the different groups h by means of the Bartlett test described earlier.

Mancova versus Manova

There may be some ambiguity as to whether we should do a manova, treating $p + c$ variables as dependent variables, or a mancova, treating c of them as covariates. When the data are from randomized groups design, or a single dependent variable, there may be little choice between the two. That is, it may not be clear which of the variables are to be treated as dependent variables and which as covariates.

It is usually safe to say that the characteristics subjects bring with them into the experiment should typically be treated as covariates. Those that are presumably affected by the treatments in an experiment should be treated as dependent variables. There are ambiguous cases, however, particularly when the groups are not randomized. Sometimes the investigator may question whether one dependent variable is affected more than some other variable with which it is associated. Such questions should be reformulated. Under these circumstances, the question is really whether the group differentiation is improved by the inclusion of the variable under consideration.

Thus we have sketched the way in which the ancova can be extended to a mancova. Perhaps it can now be seen that we can formulate mancova as a manova of the deviations of all the dependent variables from their values as predicted from the control variables. The tests for homogeneity of regression coefficients and of error variances in the various groups are also extended to the multivariate case.

SUMMARY

Manova and discriminant analysis are two closely related methods. In both, the investigator is analyzing several groups and several continuous variables and is interested in the differences among the groups on the variables. Manova, however, is an extension of anova, which means that the continuous variables are the dependent variables, even though they are taken collectively. In discriminant analysis, the roles of the sets of variables are reversed, and we seek to understand group membership in terms of differences on the variables.

In either case, the investigator may be concerned whether the groups can be differentiated at all, or about the nature of the way or ways that the groups differ. The mathematical calculations are much the same in manova and discriminant analysis. The discriminant variables are formed by using the weights that, when applied to the observed variables, maximize the ratio of among- to within-sums of squares. Maximization turns out to be an eigenvalue-eigenvector problem, and the eigenvectors provide the weights and the eigenvalues are, or are directly related to, the ratio.

The number of dimensions of differentiation is equal to p or $g - 1$, whichever is smaller, and the question of overall differentiation can be answered by considering all the dimensions at once. This test is based on the statistic called Wilks' Λ. If one major combination of variables is expected to account for the differentiation, the largest root criterion is preferable.

Although the manova may be thought of as simply testing the degree to which different contrasts among the groups are differentiated, it is also useful to think of it from a more traditional anova viewpoint. Here, the sums-of-squares-and-products matrices associated with the usual anova effects are compared to an appropriate error sums-of-squares-and-product-matrix in a way that is a direct extension of anova, except that multiple dependent variables are considered simultaneously.

The multivariate extension of anova fits in rather directly, particularly when it can be thought of as an anova of the deviations from regression of the dependent variables on the covariates.

COMPUTER APPLICATIONS

A number of computer packages offer manova and discriminant methods. Three of these are discussed below.

BMDP

The BMDP package provides a comprehensive discriminant analysis program in P7M, called Stepwise Discriminant Analysis. This is the modern version of a program that has been included in this series since the first edition of the BMD manual—virtually the paleolithic era as far as computer packages are concerned. The modern program not only has distinguished forebears; it includes just about all the features we would want in such a program.

As usual in BMDP programs, we begin by identifying the variables to be used in the analysis; the variable that identifies group membership is specified with the GROUPING IS sentence of the VARIABLE paragraph. In the GROUP paragraph the identities of the groups are specified in the CODES sentence and the names to be given to these groups are specified in the output in the NAMES sentence.

The program operates in a stepwise fashion, very similar to the multiple regression programs. That is, the program first uses one observed variable to discriminate among the groups, then adds a second, third, fourth variable, and so on, with the possibility of dropping one from the discriminant process at any stage. As in multiple regression, appropriate coding of the variables to be used for discrimination enables us to have any combination of variable-selection and hierarchical treatment of these variables. Left to its own devices, the program operates in the pure variable-selection mode. *All of the cautions and caveats that we put forth in the regression context apply here with equal, indeed greater, force.* (We say "greater" force because here we have the opportunity to capitalize on chance not only by choosing weights for the variables but also in choosing combinations of groups to be discriminated.) While we may allow the program to procede in a variable-selection mode as a matter of curiosity, to see which variables are most effective in the differentiation, it is best to use only the information at the *last step* in the analysis for inferential purposes.

As each variable is entered into the analysis, the program performs a complete analysis, including the assessment of the contribution of each variable to the differentiation. It provides Wilks' Λ, also referred to as the U-STATISTIC, along with Rao's transformation, listed as APPROXIMATE F-STATISTIC. The program does not provide the largest root criterion. However, the summary table provides the eigenvalues, so that the user can carry out the Wilks' Λ test if necessary, as well as the test for the number of dimensions. The coefficients that define the canonical variables are given as well. There is no simple way of using the eigenvalues to decide on the number of differentiation dimensions and then to perform the analysis in this reduced space as we might like. If we want the correlations of the observed variables with the canonical variates we must use the weights themselves to define new variables and then correlate these with the observed variables.

The P7M program provides extensive information with respect to classification accuracy. It provides the classification functions and a table cross-classifying the actual against the predicted group memberships. Not only is classification done for the current data, with consequent biases toward correct classification, but it also uses the jackknife procedure. That is, it uses data on $n - 1$ cases to predict the group of the nth n number of times. The cross-classification of true and predicted group membership in this mode furnishes a relatively unbiased, realistic assessment of the accuracy of the classification functions. P7M also makes explicit provision for splitting the sample into subgroups for cross-validation.

Additional features include options for using contrasts to guide the analysis. That is, where there are several groups, we may be more interested in some differentiations than others, and we may define these with the appropriate contrast vectors. There can be as many as $g - 1$ contrast vectors, although they do not do any "guiding" of the discrimination unless there are fewer than $g - 1$. There are also extensive and useful possibilities for plotting the cases in various ways.

A feature that sometimes causes some confusion to users is the one called "prior probabilities." This is relevant only to the aspect of the program that actually classifies the cases, and permits us to reflect, for example, the sizes of the respective populations of the various groups, if that is relevant. The default is to assume equal priors, that is, populations of equal size.

Manova in BMD

There are five different anova programs in BMDP and several others that do anova incidentally, but the one program that corresponds most closely to our treatment of manova is BMDP4V. P4V is a very comprehensive program that can hardly be treated in detail here, but we will try to relate the essentials to the presentation of manova in this chapter. In setting up a problem, it is necessary to distinguish between the dependent and the independent variables after they are listed in the VARIABLES paragraph. This is done by listing the independent (grouping) variables in the BETWEEN paragraph in a FACTORS ARE sentence.

A new anova concept is introduced in the WEIGHTS paragraph. Ordinarily in anova or manova the data are analyzed as they are observed, but if a variable is observational rather than manipulated, it often happens that the frequencies in the levels of the variable in the data sample do not reflect their frequencies in the population. The "WEIGHTS" paragraph allows us to take this into account; the rationale and mechanisms are explained in the BMDP manual.

The flexibility in the analysis is enormous, including a number of advanced techniques that we have not mentioned. Assuming that the reader's interest is confined to the topics covered in this text, many sentences of the ANALYSIS paragraph can be ignored when we want simply a straightforward factorial analysis with multiple dependent variables. We do need to decide whether we wish the cell means and parameter estimates printed; if so, we have EST in the ANAL paragraph. (Remember that if there are g levels of each of f factors and p dependent measures, then there are $g^f p$ means, which can be a lot to look at.) If we wish to see the eigenvalues and vectors and the within-group dispersion matrices, all of which are recommended, then we must state EVAL., VECT., DISP. The remaining sentences may be left at their default values.

The program offers additional features reflecting higher and higher values of sophistication in the user, sufficient to challenge all but the most sophisticated blackbelt in manova.

SAS

There are three discriminant function procedures in the SAS system. Two of these—DISCRIM and STEPDISC—are not very satisfactory from the point of view expressed here, at least as compared to BMDP7M. Neither of these programs will be discussed in any detail, although DISCRIM has a nice feature for testing homogeneity of covariance matrices that can be used wherever this is a matter of concern. A third procedure called CANDISC, does essentially the discriminant analyses we described. For those users whose favorite package is SAS, a combination of the GLM procedure with DISCRIM should provide a satisfactory means of doing both manova and discriminant analysis, so we will concentrate on these. Any actual classification of cases would be carried out by DISCRIM, and the interpretational parts with GLM. The GLM procedure in SAS is very general and elegant. Indeed, it can be used to perform all the analyses that we have discussed up to this point, except for the factor analysis ones.

In PROC GLM the qualitative independent variables are defined with the statement CLASSES followed by the names of the variables that form the factors in the analysis, for example, CLASSES FACA FACB. Then the analysis is specified by a MODELS statement of the form

```
MODEL Y1, Y2, Y3 = FACA FACB FACA*FACB
```

The variables to the left of the equals sign are the dependent variables, and those to the right are the effects that are to be fitted and tested—in this case, the two main effects and their interaction. Only those effects that are listed will be tested and shown. A MANOVA statement is also necessary, followed by the list of effects that are to be considered; this will generally duplicate the list to the right of the equals sign in the MODEL statement. Here it would be

```
MANOVA H = FACA FACB FACA*FACB
```

If some effect other than the within-cells matrix is to be used for the error term, then the H= list is followed by E=, specifying the error term. The sums-of-squares-and-products matrices for the effects and error will be printed if the user follows the H= list with a slash and then PRINTH PRINTE. The slash causes the listed T_E matrices to be printed, and the PRINTE statement prints T_W. The statement also causes the within-groups correlation matrix to be printed. Thus the specification of the problem discussed in this paragraph would be

```
PROC GLM;
     CLASS FACA FACB;
     MODEL Y1 Y2 Y3 = FACA FACB FACA*FACB
     MANOVA H= FACA FACB FACA*FACB / PRINTE PRINTH.
```

If the means are desired, they can be gotten by stating MEANS and following it with the list of effects for which the means are desired. For example, MEANS FACA FACB would print the marginal means for both effects but not the cell means. The univariate analyses are printed as well as the multivariate ones. The user who does not want to see them—perhaps to avoid temptation in case the overall test is not significant—can say NOUNI.

The program PROC GLM contains many other options, including one that specifies contrasts with CONTRAST and one that estimates parameters in the anova design using ESTIMATE. The former is quite straightforward, and the latter is very involved, as described in the manual.

In addition to optionally printing T_w, the various T_E matrices, and the various univariate matrices, PROC GLM prints the eigenvalues and eigenvectors for each $T_w^{-1}T_E$ and several multivariate test criteria, including two that have not been discussed here. The program also calculates the test criteria and their alpha levels. The one thing it does not do for a discriminant analysis is the stepdown tests of the individual variables.

SPSS^x

SPSS^x offers both DISCRIMINANT and MANOVA. The former is one of the more clearly described procedures, and it offers the major options and output. The user specifies the group membership variable and the discriminating variables with keywords GROUPS= and VARIABLES=. For example, if the groups are specified in FINISH and the variables are called GREQ and HOURS,

```
DISCRIMINANT GROUPS=FINISH/
    VARIABLES=GREQ,HOURS.
```

DISCRIMINANT will "predict" group membership and present a classification table. By default, the prior probabilities (population sizes) are assumed equal, but this can be changed through the PRIORS statement.

It prints by default the weights for the standardized canonical variates (called "standardized canonical discriminant function coefficients") and the within-group structure matrices. The significance of discrimination is evaluated using Wilk's Λ, but the eigenvalues are printed in case the largest root (F_0) test is used.

Variable selection methods can be used, with several optional criteria. Wilks or RAO are the two methods most directly related to the presentation here. Various cutoffs and criteria can be used to control the variable selection process.

Cross-validation can be accomplished using the SELECT subcommand, and STATISTIC 13. A variety of useful plots and tables can be printed by choosing other STATISTICS subprograms. STATISTIC 7, for example, is the test of within-group homogeneity of variance-covariance matrices.

Multivariate analysis of variance can be carried out using the MANOVA program described in Chapters 10 and 11. We merely specify the multiple dependent variables in the MANOVA statement. If these are multiple, the manova takes place, and the default output presents the major information. Useful additions include PRINT=CELLINFO(MEANS) for cell means, HOMOGENEITY=BOXM to test within-cells homogeneity, and STEPDOWN to assess the significance of variables, partialing out all others. Note that the cell information should be requested judiciously because of its potential volume.

PROBLEMS

1. Below are data on two variables for two groups. Perform the discriminant analysis, including computation of the weight vectors, $\mathbf{v}_1$, F_0, $\bar{\mathbf{z}}_1$, and $\bar{\mathbf{z}}_2$.

	Group 1								Group 2							
x_1	0	1	3	4	8	8	5	3	5	7	9	11	6	11	4	11
x_2	4	3	2	7	8	6	6	4	4	5	6	9	7	10	4	11

2. Suppose the groups in Problem 1 are further subdivided, with the first four in Group 1 falling in Group 11, the second four into Group 12; the first four in Group 2 falling in Group 21, and the last four falling in Group 22. Perform the discriminant analysis among the four groups.

3. The four groups of Problem 2 are considered to be the result of a 2 × 2 design in the obvious way: $\left[\begin{smallmatrix}11 & 12\\21 & 22\end{smallmatrix}\right]$, and x_1 and x_2 are considered dependent variables. Perform the factorial manova.

4. In the data from Problem 1, define a dichotomous Group-1-versus-Group-2 variable and "predict" it using x_1 and x_2. How do the results relate to those from Problem 1?

5. Listed below are several contrast vectors $\mathbf{c}_c$ for the data of Problem 2. Associated with each is one or more weight vectors $\mathbf{a}_c$ for the variables. Determine whether the contrast-combinations are significant at $\alpha = .05$. Explain the nature of each of the following contrast-combinations:
   (a) $\mathbf{c}_1' = (1, 1, -1, -1)$; $\mathbf{a}_1' = (1, 1)$
   (b) $\mathbf{c}_2' = (0, 1, 0, -1)$; $\mathbf{a}_2' = (1, 1)$
   (c) $\mathbf{c}_3' = (1, -1, -1, 1)$; $\mathbf{a}_3' = (.718, -.696)$
   (d) $\mathbf{c}_4' = (-2.056, 0, -1.617, .440)$; $\mathbf{a}_4' = (.808, -.589)$
   Use $F_0 = 7.78$ since Appendix Table 6 has $t = 12$ as its lowest value for $s = 2$ and $r = 2$.

ANSWERS

1. $v_1 = .61538$; $\mathbf{w}_1' = [.8944, -.4472]$; $F_0 = 4.000$; $p < .05$; $\bar{z}_1 = 1.34167$, $\bar{z}_2 = 4.02497$; $\hat{s} = 1.8284$.
   Note that the weights for the two variables are opposite in sign, even though the direction of differences in means is the same for each. This is analogous to a suppressor effect in multiple regression, and presumably reflects an occurrence that could be described by saying that the mean difference on x_2 is surprisingly small given the large difference on x_1 and the high correlation between the two.

2. $v_1 = 1.5496$; $\mathbf{w}_1' = [.80837, -.58867]$; $F_0 = 8.5228$; $p < .05$; $q_1 = 14.505$, $df = 6$, $p < .025$
   $v_2 = .3137$; $\mathbf{w}_2' = [-.37504, .92701]$; $q_2 = 3.274$, $df = 2$, $p < .25$

$$\bar{\mathbf{Z}}' = \begin{bmatrix} -.7979 & 1.3182 & 2.9349 & 1.7576 \\ 2.9580 & 3.3118 & 2.5617 & 4.4157 \end{bmatrix} \begin{matrix} s_1 = 1.2301 \\ s_2 = 1.4234 \end{matrix}$$

$$\bar{\mathbf{Z}}^{*\prime} = \begin{bmatrix} -.5999 & 1.0716 & 2.3859 & 1.4288 \\ 2.0781 & 2.3267 & 1.7998 & 3.1023 \end{bmatrix}$$

3. Row effect: $v_{1r} = .95190$; $\mathbf{w}_{1r}' = [.85080, -.52549]$; $s = 1$; $r = 2$; $t = 11$
   $F_0 = 5.2354$; $p > .05$
   Column effect: $v_{1c} = .24034$; $\mathbf{w}_{1c}' = [.09054, .99489]$
   $F_0 = 1.3219$; $p > .05$
   Interaction: $v_{1i} = .66835$; $\mathbf{w}_{1i}' = [.71790, -.69615]$;
   $F_0 = 3.6759$; $p > .05$

4. $\mathbf{b}' = -.12698, .06349$; $F_{2,13} = 4.0$ Note that $b_1 \div b_2 = w_{11} \div w_{12}$

5. (a) -12 ± 13.473. Accept the hypothesis that the average of groups 11 and 12 is the same as the average of groups 21 and 22 (that is, the same as the row effect in the manova) on the sum or total on the two dependent variables.

   (b) -4 ± 9.527. Accept the hypothesis that groups 12 and 22 do not differ on the sum of the dependent variables.

   (c) -2.872 ± 2.829. Reject (barely) the hypothesis of no interaction. Note that the combination of variables is the same as that for the canonical variate in the interaction test of Problem 3.

   (d) 7.03 ± 4.546. Reject the hypothesis that the optimal contrast among the optimally weighted variables is zero.

READINGS

Bock, R. D. (1975). *Multivariate statistical methods for the behavioral sciences.* New York: McGraw-Hill, pp. 103–64; 236–372; 395–414.

Morrison, D. F. (1976). *Multivariate statistical methods* (2nd ed). New York: McGraw-Hill, pp. 170–216; 230–46.

Tabachnick, B. G., and L. S. Fidell (1983). *Using multivariate statistics.* New York: Harper & Row, pp. 222–371.

Tatsuoka, M. M. (1970). *Discriminant analysis: The study of group differences.* Champaign, IL: Institute for ability and personality testing, pp. 1–57.

17

Canonical Correlation Analysis

ORIGIN AND APPLICATIONS

The Most Predictable Criterion

An investigator applying multiple regression is often faced with a dilemma. There are several possible dependent variables that are more or less equally defensible or important. For example, when predicting "college success" from the characteristics of student applicants, it is traditional to use grade point average as the dependent variable—that is, the definition of "success." However, we could use either freshman-year grades, or overall grades, or break them down into grades in required courses, electives, and major. Some might consider *finishing* college as the definition of success and recommend using this dichotomy, (or using the number of years required to finish) as the criterion. Or, we could ask the students how much they enjoyed college or how much they benefited from college, and use this admittedly radical criterion. Or, we could use their income the first year or the tenth year after graduation as a criterion. The alumni office might define "success" as the amount of money the student contributed to the school in later years. So there are many possible criteria or dependent variables, and our predictors in multiple regression analyses could be used to predict any or all of them.

Suppose we predict all of these variables. Some will be more predictable than others in the sense of having higher multiple R's. Which ones are more predictable may be interesting in itself, but we might consider forming a composite criterion, a weighted combination of all of them. To this end we could find some means of assessing their relative importance, and use these results to define a set of weights to apply to all the criterion scores.

Finally, we might consider whether there is a way to weight the composite so as to maximize its predictability. If nothing else, we could weight the single most predictable variable 1.0 and weight the rest of the variables 0, so it seems likely that we could pick a second dependent variable and weight it in such a way that the multiple correlation of the combination with the predictors is higher than the multiple correlation of either separate variable. Now we get excited. There must be an *optimum* set of weights for the criteria to define a composite that can then be predicted using an optimum set of weights on the predictors. The correlation between the two composites must be at least as large as the highest individual multiple, and who knows how high it might go?

By now, we may begin to have second thoughts. Isn't there a lot of opportunity to capitalize on chance here? What if all the individual multiples were nonsignificant? The correlation between the two composites might look pretty good but represent complete hash. Clearly, some inferential safeguards are desirable. Finding a way of inferring which of the variables are reliably important as members of the composite would be useful also.

Considering it further, or as we inspect the intercorrelations among the sets of variables, it may occur to us that there may be several *different* composites that are each predictable from different composites of the predictors. For example, the academic aspects of success might be expected to be predictable from academic predictors, previous grades, test scores, and the like. On the other hand, the students' own feelings about college may depend more on personal characteristics. Completing college might depend on social and economic aspects and so might the economic success factors. Thus we wonder about the extent to which these ideas of different composites predicted by different predictors are supported by the data, or what alternative relationships of this general type might exist instead.

The methods described in this chapter address all of these concerns, as well as some related problems. Canonical correlation analysis was originally developed by Harold Hotelling, a pioneer in multivariate statistics, (Hotelling, 1935, 1936) to solve the problem of combining dependent variables into a "most predictable" criterion. This type of analysis can also be used to study a variety of other types of questions, like those mentioned in the above example.

OTHER APPLICATIONS

Our first example gave a fairly clear differentiation between independent variables and dependent variables. However, we can apply canonical correlation analysis when the two sets of variables have equivalent status, as well as when one is dependent and the other independent. When a collection of variables falls into two natural classes, we may wonder about the degree and nature of the relationships between the two classes. For example, we may wonder how a battery of ability tests and collection of personality measures (or achievement scores) interrelate. (The data matrix in this study would be persons by variables.) For a set of census tracts, we could have scores for each tract on some average socioeconomic characteristics of the residents as well as scores on measures of land use, population density, and so on, and we could study the relationships between the resident characteristics and the real estate variables. (The data matrix here would be tracts by characteristics-plus-variables.) In a marketing context, we might have a number of competing products and one set of variables might represent expenditures on different types of advertising and the other set might represent sales in different segments of the market. (The data matrix would be products by characteristic-plus-sales data.) In a study employing multidimensional scaling, the projections of the stimuli on the dimensions would represent one set of variables and their average ratings on unidimensional scales (or values on objective characteristics) would represent the other set. (The data matrix would be stimuli by dimensions-plus-ratings.) In all these cases, we have a single set of entities and their values on two sets of variables, and canonical correlation analysis is applicable to each. Unfortunately, there are no well-established, routine methods of handling more than two sets of variables, except as different sets of pairs, or by combining sets to make just two large ones.

In any application, there may be several different research questions. The most primitive is the question of whether there is any between-set association at all. The corresponding null hypothesis states that the correlation is zero between all possible combinations of variables in one set and all possible combinations of variables in the other set. (This comes close to the "nullest" hypothesis that *all* intercorrelations are zero.) We can also ask, as we did at the outset of our discussion, what is the *maximum* correlation between a combination of variables in one set and a combination in the other, and what are the combinations? Other questions we can ask are, What is the number of *different* dimensions of association between the two sets? That is, is the between-set association accounted for by one combination or by several? If so, how many? What is the nature of these dimensions? What are the major variables accounting for each? How much of the total variance in one set is accounted for by its correlation with the other? Can some variables be dropped from one or both sets without disturbing the overall level of association between the sets? All these questions can be studied by means of canonical correlation analysis.

Canonical correlation analysis is also useful in a subsidiary way. For example, in the multiple-dependent-variables context, we may be genuinely interested in each individual dependent variable and how well it can be predicted. If there are a number of dependent variables, and we use multiple regression to predict each, it could turn out that only one or a few multiple Rs are significant. We can then use the inferential parts of canonical correlation analysis to provide an omnibus significance test that will provide a protection level against the possibility that all the multiple regression null hypotheses are actually true. This is much like using manova to provide a protection level when we are primarily concerned with the analyses of variance of several individual dependent variables.

CONCEPTUAL UTILITY OF CANCOR

Much of canonical correlation analysis (we'll call it *cancor* for short) will be reminiscent of methods that we have encountered previously in this text. This is no accident. At the end of this chapter we will show how cancor can be conceived of as the general multivariate case, and how all the other methods are special instances of various kinds, just as analysis of variance can be conceived as a special kind of regression analysis in which the predictors are dichotomous.

The other methods are variations on cancor in which, for example, one set has only one variable (multiple regression), or in which one set is categorical (manova). These methods differ in their orientation or in the particular type of variable involved. In fact, a computer program for canonical correlation can be coaxed into doing many different types of analysis. Each of the methods has its own special purposes and individual features, of course. However, if we stand back far enough, these individual features begin to blur, and family resemblances begin to emerge.

MAXIMIZING THE CORRELATION BETWEEN COMPOSITES

Choosing Weights

If we return to the original, most predictable criterion problem, we realize that we are optimally weighting the criteria for predictability and simultaneously optimally weighting the predictors for validity. If some sentient entity were faced with this task, it would have difficulty seeing the two halves of the problem as different. Not to be outdone by a mere sentient entity, we too should see the problem as a symmetric one. That is, what we are really looking for is one set of weights to apply to the predictors and another set to apply to the criteria such that the correlation between the two composites is as large as possible. The questions then become, "Is there a way to do this, and if so, what is it?"

It is best to begin by formulating the problem in a way that makes it as easy as possible on us. First of all, we let the matrix of scores on one set of variables be called X, n by p, rank p, and those on the other be called Y, n by q, rank q. Furthermore, we may as well let them be in standard score form because if we were to find weights for the raw scores that solved the problem we could convert them directly into weights for the standard scores and vice versa. This simplifies a good bit of the notation. Now let there be a vector of weights $\mathbf{w}_x$ $(p \times 1)$ for weighting the x's and a second one $\mathbf{w}_y$ $(q \times 1)$ for weighting the y's. Each can be used to make a composite variate, z_x and z_y, respectively:

$$\mathbf{z}_x = \mathrm{X}\mathbf{w}_x \tag{17.1}$$

$$\mathbf{z}_y = \mathrm{Y}\mathbf{w}_y \tag{17.2}$$

What we want to do is maximize the correlation between z_x and z_y. Maximizing correlations is a little troublesome because of the complications arising from the standard deviation terms in the denominator of r. In multiple regression it works out fortuitously that minimizing the error variance and maximizing the correlation are the same, but in cancor, an additional device proves necessary.

This device requires that the weights be such as to put z_x and z_y in standard score form. That is,

$$(n - 1)^{-1}\mathbf{z}_x{}'\mathbf{z}_x = (n - 1)^{-1}\mathbf{z}_y{}'\mathbf{z}_y = 1.0 \tag{17.3}$$

If we recall the relationship between the variates z_x or z_y and the observed batteries, and do a little matrix algebra, we realize that

$$\mathbf{w}_x{}'\mathrm{R}_{xx}\mathbf{w}_x = \mathbf{w}_y{}'\mathrm{R}_{yy}\mathbf{w}_y = 1.0 \tag{17.4}$$

The task, then, is to find $\mathbf{w}_x$ and $\mathbf{w}_y$ so as to maximize

$$r^* = (n - 1)^{-1}\mathbf{z}_x'\mathbf{z}_y \tag{17.5}$$

while obeying the restriction 17.3 or the equivalent 17.4.

When the equations are solved, it turns out that this too is an eigenvalue-eigenvector problem, and that there is more than one "maximum" in the mathematical sense. In fact, there are as many maxima as the smaller of the two ranks, p or q. The equations are usually formulated in terms of the sections of the supermatrix of correlations within and between sets:

$$\mathrm{R} = \begin{bmatrix} \mathrm{R}_{xx} & \mathrm{R}_{xy} \\ \mathrm{R}_{yx} & \mathrm{R}_{yy} \end{bmatrix} \tag{17.6}$$

Although we will not trace it out, maximizing Equation 17.5 under these restrictions leads to the following two expressions that are a form of characteristic equations:

$$(\mathrm{R}_{yx}\mathrm{R}_{xx}^{-1}\mathrm{R}_{xy} - r^{*^2}\mathrm{R}_{yy})\mathbf{w}_y = 0 \tag{17.7}$$

$$(\mathrm{R}_{xy}\mathrm{R}_{yy}^{-1}\mathrm{R}_{yx} - r^{*^2}\mathrm{R}_{xx})\mathbf{w}_x = 0 \tag{17.8}$$

Here, the "right" weights, $\mathbf{w}_x$ and $\mathbf{w}_y$, are the ones that have these relationships to the data.

This is a new type of characteristic equation in that we have R_{yy} or R_{xx} instead of I to be multiplied by the eigenvalue r^{*^2} and subtracted from a matrix. This poses only a temporary problem because there is a device for getting around it. We will leave it to the computers to do the getting around and simply accept the fact that the equations permit us to find r^*, $\mathbf{w}_x$, and $\mathbf{w}_y$.

The fact that this is an eigenvector-eigenvalue problem means that there is more than one solution. In fact, there are as many solutions as the ranks of X or Y, whichever is less, p or q. To simplify the discussion, we will assume that $p \geqq q$. Hence there are q different values of r^*, and each value has a corresponding $\mathbf{w}_x$ and a $\mathbf{w}_y$. The r^* may be arranged in order from largest to smallest (r_1^*, $r_2^*, \ldots, r_q^*$) and their corresponding $\mathbf{w}_x$ and $\mathbf{w}_y$ accordingly. The largest one, and its associated vectors, provides the answer to the question we started from. The vectors $\mathbf{w}_x$ and $\mathbf{w}_y$ provide the weights for the two batteries of variables that provide the largest possible correlation between the two sets of measures, and r_1^* is that correlation.

Other Combinations

Sometimes the pair of sets of weights that provide the largest correlation and the value of that correlation may provide sufficient information for the purposes of the investigator. But to answer some of our possible questions we need to know about the remaining sets of weights and their associated correlations. Each pair of weight vectors defines a pair of canonical variates, and the corresponding canonical correlation is the correlation between them.

The variates have the interesting property that all the correlations among them are zero, except the members of the same pair. This gives the matrix of correlations among the $2q$ variates the following form:

$$
R_{zz} = \left[\begin{array}{ccccccccccccccc}
1 & 0 & 0 & \cdot & \cdot & \cdot & 0 & & r_1^* & 0 & 0 & \cdot & \cdot & \cdot & 0 \\
0 & 1 & 0 & \cdot & \cdot & \cdot & 0 & & 0 & r_2^* & 0 & \cdot & \cdot & \cdot & 0 \\
0 & 0 & 1 & \cdot & \cdot & \cdot & 0 & & 0 & 0 & r_3^* & \cdot & \cdot & \cdot & 0 \\
\cdot & \cdot & \cdot & \cdot & \cdot & \cdot & \cdot & & \cdot & \cdot & \cdot & \cdot & \cdot & \cdot & \cdot \\
\cdot & \cdot & \cdot & \cdot & \cdot & \cdot & \cdot & & \cdot & \cdot & \cdot & \cdot & \cdot & \cdot & \cdot \\
\cdot & \cdot & \cdot & \cdot & \cdot & \cdot & \cdot & & \cdot & \cdot & \cdot & \cdot & \cdot & \cdot & \cdot \\
0 & 0 & 0 & \cdot & \cdot & \cdot & 1 & & 0 & 0 & 0 & 0 & 0 & 0 & r_q^* \\
 & & & & & & & & & & & & & & \\
r_1^* & 0 & 0 & \cdot & \cdot & \cdot & 0 & & 1 & 0 & 0 & \cdot & \cdot & \cdot & 0 \\
0 & r_2^* & 0 & \cdot & \cdot & \cdot & 0 & & 0 & 1 & 0 & \cdot & \cdot & \cdot & 0 \\
0 & 0 & r_3^* & \cdot & \cdot & \cdot & 0 & & 0 & 0 & 1 & \cdot & \cdot & \cdot & 0 \\
\cdot & \cdot & \cdot & \cdot & \cdot & \cdot & \cdot & & \cdot & \cdot & \cdot & \cdot & \cdot & \cdot & \cdot \\
\cdot & \cdot & \cdot & \cdot & \cdot & \cdot & \cdot & & \cdot & \cdot & \cdot & \cdot & \cdot & \cdot & \cdot \\
\cdot & \cdot & \cdot & \cdot & \cdot & \cdot & \cdot & & \cdot & \cdot & \cdot & \cdot & \cdot & \cdot & \cdot \\
0 & 0 & 0 & \cdot & \cdot & \cdot & r_q^* & & 0 & 0 & 0 & \cdot & \cdot & \cdot & 1
\end{array}\right] \tag{17.9}
$$

Thus the canonical variates are uncorrelated both within and between sets.

Inspection of the size of all the canonical correlations, supplemented by inferential methods that will be discussed later, provides the basis for deciding on the number of dimensions that the sets have in common. The inferential methods are identical in form to those that were described in the discriminant-manova context. These will be detailed below.

Structure Correlations

In a typical application, we would like to interpret the canonical variates, that is, to interpret what is being correlated across the two sets. The weights themselves, like regression weights, may be deceptive in this regard, although we will return to the topic of assessing them. The correlations of the variables with the canonical variates are often more interesting and logical in their patterns.

These correlations are called "structure correlations," much like the structure—as opposed to pattern—matrices of oblique factor loadings. In a sense, these correlations are telling us which observed variables are most like the variates. The structure correlations are often—but not always—provided by computer programs along with the weights. They may also be obtained from the matrices of weights and the original intercorrelation matrix. The correlations between the variables in the x-battery and the x-variates are

$$\mathrm{R}_{xc_x} = \mathrm{R}_{xx}\mathrm{W}_x \tag{17.10}$$

for the x-battery, and

$$\mathrm{R}_{yc_y} = \mathrm{R}_{yy}\mathrm{W}_y \tag{17.11}$$

for the y-battery. These are useful correlations for interpreting the nature of the variates within the batteries.

There is actually a second set of structure correlations, although they are rarely mentioned in this context. These are the correlations of the *variables* in one battery with the *variates* from the *other*. These correlations are useful in helping us understand which variables in one battery are associated with the variates from the other. They form the basis of what are called the "redundancy coefficients," which will be explained later. These correlations are described by the following formulas:

$$\mathrm{R}_{xc_y} = \mathrm{R}_{xy}\mathrm{W}_y \tag{17.12}$$

and

$$\mathrm{R}_{yc_x} = \mathrm{R}_{yx}\mathrm{W}_x \tag{17.13}$$

Notice that the notation needs close attention here: R_{xc_x} is the matrix of correlations of the x-variables with the canonical variates from the x battery, whereas R_{xc_y} contains the correlations of the x-variables with the y canonical variates. Of course, we should be able to compute these correlations in a more literal way, applying the weights to the observed variables and computing the various correlations. Fortunately, this is not necessary since we can take advantage of the fact that the canonical variates are already in standard score form to shortcut the process.

If R_{xc_x} and R_{xc_y} are computed, and examined closely, we find that there is a strong qualitative resemblance between the two. Examined more closely still, we find that the columns of R_{xc_x} and R_{xc_y} are proportional. Furthermore, the constants of proportionality for the different columns are the corresponding canonical correlations. The same is true for R_{yc_y} relative to R_{yc_x}. That is

$$\mathrm{R}_{xc_y} = \mathrm{R}_{xc_x}\mathrm{D}_{r*} \tag{17.14}$$

and

$$\mathrm{R}_{yc_x} = \mathrm{R}_{yc_y}\mathrm{D}_{r*} \tag{17.15}$$

where D_{r*} is a diagonal matrix containing the canonical correlations. This relationship is a consequence of Equations 17.7 and 17.8.

CANCOR VERSUS MULTIPLE REGRESSION AND COMPONENTS ANALYSIS

Conceptual Relations

As we have seen, cancor is a member of the family of methods that is a form of covariances of weighted combinations. In a way, cancor is much like a double principal components analysis. In fact, the R_{xc_x} and R_{yc_y} matrices are loadings on such components. In components analysis, the weights are chosen in order to account for as much variance as possible in the total collection of variables, and successive components are uncorrelated. In cancor, we choose *paired* weight vectors to maximize the correlation between the weighted combinations, and choose successive pairs to be uncorrelated. A closely related technique to cancor is called *interbattery* factor analysis. The purpose of this technique is to produce factors that account as accurately as possible for the correlations between variables in different batteries, rather than for all the intercorrelations among the variables, as in common factor analysis. Still another technique, called *redundancy* analysis, tries to account for variance in one battery using variates from the other battery.

Because cancor is of the weighted combinations family, we can think of it as a kind of components analysis—but not principal components—where instead of all components being uncorrelated, the $2r$ components have a certain restricted kind of correlation—that is, all correlations are zero except between the paired canonical components.

As noted at the outset, cancor is also a kind of multiple regression—in fact, a sort of multiple, multiple regression. Instead of using a multiple regression of each separate criterion in its observed form, we form a composite criterion, in fact several composite criteria, and find the regression of each on the predictors. Together, the predictors account for the same total amount of variance in the criteria as is accounted for by doing all the multiple regressions. The difference is that in multiple regressions of two criteria on a set of predictors we may find that both can be predicted fairly well, with an R^2 of say, .45 and .40. With cancor, we make two new criteria that are composites of the originals, one that we predict somewhat better and one somewhat worse than the original criteria.

Choosing One Method Over Another

The reader may wonder whether a new method is really necessary when the same purposes might perhaps be served by an old method. This is not the case, however, or is only true to a minor degree. Cancor is better suited to answer the types of questions we outlined above than is multiple regression or factor analysis, in either its components or common-factor forms. The advantage is one of precision. The other two methods can answer closely related questions, and indeed a multivariate virtuoso can often work wonders with limited resources, but cancor is more precisely directed toward certain questions.

We can sometimes answer questions of the interrelations of two sets of variables through factor analysis simply by extracting factors and inspecting the loadings to find those factors where variables from both batteries have appreciable loadings. Thus we could answer the question of whether there was variance shared between the two sets of variables. However, the question of the number of shared dimensions is more difficult to answer. While we can attempt to transform the factors in order to concentrate the loadings on as few factors as possible, we cannot be sure that this has been accomplished in an optimum way.

Another difference between factor analysis and cancor is that factor analysis tries to explain all the correlations in its common factor form or all the variance in its principal components guise, whereas cancor focuses on the interbattery relations. Concern with the within-battery information may pull the factor solution away from dimensions that might account more precisely for the between-battery correlations.

Multiple regression is also inadequate as a substitute for cancor. Even when there are several variables that are clearly criteria, and several independent predictors, separate multiple regressions may leave us with several questions. These particularly relate to whether the regression weights for the predictors are the same for predicting each of the criteria or not. Even if the regression weights are not the same, it could be that they are variations on only a couple of themes—for example, four or five different-looking sets of weights may be linear combinations of only a couple of basic ones. Note, however, that multiple regression may be answering the questions we are interested in if these linear combinations correspond with the prediction of the criteria as individual variables. If this is the case, then the often hard-to-intrepret composites furnished by cancor may only complicate matters, accompanied by some manifestation of insecurity such as wishy-washy conclusions or ego-defensive adamance.

Example

One of the empirical uses of cancor, noted earlier, involves comparing the relationship between variables of one domain with those from another. In the study in Table 17-1,* the data from the first domain consisted of two "values" inventories—Consciousness I–II, which represents the more or less traditional Western values, and Consciousness III, which represents the social-ecological values that became widespread during the 1960s and 1970s. The second domain included self-report personality inventories, consisting of scales called Normative, Authoritarian, Absorptive, and Humanistic. Chronological age was also added to the second domain to see how it would behave as predictor of the first domain.

The investigators were concerned with (1) whether there was any relationship at all between the two domains, (2) if there was a relationship, whether the two domains shared two, as opposed to only one, dimension, and (3) what the pattern of relationships was, relating variables in one domain with those in the other.

The correlation matrix for this study is shown in Table 17-1. Note that the two Consciousness variables have different patterns of correlations with the others, suggesting two somewhat separate dimensions of the relationship. A canonical correlation analysis was performed. The weights that define the canonical

*The data for this example are from a study by Krus and Tellegen (1975), and analyzed by Cliff and Krus (1976).

Table 17-1 Correlations among Seven Variables in Two Sets ($n = 58$)

	Nor	*Au*	*Ab*	*Hu*	*Age*	*C-I and II*	*C-III*
Nor	1.00	.32	−.47	.04	.08	.61	−.09
Au		1.00	−.16	.00	.00	.59	−.01
Ab			1.00	.37	−.31	−.43	.56
Hu				1.00	−.08	−.21	.41
Age					1.00	−.02	−.40
C-I and II						1.00	−.15
C-III							1.00

variates are given in Table 17-2, and the canonical correlations are given in Table 17-3. In both tables the variates are identified by Roman numerals, subscripted x or y to indicate battery of origin. The zero correlations of the noncorresponding variates are also included. All the intervariate correlations can, of course, be calculated by the formulas for correlations of weighted combinations. This is left as an exercise.

Table 17-3 shows that both canonical correlations are moderately high—.79 and .63, respectively. The first is considerably higher than any of the zero-order interbattery correlations, and the second is also higher. The existence of two largish correlations supports the notion that there are two separate dimensions of common variance between the two domains. Questions of statistical significance will be considered later.

It can be seen from Table 17-2 that one y-variate is almost entirely Consciousness I–II, whereas the other is almost entirely Consciousness III. The x-variates are more complex. The simple nature of the y-variates results from three characteristics of the data: (1) the patterns of correlations for the two y-variables are quite different; (2) y_1 and y_2 are nearly uncorrelated; and (3) there are only two y-variables. The weights for the x-variates are lower than those for the y's because there are more variables in the x-battery, and because each battery is being used to construct a variate with the same variance of 1.0.

The structure correlations are shown in Table 17-4. These are the correlations of the variables with the variates defined by weighting the variables in their own batteries, $\mathbf{R}_{xc_x}$ and $\mathbf{R}_{yc_y}$, from Equations 17.12 and 17.13, respectively. These correlations show that, as expected from the weights, the first y-variate is highly correlated with y_1 and the second with y_2. None of the x-variables correlates this highly with either x-variate. This is to be expected since there are five variables,

Table 17-2 Weights for Canonical Variates

	I_x	II_x	I_y	II_y
Nor	.449	.497	—	—
Au	.500	.314	—	—
Ab	−.304	.654	—	—
Hu	−.304	.196	—	—
Age	−.036	−.442	—	—
C-I + II	—	—	.921	.418
C-III	—	—	−.276	.973

Table 17-3 Canonical Correlations

	I_x	II_x
I_y	.79	.00
II_y	.00	.63

they do not intercorrelate highly, and no single variable stands out as a correlate of an observed y-variable, several of which have comparable interbattery correlations.

REDUNDANCY ANALYSIS

Accounting for Variance

We mentioned earlier that canonical correlation analysis is a kind of components analysis, although it is not a *principal* components analysis. It is a components analysis because it consists of weighting the observed variables to make new composite variates. There are differences, however, between canonical variates and principal components. One difference is that there are two separate sets of weights, and that the composites, instead of all being uncorrelated, have correlations of a particular kind. The paired variates from the separate batteries have maximal correlations, and all other variate correlations within and between batteries are zero.

In assessing the degree of relation between sets of measures and the number of dimensions of relation, we first use the canonical correlations. However, this is not the whole story, particularly when there are more than a few measures in each battery. We could have one or two high canonicals that are reflections of minor aspects of one or both sets of measures. On the other hand, a canonical correlation

Table 17-4 Within-Battery Structure Correlations and Variance-Accounted-For

	r_{xc_x}		r_{yc_y}		
	I_x	II_x	I_y	II_y	$\Sigma_c r_{jc}^2$
Nor	.74	.26			.62
Au	.69	.37			.61
Ab	−.70	.58			.83
Hu	−.40	.49			.40
Age	.12	−.62			.40
C-I + II			.96	.27	1.00
			−.41	.91	1.00
$\Sigma_j r_{jc}^2$	1.69	1.17	1.09	.90	
Percent	34	23	54	46	
$\Sigma_c \Sigma_j r_{jc}^2$	2.86		2.00		
Percent	57		100		

that is only moderately high could involve the major underlying dimension in both batteries, and therefore reflect a substantial relation between the two domains.

Thus investigators have sought ways of describing the overall degree of involvement of one battery with the other. This description would supplement the information provided by the canonical correlations themselves.

There are several ways of looking at this question. One simple approach is to consider separately the two sets of weights and the canonical variates they produce, asking how much of the variance in a battery as a whole, and of the individual variables it includes, is accounted for by the canonical variates constructed from that battery.

We can take advantage of the uncorrelated nature of within-battery variates to compute "variance accounted for" by squaring the structure correlations, like those in Table 17-4, summing this across the rows to see the variance accounted for in each variable by the battery's two canonical variates, and then summing down the columns to find the amount of variance in each battery that is accounted for by its first and second (and in some cases, third, fourth, or more) variates.

In Table 17-4, this information is located in the rows and columns labeled as summations. The sum of squared correlations in each column is given for each variate. Note that the first variate accounts for somewhat more of the variance in each battery than the second variate. This tends to be true in general, but in contrast to principal components, it does not need to be. Taken together, the two variates account for all the variance in the y-battery, as they must since they represent two vectors in a two-dimensional space defined by the observed variables. In the x-battery, the two variates account for 57 percent of the variance. This is relatively low, since almost any two orthogonal components would be expected to account for 40 percent of the variance in five variables.

We can see a little more of what is going on in the y-battery by looking at the right-hand column of the table. The summations in the sixth and seventh rows, referring to the variables in the y-battery, are 1.0, which merely indicates that the two components account for all the variance in these two variables, as they must. The five entries referring to the other variables show how much variance is accounted for by the two components from the x-battery. Less than half the variance of Age and of the Humanistic scale and almost all of the Absorptive scale are associated with these components while the Normative and Authoritarian scales fall between these two extremes.

The above description concerns the degree to which the variables are related to the composites formed from their own batteries. In view of the goals of cancor, the correlations of the variables with the composites formed from the *opposite* battery may be more informative. This information is given in Table 17-5, using the cross-battery correlations from Equations 17.14 and 17.15.

The correlations in Table 17.5, then, show the extent to which the individual variables in one battery are predictable from the canonical variates of the other. Because the variates are uncorrelated, summing the squares of these correlations again gives us variance accounted for, except that now the variance in one battery is accounted for by the other. In the example, we see that the first three x-battery variables have moderate correlations with the first y-variate although the sign of Absorption is opposite to the other two signs. The correlations for the x-variables with the other y-variate are smaller. Summing the squares of these correlations, we see that the first y-variate accounts for 1.06/5, or 21 percent of the variance of

Table 17-5 Between-Battery Structure Correlations and Variance-Accounted-For

	r_{yc_x}		r_{xc_y}		
	I_x	II_x	I_y	II_y	$\Sigma_c r_{jc}^2$
Nor			.59	.17	.38
Au			.55	.24	.36
Ab			−.55	.36	.43
Hu			−.31	.31	.19
Age			.09	−.40	.17
C-I and II	.76	.18			.61
C-III	−.33	.58			.44
$\Sigma_j r_{jc}^2$	.69	.36	1.05	.47	
Percent	34	18	21	9	
$\Sigma_j \Sigma_c r_{jc}^2$	1.05		1.52		
Percent	53		31		

the variables in the x-battery, whereas the second accounts for about 10 percent. Together, they account for 1.54/5, or 31 percent of the variance of the x-battery. The corresponding figures for the y-battery as accounted for by the x-variates are .69/2 (34 percent) for the first x-variate, .37/2 (18 percent) for the second, and 1.06/2 (53 percent) for both variates combined.

These total percentages—21 for the x's and 53 for the y's—are sometimes referred to as the *redundancies* of the batteries. They represent the amounts of variance in the individual batteries that can be accounted for by the opposite battery. We should note that here the x's account for more of the y's than y's account for x's; this is to be expected since there are more x's than y's. This kind of imbalance is sometimes considered a flaw in the use of redundancy analysis, although it can also be viewed as just a natural aspect of the process and the circumstances.

One further point may be noted. The sums of squared correlations in the last column of Table 17-5 are equal to the squared multiple R's of those variables with the *variables* of the other battery. That is, .17 is the R^2 of Age with the two Consciousness variables, .61 is the R^2 of Consciousness I–II with the five variables from the other battery, and so on. Thus *the redundancies are the average R^2 of the respective batteries, as predicted from the other battery.*

TRANSFORMATION OF VARIATES

Interpretability of Cancor

The investigator undertakes a multivariate analysis in order to understand what the data mean. In many applications of cancor, such understanding includes forming conclusions about the nature of the canonical variates and the nature of the underlying variables that lead to the correlations between the two batteries. The weights and the structure correlations of the variates themselves are often hard to interpret, as are the corresponding weights and structure correlations of untransformed principal components. Rotation is used with cancor for the same reasons as in components or factor analysis.

The idea of transforming the canonical variates was accepted long after its acceptance in factor analysis, and there are still restrictions on just what type of transformation is legitimate if the basic properties of cancor are to be preserved. Furthermore, while most of the nice properties of cancor are preserved, some are lost by the rotation process. Also, it is rare for transformations to be as successful as they often are in factor analysis in the sense of leading to clearly interpretable results. Nevertheless, transformation of canonical variates can be an aid in interpreting them.

Legitimate Transformations

Since the purpose of the cancor is to maximize the degree of relationship between the two sets of variates, we would like to preserve the overall degree of relationship between the two sets, even in cases where not all of the canonical variates are retained. Furthermore, having gone to the trouble of constructing *matching* variates from each battery, we would like to have a similar kind of correspondence between the transformed versions.

This turns out to be quite simple to do. Due to the uncorrelated nature of within-battery variates and their score character, any orthogonal transformation of the weight matrix W_x, such as

$$V_x = W_x T \tag{17.16}$$

leads to another set of uncorrelated, transformed variates. The correlations of these transformed variates with the original variables can be found in the usual way, or more directly, because they will be the same transformation of the original structure correlations:

$$R^*_{xc_x} = R_{xc_x}T, \text{ and } R^*_{yc_y} = R_{yc_y}T$$

If the *same* orthogonal transformation T is applied to W_y, then there is still a correspondence between the corresponding variates from each battery, because both have been transformed in the same way. Furthermore, the total predictability of one battery from the other is preserved. This preservation of predictability applies not only in the sense that the sum of the correlations between the corresponding transformed variates is the same as the sum of the canonical correlations, but also in the sense that the proportions of variance accounted for (that is, the redundancies) are the same. Thus, provided that the transformation is orthogonal, and that the same transformation is applied to both sets of variates, the degree of relation is the same for the transformed as for the untransformed data.

What we lose is the zero correlations between all but the corresponding variates from different batteries. There is now some degree of correlation between a transformed variate from one battery and transformed variates from the opposite battery besides the corresponding variate. The degree of this correlation depends on how substantial the transformation is and the degree of inequality of the canonical correlations. In the example data—where the amount of rotation is relatively small and the two canonical correlations are relatively equal—we will see that the transformed variates remain almost uncorrelated. This may be a small price to pay for the greater interpretability.

Operationally, the best approach to rotation is to obtain the structure correlations. Of the two possible sets—the within-battery correlations of Equations 17.12 and 17.13 (Table 17-4) or the cross-battery correlations of Equations 17.14 and 17.15 (Table 17-5)—we usually work with the within-battery sets. We can combine these into a single $(p + q) \times r$ matrix, where r is the number of variates to be rotated. Then we can apply some orthogonal rotation procedure such as varimax or one of its variants, or a procrustean rotation to a target hypothesis, or a combination of visual-intuitive methods. We would then use the resulting structure correlations to interpret the nature of the variates.

Example

The data in Table 17-4 provide a simple example of the procedure and its outcome. Figure 17-1 shows the configuration of points representing the variables of both sets, using their structure correlations as coordinates. A mild rotation, as indicated, separates the variates more clearly so that there is one pair where y_1 loads along with x_1, x_2, and $-x_3$, and another where y_2 loads along with x_3, x_4, and $-x_5$. The interpretation is simplified at least somewhat through the reduction in the sizes of the minor loadings. The structure is thus somewhat simplifed. The numerical values of the rotated structure correlations are given in Table 17-6.

Table 17-7 shows the correlations between the transformed canonical variates. Compared to Table 17-3, the correlation between the first pair is slightly reduced (.78 instead of .79) whereas the correlation between the second pair is correspondingly increased from .63 to .64. There is a very slight correlation (−.04) now between the noncorresponding transformed variates.

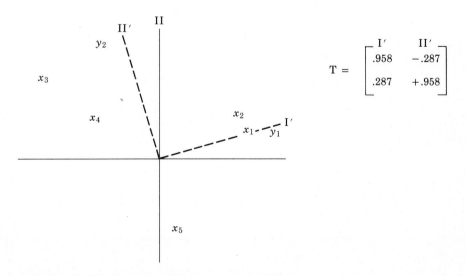

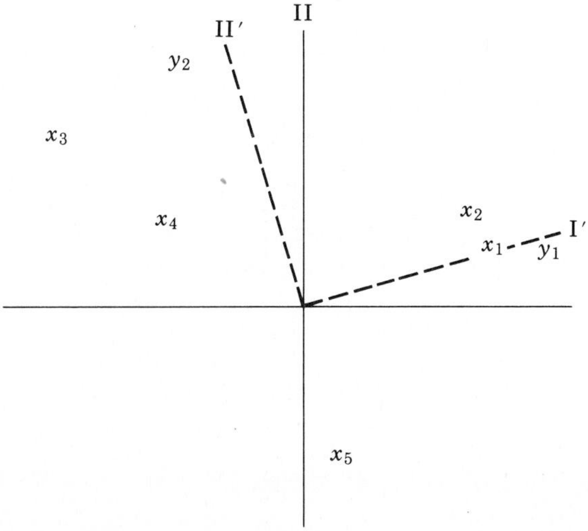

$$T = \begin{bmatrix} & I' & II' \\ & .958 & -.287 \\ & .287 & +.958 \end{bmatrix}$$

Figure 17-1 Rotation of canonical variates based on within-battery structure correlations.

Table 17-6 Transformed Within-Battery Structure Correlations

	r_{xc_x}		r_{yc_y}	
	I_x'	II_x'	I_y'	II_y'
Nor	.78	.04		
Au	.77	.15		
Ab	−.50	.76		
Hu	−.24	.59		
Age	−.06	−.63		
C-I and II			1.00	−.02
C-III			−.14	.99

Table 17-7 Correlations Between Transformed Variates

	I_x'	II_x'
I_y'	.78	−.04
II_y'	−.04	.64

Thus we see that it is possible to improve the interpretability of canonical correlation analyses, although it is necessary to carry out the rotations in a very particular way. The variates exhibit some degree of cross-correlation, although they remain uncorrelated within batteries.

STATISTICAL INFERENCE IN CANCOR

Varieties of Inferential Questions

So far, our discussion of cancor has been purely at the descriptive level, focusing on what the various numbers mean. In cancor, as everywhere else, it is necessary to remember that the obtained data are descriptive of the sample, and that what is characteristic of a sample is not necessarily characteristic of the population sampled.

Statistical inferences in cancor focus on the number and size of the canonical correlations and on testing hypotheses about the weights. As we shall see, the same questions are asked in discriminant function analysis or manova, and are answered in just about the same ways.

When we ask whether two sets of measures are irrelevant to each other we are in effect asking whether all the canonical correlations between the two are zero. Similarly, when we ask how many dimensions there are in common between the two measures, we are asking how *many* of the population canonical correlations are nonzero. Naturally, we would also like to get an accurate estimate of the size of the canonical correlations—that is, unbiased estimates of the population correlations and their confidence intervals.

Knowing the number of statistically reliable variates is not of much use if we do not know what they mean. Thus we also need to make inferences about the weights and the structural correlations as well, testing hypotheses and making confidence intervals with respect to these quantities.

As we shall see, there are routine ways to answer many of these questions. Most of these answers will be closely related to the methods from the previous chapter.

SIGNIFICANCE OF CANONICAL CORRELATIONS

The null hypothesis $\rho_1^* = \rho_2^* = \cdots = \rho_p^* = 0$ can be tested using either the generalized F-largest root or the Wilks' Λ criterion. This is because the null hypothesis is equivalent to saying $\mathbf{w}_x'\Sigma_{xy}\mathbf{w}_y = 0$ for all possible vectors $\mathbf{w}_x$ and $\mathbf{w}_y$. The details of these procedures are readily adapted from those used in the manova context.

The generalized F criterion here is expressed as

$$F_0 = \frac{t}{r}\frac{r_1^{*2}}{1 - r_1^{*2}} \tag{17.17}$$

where

$$t = n - p - q$$

$$s = \min(p, q)$$

$$r = |q - p| + 1$$

If F_0 exceeds the tabled value for the appropriate alpha level, the null hypothesis is rejected. Note that when $q = 1$, and $r_1^{*2} = R^2$, F_0 reduces to F.

The test for the number of "significant" canonical correlations is equivalent to successive tests using $\rho_m^* = \rho_{m+1}^* = \cdots = \rho_s^* = 0$ where Σr^{*2} is converted to an approximate χ^2. When $m = 1$, this is a test of the overall null hypothesis. If this hypothesis is rejected, then m is set to 2, and the test carried out again. If this, too, is rejected, then m is set to 3, and so on, until s is reached, or a null hypothesis is accepted for some value of m. If the mth null hypothesis is accepted, then the number of nonzero canonical correlations is concluded to be $m - 1$. In these tests we would use Bartlett's chi-square approximation:

$$\chi_B^2 = -[n - 1 - \tfrac{1}{2}(p + q + 1)] \sum_{j=m}^{s} \log_e (1 - r_j^{*2}) \tag{17.18}$$

for $(p - m + 1)(q - m + 1)$ degrees of freedom.

As in the manova context, whether we use Equation 17.17 or Equation 17.18 with $m = 1$ as the overall test of no association between the two batteries depends on our expectations concerning the nature of the interbattery relations. If an investigator expects a single dominant dimension, Equation 17.17 has more power; if he expects several more nearly equal ρ^*, then Equation 17.18 is preferable. Table 17-8 shows the results of these tests for the example data.

Table 17-8 Significance of Canonical Correlations

Largest Root Test

$r_1^* = .79 \quad t = 58 - 5 - 2 = 51 \quad s = \min(5, 2) = 2 \quad r = |p - q| + 1 = 4$

$$F_0 = \frac{51}{4} \frac{.79^2}{1 - .79^2} = 21.17 \quad p < .01$$

Bartlett's χ^2

i	$r_i^{*^2}$	$\ln(1 - r_m^{*^2})$	$\sum_{i=m}^{s} \ln(1 - r_m^2)$	χ^2	df^*	p
1	.6241	−.9784	−1.4841	78.657	10	<.001
2	.3969	−.5056	−.5056	26.801	4	<.001

$n - 1 - \frac{1}{2}(p + q + 1) = 53$

$df_1 = (5 - 1 + 1)(2 - 1 + 1) = 10$

$df_2 = (5 - 2 + 1)(2 - 2 + 1) = 4$

Estimating ρ^*

In multiple regression we learned that R can be a substantial overestimate of P if the number of predictors is an appreciable fraction of the number of observations. In cancor we not only have the correlation between a variable and the composite for which the weights were optimally chosen, but the correlation between two optimally weighted composites. We would therefore expect that r^* is correspondingly likely to be even more of an overestimate of ρ^* than R is of P.

This assumption is certainly accurate. Unfortunately, there does not seem to be any well-established way to correct r^* to get an unbiased estimate of ρ^*. Nevertheless, on the basis of intuitive principles, we would expect that the amount of inflation is proportional to the number of parameters estimated. Therefore, the following formula serves as a way of getting an unbiased estimate of ρ^* from r^*:

$$\hat{\rho}^{*^2} = r^{*^2} - (1 - r^{*^2}) \frac{p + q}{n - p - q - 1} \tag{17.19}$$

This formula is the same as Equation 9.3, except that p has been replaced by $p + q$ on the assumption that each additional parameter estimated will spuriously contribute one nth of the sample variance.

There do not seem to be any well-known procedures for forming confidence intervals of ρ^* either, and the problem is undoubtedly a complex one. Some rough idea might be formed by applying the Fisher z' transformation to $\hat{\rho}^*$ from Equation 17.19 and then taking $(n - p - q - 3)^{-1/2}$ as the standard error of z', assuming that the sampling distribution was normal. Unfortunately, such a procedure is likely to be valid only when it is needed least, that is, where n is very large relative to $p + q$. Perhaps the most rational conclusion is not to worry about such niceties as confidence intervals and unbiased estimates for ρ^* since no one else seems to worry.

INFERENCES ABOUT WEIGHTS

Multivariate Regression

When the two batteries of measures clearly constitute one battery of predictors and one battery of criteria, we may not really want to combine the criteria into a composite. Not only are such composites hard to interpret, but we may have genuine interest in predicting each individual criterion, and in inferences concerning what is predicting each. In such cases we may want to consider the *multiple* criteria and *multiple* prediction equations when inferences are made. Thus we need to deal, at least briefly, with what is called *multivariate regression.* This also provides a foundation for inferences about the weights in a cancor analysis.

In Chapter 9 we learned that just posing questions about regression weights can be tricky because inferences about one weight have to be separated from inferences about the others. We can expect such complexity to increase in the multivariate regression case because there are now multiple composites to consider. Thus we have to think about how to formulate the inferential questions regarding the weights for multivariate regression.

Cancor is, of course, a generalization of multiple regression; so is multivariate regression, although it is not as much of a generalization as cancor. This generalization means that there are a number of criteria, as well as a number of predictors, and that the regression of each criterion on the predictors must be found. Thus we have a set of regression weights for each dependent variable, and a general formula

$$\mathrm{B}_{xy} = \mathrm{S}_{xx}^{-1}\mathrm{S}_{xy} \tag{17.20}$$

where B is a p-predictor-by-q-criteria matrix of regression coefficients. Each column of B contains the regression weights that would have been obtained if the corresponding y-variable had been the only criterion. We also have the corresponding matrix of standard score regression weights:

$$\mathrm{B}_{xy}^{*} = \mathrm{R}_{xx}^{-1}\mathrm{R}_{xy} \tag{17.21}$$

Note that we may wish to reverse the roles of the batteries, predicting x from y, and wish that this would lead to a different set of regression coefficients, a q-by-p matrix B_{yx}, and so on. In some contexts, we may wish to consider the regression both ways. There are then $p + q$ possible multiple correlations, each with its own standard errors of estimate, and so on, one for each variable that is being predicted.

We test the hypothesis that all the multiple correlations are zero using the largest root test or Wilks' Λ. This also tests whether all the interbattery correlations are all zero, because this is the only way the multiples could be zero. Thus we can use these tests in the protection-levels sense to control the analysiswise alpha level in multivariate regression. If we need to be similarly conservative about inferences concerning the regression coefficients, we can apply the Roy system that was used in the manova-discriminant context. That is, we can make inferences by constructing confidence intervals around regression coefficients and weighted combinations of coefficients. If zero falls within the confidence interval, then we can accept the null hypothesis about that coefficient. At the same time, we must remember to be careful about just what question is being answered.

A General System for Inferences about Weights

At this point it may be useful to present the complete hierarchy of systems for inferences about regression coefficients, beginning with the simple case of an inference about a single coefficient. We will describe these systems in terms of confidence intervals around the given coefficient or combination of coefficients.

We begin with a formula that was first presented in a different form in Equation 9.7, which represents the confidence interval for a single regression coefficient:

$$b_j \pm (F_1 \hat{\sigma}_e^2 s^{jj})^{1/2} n^{-1/2} \tag{17.22}$$

This equation represents the $1 - \alpha$ confidence interval for β_j, the regression of y on x_j. Here, F_1 is the ordinary F statistic with 1 and $n - p$ degrees of freedom, and $\hat{\sigma}_e^2$ is the estimate of the error variance, that is, $ss_{dev}/(n - p)$. The term s^{jj} is the jth diagonal element of S_{xx}^{-1} and it reflects in part the predictability of x_j from the other predictors. Recall that this interval is appropriate if we are taking a parameterwise or a priori stance with respect to this particular regression coefficient; the probability is $1 - \alpha$ that this particular coefficient is indeed within the indicated interval.

If we wish *all* the coefficients to be in their c.i. with probability $1 - \alpha$, then Equation 17.22 must be modified as the Scheffé interval

$$b_j \pm (pF_p \hat{\sigma}_e^2 s^{jj})^{1/2} n^{-1/2} \tag{17.23}$$

where all the terms are the same except that F_p is the F with p rather than one numerator df, and the multiplying factor of p is introduced. Thus the c.i. of Equation 17.23 is typically much larger than Equation 17.22 by a factor of nearly $p^{1/2}$. This increase in width is because Equation 17.23 allows not only for the uncertainty about individual coefficients like 17.22 but also for combinations of coefficients, such as differences between coefficients for different variables, or for weighted sums of coefficients, and so on. In fact, there is another general expression we saw earlier as Equation 9.21

$$\mathbf{c}'\mathbf{b} \pm (pF_p \hat{\sigma}_e^2 \mathbf{c}' \mathrm{S}_{xx}^{-1} \mathbf{c})^{1/2} n^{-1/2} \tag{17.24}$$

in which $\mathbf{c}$ is a vector containing weights for combining the coefficients. The previous Equation 17.23, can be seen as a special case where the weight vector is all zeros except there is a 1.0 in the jth position, that is, just for j. If the c.i. represents a difference between two regression weights, then one of these weights has an entry in $\mathbf{c}$ of -1 and the other a weight of $+1$. There are many other possibilities, of course. If such contrasts represented by the weights are a priori, or if the alpha level is taken contrastwise, then the pF_p term in Equation 17.24 will be replaced by F_1, as in 17.22.

In the multivariate regression case, there is even more generality possible because now we can weight the criteria as well as the predictors' coefficients. We may wish to make c.i. statements not only about p regression coefficients, but about pq coefficients, and still wish to have *all* pq coefficients captured by their c.i. with probability $1 - \alpha$. If there is an a priori hypothesis about any coefficient, then Equation 17.22 still applies; the only restriction is that the error variance term must be for the appropriate dependent variable. Assuming a more general situation, one where the alpha level is to be controlled across all pq coefficients, the following formula is appropriate:

$$b_{jk} \pm \left[\frac{(n-p)}{t} rF_{0rst}\hat{\sigma}^2_{ek} s^{jj}\right]^{1/2} n^{-1/2} \tag{17.25}$$

Here, F_{0rst} is the generalized F_0 with r, s, and t parameters as in Equation 17.17. The $\hat{\sigma}^2_{ek}$ term refers to the error variance for y_k. A heavy penalty is paid here for the number of parameters that are required to be simultaneously in their c.i.

Conceivably, we may need to define a c.i. for a combination of coefficients, as in Equation 17.24 for the single-criterion case. This can be done by combining the left segment of Equation 17.25 with the right segment of Equation 17.24:

$$\mathbf{c}'\mathbf{b}_k \pm \left[\frac{(n-p)}{t} rF_{0rst}\hat{\sigma}^2_{ek}\mathbf{c}'\mathrm{S}_{xx}^{-1}\mathbf{c}\right]^{1/2} n^{-1/2} \tag{17.26}$$

Note that the vector of regression coefficients has been subscripted to identify the y-variable being predicted.

As a final step in this system, we allow for the possibility of combining columns of B as well as elements of a single column as in Equation 17.26. This may seem a bit strange; it is hard to imagine at first just what this would mean and when it might make sense. The context in which this final step does make sense is in a linear combination of the y's themselves. Thus we have

$$\mathbf{c}'\mathrm{B}\mathbf{a} \pm \left[\frac{(n-p)}{t} rF_{0rst}\mathbf{a}'\mathrm{S}_{ee}\mathbf{a}\mathbf{c}'\mathrm{S}_{xx}^{-1}\mathbf{c}\right]^{1/2} n^{-1/2} \tag{17.27}$$

as a confidence interval for a linear combination of coefficients, defined by **c**, for a linear combination of dependent variables, defined by **a**. If **a** is a vector with a 1.0 in the kth position and zero everywhere else, then Equation 17.27 reduces to Equation 17.26, just as Equation 17.26 reduces to Equation 17.25 for a single coefficient. Equation 17.27 introduces a new term, S_{ee}, which is the unbiased estimate of the covariance matrix of the deviations from regression:

$$\mathrm{S}_{ee} = \frac{n-1}{n-p}(\mathrm{S}_{yy} - \mathrm{B}'\mathrm{S}_{xx}\mathrm{B}) \tag{17.28}$$

The preceding discussion provides the complete sequence of generality for the confidence intervals for regression coefficients. Equation 17.27 is particularly relevant in the context of cancor. One obvious possibility for the **a** vector is that it can be one of the $\mathbf{w}_y$ that define a canonical variate. If it is, then, as if by magic,

$\mathbf{Bw}_y$ will equal $\mathbf{w}_x$. Thus, Equation 17.27 provides a set of c.i. for the canonical weights, and even for combinations of them (sums, differences, and so on) defined by a combining vector **c**. In the next section we will consider other methods for inferences about the canonical weights. The process of calculating c.i. is illustrated in Exhibit 17-1.

$$\underset{[s_{xx}^{jj} = 1/(1 - R_j^2)]}{\underset{\mathbf{S}_{xx}^{-1}}{\begin{bmatrix} 1.499 & -.350 & .810 & -.352 & .103 \\ & 1.116 & .022 & .009 & .035 \\ & & 1.757 & -.648 & .428 \\ & & & 1.248 & -.073 \\ & & & & 1.119 \end{bmatrix}}} \quad \underset{\mathbf{B}}{\begin{bmatrix} .431 & .137 \\ .433 & .022 \\ -.121 & .474 \\ -.191 & .209 \\ -.107 & -.247 \end{bmatrix}} \quad \underset{(s_{ee}^{kk} = 1 - R_k^2)}{\underset{\mathbf{S}_{ee}}{\begin{bmatrix} .425 & -.004 \\ -.004 & .617 \end{bmatrix}}}$$

Exhibit 17-1(a) Calculation of confidence intervals and significance of regression weights in the multivariate case. Here, the correlation matrices of Table 17-1 are treated as if they were covariance matrices. The matrix of regression weights is computed using Equation 17.21 and $\mathbf{S}_{ee}$ is computed using Equation 17.28.

b_{j1} intervals	b_{j2} intervals	$b_{j1} \pm$ c.i.	$b_{j2} \pm$ c.i.
(4.02) (.425) (1.499)/58	(4.02) (.617) (1.499)/58	*.431 ± .210	.137 ± .253
(4.02) (.425) (1.116)/58	(4.02) (.617) (1.116)/58	*.433 ± .181	.022 ± .218
(4.02) (.425) (1.757)/58	(4.02) (.617) (1.757)/58	−.121 ± .227	*.474 ± .274
(4.02) (.425) (1.248)/58	(4.02) (.617) (1.248)/58	−.191 ± .192	.209 ± .231
(4.02) (.425) (1.119)/58	(4.02) (.617) (1.119)/58	−.107 ± .182	*−.247 ± .219

Exhibit 17-1(b) Parameterwise univariate confidence intervals (Equation 17.22) for the elements of B from part (a). Note that the c.i. for b_{j1} are shorter than for the b_{j2} due to the greater predictability of y_1. Note also the effect of the R_i^2, the multiple correlations of the individual predictors, with the other *x*-variables. Starred intervals do not include zero, indicating that $\beta_{jm} = 0$ can be rejected.

b_{j1} intervals	b_{j2} intervals	$b_{j1} \pm$ c.i.	$b_{j2} \pm$ c.i.
(5) (2.39) (.425) (1.499)/58	(5) (2.39) (.617) (1.499)/58	*.431 ± .362	.137 ± .436
(5) (2.39) (.425) (1.116)/58	(5) (2.39) (.617) (1.116)/58	*.433 ± .312	.022 ± .376
(5) (2.39) (.425) (1.757)/58	(5) (2.39) (.617) (1.757)/58	−.121 ± .391	*.474 ± .472
(5) (2.39) (.425) (1.248)/58	(5) (2.39) (.617) (1.248)/58	−.191 ± .331	.209 ± .398
(5) (2.39) (.425) (1.119)/58	(5) (2.39) (.617) (1.119)/58	−.107 ± .314	−.247 ± .376

Exhibit 17-1(c) Analysiswise univariate confidence intervals, according to Equation 17.23. These are larger than those in part (b) by a factor of $(5 \times 2.39/4.02)^{1/2} = 1.724$. By this criterion, b_{52} is no longer significant and b_{32} is borderline.

Contrast	$\beta_{11} - \beta_{21}$	$\beta_{32} + \beta_{42} - \beta_{52}$	$\beta_{11} + \beta_{21} - \beta_{31} - \beta_{41} - \beta_{51}$
$\mathbf{c}'$	1 −1 0 0 0	0 0 1 1 −1	1 1 −1 −1 −1
$\mathbf{c}'\mathbf{S}_{xx}^{-1}\mathbf{c}$	3.315	2.118	4.001
$\hat{\sigma}_{\ell k}^2$	.425	.617	.425
$\mathbf{c}'\mathbf{b}$	−.002	9.30	1.283
Univariate c.i.			
"a priori"	$[(4.02)(.425)(3.315)/58]^{1/2}$ =.312	$[(4.02)(.617)(2.118)/58]^{1/2}$ =.301	.343
"post hoc"	$[(5)(2.39)(.425)(3.315)/58]^{1/2}$ =.539	$[(5)(2.39)(.617)(2.118)/58]^{1/2}$ =.519	.592
Multivariate c.i.			
$(n-p)/t$	53/51 = 1.039	1.039	1.039
$F_{0,4,2,50}$	4.03	4.03	4.03
c.i.	$[(1.039)(4)(4.03)(.425) \times (3.315)/58]^{1/2}$ (−.002) ± .638	$[(1.039)(4)(4.03)(.617) \times (2.118)/58]^{1/2}$ .930 ± .614	1.283 ± .701

Exhibit 17-1(d) Confidence intervals for combinations of coefficients. The row labeled "contrast" shows the combinations of regression coefficients that correspond to various hypotheses. The first corresponds to the difference between weights for x_1 and x_2 in predicting y_1; the second column corresponds to the pattern of signs for the larger weights in b_2; the third column follows the pattern of signs in b_1. The c' row shows the corresponding weight vectors. Then $c'S_{xx}^{-1}c$ and $\hat{\sigma}_{\ell k}^2$ are the major terms in the sampling variance of the weights. In the univariate case, the F values are the same as in part (b) and part (c). Note the wide interval for $\beta_{11} - \beta_{21}$ in both cases. In the multivariate case, F_{0rst} has parameters $r = |q - p| + 1 = 4$, $s = \min(p, q) = 2$, $t = n - p - q - 1 = 50$, and the c.i. is given by Equation 17.25.

Inferences about Canonical Weights

Inferential procedures for the weights in cancor are not widely used. This is somewhat surprising; it is true that the problem of inferences about weights is conceptually quite complex because of the double-weighting and number of coefficients that are potentially involved. Even so, the similar complexity of discriminant analysis and manova has not proved to be a barrier. Indeed, it is surprising that there has not been a more ready adaptation of the methods for deciding about the importance of variables in a discriminant analysis to the similar decisions in cancor.

In this section we will try to point out some of the parallels between discriminant analysis and cancor and also to point out how some of the questions about cancor weights can be answered by other means, given a modest amount of ingenuity.

In thinking about the "significance" of some particular canonical weight w_{jm}, we are likely to bring to the situation habits of thought from other statistical methods such as regression and manova, and to think of a null hypothesis such as $H_0{:}\omega_{jm} = 0$, where ω_{jm} is the population value of the weight of the jth variable on the mth canonical variate. The null hypothesis concerns whether x_j is irrelevant to the mth canonical variate in the population. As soon as this is stated, we realize that there are two variates involved, one variate for the battery that contains x_j and one variate for the corresponding variate in the other battery. The mth population variate in the other battery will be differently composed than the sample variate. Indeed, the mth population composite in the y-battery may be quite different from the mth variate in the sample. Thus the null hypothesis makes relatively little sense when stated this way.

A possible way out of the dilemma is to change the question slightly and treat the composite in the other battery as fixed in its definition. That is, $\mathbf{w}_{ym}$ is taken as equal to ω_{ym}. Essentially, we are saying, "suppose there is a population variate $z_m = Y\omega'_{ym}$." Now we can treat this as an ordinary variable and consider the regression of this variable on the x_j. The significance of the weights is then evaluated as if this were a regression analysis. (Naturally, we wish to be confident that the corresponding canonical correlation is itself significant before undertaking such an analysis.) These regression weights can be obtained directly because they are $r_m^*\mathbf{w}_{xm}$. It is also possible to use Equation 17.27 for the purpose of making inferences about them, setting $\mathbf{a} = \mathbf{w}_{ym}$. It is probably simpler in the long run to do the analysis in a more literal fashion, combining the y-variables using w_{ym}, using the result as a dependent variable in a regression analysis with the x's as predictors, and applying the methods used in Chapter 9. The preceding discussion then suggests a way of assessing the relevance of the jth x-variable to the mth y-variate.

We may also wonder whether x_j is relevant to *any* y-variate. We could, of course, test the significance of x_j in the manner outlined above for predicting each of the y-variates. However, if there are a number of y-variates, we may have concerns about the true alpha-level. It would also seem more elegant to have a test that would answer the question all at once, although there is no widely adopted method for doing this. It would seem that the solution would be virtually identical to testing the significance of the contribution of a single variable to the discriminant function.

Although there is no direct test of the generalized hypothesis concerning whether a predictor makes a unique contribution to the prediction of the y-variables, there is a test to determine whether a given x is unrelated to all of the y-variates. Because the y-variates are linear combinations of the y-variables, finding out whether x_j is uncorrelated with all the y-variables is equivalent to finding that it is uncorrelated with the y-variates. This test can be done by finding the multiple correlation of x_j with all the y's. If this correlation is unrelated to all of the y's, its population multiple will be zero, and we have answered the question. Thus, while there is no standard way to answer the question of whether x_j is *necessary* for predicting any variance in the y-battery, it is still possible to test the hypothesis that x_j is not *sufficient* to explain any of the y-variance.

INTERRELATIONS AMONG METHODS

Linear Models

Without exception, the methods of statistical analysis described in this text are linear or bilinear. This means that all the models for the data and methods of analysis refer to *linear* combinations of variables in the sense that the combinations are weighted sums in which the weights as well as the individual variables are in linear form. In some cases, as in cancor, there are two sets of weights, called bilinear methods. The statistical models consist of functions in which one or more derived variates are equal to the sum of products of observed variables with weights for each variable. The weights enter the model in only a linear way; there are no squared weights in the models themselves and no weights as exponents. The variables likewise enter the models in a linear fashion.* All of the methods that we have described are different ways of investigating the variance and covariances of such linear combinations.

In each case, we have a matrix of scores of the observational units—persons, countries, rats, commercial products, and so on—on a number of variables. The analysis itself involves taking a linear combination of these variables, perhaps several linear combinations.

Perhaps the simplest example is multiple regression. Here the model expresses the dependent variable as a linear function of the predictors and expresses the error variance as the variance of the deviations from these predictions. In anova, the situation is the same except that the regression is on the dichotomous dummy variables. In anova, however, we add the possibility of recombining the predictors in various ways, that is, transforming the predictors by linearly combining the dummy variables. Partial regression analysis is also based on forming linear combinations, although in partial regression there are two or more levels to the process. After finding the regression of each of several variables on the set of control variables, we treat the deviations from these regressions in a multiple regression form. (It is possible to express everything as a kind of multilevel weighting system at the cost of considerable complication in the notation.) Factorial designs in anova can be viewed as a form of partial regression analysis; indeed, where the design is nonorthogonal, we fall back on the hierarchical type of partial multiple regression in anova.

In the various forms of factor analysis, we express the observations as linear combinations of variables, but here the variables are linear combinations of derived (components) or latent (common factor) variables. In the factor analytic process, we estimate these derived or latent variables as linear combinations of the observed variables. Our interest is focused on the correlations and variances of these derived and latent variables, and the regression of the observed variables on them. In the common factors case, only these correlations and variances are determinate, not the scores on the factor variables themselves.

Discriminant analysis or manova involves both multiple groups and multiple variables. The object of the analysis is to form a linear combination of the variables to maximize the differentiation among the groups. In discriminant analysis, group membership assumes the role of the dependent variable and the continuous variables are the independent variables, whereas in manova, the roles of the dependent and independent variables are reversed. Manova also adds the possibility of factorial or other structures in the group membership—that is, linear transformation of the group membership variables.

Canonical correlation analysis also involves two sets of variables, but unlike discriminant analysis and manova, the sets of variables are continuous. The objective of cancor is to form linear combinations of both sets of variables in order to maximize the correlation between the combinations. In fact, cancor can be thought of as the general case that includes all the other analyses involving two sets of observed variables. A statistician faced with exile and allowed to take

---

*In polynomial regression, the independent variable is used to generate squared, cubed, and so on, values of itself. But once these variables are generated, there are only linear combinations of them. Indeed, we (or the computer) can ignore the fact that there is anything special about these variables. Once they are there, they are treated in a linear fashion.

along only a single computer program would slyly take one for doing cancor, since this program can be persuaded to do all the other analyses.

Organizing the Methods

Other methods considered in this text revolve around the existence of a matrix of data in which there are at least two variables and more observations than variables. All the methods we have considered can be thought of in terms of the nature of these variables and their roles in the analysis. The differentiation is a series of dichotomies.

The first dichotomy concerns whether the variables are divided into two groups or not (one-set-two-set). The other dichotomies apply only in the two-set case. The second dichotomy separates cases in which there is only one variable in a set from cases in which there are multiple variables in a set. The third dichotomy concerns whether the variables in a set are dichotomous or continuous—that is, qualitative or quantitative. The fourth distinction is between dependent and independent variables. The various methods of analysis can be defined in terms of their positions on all these dichotomies.

Classification When the variables are in a single set, there are only a few possibilities. For example, we could simply compute the correlations among the variables and look at them. Or, we could form one or more linear combinations of the variables and study their variances, covariances, and correlations. Finally, we could perform one or the other form of factor analysis of the variables.

When there are two sets of variables, there are various ways to characterize the two sets. In the simplest case—a single variable in each set—we consider whether both variables are continuous or not. If both are continuous, we have univariate regression. If one set of variables is dichotomous and the other continuous, the regression turns into the comparison of group means if the dichotomy is the independent variable and point-biserial correlation when the dichotomy is the dependent variable. The case in which both variables are dichotomous is not treated here because linear analysis of the observed variables is not very satisfactory in such cases.

If there are multiple variables in one set and a single variable in the second set, then there are several possibilities. If all variables are continuous, and the single variable is the dependent variable, then we have multiple regression. If the single variable is the independent, then the situation is statistically uninteresting. If the single dependent variable is a dichotomy, then we have two-group discriminant analysis. If the single dichotomy is the independent variable, then we have a two-group manova. Both instances of dichotomous variables can be analyzed via multiple regression. When the single variable is continuous and the multiple set is dichotomous, then we have anova.

Both multiple regression and anova can be elaborated through the introduction of control variables. In multiple regression this results in the multiple-partial correlation analyses, including the hierarchical multiple regressions. In the anova case, we have the factorial designs and other forms of experimental control of the variables. When some or all of the control variables are continuous, the result is ancova. All of these partialing analyses can be recast in terms of transformations of the predictor variables.

If there are multiple variables in both sets, then there are several corresponding possibilities. When the multiple qualitative variables are the dependent variables, we have discriminant analysis. When they are the independent variables, we have various forms of manova. When both sets of variables are continuous, we have canonical correlation analysis. Cancor can be employed either when one set of variables is dependent and the other independent or when there is no distiction in the roles of the two sets.

The reason the cancor program is a good choice for an exiled statistician is that the computer does not really notice whether the variables are dichotomous or continous; it does the same analysis either way. When there is a single variable in the set, the program will treat it as a special case of a multivariate analysis, and everything works out anyway. The calculational process may be somewhat inefficient by choosing the more elaborate multivariate model rather than multiple regression or anova, but computing costs are low and continue to decrease.

Thus all of the methods described in this text fall within the same general framework. There are, of course, a number of other recurring themes. One is the concept of degrees of freedom. This refers to the rank, or potential rank, of some matrix or the dimensionality of the corresponding vector space. Degrees of freedom are "lost" as linear constraints are placed on the spaces or matrices, and the statistical freedom of random variables is correspondingly restricted; the vectors are being confined to subspaces.

The goal of this text has been to provide some understanding of the nature of the statistical analyses that are appropriate for the multivariate data. Still, we should not lose sight of the fact that by far the most thought should go into the *gathering* of the data, not its analysis—"garbage in, garbage out." Furthermore, we can learn as much or more from simple analyses, such as plotting variables against each other, as from the most staggering computations.

SUMMARY

Cancor is a very interesting idea, offering potential answers to a number of useful and interesting empirical questions. In many instances these empirical questions can be answered directly, although sometimes they need to be reformulated in order to fit with available methodology.

To some extent, however, cancor offers more possibilities than are realized. Perhaps this is due to the very generality of the method, which makes it hard to grasp tightly. As this is being written, however, cancor methodology is undergoing rapid growth and a number of new developments. Thus there is a considerable likelihood that more of its promise will be fulfilled in the near future. Even in its present state, there are a variety of empirical purposes for which cancor is well suited.

Cancor is applied when a set of measures can be divided into two subsets on meaningful grounds, such as a set of independent and a set of dependent variables. Cancor finds two vectors of weights, one for each subset, such that the correlations between the combinations is as high as possible. The solution to finding the weights turns out to be an eigenvalue-eigenvector problem, and there are as many solutions as the rank of the smaller subset of variables. Each solution provides paired sets of weights that define canonical variates, and the correlations between canonical variates are the canonical correlations.

The canonical variates are uncorrelated within batteries, and a canonical variate has zero correlations with all the canonical variates in the opposite battery except the corresponding variate. Cancor can be used to test the hypothesis that all the between-battery correlations are zero. The number of canonical correlations that are significant, statistically or substantively, is equal to the number of dimensions shared by the two batteries.

Structure correlations between the observed variables and the canonical variates, both between and within batteries, can be calculated to help in interpretation. These can be transformed by the rotational methods of factor analysis, although the same transformation must be applied to the structure correlations of both batteries.

Canonical correlation analysis provides a general methodology, and most of the other multivariate methods are special cases of this methodology. In manova and discriminant analysis, one subset of measures is defined as dummy variables or transformations of these variables. In multiple regression, one battery contains only a single variable. In anova, one battery has only a single variable and the other consists of dummy variables or transformations of them. In partial correlation analysis and ancova, the analysis is carried out on deviations from regression on the covariates or control variables. Thus, the methods we have described can all be put in the same general framework.

COMPUTER APPLICATIONS

SAS

Using one of the standard statistical packages, canonical correlation analysis can be quite straightforward. There are relatively few options and decisions to be made, although not all the relevant information may be readily available.

The SAS procedure CANCORR has a few peculiarities that may be worth clarifying. For example, it does all the calculations in raw-score metric. Instead of the observed correlation matrix, it prints out the sums-of-squares-and-products-of-deviations matrix that we have sometimes called T_t. Thus, if one would like to know what the correlations are, it is necessary to employ PROC CORR separately, or input standarized variables. The user should also be aware that the characteristic vectors (eigenvectors) printed out are also computed in this raw-score metric. Thus the general size of the coefficients of these eigenvectors depends in a complex way on the relative sizes of the variances of the x-variables and y-variables.

The SAS subprogram prints out the canonical correlations and the corresponding values of Bartlett's chi-square statistic based on Wilks' Λ. It calculates the probability values of the statistic directly with respect to each canonical correlation. A nonsignificant χ^2 means that the null hypothesis is accepted for the corresponding ρ^* and all smaller ones.

SAS does provide the correlations of the observed variables with the canonical variates of their own batteries. You may recall that one can compute the correlations of the variables with the variates of the *opposite* battery by multiplying these within-battery correlations by the corresponding canonical correlations. From these, the redundancies are available.

One feature of the CANCORR procedure that may puzzle some users is the presence of "EDF" and "RDF" options. These allow for the possibility that the canonical correlation analysis may be done on variables which themselves are deviations from regression. That is, one may partial some control variables out of the variables involved in the cancor analysis. If so, then the tests of significance should be based on the residual degrees of freedom. The user specifies this by either entering $s + 1$, where s is the number of control variables, as the RDF, or by entering $n - s - 1$ as the EDF.

To rotate the structure correlations in SAS, we use the FACTOR procedure and input the canonical structure correlations and specify "TYPE-FACTOR."

SPSS^X

The original SPSS subprogram CANCORR was equally straightforward in its approach. Its advantage was that it would print out the correlation matrix (Statistic 2), although it did not print the structure correlations. Unfortunately, the straightforward CANCOR program is not available in SPSS^x. Canonical correlation analysis must be carried out with the all-purpose MANOVA subprogram. Stating MANOVA followed by the list of variables in the dependent battery and WITH followed by the list of variables in the predictor battery, leads to a canonical correlation analysis. However, the user must also input the subcommand PRINT= followed, rather confusingly, by DISCRIM if the relevant cancor information is to be printed.

The output is extensive, presumably due to the all-purpose MANOVA being used also as a discriminant manova and mancova program. The cancor information is included in the pages labeled EFFECT . . . WITHIN CELLS REGRESSION. Several versions of the largest root test are presented first followed by a section labeled EIGENVALUES AND CANONICAL CORRELATIONS. Note here that the eigenvalues listed are not the same as the squared canonical correlations. There is also a series of tests of the number of significant canonical correlations via Wilks' Λ. This is followed by a series of univariate F tests of the regression of each dependent variable on the variables in the independent battery.

There are also tables of the raw and standard score coefficients for the canonical variates in the dependent battery, followed by their within-battery-structure correlations. The next table reports percentages of variance accounted for by the canonical variates. The weights, structure correlations, and percentages of variance for the predictor battery follow, but care must be taken in identifying just what is what in this last table. The next section provides complete multiple regressions of each criterion against all predictors in case the user intended a multivariate regression analysis. Thus the major information is available but requires considerable care in its identification.

BMDP

The BMDP Canonical Correlation Analysis program P6M is quite straightforward in its use and interpretation and does not require any special comment. The user specifies the members of the batteries by the statement FIRST ARE, followed by the list of variables' names in one battery, and then SECOND ARE, followed by the variables' names in the second.

Output includes the complete correlation matrix, the squared multiple correlations of each variable with all the variables in the other battery, the eigenvalues and canonical correlations, Bartlett's test with alpha levels, the within-battery structure correlations, and the raw and standard score weights for the canonical variates. All are clearly labeled, except that the structure correlations are headed "canonical variable loadings."

PROBLEMS

1. For the following correlation matrix, the first two variables are in one battery and the last three in a second. Find the canonical correlations, the weights for canonical variates, and the structure correlations.

	x_1	x_2	y_1	y_2	y_3
x_1	1.00	.38	.52	.50	.20
x_2		1.00	.32	.50	.45
y_1			1.00	.40	.18
y_2				1.00	.35
y_3					1.00

2. From the results of Problem 1, find the squared multiple correlation of each variable predicted from variables in the other battery and the redundancies of each battery.

3. Assume $n = 100$, and test the canonical correlations for significance using both F_0 and Bartlett's test. Repeat the tests assuming $n = 40$.

4. Rotate the structure correlations from Problem 1 to an approximate simple structure and apply the same transformation to the weights.

5. Using the same data, suppose $s_{x_1} = 2$; $s_{x_2} = 3$; $s_{y_1} = 1$; $s_{y_2} = 2$; and $s_{y_3} = 3$. Compute the matrix of coefficients for predicting the y's from the x's. Compute the .95 c.i. assuming $n = 40$ for β_{11} and $\beta_{11} - \beta_{21}$. Assume (1) a comparisonwise stance about both x's and y's; (2) an analysiswise stance about the x's and a criterionwise stance about y_1, (that is, as if each criterion is being treated separately); and (3) an analysiswise stance about both the x's and y's.

ANSWERS

1.
$$\mathbf{W}_1 = \begin{matrix} x_1 \\ x_2 \end{matrix}\begin{bmatrix} .62994 & .87861 \\ .57333 & -.91656 \end{bmatrix} \quad \mathbf{W}_2 = \begin{matrix} y_1 \\ y_2 \\ y_3 \end{matrix}\begin{bmatrix} .45274 & .68515 \\ .59594 & -.03299 \\ .26507 & -.86015 \end{bmatrix} \quad \mathbf{R}^* = \begin{bmatrix} .69169 & 0 \\ 0 & .31632 \end{bmatrix}$$

$$\mathbf{R}_{xc_x} = \begin{bmatrix} .84780 & .53032 \\ .81270 & -.58269 \end{bmatrix} \quad \mathbf{R}_{xc_y} = \begin{bmatrix} .58641 & .16775 \\ .56213 & -.18432 \end{bmatrix}$$

$$\mathbf{R}_{yc_x} = \begin{bmatrix} .51103 & .16358 \\ .60163 & -.01897 \\ .38398 & -.23673 \end{bmatrix} \quad \mathbf{R}_{yc_y} = \begin{bmatrix} .73882 & .51712 \\ .86981 & -.05998 \\ .55514 & -.74837 \end{bmatrix}$$

2. $R^2_{x1.y} = .37201$; $R^2_{x2.y} = .34996$; $R^2_{y1.x} = .28791$; $R^2_{y2.x} = .36232$; $R^2_{y3.x} = .20348$; Redundancy of y on x = .28157; Redundancy of x on y = .36098

3. $n = 100$: $F_0 = 43.571$, $\alpha << .01$; $\chi^2_6 = 72.608$; $\alpha << .005$; $\chi^2_2 = 10.121$, $\alpha < .01$. $n = 40$: $F_0 = 16.052$, $\alpha < .01$; $\chi^2_6 = 27.228$; $\alpha < .005$; $\chi^2_2 = 3.795$, $\alpha > .10$; Tests of hypothesis of independence of variables in two batteries are rejected with either sample size. Both canonical correlations are significant if $n = 100$, but only one is significant if $n = 40$.

Using the transformation matrix,

$$\mathrm{T} = \begin{array}{c} \\ c_1 \\ c_2 \end{array}\begin{array}{c} \begin{array}{cc} c_1' & c_2' \end{array} \\ \begin{bmatrix} .7962 & .6051 \\ .6051 & -.7962 \end{bmatrix} \end{array}$$

4.

$$\begin{array}{c} \\ x_1 \\ x_2 \end{array}\begin{array}{c} \begin{array}{cc} c_1' & c_2' \end{array} \\ \begin{bmatrix} .9959 & .0908 \\ .2945 & .9557 \end{bmatrix} \end{array}\quad\begin{array}{c} \begin{array}{cc} c_1' & c_2' \end{array} \\ \begin{bmatrix} 1.0332 & -.3183 \\ -.0981 & 1.0766 \end{bmatrix} \end{array}$$

$$\begin{array}{c} y_1 \\ y_2 \\ y_3 \\ \end{array}\begin{array}{c} \begin{bmatrix} .9011 & .0353 \\ .6562 & .5741 \\ -.0108 & .9317 \end{bmatrix} \\ \mathrm{R}_c' \end{array}\quad\begin{array}{c} \begin{bmatrix} .7750 & -.2715 \\ .4545 & .3869 \\ -.3084 & .8452 \end{bmatrix} \\ \mathrm{W}_c' \end{array}$$

5. $\mathrm{B} = \begin{bmatrix} .2328 & .3624 & .0509 \\ .0477 & .2417 & .4373 \end{bmatrix}$ (1) $.2328 \pm .1501$; $.1851 \pm .2089$;

(2) $.2329 \pm .1891$; $.1851 \pm .2632$; (3) $.2328 \pm .2659$; $.1851 \pm .3701$

READINGS

Cooley, W. W., and P. R. Lohnes (1971). *Multivariate data analysis.* New York: Wiley, pp. 168–93.

Levine, M.S. (1977). *Canonical analysis and factor comparison.* Beverly Hills, CA: Sage, pp. 7–36.

Tabachnick, B. G., and L. S. Fidell (1983). *Using multivariate statistics.* New York: Harper and Row, pp. 146–72.

APPENDIX

Table 1 shows the proportion of the normal distribution that falls below the given z-values. The z-values that correspond to one-tailed alpha levels are found by subtracting the alpha level from one. For example, $1.0 - .05 = .95$, so the corresponding z-value falls between 1.64 and 1.65. For two-tailed alphas, divide α by two and subtract from one as above. For example, $1.0 - (.05/2) = .975$, so the corresponding z-value is 1.96.

In Table 2, "$1 - \alpha$" is the probability of finding a χ^2 that is at least as large as the tabled value with the given df.

Table 6 is for the generalized F_0 statistic that is used in the largest root test in multivariate analyses:

$$F_0 = \frac{t}{r} \lambda_1$$

The parameters r, s, and t refer to degrees of freedom, or the rank of a matrix of deviations. Note that the F_0 for different values of s are on different pages.

In discriminant analysis, $r = |g - 1 - p|$ where g is the number of groups and p the number of observed variables; $t = n - p - g$ where n is the total number of cases in the groups to be discriminated; $s = g - 1$ or p, whichever is smaller. In manova, $g - 1$ is replaced by the number of degrees of freedom for the effect. In cancor, $r = |p - g|$ where p is the number of variables in one battery and q the number of variables in the other; s equals p or q, whichever is smaller; $t = n - p - q - 1$. Entries in the table were converted by Bock (1975) from Pillai's (1967) tables.

Table 1 Cumulative Normal Distribution Function

z	*0.00*	*0.01*	*0.02*	*0.03*	*0.04*	*0.05*	*0.06*	*0.07*	*0.08*	*0.09*
0.0	0.5000	0.5040	0.5080	0.5120	0.5160	0.5199	0.5239	0.5279	0.5319	0.5359
0.1	0.5398	0.5438	0.5478	0.5517	0.5557	0.5596	0.5636	0.5675	0.5714	0.5753
0.2	0.5793	0.5832	0.5871	0.5910	0.5948	0.5987	0.6026	0.6064	0.6103	0.6141
0.3	0.6179	0.6217	0.6255	0.6293	0.6331	0.6368	0.6406	0.6443	0.6480	0.6517
0.4	0.6554	0.6591	0.6628	0.6664	0.6700	0.6736	0.6772	0.6808	0.6844	0.6879
0.5	0.6915	0.6950	0.6985	0.7019	0.7054	0.7088	0.7123	0.7157	0.7190	0.7224
0.6	0.7257	0.7291	0.7324	0.7357	0.7389	0.7422	0.7454	0.7486	0.7517	0.7549
0.7	0.7580	0.7611	0.7642	0.7673	0.7704	0.7734	0.7764	0.7794	0.7823	0.7852
0.8	0.7881	0.7910	0.7939	0.7967	0.7995	0.8023	0.8051	0.8078	0.8106	0.8133
0.9	0.8159	0.8186	0.8212	0.8238	0.8264	0.8289	0.8315	0.8340	0.8365	0.8389
1.0	0.8413	0.8438	0.8461	0.8485	0.8508	0.8531	0.8554	0.8577	0.8599	0.8621
1.1	0.8643	0.8665	0.8686	0.8708	0.8729	0.8749	0.8770	0.8790	0.8810	0.8830
1.2	0.8849	0.8869	0.8888	0.8907	0.8925	0.8944	0.8962	0.8980	0.8997	0.9015
1.3	0.9032	0.9049	0.9066	0.9082	0.9099	0.9115	0.9131	0.9147	0.9162	0.9177
1.4	0.9192	0.9207	0.9222	0.9236	0.9251	0.9265	0.9279	0.9292	0.9306	0.9319
1.5	0.9332	0.9345	0.9357	0.9370	0.9382	0.9394	0.9406	0.9418	0.9429	0.9441
1.6	0.9452	0.9463	0.9474	0.9484	0.9495	0.9505	0.9515	0.9525	0.9535	0.9545
1.7	0.9554	0.9564	0.9573	0.9582	0.9591	0.9599	0.9608	0.9616	0.9625	0.9633
1.8	0.9641	0.9649	0.9656	0.9664	0.9671	0.9678	0.9686	0.9693	0.9699	0.9706
1.9	0.9713	0.9719	0.9726	0.9732	0.9738	0.9744	0.9750	0.9756	0.9761	0.9767
2.0	0.9772	0.9778	0.9783	0.9788	0.9793	0.9798	0.9803	0.9808	0.9812	0.9817
2.1	0.9821	0.9826	0.9830	0.9834	0.9838	0.9842	0.9846	0.9850	0.9854	0.9857
2.2	0.9861	0.9864	0.9868	0.9871	0.9875	0.9878	0.9881	0.9884	0.9887	0.9890
2.3	0.9893	0.9896	0.9898	0.9901	0.9904	0.9906	0.9909	0.9911	0.9913	0.9916
2.4	0.9918	0.9920	0.9922	0.9925	0.9927	0.9929	0.9931	0.9932	0.9934	0.9936
2.5	0.9938	0.9940	0.9941	0.9943	0.9945	0.9946	0.9948	0.9949	0.9951	0.9952
2.6	0.9953	0.9955	0.9956	0.9957	0.9959	0.9960	0.9961	0.9962	0.9963	0.9964
2.7	0.9965	0.9966	0.9967	0.9968	0.9969	0.9970	0.9971	0.9972	0.9973	0.9974
2.8	0.9974	0.9975	0.9976	0.9977	0.9977	0.9978	0.9979	0.9979	0.9980	0.9981
2.9	0.9981	0.9982	0.9982	0.9983	0.9984	0.9984	0.9985	0.9985	0.9986	0.9986
3.0	0.9987	0.9987	0.9987	0.9988	0.9988	0.9989	0.9989	0.9989	0.9990	0.9990
3.1	0.9990	0.9991	0.9991	0.9991	0.9992	0.9992	0.9992	0.9992	0.9993	0.9993
3.2	0.9993	0.9993	0.9994	0.9994	0.9994	0.9994	0.9994	0.9995	0.9995	0.9995
3.3	0.9995	0.9995	0.9995	0.9996	0.9996	0.9996	0.9996	0.9996	0.9996	0.9997
3.4	0.9997	0.9997	0.9997	0.9997	0.9997	0.9997	0.9997	0.9997	0.9997	0.9998

z_α	1.282	1.645	1.960	2.326	2.576	3.090	3.291	3.891	4.417
$1 - \alpha = \Phi(z_\alpha)$	0.90	0.95	0.975	0.99	0.995	0.999	0.9995	0.99995	0.999995
2α	0.20	0.10	0.05	0.02	0.01	0.002	0.001	0.0001	0.00001

Reproduced by permission from A. M. Mood, *Introduction to the Theory of Statistics*, McGraw-Hill Book Company, New York, 1950.

Table 2 Percentage Points of the Chi-Squared Distribution

$1-\alpha$ / *df*	*0.005*	*0.010*	*0.025*	*0.050*	*0.100*	*0.250*
1	0.0^4393	0.0^3157	0.0^3982	0.0^2393	0.0158	0.102
2	0.0100	0.0201	0.0506	0.103	0.211	0.575
3	0.0717	0.115	0.216	0.352	0.584	1.21
4	0.207	0.297	0.484	0.711	1.06	1.92
5	0.412	0.554	0.831	1.15	1.61	2.67
6	0.676	0.872	1.24	1.64	2.20	3.45
7	0.989	1.24	1.69	2.17	2.83	4.25
8	1.34	1.65	2.18	2.73	3.49	5.07
9	1.73	2.09	2.70	3.33	4.17	5.90
10	2.16	2.56	3.25	3.94	4.87	6.74
11	2.60	3.05	3.82	4.57	5.58	7.58
12	3.07	3.57	4.40	5.23	6.30	8.44
13	3.57	4.11	5.01	5.89	7.04	9.30
14	4.07	4.66	5.63	6.57	7.79	10.2
15	4.60	5.23	6.26	7.26	8.55	11.0
16	5.14	5.81	6.91	7.96	9.31	11.9
17	5.70	6.41	7.56	8.67	10.1	12.8
18	6.26	7.01	8.23	9.39	10.9	13.7
19	6.84	7.63	8.91	10.1	11.7	14.6
20	7.43	8.26	9.59	10.9	12.4	15.5
21	8.03	8.90	10.3	11.6	13.2	16.3
22	8.64	9.54	11.0	12.3	14.0	17.2
23	9.26	10.2	11.7	13.1	14.8	18.1
24	9.89	10.9	12.4	13.8	15.7	19.0
25	10.5	11.5	13.1	14.6	16.5	19.9
26	11.2	12.2	13.8	15.4	17.3	20.8
27	11.8	12.9	14.6	16.2	18.1	21.7
28	12.5	13.6	15.3	16.9	18.9	22.7
29	13.1	14.3	16.0	17.7	19.8	23.6
30	13.8	15.0	16.8	18.5	20.6	24.5

Reproduced by permission from Catherine M. Thompson: Tables of percentage points of the incomplete beta function and of the chi-square distribution, *Biometrika,* vol. 32 (1941), pp. 187–191.

Table 2 Percentage Points of the Chi-Squared Distribution *(continued)*

$1-\alpha$ / *df*	*0.500*	*0.750*	*0.900*	*0.950*	*0.975*	*0.990*	*0.995*
1	0.455	1.32	2.71	3.84	5.02	6.63	7.88
2	1.39	2.77	4.61	5.99	7.38	9.21	10.6
3	2.37	4.11	6.25	7.81	9.35	11.3	12.8
4	3.36	5.39	7.78	9.49	11.1	13.3	14.9
5	4.35	6.63	9.24	11.1	12.8	15.1	16.7
6	5.35	7.84	10.6	12.6	14.4	16.8	18.5
7	6.35	9.04	12.0	14.1	16.0	18.5	20.3
8	7.34	10.2	13.4	15.5	17.5	20.1	22.0
9	8.34	11.4	14.7	16.9	19.0	21.7	23.6
10	9.34	12.5	16.0	18.3	20.5	23.2	25.2
11	10.3	13.7	17.3	19.7	21.9	24.7	26.8
12	11.3	14.8	18.5	21.0	23.3	26.2	28.3
13	12.3	16.0	19.8	22.4	24.7	27.7	29.8
14	13.3	17.1	21.1	23.7	26.1	29.1	31.3
15	14.3	18.2	22.3	25.0	27.5	30.6	32.8
16	15.3	19.4	23.5	26.3	28.8	32.0	34.3
17	16.3	20.5	24.8	27.6	30.2	33.4	35.7
18	17.3	21.6	26.0	28.9	31.5	34.8	37.2
19	18.3	22.7	27.2	30.1	32.9	36.2	38.6
20	19.3	23.8	28.4	31.4	34.2	37.6	40.0
21	20.3	24.9	29.6	32.7	35.5	38.9	41.4
22	21.3	26.0	30.8	33.9	36.8	40.3	42.8
23	22.3	27.1	32.0	35.2	38.1	41.6	44.2
24	23.3	28.2	33.2	36.4	39.4	43.0	45.6
25	24.3	29.3	34.4	37.7	40.6	44.3	46.9
26	25.3	30.4	35.6	38.9	41.9	45.6	48.3
27	26.3	31.5	36.7	40.1	43.2	47.0	49.6
28	27.3	32.6	37.9	41.3	44.5	48.3	51.0
29	28.3	33.7	39.1	42.6	45.7	49.6	52.3
30	29.3	34.8	40.3	43.8	47.0	50.9	53.7

Reproduced by permission from Catherine M. Thompson: Tables of percentage points of the incomplete beta function and of the chi-square distribution, *Biometrika*, vol. 32 (1941), pp. 187–191.

Table 3 One-tailed Percentage Points of the *t* Distribution

$1-\alpha$ / *df*	*0.75*	*0.90*	*0.95*	*0.975*	*0.99*	*0.995*	*0.9995*
1	1.000	3.078	6.314	12.706	31.821	63.657	636.619
2	0.816	1.886	2.920	4.303	6.965	9.925	31.598
3	0.765	1.638	2.353	3.182	4.541	5.841	12.941
4	0.741	1.533	2.132	2.776	3.747	4.604	8.610
5	0.727	1.476	2.015	2.571	3.365	4.032	6.859
6	0.718	1.440	1.943	2.447	3.143	3.707	5.959
7	0.711	1.415	1.895	2.365	2.998	3.499	5.405
8	0.706	1.397	1.860	2.306	2.896	3.355	5.041
9	0.703	1.383	1.833	2.262	2.821	3.250	4.781
10	0.700	1.372	1.812	2.228	2.764	3.169	4.587
11	0.697	1.363	1.796	2.201	2.718	3.106	4.437
12	0.695	1.356	1.782	2.179	2.681	3.055	4.318
13	0.694	1.350	1.771	2.160	2.650	3.012	4.221
14	0.692	1.345	1.761	2.145	2.624	2.977	4.140
15	0.691	1.341	1.753	2.131	2.602	2.947	4.073
16	0.690	1.337	1.746	2.120	2.583	2.921	4.015
17	0.689	1.333	1.740	10	2.567	2.898	3.965
18	0.688	1.330	1.734	2.101	2.552	2.878	3.922
19	0.688	1.328	1.729	2.093	2.539	2.861	3.883
20	0.687	1.325	1.725	2.086	2.528	2.845	3.850
21	0.686	1.323	1.721	2.080	2.518	2.831	3.819
22	0.686	1.321	1.717	2.074	2.508	2.819	3.792
23	0.685	1.319	1.714	2.069	2.500	2.807	3.767
24	0.685	1.318	1.711	2.064	2.492	2.797	3.745
25	0.684	1.316	1.708	2.060	2.485	2.787	3.725
26	0.684	1.315	1.706	2.056	2.479	2.779	3.707
27	0.684	1.314	1.703	2.052	2.473	2.771	3.690
28	0.683	1.313	1.701	2.048	2.467	2.763	3.674
29	0.683	1.311	1.699	2.045	2.462	2.756	3.659
30	0.683	1.310	1.697	2.042	2.457	2.750	3.646
40	0.681	1.303	1.684	2.021	2.423	2.704	3.551
60	0.679	1.296	1.671	2.000	2.390	2.660	3.460
120	0.677	1.289	1.658	1.980	2.358	2.617	3.373
∞	0.674	1.282	1.645	1.960	2.326	2.576	3.291

Adapted from Table III of R. A. Fisher and F. Yates, *Statistical Tables for Biological, Agricultural, and Medical Research,* published by Oliver & Boyd Ltd., Edinburgh.

Table 4 Critical Values of the *F* Distribution
Points for the Distribution of *F* for α = .05 and **.01**

	df_1 degrees of freedom (for greater mean square)											
df_2	*1*	*2*	*3*	*4*	*5*	*6*	*7*	*8*	*9*	*10*	*11*	*12*
1	161	200	216	225	230	234	237	239	241	242	243	244
	4,052	**4,999**	**5,403**	**5,625**	**5,764**	**5,859**	**5,928**	**5,981**	**6,022**	**6,056**	**6,082**	**6,106**
2	18.51	19.00	19.16	19.25	19.30	19.33	19.36	19.37	19.38	19.39	19.40	19.41
	98.49	**99.00**	**99.17**	**99.25**	**99.30**	**99.33**	**99.34**	**99.36**	**99.38**	**99.40**	**99.41**	**99.42**
3	10.13	9.55	9.28	9.12	9.01	8.94	8.88	8.84	8.81	8.78	8.76	8.74
	34.12	**30.82**	**29.46**	**28.71**	**28.24**	**27.91**	**27.67**	**27.49**	**27.34**	**27.23**	**27.13**	**27.05**
4	7.71	6.94	6.59	6.39	6.26	6.16	6.09	6.04	6.00	5.96	5.93	5.91
	21.20	**18.00**	**16.69**	**15.98**	**15.52**	**15.21**	**14.98**	**14.80**	**14.66**	**14.54**	**14.45**	**14.37**
5	6.61	5.79	5.41	5.19	5.05	4.95	4.88	4.82	4.78	4.74	4.70	4.68
	16.26	**13.27**	**12.06**	**11.39**	**10.97**	**10.67**	**10.45**	**10.27**	**10.15**	**10.05**	**9.96**	**9.89**
6	5.99	5.14	4.76	4.53	4.39	4.28	4.21	4.15	4.10	4.06	4.03	4.00
	13.74	**10.92**	**9.78**	**9.15**	**8.75**	**8.47**	**8.26**	**8.10**	**7.98**	**7.87**	**7.79**	**7.72**
7	5.59	4.74	4.35	4.12	3.97	3.87	3.79	3.73	3.68	3.63	3.60	3.57
	12.25	**9.55**	**8.45**	**7.85**	**7.46**	**7.19**	**7.00**	**6.84**	**6.71**	**6.62**	**6.54**	**6.47**
8	5.32	4.46	4.07	3.84	3.69	3.58	3.50	3.44	3.39	3.34	3.31	3.28
	11.26	**8.65**	**7.59**	**7.01**	**6.63**	**6.37**	**6.19**	**6.03**	**5.91**	**5.82**	**5.74**	**5.67**
9	5.12	4.26	3.86	3.63	3.48	3.37	3.29	3.23	3.18	3.13	3.10	3.07
	10.56	**8.02**	**6.99**	**6.42**	**6.06**	**5.80**	**5.62**	**5.47**	**5.35**	**5.26**	**5.18**	**5.11**
10	4.96	4.10	3.71	3.48	3.33	3.22	3.14	3.07	3.02	2.97	2.94	2.91
	10.04	**7.56**	**6.55**	**5.99**	**5.64**	**5.39**	**5.21**	**5.06**	**4.95**	**4.85**	**4.78**	**4.71**
11	4.84	3.98	3.59	3.36	3.20	3.09	3.01	2.95	2.90	2.86	2.82	2.79
	9.65	**7.20**	**6.22**	**5.67**	**5.32**	**5.07**	**4.88**	**4.74**	**4.63**	**4.54**	**4.46**	**4.40**
12	4.75	3.88	3.49	3.26	3.11	3.00	2.92	2.85	2.80	2.76	2.72	2.69
	9.33	**6.93**	**5.95**	**5.41**	**5.06**	**4.82**	**4.65**	**4.50**	**4.39**	**4.30**	**4.22**	**4.16**
13	4.67	3.80	3.41	3.18	3.02	2.02	2.84	2.77	2.72	2.67	2.63	2.60
	9.07	**6.70**	**5.74**	**5.20**	**4.86**	**4.62**	**4.44**	**4.30**	**4.19**	**4.10**	**4.02**	**3.96**

The function, $F = \alpha$ with exponent 2*s*, is computed in part from Fisher's table VI (7). Additional entries are by interpolation, mostly graphical.

Source: Adapted from George W. Snedecor, *Statistical Methods* (Ames, Iowa: Iowa State College Press, 1946), pp. 222–225.

df₁ degrees of freedom (for greater mean square)												
14	*16*	*20*	*24*	*30*	*40*	*50*	*75*	*100*	*200*	*500*	∞	df_2
245	246	248	249	250	251	252	253	253	254	254	254	1
6,142	**6,169**	**6,208**	**6,234**	**6,258**	**6,286**	**6,302**	**6,323**	**6,334**	**6,352**	**6,361**	**6,366**	
19.42	19.43	19.44	19.45	19.46	19.47	19.47	19.48	19.49	19.49	19.50	19.50	2
99.43	**99.44**	**99.45**	**99.46**	**99.47**	**99.48**	**99.48**	**99.49**	**99.49**	**99.49**	**99.50**	**99.50**	
8.71	8.69	8.66	8.64	8.62	8.60	8.58	8.57	8.56	8.54	8.54	8.53	3
26.92	**26.83**	**26.69**	**26.60**	**26.50**	**26.41**	**26.35**	**26.27**	**26.23**	**26.18**	**26.14**	**26.12**	
5.87	5.84	5.80	5.77	5.74	5.71	5.70	5.68	5.66	5.65	5.64	5.63	4
14.24	**14.15**	**14.02**	**13.93**	**13.83**	**13.74**	**13.69**	**13.61**	**13.57**	**13.52**	**13.48**	**13.46**	
4.64	4.60	4.56	4.53	4.50	4.46	4.44	4.42	4.40	4.38	4.37	4.36	5
9.77	**9.68**	**9.55**	**9.47**	**9.38**	**9.29**	**9.24**	**9.17**	**9.13**	**9.07**	**9.04**	**9.02**	
3.96	3.92	3.87	3.84	3.81	3.77	3.75	3.72	3.71	3.69	3.68	3.67	6
7.60	**7.52**	**7.39**	**7.31**	**7.23**	**7.14**	**7.09**	**7.02**	**6.99**	**6.94**	**6.90**	**6.88**	
3.52	3.49	3.44	3.41	3.38	3.34	3.32	3.29	3.28	3.25	3.24	3.23	7
6.35	**6.27**	**6.15**	**6.07**	**5.98**	**5.90**	**5.85**	**5.78**	**5.75**	**5.70**	**5.67**	**5.65**	
3.23	3.20	3.15	3.12	3.08	3.05	3.03	3.00	2.98	2.96	2.94	2.93	8
5.56	**5.48**	**5.36**	**5.28**	**5.20**	**5.11**	**5.06**	**5.00**	**4.96**	**4.91**	**4.88**	**4.86**	
3.02	2.98	2.93	2.90	2.86	2.82	2.80	2.77	2.76	2.73	2.72	2.71	9
5.00	**4.92**	**4.80**	**4.73**	**4.64**	**4.56**	**4.51**	**4.45**	**4.41**	**4.36**	**4.33**	**4.31**	
2.86	2.82	2.77	2.74	2.70	2.67	2.64	2.61	2.59	2.56	2.55	2.54	10
4.60	**4.52**	**4.41**	**4.33**	**4.25**	**4.17**	**4.12**	**4.05**	**4.01**	**3.96**	**3.93**	**3.91**	
2.74	2.70	2.65	2.61	2.57	2.53	2.50	2.47	2.45	2.42	2.41	2.40	11
4.29	**4.21**	**4.10**	**4.02**	**3.94**	**3.86**	**3.80**	**3.74**	**3.70**	**3.66**	**3.62**	**3.60**	
2.64	2.60	2.54	2.50	2.46	2.42	2.40	2.36	2.35	2.32	2.31	2.30	12
4.05	**3.98**	**3.86**	**3.78**	**3.70**	**3.61**	**3.56**	**3.49**	**3.46**	**3.41**	**3.38**	**3.36**	
2.55	2.51	2.46	2.42	2.38	2.34	2.32	2.28	2.26	2.24	2.22	2.21	13
3.85	**3.78**	**3.67**	**3.59**	**3.51**	**3.42**	**3.37**	**3.30**	**3.27**	**3.21**	**3.18**	**3.16**	

Table 4 Critical Values of the *F* Distribution
Points for the Distribution of *F* for $\alpha = .05$ and **.01** *(continued)*

	df₁ degrees of freedom (for greater mean square)											
df_2	*1*	*2*	*3*	*4*	*5*	*6*	*7*	*8*	*9*	*10*	*11*	*12*
14	4.60	3.74	3.34	3.11	2.96	2.85	2.77	2.70	2.65	2.60	2.56	2.53
	8.86	**6.51**	**5.56**	**5.03**	**4.69**	**4.46**	**4.28**	**4.14**	**4.03**	**3.94**	**3.86**	**3.80**
15	4.54	3.68	3.29	3.06	2.90	2.79	2.70	2.64	2.59	2.55	2.51	2.48
	8.68	**6.36**	**5.42**	**4.89**	**4.56**	**4.32**	**4.14**	**4.00**	**3.89**	**3.80**	**3.73**	**3.67**
16	4.49	3.63	3.24	3.01	2.85	2.74	2.66	2.59	2.54	2.49	2.45	2.42
	8.53	**6.23**	**5.29**	**4.77**	**4.44**	**4.20**	**4.03**	**3.89**	**3.78**	**3.69**	**3.61**	**3.55**
17	4.45	3.59	3.20	2.96	2.81	2.70	2.62	2.55	2.50	2.45	2.41	2.38
	8.40	**6.11**	**5.18**	**4.67**	**4.34**	**4.10**	**3.93**	**3.79**	**3.68**	**3.59**	**3.52**	**3.45**
18	4.41	3.55	3.16	2.93	2.77	2.66	2.58	2.51	2.46	2.41	2.37	2.34
	8.28	**6.01**	**5.09**	**4.58**	**4.25**	**4.01**	**3.85**	**3.71**	**3.60**	**3.51**	**3.44**	**3.37**
19	4.38	3.52	3.13	2.90	2.74	2.63	2.55	2.48	2.43	2.38	2.34	2.31
	8.18	**5.93**	**5.01**	**4.50**	**4.17**	**3.94**	**3.77**	**3.63**	**3.52**	**3.43**	**3.36**	**3.30**
20	4.35	3.49	3.10	2.87	2.71	2.60	2.52	2.45	2.40	2.35	2.31	2.28
	8.10	**5.85**	**4.94**	**4.43**	**4.10**	**3.87**	**3.71**	**3.56**	**3.45**	**3.37**	**3.30**	**3.23**
21	4.32	3.47	3.07	2.84	2.68	2.57	2.49	2.42	2.37	2.32	2.28	2.25
	8.02	**5.78**	**4.87**	**4.37**	**4.04**	**3.81**	**3.65**	**3.51**	**3.40**	**3.31**	**3.24**	**3.17**
22	4.30	3.44	3.05	2.82	2.66	2.55	2.47	2.40	2.35	2.30	2.26	2.23
	7.94	**5.72**	**4.82**	**4.31**	**3.99**	**3.76**	**3.59**	**3.45**	**3.35**	**3.26**	**3.18**	**3.12**
23	4.28	3.42	3.03	2.80	2.64	2.53	2.45	2.38	2.32	2.28	2.24	2.20
	7.88	**5.66**	**4.76**	**4.26**	**3.94**	**3.71**	**3.54**	**3.41**	**3.30**	**3.21**	**3.14**	**3.07**
24	4.26	3.40	3.01	2.78	2.62	2.51	2.43	2.36	2.30	2.26	2.22	2.18
	7.82	**5.61**	**4.72**	**4.22**	**3.90**	**3.67**	**3.50**	**3.36**	**3.25**	**3.17**	**3.09**	**3.03**
25	4.24	3.38	2.99	2.76	2.60	2.49	2.41	2.34	2.28	2.24	2.20	2.16
	7.77	**5.57**	**4.68**	**4.18**	**3.86**	**3.63**	**3.46**	**3.32**	**3.21**	**3.13**	**3.05**	**2.99**
26	4.22	3.37	2.98	2.74	2.59	2.47	2.39	2.32	2.27	2.22	2.18	2.15
	7.72	**5.53**	**4.64**	**4.14**	**3.82**	**3.59**	**3.42**	**3.29**	**3.17**	**3.09**	**3.02**	**2.96**

(continued)

df₁ degrees of freedom (for greater mean square)												
14	*16*	*20*	*24*	*30*	*40*	*50*	*75*	*100*	*200*	*500*	∞	df_2
2.48	2.44	2.39	2.35	2.31	2.27	2.24	2.21	2.19	2.16	2.14	2.13	14
3.70	**3.62**	**3.51**	**3.43**	**3.34**	**3.26**	**3.21**	**3.14**	**3.11**	**3.06**	**3.02**	**3.00**	
2.43	2.39	2.33	2.29	2.25	2.21	2.18	2.15	2.12	2.10	2.08	2.07	15
3.56	**3.48**	**3.36**	**3.29**	**3.20**	**3.12**	**3.07**	**3.00**	**2.97**	**2.92**	**2.89**	**2.87**	
2.37	2.33	2.28	2.24	2.20	2.16	2.13	2.09	2.07	2.04	2.02	2.01	16
3.45	**3.37**	**3.25**	**3.18**	**3.10**	**3.01**	**2.96**	**2.89**	**2.86**	**2.80**	**2.77**	**2.75**	
2.33	2.29	2.23	2.19	2.15	2.11	2.08	2.04	2.02	1.99	1.97	1.96	17
3.35	**3.27**	**3.16**	**3.08**	**3.00**	**2.92**	**2.86**	**2.79**	**2.76**	**2.70**	**2.67**	**2.65**	
2.29	2.25	2.19	2.15	2.11	2.07	2.04	2.00	1.98	1.95	1.93	1.92	18
3.27	**3.19**	**3.07**	**3.00**	**2.91**	**2.83**	**2.78**	**2.71**	**2.68**	**2.62**	**2.59**	**2.57**	
2.26	2.21	2.15	2.11	2.07	2.02	2.00	1.96	1.94	1.91	1.90	1.88	19
3.19	**3.12**	**3.00**	**2.92**	**2.84**	**2.76**	**2.70**	**2.63**	**2.60**	**2.54**	**2.51**	**2.49**	
2.23	2.18	2.12	2.08	2.04	1.99	1.96	1.92	1.90	1.87	1.85	1.84	20
3.13	**3.05**	**2.94**	**2.86**	**2.77**	**2.69**	**2.63**	**2.56**	**2.53**	**2.47**	**2.44**	**2.42**	
2.20	2.15	2.09	2.05	2.00	1.96	1.93	1.89	1.87	1.84	1.82	1.81	21
3.07	**2.99**	**2.88**	**2.80**	**2.72**	**2.63**	**2.58**	**2.51**	**2.47**	**2.42**	**2.38**	**2.36**	
2.18	2.13	2.07	2.03	1.98	1.93	1.91	1.87	1.84	1.81	1.80	1.78	22
3.02	**2.94**	**2.83**	**2.75**	**2.67**	**2.58**	**2.53**	**2.46**	**2.42**	**2.37**	**2.33**	**2.31**	
2.14	2.10	2.04	2.00	1.96	1.91	1.88	1.84	1.82	1.79	1.77	1.76	23
2.97	**2.89**	**2.78**	**2.70**	**2.62**	**2.53**	**2.48**	**2.41**	**2.37**	**2.32**	**2.28**	**2.26**	
2.13	2.09	2.02	1.98	1.94	1.89	1.86	1.82	1.80	1.76	1.74	1.73	24
2.93	**2.85**	**2.74**	**2.66**	**2.58**	**2.49**	**2.44**	**2.36**	**2.33**	**2.27**	**2.23**	**2.21**	
2.11	2.06	2.00	1.96	1.92	1.87	1.84	1.80	1.77	1.74	1.72	1.71	25
2.89	**2.81**	**2.70**	**2.62**	**2.54**	**2.45**	**2.40**	**2.32**	**2.29**	**2.23**	**2.19**	**2.17**	
2.10	2.05	1.99	1.95	1.90	1.85	1.82	1.78	1.76	1.72	1.70	1.69	26
2.86	**2.77**	**2.66**	**2.58**	**2.50**	**2.41**	**2.36**	**2.28**	**2.25**	**2.19**	**2.15**	**2.13**	

Table 4 Critical Values of the *F* Distribution
Points for the Distribution of *F* for α = .05 and **.01** *(continued)*

	df_1 degrees of freedom (for greater mean square)											
df_2	*1*	*2*	*3*	*4*	*5*	*6*	*7*	*8*	*9*	*10*	*11*	*12*
27	4.21	3.35	2.96	2.73	2.57	2.46	2.37	2.30	2.25	2.20	2.16	2.13
	7.68	**5.49**	**4.60**	**4.11**	**3.79**	**3.56**	**3.39**	**3.26**	**3.14**	**3.06**	**2.98**	**2.93**
28	4.20	3.34	2.95	2.71	2.56	2.44	2.36	2.29	2.24	2.19	2.15	2.12
	7.64	**5.45**	**4.57**	**4.07**	**3.76**	**3.53**	**3.36**	**3.23**	**3.11**	**3.03**	**2.95**	**2.90**
29	4.18	3.33	2.93	2.70	2.54	2.43	2.35	2.28	2.22	2.18	2.14	2.10
	7.60	**5.42**	**4.54**	**4.04**	**3.73**	**3.50**	**3.33**	**3.20**	**3.08**	**3.00**	**2.92**	**2.87**
30	4.17	3.32	2.92	2.69	2.53	2.42	2.34	2.27	2.21	2.16	2.12	2.09
	7.56	**5.39**	**4.51**	**4.02**	**3.70**	**3.47**	**3.30**	**3.17**	**3.06**	**2.98**	**2.90**	**2.84**
32	4.15	3.30	2.90	2.67	2.51	2.40	2.32	2.25	2.19	2.14	2.10	2.07
	7.50	**5.34**	**4.46**	**3.97**	**3.66**	**3.42**	**3.25**	**3.12**	**3.01**	**2.94**	**2.86**	**2.80**
34	4.13	3.28	2.88	2.65	2.49	2.38	2.30	2.23	2.17	2.12	2.08	2.05
	7.44	**5.29**	**4.42**	**3.93**	**3.61**	**3.38**	**3.21**	**3.08**	**2.97**	**2.89**	**2.82**	**2.76**
36	4.11	3.26	2.86	2.63	2.48	2.36	2.28	2.21	2.15	2.10	2.06	2.03
	7.39	**5.25**	**4.38**	**3.89**	**3.58**	**3.35**	**3.18**	**3.04**	**2.94**	**2.86**	**2.78**	**2.72**
38	4.10	3.25	2.85	2.62	2.46	2.35	2.26	2.19	2.14	2.09	2.05	2.02
	7.35	**5.21**	**4.34**	**3.86**	**3.54**	**3.32**	**3.15**	**3.02**	**2.91**	**2.82**	**2.75**	**2.69**
40	4.08	3.23	2.84	2.61	2.45	2.34	2.25	2.18	2.12	2.07	2.04	2.00
	7.31	**5.18**	**4.31**	**3.83**	**3.51**	**3.29**	**3.12**	**2.99**	**2.88**	**2.80**	**2.73**	**2.66**
42	4.07	3.22	2.83	2.59	2.44	2.32	2.24	2.17	2.11	2.06	2.02	1.99
	7.27	**5.15**	**4.29**	**3.80**	**3.49**	**3.26**	**3.10**	**2.96**	**2.86**	**2.77**	**2.70**	**2.64**
44	4.06	3.21	2.82	2.58	2.43	2.31	2.23	2.16	2.10	2.05	2.01	1.98
	7.24	**5.12**	**4.26**	**3.78**	**3.46**	**3.24**	**3.07**	**2.94**	**2.84**	**2.75**	**2.68**	**2.62**
46	4.05	3.20	2.81	2.57	2.42	2.30	2.22	2.14	2.09	2.04	2.00	1.97
	7.21	**5.10**	**4.24**	**3.76**	**3.44**	**3.22**	**3.05**	**2.92**	**2.82**	**2.73**	**2.66**	**2.60**
48	4.04	3.19	2.80	2.56	2.41	2.30	2.21	2.14	2.08	2.03	1.99	1.96
	7.19	**5.08**	**4.22**	**3.74**	**3.42**	**3.20**	**3.04**	**2.90**	**2.80**	**2.71**	**2.64**	**2.58**

df₁ degrees of freedom (for greater mean square)												
14	*16*	*20*	*24*	*30*	*40*	*50*	*75*	*100*	*200*	*500*	∞	df_2
2.08	2.03	1.97	1.93	1.88	1.84	1.80	1.76	1.74	1.71	1.68	1.67	27
2.83	**2.74**	**2.63**	**2.55**	**2.47**	**2.38**	**2.33**	**2.25**	**2.21**	**2.16**	**2.12**	**2.10**	
2.06	2.02	1.96	1.91	1.87	1.81	1.78	1.75	1.72	1.69	1.67	1.65	28
2.80	**2.71**	**2.60**	**2.52**	**2.44**	**2.35**	**2.30**	**2.22**	**2.18**	**2.13**	**2.09**	**2.06**	
2.05	2.00	1.94	1.90	1.85	1.80	1.77	1.73	1.71	1.68	1.65	1.64	29
2.77	**2.68**	**2.57**	**2.49**	**2.41**	**2.32**	**2.27**	**2.19**	**2.15**	**2.10**	**2.06**	**2.03**	
2.04	1.99	1.93	1.89	1.84	1.79	1.76	1.72	1.69	1.66	1.64	1.62	30
2.74	**2.66**	**2.55**	**2.47**	**2.38**	**2.29**	**2.24**	**2.16**	**2.13**	**2.07**	**2.03**	**2.01**	
2.02	1.97	1.91	1.86	1.82	1.76	1.74	1.69	1.67	1.64	1.61	1.59	32
2.70	**2.62**	**2.51**	**2.42**	**2.34**	**2.25**	**2.20**	**2.12**	**2.08**	**2.02**	**1.98**	**1.96**	
2.00	1.95	1.89	1.84	1.80	1.74	1.71	1.67	1.64	1.61	1.59	1.57	34
2.66	**2.58**	**2.47**	**2.38**	**2.30**	**2.21**	**2.15**	**2.08**	**2.04**	**1.98**	**1.94**	**1.91**	
1.98	1.93	1.87	1.82	1.78	1.72	1.69	1.65	1.62	1.59	1.56	1.55	36
2.62	**2.54**	**2.43**	**2.35**	**2.26**	**2.17**	**2.12**	**2.04**	**2.00**	**1.94**	**1.90**	**1.87**	
1.96	1.92	1.85	1.80	1.76	1.71	1.67	1.63	1.60	1.57	1.54	1.53	38
2.59	**2.51**	**2.40**	**2.32**	**2.22**	**2.14**	**2.08**	**2.00**	**1.97**	**1.90**	**1.86**	**1.84**	
1.95	1.90	1.84	1.79	1.74	1.69	1.66	1.61	1.59	1.55	1.53	1.51	40
2.56	**2.49**	**2.37**	**2.29**	**2.20**	**2.11**	**2.05**	**1.97**	**1.94**	**1.88**	**1.84**	**1.81**	
1.94	1.89	1.82	1.78	1.73	1.68	1.64	1.60	1.57	1.54	1.51	1.49	42
2.54	**2.46**	**2.35**	**2.26**	**2.17**	**2.08**	**2.02**	**1.94**	**1.91**	**1.85**	**1.80**	**1.78**	
1.92	1.88	1.81	1.76	1.72	1.66	1.63	1.58	1.56	1.52	1.50	1.48	44
2.52	**2.44**	**2.32**	**2.24**	**2.15**	**2.06**	**2.00**	**1.92**	**1.88**	**1.82**	**1.78**	**1.75**	
1.91	1.87	1.80	1.75	1.71	1.65	1.62	1.57	1.54	1.51	1.48	1.46	46
2.50	**2.42**	**2.30**	**2.22**	**2.13**	**2.04**	**1.98**	**1.90**	**1.86**	**1.80**	**1.76**	**1.72**	
1.90	1.86	1.79	1.74	1.70	1.64	1.61	1.56	1.53	1.50	1.47	1.45	48
2.48	**2.40**	**2.28**	**2.20**	**2.11**	**2.02**	**1.96**	**1.88**	**1.84**	**1.78**	**1.73**	**1.70**	

Table 4 Critical Values of the *F* Distribution
Points for the Distribution of *F* for α = .05 and **.01** *(continued)*

df_2	*df₁ degrees of freedom (for greater mean square)* 1	2	3	4	5	6	7	8	9	10	11	12
50	4.03 **7.17**	3.18 **5.06**	2.79 **4.20**	2.56 **3.72**	2.40 **3.41**	2.29 **3.18**	2.20 **3.02**	2.13 **2.88**	2.07 **2.78**	2.02 **2.70**	1.98 **2.62**	1.95 **2.56**
55	4.02 **7.12**	3.17 **5.01**	2.78 **4.16**	2.54 **3.68**	2.38 **3.37**	2.27 **3.15**	2.18 **2.98**	2.11 **2.85**	2.05 **2.75**	2.00 **2.66**	1.97 **2.59**	1.93 **2.53**
60	4.00 **7.08**	3.15 **4.98**	2.76 **4.13**	2.52 **3.65**	2.37 **3.34**	2.25 **3.12**	2.17 **2.95**	2.10 **2.82**	2.04 **2.72**	1.99 **2.63**	1.95 **2.56**	1.92 **2.50**
65	3.99 **7.04**	3.14 **4.95**	2.75 **4.10**	2.51 **3.62**	2.36 **3.31**	2.24 **3.09**	2.15 **2.93**	2.08 **2.79**	2.02 **2.70**	1.98 **2.61**	1.94 **2.54**	1.90 **2.47**
70	3.98 **7.01**	3.13 **4.92**	2.74 **4.08**	2.50 **3.60**	2.35 **3.29**	2.23 **3.07**	2.14 **2.91**	2.07 **2.77**	2.01 **2.67**	1.97 **2.59**	1.93 **2.51**	1.89 **2.45**
80	3.96 **6.96**	3.11 **4.88**	2.72 **4.04**	2.48 **3.56**	2.33 **3.25**	2.21 **3.04**	2.12 **2.87**	2.05 **2.74**	1.99 **2.64**	1.95 **2.55**	1.91 **2.48**	1.88 **2.41**
100	3.94 **6.90**	3.09 **4.82**	2.70 **3.98**	2.46 **3.51**	2.30 **3.20**	2.19 **2.99**	2.10 **2.82**	2.03 **2.69**	1.97 **2.59**	1.92 **2.51**	1.88 **2.43**	1.85 **2.36**
125	3.92 **6.84**	3.07 **4.78**	2.68 **3.94**	2.44 **3.47**	2.29 **3.17**	2.17 **2.95**	2.08 **2.79**	2.01 **2.65**	1.95 **2.56**	1.90 **2.47**	1.86 **2.40**	1.83 **2.33**
150	3.91 **6.81**	3.06 **4.75**	2.67 **3.91**	2.43 **3.44**	2.27 **3.14**	2.16 **2.92**	2.07 **2.76**	2.00 **2.62**	1.94 **2.53**	1.89 **2.44**	1.85 **2.37**	1.82 **2.30**
200	3.89 **6.76**	3.04 **4.71**	2.65 **3.88**	2.41 **3.41**	2.26 **3.11**	2.14 **2.90**	2.05 **2.73**	1.98 **2.60**	1.92 **2.50**	1.87 **2.41**	1.83 **2.34**	1.80 **2.28**
400	3.86 **6.70**	3.02 **4.66**	2.62 **3.83**	2.39 **3.36**	2.23 **3.06**	2.12 **2.85**	2.03 **2.69**	1.96 **2.55**	1.90 **2.46**	1.85 **2.37**	1.81 **2.29**	1.78 **2.23**
1000	3.85 **6.66**	3.00 **4.62**	2.61 **3.80**	2.38 **3.34**	2.22 **3.04**	2.10 **2.82**	2.02 **2.66**	1.95 **2.53**	1.89 **2.43**	1.84 **2.34**	1.80 **2.26**	1.76 **2.20**
∞	3.84 **6.64**	2.99 **4.60**	2.60 **3.78**	2.37 **3.32**	2.21 **3.02**	2.09 **2.80**	2.01 **2.64**	1.94 **2.51**	1.88 **2.41**	1 .83 **2.32**	1.79 **2.24**	1.75 **2.18**

df₁ degrees of freedom (for greater mean square)												
14	*16*	*20*	*24*	*30*	*40*	*50*	*75*	*100*	*200*	*500*	∞	df_2
1.90 **2.46**	1.85 **2.39**	1.78 **2.26**	1.74 **2.18**	1.69 **2.10**	1.63 **2.00**	1.60 **1.94**	1.55 **1.86**	1.52 **1.82**	1.48 **1.76**	1.46 **1.71**	1.44 **1.68**	50
1.88 **2.43**	1.83 **2.35**	1.76 **2.23**	1.72 **2.15**	1.67 **2.06**	1.61 **1.96**	1.58 **1.90**	1.52 **1.82**	1.50 **1.78**	1.46 **1.71**	1.43 **1.66**	1.41 **1.64**	55
1.86 **2.40**	1.81 **2.32**	1.75 **2.20**	1.70 **2.12**	1.65 **2.03**	1.59 **1.93**	1.56 **1.87**	1.50 **1.79**	1.48 **1.74**	1.44 **1.68**	1.41 **1.63**	1.39 **1.60**	60
1.85 **2.37**	1.80 **2.30**	1.73 **2.18**	1.68 **2.09**	1.63 **2.00**	1.57 **1.90**	1.54 **1.84**	1.49 **1.76**	1.46 **1.71**	1.42 **1.64**	1.39 **1.60**	1.37 **1.56**	65
1.84 **2.35**	1.79 **2.28**	1.72 **2.15**	1.67 **2.07**	1.62 **1.98**	1.56 **1.88**	1.53 **1.82**	1.47 **1.74**	1.45 **1.69**	1.40 **1.62**	1.37 **1.56**	1.35 **1.53**	70
1.82 **2.32**	1.77 **2.24**	1.70 **2.11**	1.65 **2.03**	1.60 **1.94**	1.54 **1.84**	1.51 **1.78**	1.45 **1.70**	1.42 **1.65**	1.38 **1.57**	1.35 **1.52**	1.32 **1.49**	80
1.79 **2.26**	1.75 **2.19**	1.68 **2.06**	1.63 **1.98**	1.57 **1.89**	1.51 **1.79**	1.48 **1.73**	1.42 **1.64**	1.39 **1.59**	1.34 **1.51**	1.30 **1.46**	1.28 **1.43**	100
1.77 **2.23**	1.72 **2.15**	1.65 **2.03**	1.60 **1.94**	1.55 **1.85**	1.49 **1.75**	1.45 **1.68**	1.39 **1.59**	1.36 **1.54**	1.31 **1.46**	1.27 **1.40**	1.25 **1.37**	125
1.76 **2.20**	1.71 **2.12**	1.64 **2.00**	1.59 **1.91**	1.54 **1.83**	1.47 **1.72**	1.44 **1.66**	1.37 **1.56**	1.34 **1.51**	1.29 **1.43**	1.25 **1.37**	1.22 **1.33**	150
1.74 **2.17**	1.69 **2.09**	1.62 **1.97**	1.57 **1.88**	1.52 **1.79**	1.45 **1.69**	1.42 **1.62**	1.35 **1.53**	1.32 **1.48**	1.26 **1.39**	1.22 **1.33**	1.19 **1.28**	200
1.72 **2.12**	1.67 **2.04**	1.60 **1.92**	1.54 **1.84**	1.49 **1.74**	1.42 **1.64**	1.38 **1.57**	1.32 **1.47**	1.28 **1.42**	1.22 **1.32**	1.16 **1.24**	1.13 **1.19**	400
1.70 **2.09**	1.65 **2.01**	1.58 **1.89**	1.53 **1.81**	1.47 **1.71**	1.41 **1.61**	1.36 **1.54**	1.30 **1.44**	1.26 **1.38**	1.19 **1.28**	1.18 **1.19**	1.08 **1.11**	1000
1.69 **2.07**	1.64 **1.99**	1.57 **1.87**	1.52 **1.79**	1.46 **1.69**	1.40 **1.59**	1.35 **1.52**	1.28 **1.41**	1.24 **1.36**	1.17 **1.25**	1.11 **1.15**	1.00 **1.00**	∞

Table 5 Table of z' for Values of r from .00 to 1.00

$$z_r' = \frac{1}{2}[\log_e(1 + r) - \log_e(1 - r)] = 1.1513 \left[\log \frac{1 + r}{1 - r}\right]$$

r from zero to 0.89 by hundredths

r	.00	.01	.02	.03	.04	.05	.06	.07	.08	.09
.0	.0000	.0100	.0200	.0300	.0400	.0500	.0601	.0701	.0802	.0902
.1	.1003	.1104	.1206	.1307	.1409	.1511	.1614	.1717	.1820	.1923
.2	.2027	.2132	.2237	.2342	.2448	.2554	.2661	.2769	.2877	.2986
.3	.3095	.3205	.3316	.3428	.3541	.3654	.3769	.3884	.4001	.4118
.4	.4236	.4356	.4477	.4599	.4722	.4847	.4973	.5101	.5230	.5361
.5	.5493	.5627	.5763	.5901	.6042	.6184	.6328	.6475	.6625	.6777
.6	.6931	.7089	.7250	.7414	.7582	.7753	.7928	.8107	.8291	.8480
.7	.8673	.8872	.9076	.9287	.9505	.9730	.9962	1.0203	1.0454	1.0714
.8	1.0986	1.1270	1.1568	1.1881	1.2212	1.2562	1.2933	1.3331	1.3758	1.4219

r from 0.900 to 0.999 by thousandths

r	.000	.001	.002	.003	.004	.005	.006	.007	.008	.009
.90	1.4722	1.4775	1.4828	1.4882	1.4937	1.4992	1.5047	1.5103	1.5160	1.5217
.91	1.5275	1.5334	1.5393	1.5453	1.5513	1.5574	1.5636	1.5698	1.5762	1.5826
.92	1.5890	1.5956	1.6022	1.6089	1.6157	1.6226	1.6296	1.6366	1.6438	1.6510
.93	1.6584	1.6658	1.6734	1.6811	1.6888	1.6967	1.7047	1.7129	1.7211	1.7295
.94	1.7380	1.7467	1.7555	1.7645	1.7736	1.7828	1.7923	1.8019	1.8117	1.8216
.95	1.8318	1.8421	1.8527	1.8635	1.8745	1.8857	1.8972	1.9090	1.9210	1.9333
.96	1.9459	1.9588	1.9721	1.9857	1.9996	2.0139	2.0287	2.0439	2.0595	2.0756
.97	2.0923	2.1095	2.1273	2.1457	2.1649	2.1847	2.2054	2.2269	2.2494	2.2729
.98	2.2976	2.3235	2.3507	2.3796	2.4101	2.4427	2.4774	2.5147	2.5550	2.5987
.99	2.6467	2.6996	2.7587	2.8257	2.9031	2.9945	3.1063	3.2504	3.4534	3.8002

Adapted by permission from A. E. Treloar, *Elements of Statistical Reasoning* (New York: John Wiley & Sons, 1939).

Table 6 The Generalized F Statistic, F_0

$s = 1$ $\alpha = .05$

t	r: 1	2	4	6	8	10	12	16	22	32
1*	161.00	200.00	225.00	234.00	239.00	242.00	244.00	246.00	249.00	250.00
2	18.50	19.00	19.20	19.30	19.40	19.40	19.40	19.40	19.50	19.50
3	10.10	9.55	9.12	8.94	8.85	8.79	8.74	8.69	8.65	8.61
4	7.71	6.94	6.39	6.16	6.04	5.96	5.91	5.84	5.79	5.74
5	6.61	5.79	5.19	4.95	4.82	4.74	4.68	4.60	4.54	4.49
6	5.99	5.14	4.53	4.28	4.15	4.06	4.00	3.92	3.86	3.80
7	5.59	4.74	4.12	3.87	3.73	3.64	3.57	3.49	3.43	3.37
8	5.32	4.46	3.84	3.58	3.44	3.35	3.28	3.20	3.13	3.07
9	5.12	4.26	3.63	3.37	3.23	3.14	3.07	2.99	2.92	2.85
10	4.96	4.10	3.48	3.22	3.07	2.98	2.91	2.83	2.75	2.69
11	4.84	3.98	3.36	3.09	2.95	2.85	2.79	2.70	2.63	2.56
12	4.75	3.89	3.26	3.00	2.85	2.75	2.69	2.60	2.52	2.46
13	4.67	3.81	3.18	2.92	2.77	2.67	2.60	2.51	2.44	2.37
14	4.60	3.74	3.11	2.85	2.70	2.60	2.53	2.44	2.37	2.30
15	4.54	3.68	3.06	2.79	2.64	2.54	2.48	2.38	2.31	2.24
16	4.49	3.63	3.01	2.74	2.59	2.49	2.42	2.33	2.25	2.18
17	4.45	3.59	2.96	2.70	2.55	2.45	2.38	2.29	2.21	2.14
18	4.41	3.55	2.93	2.66	2.51	2.41	2.34	2.25	2.17	2.10
19	4.38	3.52	2.90	2.63	2.48	2.38	2.31	2.21	2.13	2.06
20	4.35	3.49	2.87	2.60	2.45	2.35	2.28	2.18	2.10	2.03
22	4.30	3.44	2.82	2.55	2.40	2.30	2.23	2.13	2.05	1.97
24	4.26	3.40	2.78	2.51	2.36	2.25	2.18	2.09	2.00	1.93
26	4.23	3.37	2.74	2.47	2.32	2.22	2.15	2.05	1.97	1.89
28	4.20	3.34	2.71	2.45	2.29	2.19	2.12	2.02	1.93	1.86
30	4.17	3.32	2.69	2.42	2.27	2.16	2.09	1.99	1.91	1.83
32	4.15	3.29	2.67	2.40	2.24	2.14	2.07	1.97	1.88	1.81
34	4.13	3.28	2.65	2.38	2.23	2.12	2.05	1.95	1.86	1.79
36	4.11	3.26	2.63	2.36	2.21	2.11	2.03	1.93	1.85	1.77
38	4.10	3.24	2.62	2.35	2.19	2.09	2.02	1.92	1.83	1.75
40	4.08	3.23	2.61	2.34	2.18	2.08	2.00	1.90	1.81	1.73
42	4.07	3.22	2.59	2.32	2.17	2.06	1.99	1.89	1.80	1.72
44	4.06	3.21	2.58	2.31	2.16	2.05	1.98	1.88	1.79	1.71
46	4.05	3.20	2.57	2.30	2.15	2.04	1.97	1.87	1.78	1.70
48	4.04	3.19	2.57	2.29	2.14	2.03	1.96	1.86	1.77	1.69
50	4.03	3.18	2.56	2.29	2.13	2.03	1.95	1.85	1.76	1.68
55	4.02	3.16	2.54	2.27	2.11	2.01	1.93	1.83	1.74	1.66
60	4.00	3.15	2.53	2.25	2.10	1.99	1.92	1.82	1.72	1.64
65	3.99	3.14	2.51	2.24	2.08	1.98	1.90	1.80	1.71	1.62
70	3.98	3.13	2.50	2.23	2.07	1.97	1.89	1.79	1.70	1.61
80	3.96	3.11	2.49	2.21	2.06	1.95	1.88	1.77	1.68	1.59
90	3.95	3.10	2.47	2.20	2.04	1.94	1.86	1.76	1.66	1.57
100	3.94	3.09	2.46	2.19	2.03	1.93	1.85	1.75	1.65	1.56
125	3.92	3.07	2.44	2.17	2.01	1.91	1.83	1.72	1.63	1.54
150	3.90	3.06	2.43	2.16	2.00	1.89	1.82	1.71	1.61	1.52
200	3.89	3.04	2.42	2.14	1.98	1.88	1.80	1.69	1.60	1.50
300	3.87	3.03	2.40	2.13	1.97	1.86	1.78	1.68	1.58	1.48
500	3.86	3.01	2.39	2.12	1.96	1.85	1.77	1.66	1.56	1.47
1,000	3.85	3.00	2.38	2.11	1.95	1.84	1.76	1.65	1.55	1.46
2,000	3.84	3.00	2.37	2.10	1.94	1.83	1.75	1.64	1.54	1.44

*Entries in this row should be multiplied by 10, for example, 1,610.0 instead of 161.00.

Table 6 The Generalized *F* Statistic *(continued)*

$s = 1$ $\alpha = .01$

t	*r:* 1	2	4	6	8	10	12	16	22	32
1*	405.00	500.00	563.00	586.00	598.00	606.00	611.00	617.00	622.00	627.00
2	98.50	99.00	99.20	99.30	99.40	99.40	99.40	99.40	99.50	99.50
3	34.10	30.80	28.70	27.90	27.50	27.20	27.10	26.80	26.60	26.50
4	21.20	18.00	16.00	15.20	14.80	14.50	14.40	14.20	14.00	13.80
5	16.30	13.30	11.40	10.70	10.30	10.10	9.89	9.68	9.51	9.36
6	13.70	10.90	9.15	8.47	8.10	7.87	7.72	7.52	7.35	7.21
7	12.20	9.55	7.85	7.19	6.84	6.62	6.47	6.27	6.11	5.97
8	11.30	8.65	7.01	6.37	6.03	5.81	5.67	5.48	5.32	5.18
9	10.60	8.02	6.42	5.80	5.47	5.26	5.11	4.92	4.77	4.63
10	10.00	7.56	5.99	5.39	5.06	4.85	4.71	4.52	4.36	4.23
11	9.65	7.21	5.67	5.07	4.74	4.54	4.40	4.21	4.06	3.92
12	9.33	6.93	5.41	4.82	4.50	4.30	4.16	3.97	3.82	3.68
13	9.07	6.70	5.21	4.62	4.30	4.10	3.96	3.78	3.62	3.49
14	8.86	6.51	5.04	4.46	4.14	3.94	3.80	3.62	3.46	3.33
15	8.68	6.36	4.89	4.32	4.00	3.80	3.67	3.49	3.33	3.19
16	8.53	6.23	4.77	4.20	3.89	3.69	3.55	3.37	3.22	3.08
17	8.40	6.11	4.67	4.10	3.79	3.59	3.46	3.27	3.12	2.98
18	8.29	6.01	4.58	4.01	3.71	3.51	3.37	3.19	3.03	2.90
19	8.18	5.93	4.50	3.94	3.63	3.43	3.30	3.12	2.96	2.82
20	8.10	5.85	4.43	3.87	3.56	3.37	3.23	3.05	2.90	2.76
22	7.95	5.72	4.31	3.76	3.45	3.26	3.12	2.94	2.78	2.65
24	7.82	5.61	4.22	3.67	3.36	3.17	3.03	2.85	2.70	2.56
26	7.72	5.53	4.14	3.59	3.29	3.09	2.96	2.78	2.62	2.48
28	7.64	5.45	4.07	3.53	3.23	3.03	2.90	2.72	2.56	2.42
30	7.56	5.39	4.02	3.47	3.17	2.98	2.84	2.66	2.51	2.37
32	7.50	5.34	3.97	3.43	3.13	2.93	2.80	2.62	2.46	2.32
34	7.44	5.29	3.93	3.39	3.09	2.89	2.76	2.58	2.42	2.28
36	7.40	5.25	3.89	3.35	3.05	2.86	2.72	2.54	2.38	2.24
38	7.35	5.21	3.86	3.32	3.02	2.83	2.69	2.51	2.35	2.21
40	7.31	5.18	3.83	3.29	2.99	2.80	2.66	2.48	2.33	2.18
42	7.28	5.15	3.80	3.27	2.97	2.78	2.64	2.46	2.30	2.16
44	7.25	5.12	3.78	3.24	2.95	2.75	2.62	2.44	2.28	2.13
46	7.22	5.10	3.76	3.22	2.93	2.73	2.60	2.42	2.26	2.11
48	7.19	5.08	3.74	3.20	2.91	2.72	2.58	2.40	2.24	2.10
50	7.17	5.06	3.72	3.19	2.89	2.70	2.56	2.38	2.22	2.08
55	7.12	5.01	3.68	3.15	2.85	2.66	2.53	2.34	2.18	2.04
60	7.08	4.98	3.65	3.12	2.82	2.63	2.50	2.31	2.15	2.01
65	7.04	4.95	3.62	3.09	2.80	2.61	2.47	2.29	2.13	1.98
70	7.01	4.92	3.60	3.07	2.78	2.59	2.45	2.27	2.11	1.97
80	6.96	4.88	3.56	3.04	2.74	2.55	2.42	2.23	2.07	1.92
90	6.93	4.85	3.54	3.01	2.72	2.52	2.39	2.21	2.04	1.90
100	6.90	4.82	3.51	2.99	2.69	2.50	2.37	2.19	2.02	1.87
125	6.84	4.78	3.47	2.95	2.66	2.47	2.33	2.15	1.98	1.83
150	6.81	4.75	3.45	2.92	2.63	2.44	2.31	2.12	1.96	1.81
200	6.76	4.71	3.41	2.89	2.60	2.41	2.27	2.09	1.93	1.77
300	6.72	4.68	3.38	2.86	2.57	2.38	2.24	2.06	1.89	1.74
500	6.69	4.65	3.36	2.84	2.55	2.36	2.22	2.04	1.87	1.72
1,000	6.66	4.63	3.34	2.82	2.53	2.34	2.20	2.02	1.85	1.70
2,000	6.63	4.61	3.32	2.80	2.51	2.32	2.18	2.00	1.83	1.68

*Entries in this row should be multiplied by 10, for example, 4,050.0 instead of 405.00.

$s = 2$ $\alpha = .05$

t	r: 1	2	4	6	8	10	12	16	22	32
12	12.23	7.78	5.59	4.81	4.40	4.15	3.98	3.76	3.58	3.42
14	11.87	7.54	5.39	4.62	4.22	3.98	3.81	3.59	3.41	3.25
16	11.50	7.29	5.19	4.44	4.05	3.80	3.63	3.42	3.24	3.08
18	11.14	7.05	4.99	4.25	3.87	3.62	3.46	3.25	3.07	2.91
20	10.78	6.81	4.79	4.07	3.69	3.45	3.29	3.08	2.90	2.74
25	10.24	6.44	4.49	3.79	3.42	3.19	3.03	2.82	2.64	2.48
30	9.95	6.24	4.33	3.64	3.27	3.04	2.88	2.68	2.50	2.34
35	9.74	6.10	4.22	3.53	3.17	2.94	2.78	2.58	2.40	2.24
40	9.59	6.00	4.14	3.46	3.10	2.87	2.71	2.51	2.33	2.17
45	9.48	5.93	4.07	3.40	3.04	2.81	2.66	2.45	2.27	2.11
50	9.39	5.87	4.03	3.35	3.00	2.77	2.61	2.41	2.23	2.07
60	9.27	5.78	3.95	3.29	2.93	2.71	2.55	2.35	2.17	2.00
70	9.19	5.72	3.91	3.25	2.89	2.67	2.51	2.31	2.13	1.96
80	9.12	5.68	3.87	3.21	2.86	2.63	2.48	2.27	2.09	1.93
90	9.07	5.65	3.84	3.19	2.83	2.61	2.45	2.25	2.07	1.90
100	9.04	5.62	3.83	3.17	2.82	2.59	2.44	2.23	2.05	1.89
150	8.91	5.53	3.75	3.10	2.75	2.53	2.37	2.17	1.98	1.82
200	8.84	5.49	3.72	3.07	2.72	2.50	2.34	2.14	1.95	1.78
300	8.79	5.45	3.69	3.04	2.69	2.47	2.31	2.11	1.92	1.75
500	8.74	5.42	3.66	3.02	2.67	2.45	2.29	2.09	1.90	1.73
1,000	8.71	5.39	3.64	3.00	2.65	2.43	2.27	2.07	1.88	1.71
2,000	8.69	5.38	3.63	2.99	2.64	2.42	2.27	2.06	1.87	1.70

$s = 2$ $\alpha = .01$

t	r: 1	2	4	6	8	10	12	16	22	32
12	20.36	12.57	8.75	7.40	6.70	6.27	5.98	5.61	5.30	5.03
14	19.50	12.01	8.31	7.01	6.33	5.91	5.63	5.27	4.96	4.70
16	18.63	11.45	7.88	6.61	5.96	5.55	5.28	4.92	4.62	4.37
18	17.77	10.89	7.44	6.22	5.59	5.19	4.92	4.58	4.29	4.04
20	16.90	10.32	7.00	5.83	5.21	4.83	4.57	4.24	3.95	3.71
25	15.64	9.50	6.37	5.26	4.67	4.31	4.06	3.74	3.46	3.23
30	14.98	9.07	6.04	4.96	4.39	4.04	3.79	3.48	3.21	2.97
35	14.52	8.77	5.81	4.75	4.20	3.85	3.61	3.30	3.03	2.80
40	14.20	8.56	5.65	4.61	4.06	3.72	3.48	3.17	2.91	2.68
45	13.96	8.40	5.53	4.50	3.96	3.62	3.39	3.08	2.82	2.59
50	13.77	8.28	5.44	4.42	3.88	3.54	3.31	3.01	2.75	2.51
60	13.50	8.10	5.30	4.30	3.77	3.43	3.20	2.90	2.64	2.41
70	13.33	7.99	5.22	4.22	3.69	3.36	3.13	2.83	2.57	2.34
80	13.17	7.89	5.14	4.15	3.63	3.30	3.07	2.77	2.51	2.28
90	13.08	7.83	5.10	4.11	3.59	3.26	3.04	2.74	2.48	2.25
100	13.01	7.78	5.06	4.08	3.56	3.23	3.01	2.71	2.45	2.22
150	12.74	7.60	4.92	3.96	3.45	3.12	2.90	2.60	2.34	2.11
200	12.61	7.52	4.86	3.90	3.39	3.07	2.85	2.55	2.30	2.06
300	12.49	7.44	4.80	3.85	3.34	3.02	2.80	2.51	2.25	2.01
500	12.40	7.38	4.76	3.81	3.31	2.99	2.76	2.47	2.21	1.98
1,000	12.32	7.33	4.72	3.78	3.27	2.95	2.73	2.44	2.18	1.95
2,000	12.29	7.31	4.70	3.76	3.26	2.94	2.72	2.43	2.17	1.93

Table 6 The Generalized *F* Statistic *(continued)*

$s = 3$ $\alpha = .05$

t	r: 1	2	4	6	8	10	12	16	22	32
12	20.28	12.12	8.08	6.69	5.98	5.54	5.25	4.88	4.57	4.31
14	19.55	11.66	7.74	6.39	5.70	5.28	4.99	4.63	4.32	4.06
16	18.82	11.20	7.40	6.09	5.42	5.01	4.73	4.37	4.08	3.82
18	18.09	10.74	7.06	5.79	5.14	4.74	4.47	4.12	3.83	3.58
20	17.36	10.29	6.72	5.49	4.86	4.47	4.21	3.87	3.59	3.34
25	16.28	9.60	6.22	5.05	4.44	4.07	3.82	3.49	3.22	2.98
30	15.69	9.23	5.95	4.81	4.22	3.85	3.61	3.29	3.02	2.78
35	15.28	8.97	5.76	4.64	4.06	3.70	3.46	3.14	2.88	2.64
40	14.98	8.78	5.62	4.52	3.95	3.59	3.35	3.04	2.77	2.54
45	14.76	8.64	5.52	4.43	3.86	3.51	3.27	2.96	2.70	2.47
50	14.59	8.53	5.44	4.36	3.80	3.45	3.21	2.90	2.64	2.41
60	14.34	8.37	5.32	4.25	3.70	3.35	3.12	2.81	2.55	2.32
70	14.17	8.27	5.25	4.19	3.64	3.30	3.06	2.76	2.49	2.26
80	14.03	8.18	5.18	4.13	3.58	3.24	3.01	2.70	2.44	2.21
90	13.94	8.12	5.14	4.10	3.55	3.21	2.98	2.67	2.41	2.18
100	13.88	8.08	5.11	4.07	3.52	3.19	2.95	2.65	2.39	2.15
150	13.62	7.92	4.99	3.96	3.42	3.09	2.86	2.56	2.30	2.06
200	13.50	7.84	4.93	3.91	3.38	3.04	2.81	2.51	2.25	2.02
300	13.38	7.77	4.88	3.86	3.33	3.00	2.77	2.47	2.21	1.98
500	13.29	7.71	4.84	3.83	3.30	2.97	2.74	2.44	2.18	1.94
1,000	13.22	7.66	4.80	3.80	3.27	2.94	2.71	2.42	2.15	1.92
2,000	13.19	7.64	4.79	3.78	3.26	2.93	2.70	2.40	2.14	1.90

$s = 3$ $\alpha = .01$

t	r: 1	2	4	6	8	10	12	16	22	32
12	31.94	18.81	12.30	10.08	8.94	8.25	7.79	7.19	6.70	6.28
14	30.35	17.83	11.61	9.48	8.39	7.73	7.28	6.71	6.24	5.83
16	28.76	16.85	10.92	8.88	7.84	7.21	6.78	6.23	5.77	5.38
18	27.17	15.88	10.22	8.28	7.29	6.68	6.27	5.75	5.31	4.93
20	25.59	14.90	9.53	7.69	6.74	6.16	5.77	5.26	4.84	4.48
25	23.28	13.48	8.53	6.82	5.94	5.40	5.04	4.57	4.17	3.83
30	22.08	12.74	8.01	6.37	5.53	5.01	4.66	4.21	3.83	3.50
35	21.26	12.23	7.65	6.07	5.25	4.75	4.40	3.96	3.59	3.27
40	20.69	11.88	7.41	5.86	5.06	4.56	4.23	3.79	3.42	3.10
45	20.26	11.62	7.22	5.70	4.91	4.42	4.09	3.66	3.30	2.98
50	19.92	11.41	7.08	5.57	4.80	4.32	3.99	3.57	3.20	2.89
60	19.43	11.11	6.87	5.39	4.63	4.16	3.84	3.42	3.06	2.75
70	19.12	10.92	6.74	5.28	4.53	4.06	3.74	3.33	2.98	2.66
80	18.85	10.75	6.62	5.18	4.44	3.98	3.66	3.25	2.90	2.59
90	18.69	10.65	6.55	5.12	4.38	3.92	3.61	3.20	2.85	2.54
100	18.56	10.57	6.49	5.07	4.34	3.88	3.57	3.16	2.81	2.50
150	18.08	10.28	6.29	4.90	4.18	3.73	3.42	3.02	2.67	2.37
200	17.85	10.14	6.19	4.82	4.10	3.66	3.35	2.95	2.61	2.30
300	17.64	10.01	6.10	4.74	4.03	3.59	3.29	2.89	2.55	2.24
500	17.48	9.91	6.03	4.68	3.98	3.54	3.24	2.84	2.50	2.19
1,000	17.34	9.83	5.98	4.63	3.93	3.50	3.20	2.80	2.46	2.16
2,000	17.28	9.79	5.95	4.61	3.91	3.48	3.18	2.78	2.45	2.14

(continued)

$s = 4$ $\alpha = .05$

t	*r*: 1	2	4	6	8	10	12	16	22	32
12	29.37	16.97	10.80	8.70	7.64	7.00	6.57	6.02	5.57	5.19
14	28.15	16.23	10.29	8.27	7.25	6.63	6.21	5.68	5.24	4.87
16	26.94	15.50	9.78	7.84	6.86	6.26	5.85	5.34	4.92	4.56
18	25.72	14.76	9.28	7.41	6.46	5.89	5.50	5.00	4.59	4.24
20	24.51	14.03	8.77	6.98	6.07	5.51	5.14	4.66	4.27	3.93
25	22.70	12.94	8.02	6.34	5.49	4.96	4.61	4.16	3.78	3.46
30	21.72	12.34	7.61	6.00	5.17	4.67	4.32	3.89	3.52	3.20
35	21.04	11.93	7.33	5.76	4.95	4.46	4.12	3.70	3.33	3.02
40	20.56	11.64	7.13	5.59	4.80	4.31	3.98	3.56	3.20	2.90
45	20.19	11.42	6.98	5.46	4.68	4.20	3.87	3.46	3.10	2.80
50	19.90	11.25	6.86	5.36	4.59	4.11	3.79	3.38	3.03	2.72
60	19.49	10.99	6.69	5.21	4.45	3.99	3.67	3.26	2.91	2.61
70	19.22	10.83	6.58	5.12	4.37	3.90	3.59	3.18	2.84	2.54
80	18.98	10.69	6.48	5.03	4.29	3.83	3.52	3.12	2.78	2.47
90	18.84	10.60	6.42	4.98	4.24	3.79	3.48	3.08	2.74	2.44
100	18.73	10.53	6.37	4.94	4.21	3.75	3.44	3.05	2.70	2.40
150	18.31	10.28	6.20	4.79	4.07	3.63	3.32	2.92	2.59	2.29
200	18.10	10.15	6.11	4.72	4.01	3.56	3.26	2.87	2.53	2.23
300	17.92	10.04	6.04	4.66	3.95	3.51	3.20	2.82	2.48	2.18
500	17.77	9.96	5.98	4.61	3.90	3.46	3.16	2.77	2.44	2.14
1,000	17.65	9.88	5.93	4.57	3.86	3.42	3.13	2.74	2.40	2.10
2,000	17.60	9.85	5.91	4.55	3.84	3.41	3.11	2.72	2.39	2.09

$s = 4$ $\alpha = .01$

t	*r*: 1	2	4	6	8	10	12	16	22	32
12	44.96	25.74	16.19	12.95	11.32	10.33	9.66	8.82	8.12	7.53
14	42.46	24.26	15.19	12.12	10.57	9.62	8.99	8.19	7.52	6.96
16	39.96	22.77	14.19	11.29	9.82	8.92	8.32	7.56	6.92	6.39
18	37.46	21.28	13.20	10.46	9.07	8.22	7.65	6.93	6.33	5.82
20	34.95	19.80	12.20	9.62	8.31	7.52	6.98	6.30	5.73	5.24
25	31.33	17.65	10.77	8.42	7.23	6.51	6.01	5.39	4.87	4.42
30	29.47	16.54	10.03	7.81	6.68	5.99	5.52	4.93	4.43	4.00
35	28.19	15.79	9.53	7.40	6.31	5.64	5.19	4.61	4.13	3.71
40	27.30	15.26	9.18	7.11	6.05	5.40	4.96	4.39	3.92	3.51
45	26.64	14.87	8.92	6.89	5.85	5.21	4.78	4.23	3.76	3.36
50	26.12	14.56	8.72	6.72	5.70	5.07	4.65	4.10	3.64	3.24
60	25.37	14.12	8.42	6.48	5.48	4.87	4.45	3.92	3.50	3.08
70	24.90	13.84	8.24	6.33	5.34	4.74	4.33	3.80	3.36	2.97
80	24.48	13.59	8.08	6.19	5.22	4.63	4.22	3.70	3.26	2.87
90	24.23	13.45	7.98	6.11	5.15	4.56	4.16	3.64	3.20	2.82
100	24.03	13.33	7.90	6.04	5.09	4.50	4.10	3.59	3.16	2.77
150	23.30	12.89	7.62	5.81	4.88	4.31	3.91	3.41	2.98	2.60
200	22.95	12.69	7.48	5.69	4.78	4.21	3.83	3.33	2.90	2.52
300	22.63	12.50	7.36	5.59	4.68	4.13	3.74	3.25	2.83	2.45
500	22.38	12.35	7.26	5.51	4.61	4.06	3.68	3.19	2.77	2.39
1,000	22.18	12.23	7.18	5.45	4.55	4.00	3.63	3.14	2.72	2.35
2,000	22.08	12.17	7.14	5.42	4.53	3.98	3.60	3.12	2.70	2.33

Table 6 The Generalized *F* Statistic *(continued)*

$s = 5$ $\alpha = .05$

t	*r:* 1	2	4	6	8	10	12	16	22	32
12	39.52	22.33	13.76	10.88	9.42	8.54	7.95	7.21	6.59	6.07
14	37.70	21.26	13.06	10.30	8.90	8.06	7.49	6.78	6.18	5.68
16	35.87	20.19	12.36	9.71	8.38	7.57	7.03	6.34	5.77	5.29
18	34.05	19.12	11.65	9.13	7.86	7.08	6.57	5.91	5.36	4.90
20	32.23	18.05	10.95	8.55	7.34	6.60	6.10	5.48	4.96	4.51
25	29.53	16.47	9.91	7.69	6.57	5.88	5.42	4.83	4.35	3.93
30	28.06	15.61	9.35	7.23	6.15	5.49	5.05	4.49	4.02	3.62
35	27.05	15.01	8.96	6.90	5.86	5.22	4.79	4.25	3.79	3.40
40	26.33	14.60	8.68	6.68	5.66	5.03	4.61	4.08	3.63	3.24
45	25.79	14.28	8.47	6.50	5.50	4.89	4.48	3.95	3.50	3.12
50	25.37	14.03	8.31	6.37	5.38	4.78	4.37	3.85	3.41	3.03
60	24.75	13.66	8.07	6.17	5.20	4.61	4.21	3.70	3.27	2.89
70	24.35	13.43	7.92	6.05	5.09	4.51	4.11	3.60	3.18	2.80
80	24.00	13.23	7.79	5.94	4.99	4.42	4.02	3.52	3.10	2.73
90	23.79	13.11	7.71	5.87	4.93	4.36	3.97	3.47	3.05	2.68
100	23.62	13.00	7.64	5.82	4.88	4.31	3.93	3.43	3.01	2.64
150	23.00	12.64	7.40	5.62	4.71	4.15	3.77	3.28	2.86	2.50
200	22.70	12.46	7.29	5.52	4.62	4.07	3.69	3.21	2.80	2.43
300	22.42	12.30	7.18	5.44	4.54	4.00	3.62	3.14	2.73	2.37
500	22.21	12.18	7.10	5.37	4.48	3.94	3.57	3.09	2.68	2.32
1,000	22.03	12.07	7.03	5.31	4.43	3.89	3.52	3.05	2.64	2.28
2,000	21.95	12.03	7.00	5.29	4.41	3.87	3.50	3.03	2.62	2.26

$s = 5$ $\alpha = .01$

t	*r:* 1	2	4	6	8	10	12	16	22	32
12	59.47	33.41	20.42	16.05	13.85	12.52	11.63	10.50	9.58	8.80
14	55.86	31.32	19.08	14.96	12.88	11.62	10.78	9.72	8.84	8.10
16	52.26	29.23	17.74	13.86	11.91	10.72	9.93	8.93	8.10	7.40
18	48.66	27.15	16.39	12.77	10.93	9.83	9.08	8.14	7.36	6.70
20	45.05	25.06	15.05	11.67	9.96	8.93	8.23	7.35	6.62	6.01
25	39.86	22.05	13.12	10.10	8.57	7.64	7.02	6.22	5.56	5.00
30	37.20	20.52	12.14	9.30	7.86	6.99	6.40	5.65	5.03	4.50
35	35.38	19.47	11.46	8.76	7.38	6.54	5.98	5.26	4.66	4.15
40	34.12	18.74	11.00	8.38	7.05	6.24	5.69	4.99	4.41	3.91
45	33.18	18.20	10.65	8.10	6.80	6.01	5.47	4.79	4.22	3.73
50	32.45	17.78	10.39	7.88	6.61	5.83	5.31	4.64	4.07	3.59
60	31.39	17.16	10.00	7.57	6.33	5.57	5.06	4.41	3.86	3.39
70	30.72	16.78	9.76	7.37	6.15	5.41	4.91	4.27	3.73	3.26
80	30.14	16.44	9.54	7.19	6.00	5.27	4.78	4.15	3.61	3.15
90	29.79	16.24	9.41	7.09	5.91	5.19	4.70	4.07	3.54	3.08
100	29.50	16.08	9.31	7.01	5.83	5.12	4.63	4.01	3.49	3.03
150	28.47	15.49	8.93	6.70	5.57	4.87	4.40	3.79	3.28	2.83
200	27.98	15.20	8.75	6.56	5.44	4.75	4.29	3.69	3.18	2.74
300	27.53	14.94	8.59	6.43	5.32	4.64	4.18	3.59	3.09	2.65
500	27.18	14.75	8.46	6.32	5.23	4.56	4.11	3.52	3.02	2.58
1,000	26.90	14.58	8.36	6.24	5.16	4.49	4.04	3.46	2.97	2.53
2,000	26.77	14.51	8.31	6.20	5.12	4.46	4.01	3.43	2.94	2.50

(continued)

$s = 6$ $\alpha = .05$

t	r: 1	2	4	6	8	10	12	16	22	32
12	50.73	28.22	16.99	13.22	11.32	10.18	9.41	8.45	7.65	6.99
14	48.18	26.76	16.06	12.46	10.66	9.57	8.83	7.91	7.15	6.52
16	45.63	25.29	15.13	11.71	9.99	8.95	8.26	7.38	6.66	6.05
18	43.09	23.83	14.20	10.96	9.33	8.34	7.68	6.85	6.16	5.58
20	40.54	22.37	13.27	10.21	8.67	7.73	7.10	6.31	5.66	5.10
25	36.77	20.20	11.90	9.10	7.69	6.83	6.25	5.53	4.92	4.41
30	34.73	19.04	11.16	8.50	7.16	6.34	5.80	5.10	4.52	4.03
35	33.32	18.23	10.65	8.09	6.79	6.01	5.48	4.81	4.25	3.77
40	32.33	17.66	10.28	7.80	6.54	5.77	5.25	4.60	4.05	3.58
45	31.58	17.23	10.01	7.58	6.34	5.59	5.08	4.44	3.90	3.44
50	30.99	16.89	9.80	7.40	6.19	5.45	4.95	4.32	3.79	3.33
60	30.13	16.40	9.49	7.15	5.97	5.24	4.76	4.14	3.62	3.17
70	29.58	16.08	9.29	6.99	5.82	5.11	4.63	4.02	3.51	3.07
80	29.10	15.81	9.11	6.85	5.70	5.00	4.53	3.92	3.41	2.97
90	28.81	15.64	9.01	6.76	5.62	4.93	4.46	3.86	3.36	2.92
100	28.57	15.50	8.92	6.69	5.56	4.87	4.41	3.81	3.31	2.87
150	27.71	15.01	8.61	6.44	5.34	4.67	4.21	3.63	3.14	2.71
200	27.30	14.77	8.46	6.32	5.23	4.57	4.12	3.54	3.05	2.63
300	26.92	14.56	8.32	6.21	5.14	4.48	4.03	3.46	2.98	2.55
500	26.63	14.39	8.22	6.13	5.06	4.41	3.97	3.40	2.92	2.50
1,000	26.38	14.25	8.13	6.05	5.00	4.35	3.91	3.35	2.87	2.45
2,000	26.27	14.18	8.09	6.02	4.97	4.32	3.89	3.32	2.85	2.43

$s = 6$ $\alpha = .01$

t	r: 1	2	4	6	8	10	12	16	22	32
12	75.49	41.82	25.02	19.39	16.56	14.85	13.71	12.27	11.08	10.09
14	70.59	39.04	23.29	18.00	15.34	13.74	12.67	11.31	10.20	9.26
16	65.70	36.26	21.55	16.61	14.13	12.63	11.62	10.36	9.31	8.43
18	60.80	33.48	19.81	15.22	12.91	11.52	10.58	9.40	8.43	7.60
20	55.91	30.69	18.08	13.83	11.70	10.40	9.54	8.44	7.54	6.78
25	48.87	26.70	15.58	11.84	9.96	8.81	8.05	7.08	6.27	5.59
30	45.29	24.67	14.33	10.84	9.08	8.01	7.30	6.39	5.64	4.99
35	42.85	23.29	13.47	10.16	8.48	7.47	6.79	5.92	5.20	4.59
40	41.17	22.34	12.88	9.69	8.07	7.09	6.43	5.60	4.90	4.31
45	39.91	21.62	12.43	9.33	7.77	6.81	6.17	5.36	4.68	4.10
50	38.93	21.07	12.09	9.06	7.53	6.60	5.97	5.17	4.50	3.93
60	37.51	20.27	11.60	8.67	7.19	6.28	5.68	4.90	4.25	3.70
70	36.63	19.77	11.29	8.42	6.97	6.09	5.49	4.73	4.10	3.55
80	35.85	19.33	11.01	8.21	6.78	5.92	5.33	4.59	3.96	3.42
90	35.38	19.06	10.85	8.08	6.67	5.81	5.24	4.50	3.88	3.34
100	35.00	18.85	10.72	7.97	6.58	5.73	5.16	4.43	3.81	3.28
150	33.63	18.07	10.24	7.59	6.25	5.43	4.87	4.17	3.57	3.05
200	32.97	17.71	10.02	7.41	6.09	5.29	4.74	4.04	3.45	2.94
300	32.38	17.37	9.81	7.25	5.95	5.16	4.62	3.93	3.35	2.84
500	31.93	17.11	9.65	7.12	5.84	5.06	4.52	3.84	3.27	2.76
1,000	31.55	16.90	9.52	7.02	5.75	4.97	4.44	3.77	3.20	2.70
2,000	31.37	16.80	9.46	6.97	5.71	4.93	4.41	3.74	3.17	2.67

Table 6 The Generalized *F* Statistic *(continued)*

$s = 7$ $\alpha = .05$

t	r: 1	2	4	6	8	10	12	16	22	32
12	62.95	34.62	20.47	15.73	13.35	11.91	10.95	9.75	8.75	7.92
14	59.57	32.71	19.29	14.79	12.53	11.17	10.25	9.11	8.15	7.37
16	56.19	30.80	18.10	13.85	11.71	10.42	9.55	8.47	7.56	6.81
18	52.81	28.89	16.92	12.90	10.89	9.67	8.85	7.83	6.97	6.26
20	49.44	26.98	15.74	11.96	10.06	8.92	8.15	7.18	6.38	5.71
25	44.44	24.15	13.99	10.57	8.85	7.81	7.12	6.24	5.51	4.89
30	41.75	22.63	13.05	9.83	8.20	7.22	6.56	5.73	5.04	4.45
35	39.88	21.58	12.40	9.31	7.75	6.81	6.18	5.38	4.71	4.14
40	38.56	20.83	11.94	8.95	7.44	6.52	5.91	5.13	4.48	3.93
45	37.56	20.27	11.59	8.67	7.20	6.30	5.70	4.94	4.31	3.76
50	36.78	19.83	11.32	8.46	7.01	6.13	5.54	4.79	4.17	3.63
60	35.63	19.19	10.93	8.14	6.74	5.88	5.31	4.58	3.97	3.45
70	34.92	18.79	10.68	7.95	6.57	5.73	5.16	4.44	3.84	3.33
80	34.29	18.43	10.46	7.77	6.41	5.59	5.03	4.32	3.73	3.22
90	33.92	18.22	10.33	7.67	6.32	5.50	4.95	4.25	3.66	3.16
100	33.60	18.04	10.22	7.58	6.25	5.43	4.89	4.19	3.61	3.10
150	32.43	17.40	9.83	7.27	5.97	5.19	4.65	3.98	3.41	2.91
200	31.95	17.10	9.63	7.12	5.84	5.07	4.54	3.87	3.31	2.82
300	31.42	16.81	9.46	6.98	5.72	4.96	4.44	3.78	3.22	2.73
500	31.03	16.59	9.33	6.87	5.63	4.87	4.36	3.70	3.15	2.67
1,000	30.71	16.41	9.21	6.78	5.55	4.80	4.29	3.64	3.09	2.61
2,000	30.58	16.33	9.16	6.74	5.52	4.77	4.26	3.61	3.07	2.59

$s = 7$ $\alpha = .01$

t	r: 1	2	4	6	8	10	12	16	22	32
12	93.02	50.98	30.00	22.97	19.45	17.32	15.90	14.12	12.65	11.42
14	86.65	47.41	27.82	21.26	17.97	15.98	14.65	12.98	11.61	10.46
16	80.28	43.85	25.64	19.55	16.48	14.64	13.40	11.85	10.57	9.49
18	73.90	40.28	23.47	17.83	15.00	13.29	12.15	10.71	9.53	8.53
20	67.53	36.71	21.29	16.12	13.52	11.95	10.90	9.58	8.48	7.56
25	58.39	31.60	18.18	13.67	11.40	10.03	9.11	7.95	7.00	6.18
30	53.78	29.02	16.61	12.44	10.34	9.07	8.22	7.14	6.25	5.50
35	50.63	27.26	15.54	11.61	9.62	8.42	7.61	6.59	5.75	5.03
40	48.46	26.05	14.81	11.03	9.12	7.97	7.19	6.21	5.40	4.70
45	46.84	25.15	14.26	10.60	8.75	7.64	6.88	5.93	5.14	4.46
50	45.58	24.45	13.84	10.27	8.47	7.38	6.64	5.71	4.94	4.28
60	43.76	23.43	13.23	9.79	8.05	7.00	6.29	5.40	4.65	4.01
70	42.63	22.80	12.85	9.49	7.80	6.77	6.08	5.20	4.47	3.84
80	41.63	22.25	12.51	9.23	7.57	6.56	5.89	5.03	4.31	3.69
90	41.04	21.92	12.31	9.07	7.44	6.44	5.77	4.92	4.21	3.60
100	40.55	21.64	12.14	8.94	7.33	6.34	5.68	4.84	4.13	3.53
150	38.80	20.67	11.56	8.48	6.93	5.98	5.35	4.54	3.85	3.27
200	37.96	20.21	11.28	8.27	6.74	5.82	5.19	4.39	3.71	3.14
300	37.21	19.78	11.02	8.07	6.57	5.66	5.05	4.26	3.60	3.03
500	36.63	19.46	10.83	7.92	6.44	5.54	4.93	4.16	3.51	2.94
1,000	36.15	19.19	10.67	7.79	6.33	5.44	4.84	4.08	3.43	2.87
2,000	35.92	19.07	10.59	7.73	6.28	5.40	4.80	4.04	3.40	2.83

(continued)

$s = 8$ $\alpha = .05$

t	*r:* 1	2	4	6	8	10	12	16	22	32
12	76.34	41.58	24.22	18.41	15.51	13.74	12.57	11.10	9.90	8.88
14	71.99	39.16	22.75	17.26	14.51	12.85	11.74	10.34	9.21	8.24
16	67.65	36.73	21.28	16.11	13.52	11.95	10.90	9.59	8.52	7.60
18	63.30	34.31	19.81	14.96	12.53	11.05	10.07	8.83	7.82	6.96
20	58.95	31.88	18.35	13.81	11.53	10.15	9.24	8.08	7.13	6.32
25	52.53	28.31	16.18	12.11	10.07	8.83	8.01	6.97	6.11	5.38
30	49.09	26.39	15.02	11.21	9.29	8.13	7.35	6.37	5.56	4.87
35	46.70	25.06	14.21	10.58	8.75	7.64	6.89	5.96	5.18	4.52
40	45.00	24.12	13.65	10.13	8.36	7.30	6.57	5.67	4.91	4.27
45	43.74	23.42	13.22	9.80	8.08	7.03	6.33	5.45	4.71	4.08
50	42.77	22.86	12.89	9.54	7.85	6.83	6.14	5.28	4.55	3.94
60	41.29	22.60	12.40	9.16	7.52	6.53	5.87	5.02	4.32	3.72
70	40.37	21.55	12.10	8.92	7.32	6.35	5.69	4.87	4.17	3.58
80	39.56	21.10	11.83	8.71	7.13	6.18	5.54	4.72	4.04	3.46
90	39.08	20.83	11.67	8.58	7.02	6.08	5.45	4.64	3.97	3.39
100	38.68	20.61	11.53	8.47	6.93	6.00	5.37	4.57	3.90	3.33
150	37.23	19.80	11.50	8.10	6.61	5.70	5.09	4.32	3.67	3.11
200	36.56	19.42	10.82	7.91	6.45	5.56	4.96	4.20	3.56	3.01
300	35.94	19.06	10.60	7.75	6.31	5.43	4.84	4.09	3.46	2.91
500	35.42	18.79	10.43	7.62	6.19	5.33	4.75	4.00	3.38	2.83
1,000	35.01	18.56	10.30	7.51	6.10	5.25	4.67	3.93	3.31	2.77
2,000	34.84	18.46	10.24	7.46	6.06	5.21	4.63	3.90	3.28	2.74

$s = 8$ $\alpha = .01$

t	*r:* 1	2	4	6	8	10	12	16	22	32
12	112.06	60.90	35.34	26.80	22.52	19.94	18.22	16.05	14.28	12.79
14	104.03	56.45	32.68	24.73	20.75	18.35	16.74	14.73	13.07	11.68
16	96.00	52.01	30.02	22.66	18.98	16.75	15.27	13.40	11.87	10.58
18	87.97	47.56	27.36	20.60	17.21	15.16	13.79	12.08	10.66	9.47
20	79.94	43.12	24.70	18.53	15.43	13.57	12.32	10.75	9.46	8.37
25	68.44	36.76	20.90	15.58	12.91	11.30	10.22	8.86	7.74	6.79
30	62.66	33.57	18.99	14.11	11.65	10.17	9.17	7.92	6.88	6.00
35	58.72	31.39	17.70	13.10	10.79	9.40	8.46	7.28	6.30	5.47
40	56.01	29.90	16.81	12.42	10.20	8.87	7.97	6.84	5.90	5.10
45	53.98	28.78	16.14	11.90	9.77	8.47	7.61	6.51	5.60	4.83
50	52.42	27.92	15.63	11.51	9.43	8.17	7.33	6.26	5.37	4.62
60	50.16	26.67	14.89	10.93	8.94	7.73	6.92	5.90	5.04	4.31
70	48.75	25.90	14.43	10.58	8.64	7.46	6.67	5.67	4.83	4.12
80	47.51	25.21	14.02	10.26	8.37	7.22	6.45	5.47	4.65	3.95
90	46.77	24.81	13.78	10.08	8.21	7.08	6.32	5.35	4.55	3.85
100	46.16	24.47	13.58	9.92	8.08	6.96	6.21	5.25	4.46	3.77
150	43.99	23.28	12.88	9.38	7.61	6.54	5.82	4.90	4.14	3.48
200	42.96	22.71	12.54	9.12	7.39	6.34	5.64	4.74	3.99	3.34
300	42.02	22.19	12.24	8.88	7.19	6.16	5.47	4.59	3.85	3.21
500	41.30	21.80	12.00	8.70	7.04	6.02	5.34	4.47	3.74	3.11
1,000	40.70	21.47	11.81	8.55	6.91	5.91	5.24	4.38	3.66	3.03
2,000	40.43	21.32	11.71	8.48	6.85	5.86	5.19	4.33	3.62	2.99

REFERENCES

Ayres, F. (1962). *Theory and problems of matrices.* New York: Schaum.

Bartlett, M. S. (1947). Multivariate analysis. *Journal of the Royal Statistical Society,* Series B, *9,* 39–52, 176–197.

Bentler, P. M. (1985). *Theory and implementation of EQS, a structural equations program.* Los Angeles: BMDP Statistical Software.

Bock, R. D. (1975). *Multivariate statistical methods for behavioral sciences.* New York: McGraw-Hill.

Box, G. E. P. (1949). A general distribution theory for a class of likelihood criteria. *Biometrika, 36,* 317–346.

Browne, M. W. (1975a). Predictive validity of a linear regression equation. *British Journal of Mathematical and Statistical Psychology, 28,* 79–87.

Browne, M. W. (1975b). A comparison of single sample and cross-validation methods for estimating the mean squared error of prediction in a multiple linear regression. *British Journal of Mathematical and Statistical Psychology, 28,* 112–120.

Cattin, P. (1980). Note on the estimation of the squared cross-validated multiple correlation of a regression model. *Psychological Bulletin, 87,*63–65.

Cliff, N. (1966). Orthogonal rotation to congruence. *Psychometrika, 31,* 33–42.

Cliff, N. (1970). The relation between sample and population eigenvectors. *Psychometrika,* 35, 163–178.

Cliff, N. (1982). What is and isn't measurement? In G. Keren (Ed.), *Statistical and methodological issues in psychology and social sciences research.* Englewood Cliffs, NJ: Erlbaum.

Cliff, N. (1983). Some cautions concerning the application of causal modeling methods. *Multivariate Behavioral Research, 18,* 115–126.

Cliff, N. (1985). The eigenvalues-greater-than-one rule and the reliability of components. Unpublished manuscript.

Cliff, N., & D. J. Krus (1976). Interpretation of canonical analysis: Rotated versus unrotated solutions. *Psychometrika, 41,* 35–42.

Cochran, W. C., & G. M. Cox (1957). *Experimental designs* (2nd ed.). New York: Wiley.

Cooley, W. W., & P. R. Lohnes (1971). *Multivariate data analysis.* New York: Wiley.

Cramer, E. M., & M. I. Appelbaum (1980). Nonorthogonal analysis of variance—once again. *Psychological Bulletin, 87,* 51–57.

Darlington, R. B. (1968). Multiple regression in psychological research and practice. *Psychological Bulletin, 69,* 161–182.

Darlington, R. B. (1978). Reduced variance regression. *Psychological Bulletin, 85,* 1238–1255.

Darlington, R. B., & C. M. Boyce (1982). Ridge and other new varieties of regression. In G. Keren (Ed.), *Statistical and methodological issues in psychology and social sciences research.* Hillsdale, NJ: Erlbaum.

Dixon, W. J. (Ed.) (1981). *BMDP statistical software, 1981.* Berkeley, CA: University of California.

Fisher, R. A. (1951). *The design of experiments.* Edinburgh, Oliver and Boyd.

Freedman, D. A. (1983). A note on screening regression equations. *American Statistician, 37,* 152–155.

Gorsuch, R. L. (1974). *Factor analysis.* Philadelphia: Saunders.

Green, B. F. (1977). Parameter sensitivity in multivariate methods. *Multivariate Behavioral Research, 12,* 263–288.

Green, P. E., & J. D. Carroll (1976). *Mathematical tools for applied multivariate analysis.* New York: Academic Press.

Guttman, L. (1954). Some necessary conditions for common factor analysis. *Psychometrika, 19,* 149–161.

Harman, H. H. (1976). *Modern factor analysis, 3rd ed.* Chicago: University of Chicago.

Harman, H. H., & W. H. Jones (1966). Factor analysis by minimizing residuals. *Psychometrika, 31,* 351–368.

Hays, W. B. (1982). *Statistics* (3rd ed.). New York: Holt, Rinehart, and Winston.

Heath, H. (1952). A factor analysis of women's measurements taken for garment and pattern construction. *Psychometrika, 17,* 87–100.

Hocking, R. R. (1976). The analysis and selection of variables in linear regression. *Biometrics, 32,* 1–44.

Horst, P. (1963). *Matrix algebra for social scientists.* New York: Holt, Rinehart, and Winston.

Horst, P, & A. L. Edwards (1982). Analysis of non-orthogonal designs: The 2^k factorial experiment. *Psychological Bulletin, 91,* 190–192.

Hotelling, H. (1935). The most predictable criterion. *Journal of Educational Psychology, 26,* 139–142.

Hotelling, H. (1936). Relations between two sets of variates. *Biometrika, 28,* 321–377.

Jöreskog, K. G., & D. Sorbom (1982). *LISREL V: Analysis of linear structural relationships by maximum likelihood and least squares.* Uppsala, Sweden: University of Uppsala.

Kaiser, H. F. (1961). A note of Guttman's lower bound for the number of common factors. *British Journal of Mathematical and Statistical Psychology, 14,* 1.

Kaiser, H. F. (1970). A second generation Little Jiffy. *Psychometrika, 35,* 401–415.

Kempthorne, O. (1950). *Design and analysis of experiments.* New York: Wiley.

Keren, G. (1982). A balanced approach to unbalanced designs. In G. Keren (Ed.), *Statistical and methodological issues in psychology and social sciences research.* Hillsdale, NJ: Erlbaum.

Kerlinger, F. N., & E. J. Pedhazur (1973). *Multiple regression in behavioral research.* New York: Holt, Rinehart, and Winston.

Kim, J-O., & C. W. Mueller (1978a). *Introduction to factor analysis.* Beverly Hills, CA: Sage.

Kim, J-O., & C. W. Mueller (1978b). *Factor analysis: statistical methods and practical issues.* Beverly Hills, CA: Sage.

Kirk, R. E. (1982). *Experimental design: Procedures for the behavioral sciences.* Belmont, CA: Brooks/Cole.

Krantz, D. H. (1974). Measurement structures and psychological laws. *Science, 175,* 1427–1435.

Krantz, D. H., R. D. Luce, S. Suppes, & A. Tversky (1971). *Foundations of measurement.* New York: Academic Press.

Krus, D. J., & A. Tellegen. (1975). Consciousness III: fact or fiction. *Psychological Reports, 36,* 23–30.

Levine, M. S. (1977). *Canonical analysis and factor comparison.* Beverly Hills, CA: Sage.

Lindeman, R. H., P. F. Merenda, & R. Z. Gold (1980). *Introduction to bivariate and multivariate analysis.* Glenview, IL: Scott, Foresman.

Linn, R. L., & C. E. Werts (1982). Measurement error in regression. In G. Keren (Ed.), *Statistical and methodological issues in Psychology and social science research.* Hillsdale, NJ: Erlbaum.

Long, J. S. (1983a). *Confirmatory factor analysis.* Beverly Hills, CA: Sage.

Long, J. S. (1983b). *Covariance structure models: An introduction to LISREL.* Beverly Hills, CA: Sage.

McDonald, R. P. (1978). A simple comprehensive model for the analysis of covariance structures. *British Journal of Mathematical and Statistical Psychology, 31,* 59–72.

McDonald, R. P. (1980). A simple comprehensive model for the analysis of covariance structures. Some remarks on applications. *British Journal of Mathematical and Statistical Psychology, 33,* 161–183.

McDonald, R. P. (1985). *Factor analysis and related methods.* Hillsdale, NJ: Erlbaum.

McDonald, R. P., & S. A. Mulaik (1979). Determinacy of common factors: A non-technical review. *Psychological Bulletin, 86,* 297–306.

Miller, A. J. (1984). Selection of subsets of regression variables. *Journal of the Royal Statistical Association,* Series A, *147,* 389–425.

Mills, G. (1969). *Introduction of linear algebra: A primer for social scientists.* Chicago: Aldine.

Morris, J. D. (1982). Ridge regression and some alternative weighting techniques. *Psychological Bulletin, 91,* 203–210.

Morrison, D. F. (1976). *Multivariate statistical methods* (2nd ed.). New York: McGraw-Hill.

Mulaik, S. A. (1972). *The foundations of factor analysis.* New York: McGraw-Hill.

Myers, J. L. (1979). *Fundamentals of experimental design.* Boston: Allyn & Bacon.

Nie, N. H., C. H. Hull, J. G. Jenkins, K. Steinbrenner, & D. H. Bent (1975). *SPSS: Statistical package for the social sciences* (2nd ed.). New York: McGraw-Hill.

Overall, J. E., & C. J. Klett (1972). *Applied multivariate analysis.* New York: McGraw-Hill.

Overall, J. E., D. M. Lee, & C. W. Hornick (1981). Comparison of two strategies for analysis of variance in nonorthogonal designs. *Psychological Bulletin, 90,* 367–375.

Pedhazur, E. (1982). *Multiple regression in behavioral research.* New York: Holt, Rinehart, and Winston.

Rao, C. R. (1951). An asymptotic expansion of the distribution of Wilks' criterion. *Bulletin of the International Statistical Institute, 33,* 177–180.

Roy, S. N. (1957). *Some aspects of multivariate analysis.* New York: Wiley.

Rozeboom, W. W. (1978). Estimation of cross-validated multiple correlation: A clarification. *Psychological Bulletin, 85,* 1348–1351.

SAS Institute, Inc. (1979). *SAS User's guide, 1979 Edition.* Raleigh, NC: SAS Institute.

Spearman, C. (1904). General intelligence, objectively measured. *American Journal of Psychology, 38,* 201–293.

SPSS, Inc. (1983). *SPSS^x User's Guide.* New York: McGraw-Hill.

Steiger, J. H., & M. W. Browne (1984). The comparison of interdependent correlations between optimal linear composites. *Psychometrika, 49,* 11–24.

Tabachnick, B. G., & L. S. Fidell (1983). *Using multivariate statistics.* New York: Harper & Row.

Tatsuoka, M. M. (1970). *Discriminant analysis: The study of group differences.* Champaign, IL: Institute for Ability and Personality Testing.

Tatsuoka, M. M. (1971a). *Multivariate analysis: Techniques for educational and psychological research.* New York: Wiley.

Tatsuoka, M. M. (1971b). *Significance tests: Univariate and multivariate.* Champaign, IL: Institute for Ability and Personality Testing.

Tatsuoka, M. M. (1975). *The general linear model: A "new" trend in analysis of variance.* Champaign, IL: Institute for Personality and Ability Testing.

Tatsuoka, M. M. (1976). *Validation studies: The use of multiple regression equations.* Champaign, IL: Institute for Personality and Ability Testing.

van de Geer, J. P. (1971). *Introduction to multivariate analysis for the social sciences.* San Francisco: Freeman.

Velicer, W. F., A. C. Peacock, & D. N. Jackson (1982). A comparison of component and factor patterns: A Monte Carlo approach. *Multivariate Behavioral Research, 17,* 371–388.

Wilcox, R. R. (1987). New designs in the analysis of variance. *Annual Review of Psychology, 38,* 27–60.

Wilkinson, L., & G. E. Dallal (1981). Tests of significance in forward selection regression with an F-to-enter stopping rule. *Technometrics, 23,* 377–380.

Wilks, S. S. (1932). Certain generalizations in the analysis of variance. *Biometrika, 24,* 471–494.

Winer, B. J. (1971). *Statistical principles in experimental design.* New York: McGraw-Hill.

Zwick, W. R., & W. F. Velicer (1982). Factors influencing four rules for determining the number of components to retain. *Multivariate Behavioral Research, 17,* 253–269.

Zwick, W. R., & W. F. Velicer (1986). A comparison of five rules for determining the number of components to retain. *Psychological Bulletin, 99,* 432–442.

Index

D

E

F

G

H

I

J

K

L

N

O

P

Q

R

S

A 7
B 8
C 9
D 0
E 1
F 2
G 3
H 4
I 5
J 6